Communications in Computer and Information Science   **693**

Commenced Publication in 2007
Founding and Former Series Editors:
Alfredo Cuzzocrea, Orhun Kara, Dominik Ślęzak, and Xiaokang Yang

José Braz · Nadia Magnenat-Thalmann
Paul Richard · Lars Linsen
Alexandru Telea · Sebastiano Battiato
Francisco Imai (Eds.)

Computer Vision, Imaging and Computer Graphics Theory and Applications

11th International Joint Conference, VISIGRAPP 2016
Rome, Italy, February 27–29, 2016
Revised Selected Papers

Springer

Editors
José Braz
Escola Superior de Tecnologia do IPS
Setúbal
Portugal

Nadia Magnenat-Thalmann
MiraLab
University of Geneva
Carouge
Switzerland

Paul Richard
LISA - ISTIA
University of Angers
Angers
France

Lars Linsen
Department of Computer Science
  and Electrical Engineering
Jacobs University
Bremen
Germany

Alexandru Telea
University of Groningen
Groningen
The Netherlands

Sebastiano Battiato
Università di Catania
Catania
Italy

Francisco Imai
Research Innovation Center
Canon U.S.A. Inc.
San Jose, CA
USA

ISSN 1865-0929   ISSN 1865-0937 (electronic)
Communications in Computer and Information Science
ISBN 978-3-319-64869-9   ISBN 978-3-319-64870-5 (eBook)
DOI 10.1007/978-3-319-64870-5

Library of Congress Control Number: 2017948186

Printed on acid-free paper

This Springer imprint is published by Springer Nature
The registered company is Springer International Publishing AG
The registered company address is: Gewerbestrasse 11, 6330 Cham, Switzerland

Preface

This book includes the extended versions of selected papers from VISIGRAPP 2016, the International Joint Conference on Computer Vision, Imaging and Computer Graphics Theory and Applications, which was held in Rome, Italy, during February 27–29, 2016. The conference was organized by the Institute for Systems and Technologies of Information, Control and Communication (INSTICC), in cooperation with ACM SIGGRAPH, AFIG and Eurographics and technically co-sponsored by the IEEE Computer Society, IEEE VGMT, and IEEE TCMC. VISIGRAPP comprises three conferences, namely, the International Conference on Computer Vision Theory and Applications (VISAPP), the International Conference on Computer Graphics Theory and Applications (GRAPP), and the International Conference on Information Visualization Theory and Applications (IVAPP).

In total 338 paper submissions were received for VISIGRAPP from more than 45 countries. After a rigorous double-blind evaluation, only 24% of the papers were accepted and published as full papers. These numbers show that our conference is aiming for the highest scientific standards, and that it can now be considered a well-established venue for researchers in the broad fields of computer vision, image analysis, computer graphics, and information visualization. From the set of full papers, 28 were selected for inclusion into this book. The selection process was based on quantitative and qualitative evaluation results provided by the Program Committee as well as the feedback on paper presentations provided by the session chairs during the conference. After selection, the accepted papers were further revised and extended by the authors. Our gratitude goes to all contributors and reviewers, without whom this book would not have been possible. Apart from the full papers, 27% of the papers were accepted for short presentations and 19% accepted for poster presentations. However, these works were not considered for the present book selection process. We do not expect that each individual reader is equally interested in all of the selected VISIGRAPP papers. However, the diversity of these papers makes it very likely that all readers can find something of interest in this selection.

As VISAPP 2016 constitutes the largest part of VISIGRAPP with 244 submissions, we decided to select and integrate 14 extended full papers aiming to cover different aspects and areas related to computer vision, such as image formation and pre-processing, image and video analysis and understanding, motion tracking, stereo vision as well as diverse computer vision applications and services. Here, we would like to mention that when we selected the papers from VISAPP for this book, our intention was to cover and highlight research from different areas and subareas related to computer vision. These papers were mainly competing with other VISAPP papers having similar content, and therefore we would like to point out that other high-quality papers accepted at the conference could have been integrated into this book if we would have enough space for them.

Concerning GRAPP 2016, 48 papers were submitted, and we decided to include eight extended full papers in this book. We tried to cover the main areas of computer graphics to make the content of the book similar to the research addressed in the conference.

The selected IVAPP 2016 papers are not only excellent representatives of the field of information visualization, but also form a balanced representation of the field itself. Above all, they are almost as diverse and exciting as the field of information visualization.

VISIGRAPP 2016 also included four invited keynote lectures, presented by internationally renowned researchers, whom we would like to thank for their contribution to the overall quality of the conference. These are in alphabetical order: Gilles Bailly (CNRS; Telecom ParisTech; University Paris-Saclay, France), Torsten Moeller (University of Vienna, Austria), Vittorio Murino (Istituto Italiano di Tecnologia, Italy), and Dieter Schmalstieg (Institute for Com-puter Graphics and Vision at Graz University of Technology, Austria).

We wish to thank all those who supported VISIGRAPP and helped to organize the conference. On behalf of the conference Organizing Committee, we would like to especially thank the authors, whose work was the essential part of the conference and contributed to a very successful event. We would also like to thank the members of the Program Committee, whose expertise and diligence were instrumental in ensuring the quality of the final contributions. We also wish to thank all the members of the Organizing Committee, whose work and commitment was invaluable. Last but not least, we would like to thank Springer for their collaboration and help in getting this book to print.

February 2017

José Braz
Nadia Magnenat-Thalmann
Paul Richard
Lars Linsen
Alexandru Telea
Sebastiano Battiato
Francisco Imai

Organization

Conference Chair

Jose Braz   Escola Superior de Tecnologia de Setúbal, Portugal

Program Co-chairs

VISAPP

Sebastiano Battiato   University of Catania, Italy
Francisco Imai   Canon U.S.A. Inc., Innovation Center, USA

IVAPP

Lars Linsen   Jacobs University, Bremen, Germany
Alexandru Telea   University of Groningen, The Netherlands

GRAPP

Nadia Magnenat-Thalmann   NTU, Singapore and MIRALab, University of Geneva, Switzerland
Paul Richard   University of Angers, France

VISAPP Program Committee

Amr Abdel-Dayem   Laurentian University, Canada
Vicente Alarcon-Aquino   Universidad de las Americas Puebla, Mexico
Mokhled S. Al-Tarawneh   Mu'tah University, Jordan
Matthew Antone   Massachusetts Institute of Technology, USA
Pantelis Asvestas   Technological Educational Institute of Athens, Greece
Hichem Bannour   Marin Software, France
Arrate Muñoz Barrutia   Universidad Carlos III de Madrid, Spain
Giuseppe Baruffa   University of Perugia, Italy
Ardhendu Behera   Edge Hill University, UK
Saeid Belkasim   Georgia State University, USA
Fabio Bellavia   Università degli studi di Firenze (presso il prof. Carlo Colombo), Italy
Olga Bellon   IMAGO Research Group - Universidade Federal do Paraná, Brazil

Neil Bergmann	University of Queensland, Australia
Du Bo	Wuhan University, China
Adrian Bors	University of York, UK
Giosue Lo Bosco	University of Palermo, Italy
Marius Brezovan	University of Craiova, Romania
Valentin Brimkov	State University of New York, USA
Alfred Bruckstein	Technion, Israel
Arcangelo R. Bruna	STMicroelectronics, Italy
Xianbin Cao	Beihang University, China
Alice Caplier	GIPSA-lab, France
Franco Alberto Cardillo	Consiglio Nazionale delle Ricerche, Italy
M. Emre Celebi	Louisiana State University in Shreveport, USA
Chee Seng Chan	University of Malaya, Malaysia
Satish Chand	Netaji Subhas Institute of Technology and Jawaharlal Nehru University Delhi, India
Chin-Chen Chang	Feng Chia University, Taiwan
Samuel Cheng	University of Oklahoma, USA
Laurent Cohen	University of Paris Dauphine, France
Sara Colantonio	ISTI-CNR, Italy
David Connah	Université François Rabelais Tours, France
Donatello Conte	Université François Rabelais Tours, France
Guido de Croon	Delft University of Technology, The Netherlands
Dima Damen	University of Bristol, UK
Roy Davies	Royal Holloway, University of London, UK
Kenneth Dawson-Howe	Trinity College Dublin, Ireland
Joachim Denzler	Friedrich Schiller University of Jena, Germany
Thomas M. Deserno	Aachen University of Technology (RWTH), Germany
Michel Devy	LAAS-CNRS, France
Sotirios Diamantas	Athens Information Technology, Greece
Yago Diez	Tohoku University, Japan
Jana Dittmann	Otto-von-Guericke-Universität Magdeburg, Germany
Anastasios Drosou	Centre for Research and Technology, Hellas, Greece
Mahmoud El-Sakka	The University of Western Ontario, Canada
Grigori Evreinov	University of Tampere, Finland
Shu-Kai S. Fan	National Taipei University of Technology, Taiwan
Giovanni Maria Farinella	Università di Catania, Italy
Jean-Baptiste Fasquel	University of Angers, France
Aaron Fenster	Robarts Research Institute, Canada
Gernot A. Fink	TU Dortmund, Germany
David Fofi	Le2i, France
Tyler Folsom	QUEST Integrated Inc., USA
Gian Luca Foresti	Unversity of Udine, Italy
Mohamed M. Fouad	Military Technical College, Egypt
Antonios Gasteratos	Democritus University of Thrace, Greece
Claudio Gennaro	CNR, Italy
Herman Gomes	Federal University of Campina Grande, Brazil

Amr Goneid — The American University in Cairo, Egypt
Manuel González-Hidalgo — Balearic Islands University, Spain
Bernard Gosselin — University of Mons, Belgium
Nikos Grammalidis — Centre of Research and Technology Hellas, Greece
Costantino Grana — Università degli Studi di Modena e Reggio Emilia, Italy
Christos Grecos — Central Washington University, USA
Benjamin Guthier — Universität Mannheim, Germany
Hsi-Chin Hsin — National United University, Taiwan
Hui-Yu Huang — National Formosa University, Taiwan
Céline Hudelot — Ecole Centrale de Paris, France
Chih-Cheng Hung — Kennesaw State University, USA
Laura Igual — Universitat de Barcelona, Spain
Jiri Jan — University of Technology Brno, Czech Republic
Tatiana Jaworska — Polish Academy of Sciences, Poland
Xiuping Jia — University of New South Wales, Australia
Xiaoyi Jiang — University of Münster, Germany
Zhong Jin — Nanjing University of Science and Technology, China
Leo Joskowicz — The Hebrew University of Jerusalem, Israel
Joni-Kristian Kamarainen — Tampere University of Technology, Finland
Martin Kampel — Vienna University of Technology, Austria
Mohan Kankanhalli — National University of Singapore, Singapore
Sehwan Kim — WorldViz LLC, USA
Nahum Kiryati — Tel Aviv University, Israel
Syoji Kobashi — University of Hyogo, Japan
Seong Kong — Sejong University, South Korea
Mario Köppen — Kyushu Institute of Technology, Japan
Andreas Koschan — University of Tennessee, USA
Dimitrios Kosmopoulos — University of Patras, Greece
Constantine Kotropoulos — Aristotle University of Thessaloniki, Greece
Jorma Laaksonen — Aalto University School of Science, Finland
Agata Lapedriza — Universitat Oberta de Catalunya, Spain
Slimane Larabi — U.S.T.H.B. University, Algeria
Mónica G. Larese — CIFASIS-CONICET, National University of Rosario, Argentina
Denis Laurendeau — Laval University, Canada
Sébastien Lefèvre — Université de Bretagne Sud, France
Daw-Tung Dalton Lin — National Taipei University, Taiwan
Huei-Yung Lin — National Chung Cheng University, Taiwan
Luis Jiménez Linares — University of de Castilla-La Mancha, Spain
Nicolas Loménie — Université Paris Descartes, France
Angeles López — Universitat Jaume I, Spain
Ilias Maglogiannis — University of Piraeus, Greece
Baptiste Magnier — LGI2P de l'Ecole des Mines d'ALES, France
András Majdik — MTA SZTAKI, Hungary
Lucio Marcenaro — University of Genova, Italy
Emmanuel Marilly — NOKIA - Bell Labs, France

Salvatore Vitabile	University of Palermo, Italy
Vassilios Vonikakis	Advanced Digital Sciences Center (ADSC), Singapore
Frank Wallhoff	Jade University of Applied Science, Germany
Tao Wang	BAE Systems, USA
Wen-June Wang	National Central University, Taiwan
Toyohide Watanabe	Nagoya Industrial Science Research Institute, Japan
Joost van de Weijer	Autonomous University of Barcelona, Spain
Quan Wen	University of Electronic Science and Technology of China, China
Andrew Willis	University of North Carolina at Charlotte, USA
Christian Wöhler	TU Dortmund University, Germany
Stefan Wörz	University of Heidelberg, Germany
Pingkun Yan	Chinese Academy of Sciences, China
Guoan Yang	Xian Jiaotong University, China
Lara Younes	CentraleSupelec, France
Hongfeng Yu	University of Nebraska - Lincoln, USA
Shan Yu	French National Institute for Research in Computer Science and Control, USA
Yizhou Yu	University of Illinois, USA
Qieshi Zhang	Waseda University, Japan
Jianmin Zheng	Nanyang Technological University, Singapore
Huiyu Zhou	Queen's University Belfast, UK
Yun Zhu	UCSD, USA
Zhigang Zhu	City College of New York, USA
Peter Zolliker	Empa, Swiss Federal Laboratories for Materials Science and Technology, Switzerland
Ju Jia (Jeffrey) Zou	University of Western Sydney, Australia
Tatjana Zrimec	University New South Wales, Slovenia

VISAPP Additional Reviewers

Andrea Albarelli	Università Ca' Foscari Venezia, Italy
Filippo Bergamasco	Università Ca' Foscari di Venezia, Italy
Liyan Chen	Tongji University, China
Luca Cosmo	Università Ca' Foscari Venezia, Italy
Mario D'Acunto	Italian National Research Council (CNR), Italy
João Fonseca	Life and Health Sciences Research Institute (ICVS), Portugal
Weifeng Ge	The University of Hong Kong, SAR China
Ioannis Kapsouras	Aristotle University of Thessaloniki, Greece
Jatuporn Leksut	University of Southern California, USA
Guanbin Li	HKU, Hong Kong, SAR China
Wei Li	CUNY City College, USA
Filippo Milotta	University of Catania, Italy
Pedro Morais	ICVS 3B's, Portugal

Greg Olmschenk	The Graduate Center of the City University of New York, USA
Axel Plinge	TU Dortmund, Germany
Christopher Pramerdorfer	TU Wien, Austria
Sandro Queirós	Life and Health Sciences Research Institute (ICVS), Portugal
Marco Righi	Institute of Information Science and Technologies, Italy
Sebastian Sudholt	Technische Universität Dortmund, Germany
Hao Tang	City University of New York (BMCC), USA
Lei Yang	The University of Oklahoma, USA
Flavio Zavan	UFPR, Brazil
Dayong Zhou	8 Rivers, USA

IVAPP Program Committee

Wolfgang Aigner	St. Poelten University of Applied Sciences, Austria
Vladan Babovic	National University of Singapore, Singapore
Rita Borgo	King's College London, UK
David Borland	University of North Carolina at Chapel Hill, USA
Massimo Brescia	Istituto Nazionale di AstroFisica, Italy
Ross Brown	Queensland University of Technology, Brisbane, Australia
Maria Beatriz Carmo	Universidade de Lisboa, Portugal
Guoning Chen	University of Houston, USA
Lieu-Hen Chen	National Chi Nan University, Taiwan
Christoph Dalitz	Niederrhein University of Applied Sciences, Germany
Mihaela Dinsoreanu	Technical University of Cluj-Napoca, Romania
Georgios Dounias	University of the Aegean, Greece
Achim Ebert	University of Kaiserslautern, Germany
Ronak Etemadpour	Oklahoma State University, USA
Chi-Wing Fu	The Chinese University of Hong Kong, Hong Kong, SAR China
Mohammad Ghoniem	Luxembourg Institute of Science and Technology, Luxembourg
Martin Graham	University of Edinburgh, UK
Theresia Gschwandtner	Vienna University of Technology, Austria
Alfred Inselberg	Tel Aviv University, Israel
Mark W. Jones	Swansea University, UK
Jessie Kennedy	Edinburgh Napier University, UK
Andreas Kerren	Linnaeus University, Sweden
Sehwan Kim	WorldViz LLC, USA
Jörn Kohlhammer	Fraunhofer Institute for Computer Graphics Research, Germany
Martin Kraus	Aalborg University, Denmark
Simone Kriglstein	Vienna University of Technology, Austria
Denis Lalanne	University of Fribourg, Switzerland

Jie Liang	Peking University, China
Chun-Cheng Lin	National Chiao Tung University, Taiwan
Giuseppe Liotta	University of Perugia, Italy
Krešimir Matkovic	VRVis Research Center, Austria
Luis Gustavo Nonato	Universidade de Sao Paulo, Brazil
Steffen Oeltze-Jafra	Innovation Center Computer Assisted Surgery (Iccas), University of Leipzig, Germany
Benoît Otjacques	Luxembourg Institute of Science and Technology (LIST), Luxembourg
Philip J. Rhodes	University of Mississippi, USA
Adrian Rusu	Fairfield University, USA
Ignaz Rutter	Karlsruher Institut für Technologie, Germany
Giuseppe Santucci	University of Roma, Italy
Angel Sappa	Computer Vision Center, Spain
Tobias Schreck	Graz University of Technology, Austria
Hans-Joerg Schulz	Fraunhofer-Institut für Graphische Datenverarbeitung IGD, Germany
Heidrun Schumann	University of Rostock, Germany
Michael Sedlmair	University of Vienna, Austria
Marc Streit	Johannes Kepler Universität Linz, Austria
Yasufumi Takama	Tokyo Metropolitan University, Japan
Ying Tan	Peking University, China
Sidharth Thakur	Renaissance Computing Institute (RENCI), USA
Roberto Theron	Universidad de Salamanca, Spain
Christian Tominski	University of Rostock, Germany
Cagatay Turkay	City University London, UK
Günter Wallner	University of Applied Arts Vienna, Austria
Daniel Weiskopf	Universität Stuttgart, Germany
Kai Xu	Middlesex University, UK
Hsu-Chun Yen	National Taiwan University, Taiwan
Hongfeng Yu	University of Nebraska - Lincoln, USA
Xiaoru Yuan	Peking University, China

IVAPP Additional Reviewers

Francesco Bongiovanni	Luxembourg Institute of Science and Technology, Luxembourg
Christian Bors	TU Wien, Austria
Davide Ceneda	Vienna University of Technology, Austria
Florian Evequoz	University of Fribourg, Switzerland
Kostiantyn Kucher	ISOVIS Group, Linnaeus University, Sweden
Roger Almeida Leite	TU Wien, Brazil
Yunai Wang	Shenzhen Institutes of Advanced Technology, China
Wei Zeng	Singapore-ETH Centre, Singapore

GRAPP Program Committee

Francisco Abad	Universidad Politécnica de Valencia, Spain
Marco Agus	King Abdullah University of Science and Technology, Saudi Arabia
Marco Ament	Karlsruhe Institute of Technology, Germany
Lilian Aveneau	University of Poitiers, France
Francesco Banterle	Visual Computing Lab, Italy
Hu-Jun Bao	Zhejiang University, China
Thomas Bashford-Rogers	University of Warwick, UK
Dominique Bechmann	CNRS-Université de Strasbourg, France
Rafael Bidarra	Delft University of Technology, The Netherlands
Venceslas Biri	University Paris Est, France
Fernando Birra	UNL, Portugal
Kristopher J. Blom	Virtual Human Technologies, Czech Republic
Carles Bosch	Eurecat, Spain
Stephen Brooks	Dalhousie University, Canada
Cédric Buche	LabSticc-CERV, France
Maria Beatriz Carmo	Universidade de Lisboa, Portugal
L.G. Casado	University of Almeria, Spain
Eva Cerezo	University of Zaragoza, Spain
Raphaelle Chaine	LIRIS Université Lyon 1, France
Teresa Chambel	Lasige, University of Lisbon, Portugal
Antoni Chica	Universitat Politecnica de Catalunya, Spain
Hwan-gue Cho	Pusan National University, Korea, Republic of
Miguel Chover	Universitat Jaume I, Spain
Marc Christie	IRISA/Inria Rennes, France
Ana Paula Cláudio	BioISI, Universidade de Lisboa, Portugal
Sabine Coquillart	Inria, France
António Cardoso Costa	ISEP, Portugal
Carsten Dachsbacher	Karlsruhe Institute of Technology, Germany
Juan José Jiménez Delgado	Universidad de Jaen, Spain
Miguel Sales Dias	Microsoft Language Development Center/ISCTE-Lisbon University Institute, Portugal
Paulo Dias	Universidade de Aveiro, Portugal
John Dingliana	Trinity College Dublin, Ireland
Thierry Duval	IMT Atlantique, France
Arjan Egges	Utrecht University, The Netherlands
Elmar Eisemann	Delft University of Technology, The Netherlands
Marius Erdt	Fraunhofer IDM@NTU, Singapore
Bianca Falcidieno	Consiglio Nazionale delle Ricerche, Italy
Petros Faloutsos	York University, Canada
Jean-Philippe Farrugia	LIRIS Lab, France
Francisco R. Feito	University of Jaén, Spain
Jie-Qing Feng	State Key Lab of CAD&CG, Zijingang Campus, Zhejiang University, China

Luiz Henrique de Figueiredo	Impa, Brazil
Mauro Figueiredo	Universidade do Algarve, Portugal
Fabian Di Fiore	Hasselt University, Belgium
Cedric Fleury	Université de Paris-Sud, France
Carla Maria Dal Sasso Freitas	Universidade Federal do Rio Grande do Sul, Brazil
Ioannis Fudos	University of Ioannina, Greece
Alejandro García-Alonso	University of the Basque Country, Spain
Miguel Gea	University of Granada, Spain
Djamchid Ghazanfarpour	Xlim Laboratory (UMR CNRS 7252) - University of Limoges, France
Stephane Gobron	HES-SO/HE-Arc/ISIC, Switzerland
Alexandrino Gonçalves	Polytechnic Institute of Leiria, Portugal
Laurent Grisoni	University of Lille Science and Technologies, France
Jerome Grosjean	LSIIT, France
James Hahn	George Washington University, USA
Vlastimil Havran	Czech Technical University in Prague, Czech Republic
Nancy Hitschfeld	University of Chile, Chile
Ludovic Hoyet	Inria Rennes - Centre Bretagne Atlantique, France
Andres Iglesias	University of Cantabria, Spain
Insung Ihm	Sogang University, South Korea
Alex Pappachen James	Nazarbayev University, Kazakhstan
Jean-Pierre Jessel	IRIT, Paul Sabatier University, Toulouse, France
Xiaogang Jin	Zhejiang University, China
Robert Joan-Arinyo	Universitat Politecnica de Catalunya, Spain
Chris Joslin	Carleton University, Canada
Josef Kohout	University of West Bohemia, Czech Republic
Torsten Kuhlen	RWTH Aachen University, Germany
Richard Kulpa	Université Rennes 2, France
Won-sook Lee	University of Ottawa, Canada
Miguel Leitão	ISEP, Portugal
Alejandro León	University of Granada, Spain
Frederick Li	University of Durham, UK
Hélio Lopes	PUC-Rio, Brazil
Pedro Faria Lopes	ISCTE-IUL, Portugal
Jorge Lopez-Moreno	Universidad Rey Juan Carlos, Spain
Joaquim Madeira	University of Aveiro, Portugal
Luís Magalhães	University of Minho, Portugal
Michael Manzke	Trinity College Dublin, Ireland
Ricardo Marroquim	Rio de Janeiro Federal University, Brazil
Belen Masia	Universidad de Zaragoza, Spain
Oliver Mattausch	University of Zurich, Switzerland
Nelson Max	University of California, USA
Daniel Meneveaux	University of Poitiers, France
Stéphane Mérillou	University of Limoges, France

Ramon Molla	Universitat Politècnica de València, Spain
David Mould	Carleton University, Canada
João Paulo Moura	Universidade de Trás-os-Montes e Alto Douro, Portugal
Georgios Papaioannou	Athens University of Economics and Business, Greece
Alexander Pasko	Bournemouth University, UK
Giuseppe Patané	CNR, Italian National Research Council, Italy
Daniel Patel	University of Bergen, Norway
Nuria Pelechano	Universitat Politecnica de Catalunya, Spain
João Pereira	Instituto Superior de Engenharia do Porto, Portugal
João Madeiras Pereira	INESC-ID/IST, Portugal
Sinésio Pesco	PUC-Rio Institute, Brazil
Christopher Peters	KTH Royal Institute of Technology, Sweden
Ruggero Pintus	CRS4 - Center for Advanced Studies, Research and Development in Sardinia, Italy
Anna Puig	University of Barcelona, Spain
Inmaculada Remolar	Universitat Jaume I, Spain
Mickael Ribardière	University of Poitiers, XLIM, France
María Cecilia Rivara	Universidad de Chile, Chile
Inmaculada Rodríguez	University of Barcelona, Spain
Przemyslaw Rokita	Warsaw University of Technology, Poland
Teresa Romão	Faculdade de Ciências e Tecnologia/Universidade de Nova Lisboa, Portugal
Isaac Rudomin	BSC, Spain
Holly Rushmeier	Yale University, USA
Luis Paulo Santos	Universidade do Minho, Portugal
Basile Sauvage	University of Strasbourg, France
Mateu Sbert	Tianjin University/Universitat de Girona, Spain
Rafael J. Segura	Universidad de Jaen, Spain
Ari Shapiro	University of Southern California, USA
Frutuoso Silva	University of Beira Interior, Portugal
A. Augusto Sousa	FEUP/INESC Porto, Portugal
Ching-Liang Su	Da Yeh University, India
Jie Tang	Nanjing University, China
Domenico Tegolo	Università degli Studi di Palermo, Italy
Matthias Teschner	University of Freiburg, Germany
Daniel Thalmann	Nanyang Technological University, Singapore
Juan Carlos Torres	Universidad de Granada, Spain
Torsten Ullrich	Fraunhofer Austria Research, Austria
Anna Ursyn	University of Northern Colorado, USA
Luiz Velho	IMPA, Instituto de Matematica Pura e Aplicada, Brazil
Andreas Weber	University of Bonn, Germany
Daniel Weiskopf	Universität Stuttgart, Germany
Burkhard Wuensche	University of Auckland, New Zealand
Lihua You	Bournemouth University, UK

GRAPP Additional Reviewers

Marco Livesu	CNR IMATI, Italy
Lis Ingrid Roque	Polytechnic Institute - Rio de Janeiro State University, Brazil
Mohamad Salimian	Dalhousie Univesity, Canada
Simona Schiavi	Inria, Ecole Polytechnique, France
Sylvain Thery	ICube, France
Myrthe Tielman	Delft University of Technology, The Netherlands

Invited Speakers

Torsten Moeller	University of Vienna, Austria
Vittorio Murino	Istituto Italiano di Tecnologia, Italy
Dieter Schmalstieg	Institute for Computer Graphics and Vision at Graz University of Technology, Austria
Gilles Bailly	CNRS, Telecom ParisTech, University Paris-Saclay, France

Contents

Invited Paper

Groups and Crowds: Behaviour Analysis of People Aggregations   3
Sadegh Mohammadi, Francesco Setti, Alessandro Perina,
Marco Cristani, and Vittorio Murino

Computer Graphics Theory and Applications

Real-Time Contour Image Vectorization on GPU   35
Xiaoliang Xiong, Jie Feng, and Bingfeng Zhou

Screen Space Curvature and Ambient Occlusion   51
Martin Prantl, Libor Váša, and Ivana Kolingerová

Multi-Class Error-Diffusion with Blue-Noise Property and Its Application . . .   72
Xiaoliang Xiong, Haoli Fan, Jie Feng, Zhihong Liu, and Bingfeng Zhou

Copula Eigenfaces with Attributes: Semiparametric Principal Component
Analysis for a Combined Color, Shape and Attribute Model   95
Bernhard Egger, Dinu Kaufmann, Sandro Schönborn, Volker Roth,
and Thomas Vetter

Representing Shapes of 2D Point Sets by Straight Outlines   113
Dirk Feldmann and Melanie Pohl

Sketching 2D Character Animation Using a Data-Assisted Interface   135
Priyanka Patel, Heena Gupta, and Parag Chaudhuri

Skin Deformation Methods for Interactive Character Animation   153
Nadine Abu Rumman and Marco Fratarcangeli

Appealing Avatars from 3D Body Scans: Perceptual Effects of Stylization   175
Reuben Fleming, Betty J. Mohler, Javier Romero, Michael J. Black,
and Martin Breidt

Information Visualization Theory and Applications

On the Visualization of Hierarchical Relations and Tree Structures
with TagSpheres   199
Stefan Jänicke and Gerik Scheuermann

Visual Analysis of Character and Plot Information Extracted
from Narrative Text.   220
   Markus John, Steffen Lohmann, Steffen Koch, Michael Wörner,
   and Thomas Ertl

Visual Querying of Semantically Enriched Movement Data   242
   Florian Haag, Robert Krüger, and Thomas Ertl

Correlation Coordinate Plots: Efficient Layouts for Correlation Tasks   264
   Hoa Nguyen and Paul Rosen

Analysis and Comparison of Feature-Based Patterns in Urban
Street Networks   287
   Lin Shao, Sebastian Mittelstädt, Ran Goldblatt, Itzhak Omer,
   Peter Bak, and Tobias Schreck

Swarm-Based Edge Bundling Applied to Flow Mapping   310
   Evgheni Polisciuc and Penousal Machado

Computer Vision Theory and Applications

Relative Pose Estimation from Straight Lines Using Optical
Flow-Based Line Matching and Parallel Line Clustering   329
   Naja von Schmude, Pierre Lothe, Jonas Witt, and Bernd Jähne

A Detailed Description of Direct Stereo Visual Odometry Based on Lines . . .   353
   Thomas Holzmann, Friedrich Fraundorfer, and Horst Bischof

Consumer-Level Virtual Reality Motion Capture.   374
   Catarina Runa Miranda and Verónica Costa Orvalho

Ground-Truth Tracking Data Generation Using Rotating
Real-World Objects.   395
   Zoltán Pusztai and Levente Hajder

The Sliced Pineapple Grid Feature for Predicting Grasping Affordances.   418
   Mikkel Tang Thomsen, Dirk Kraft, and Norbert Krüger

Extending Guided Image Filtering for High-Dimensional Signals   439
   Shu Fujita and Norishige Fukushima

Exemplar-Based Image Inpainting Using an Affine Invariant
Similarity Measure   454
   Vadim Fedorov, Pablo Arias, Gabriele Facciolo, and Coloma Ballester

Real-Time Visual Odometry by Patch Tracking Using GPU-Based
Perspective Calibration   475
  Rafael F.V. Saracchini, Carlos A. Catalina, Rodrigo Minetto,
  and Jorge Stolfi

Adaptive Non-local Means Using Weight Thresholding   493
  Asif Khan and Mahmoud R. El-Sakka

How Good Can a Face Identifier Be Without Learning?   515
  Yang Zhong, Anders Hedman, and Haibo Li

Object Tracking Guided by Segmentation Reliability Measures
and Local Features   534
  Cristian M. Orellana and Marcos D. Zuniga

Affordance Origami: Unfolding Agent Models for Hierarchical
Affordance Prediction   555
  Viktor Seib, Malte Knauf, and Dietrich Paulus

From Occlusion to Global Depth Order, a Monocular Approach   575
  Babak Rezaeirowshan, Coloma Ballester, and Gloria Haro

Infinite, Sparse 3D Modelling Volumes   593
  Eugen Funk and Anko Börner

Author Index   607

Invited Paper

Groups and Crowds: Behaviour Analysis of People Aggregations

Sadegh Mohammadi[1]([✉]), Francesco Setti[2], Alessandro Perina[3],
Marco Cristani[2], and Vittorio Murino[1,2]

[1] Pattern Analysis and Computer Vision (PAVIS), Istituto Italiano di Tecnologia,
Genova, Italy
{sadegh.mohammadi,vittorio.murino}@iit.it
[2] Department of Computer Science, University of Verona, Verona, Italy
{francesco.setti,marco.cristani,vittorio.murino}@univar.it
[3] Microsoft Corp, WDG Core Data Science, Redmond, USA
alperina@microsoft.com

Abstract. Automatic analysis of human behavior in social environment is a key topic for the computer vision community, with applications in security and video surveillance. While human behavior at an individual (single person) level has been widely studied in the past years, analysis of groups and crowd behavior, is still at a preliminary stage, with room for new approaches to emerge. Recently, there has been significant research effort dedicated to the development of automated computer vision techniques, intended to enhance safety of our societies by monitoring human behaviors and their actions in *groups* and *crowd* level. In particular, groups are usually formed by number of people who gathered for private meeting, birthday party, or wedding, while we consider crowd as huge number of people are gathered together to participate for a national or religious event, or protest due to some dissatisfaction. In this chapter, we will provide a broad overview on proposed approaches on human behavior analysis in group and crowd level, as well as, a detailed of some most recent state-of-the-art methods along with extensive experiments and comparison.

1 Introduction

Analyzing the visual content of scenes in videos is increasingly becoming an active research area in computer vision, due to its growing demand in security and surveillance applications. The content of a video captured by surveillance cameras can be potentially monitored by expert personnel for retaining public safety and reducing social crimes in crowded places such as airports, stadiums and malls. However, this is drastically limited by the scarcity of trained personnel and the natural limitation of human attention capabilities to monitor a huge amount of videos simultaneously filmed by multiple surveillance cameras [21]. This hurdle has motivated vision communities to develop methods for automated analysis of crowd scenes recorded by surveillance cameras.

© Springer International Publishing AG 2017
J. Braz et al. (Eds.): VISIGRAPP 2016, CCIS 693, pp. 3–32, 2017.
DOI: 10.1007/978-3-319-64870-5_1

Hitherto, numerous computer vision techniques have been successfully developed to detect and understand human activities in video data [45]. These techniques could be divided in two fundamental types: those which understand the social dynamics occurring in the scene by lying on the fine analysis of each single individual in terms of fine-grained cues (gestures, head pose orientation, feet orientation etc.), and those that operate on more crowded scenarios, where the number of individuals is so high that a per-pedestrian robust analysis cannot be envisaged. In the first case, one of the most intriguing analysis is the number of people who formed the groups within a scene and their activities. *Detection of groups of interacting people* is a very interesting and useful task in many modern technologies, with application fields spanning from video-surveillance to social robotics.

In the case that per-person fine analysis cannot be carried out, another branch of approaches should be taken into account. For example, in the case of highly crowded scenarios (more than 30 people), person detection, tracking, gesture recognition and other techniques for fine-grained analysis are often degraded by the presence of severe occlusions, cluttered background, low quality of surveillance data and, most importantly, by the complex interplays among people involved in crowd [28].

That has opened up a new broad research line which is generally referred to as *crowd scene analysis* in the computer vision literature [33, 64].

In this chapter, we analyze the very last state of the art related to the group analysis, together with the latest results in terms of crowd scene analysis.

A group can be broadly understood as a social unit comprising several members who stand in status and relationships with one another [5]. However, there are many kinds of groups, that differ in dimension (small groups or crowds), durability (ephemeral, ad hoc or stable groups), in/formality of organization, degree of sense of belonging, level of physical dispersion, etc. [6] (see the literature review in the next section). In this article, we build from the concepts of sociological analysis and we focus on free-standing conversational groups (FCGs), or small ensembles of co-present persons engaged in ad hoc focused encounters [6–8]. FCGs represent crucial social situations, and one of the most fundamental bases of dynamic sociality: these facts make them a crucial target for the modern automated monitoring and profiling strategies which have started to appear in the literature in the last three years [3, 9–14]. FCGs emerge during many and diverse social occasions, such as a party, a social dinner, a coffee break, a visit in a museum, a day at the seaside, a walk in the city plaza or at the mall; more generally, when people spontaneously decide to be in each others immediate presence to interact with one another. For these reasons, FCGs are fundamental social entities, whose automated analysis may bring to a novel level of activity and behavior analysis.

In a FCG, people communicate to the other participants, among and above all the rest, what they think they are doing together, what they regard as the activity at hand. And they do so not only, and perhaps not so much, by talking, but also, and as much, by exploiting non-verbal modalities of expression, also called social signals [23], among which positional and orientational forms play a crucial role (cf. also [7], p. 11). In fact, the spatial position and orientation of people define one of the most important proxemic notions which describe an FCG, that is, Adam Kendons Facing Formation, mostly known as F-formation.

Detecting free-standing conversational groups is useful in many contexts. In video-surveillance, automatically understanding the network of social relationships observed in an ecological scenario may result beneficial for advanced suspect profiling, improving and automatizing SPOT (Screening Passengers by Observation Technique) protocols [26], which nowadays are performed uniquely by human operators. In this chapter we analyze one of the latest technique for group detection, acting on single images acquired by a monocular camera, which operates on positional and orientational information of the individuals in the scene. Unlike previous approaches, the methodology is a direct formulation of the sociological principles (proximity, orientation and ease of access) concerning F-formations.

However, the aforementioned approaches is useful for moderate crowd scenarios, where the people are segmentable, and we can track them within a frame of video. Whilst, this is not a case for crowd scenes. Therefore, crowd scene analysis has recently attracted intense attention from the vision community. In particular, proposed method in this area can be categorized into three topics, including (1) crowd density estimation and people counting, (2) tracking in crowd, and (3) modeling crowd behaviors [21]. Estimating the number of people in a crowd is the foremost stage for several real-world applications such as safety control, monitoring public transportation, crowd rendering for animation and crowd simulation for urban planning. Despite many significant works in this area [13,17], automated crowd density estimation still remains an open problem in computer vision due to extreme occlusions and visual ambiguities of human appearance in crowd images [50]. Tracking individuals (or objects) in crowd scenes is another challenging task [48,56]: other than severe occlusions, cluttered background and pattern deformations, which are common difficulties in visual object tracking, the efficiency of crowd trackers is largely dependent on crowd density and dynamics, people social interactions as well as the crowd's psychological characteristics [2][1].

The primary goal of modeling crowd behaviors is to identify abnormal events such as riot, panic and violence in crowd scenes [29]. Despite recent success in this research field, detecting crowd abnormalities still remain an open and very challenging problem. The biggest issue of crowd anomaly detection lies in the definition of abnormality as it is strongly context dependent [25,37]. For example, riding a bike in a street is a normal action, whereas it is considered abnormal in another scene with a different context such as a park or sidewalk. Similarly, people gathering for a social event is a normal event, while same gathering at the same place to "protest against a law" is an abnormal event. Another challenge stems from the lack of adequate training samples to learn a well-generalized crowd model. This drastically degrades the generalization power of current crowd

---

[1] Readers are referred to [33,50] for a full treatment on the tasks of crowd tracking and density estimation.

models, since they are not capable of capturing the large intra-class variations of crowd behaviors [21].

Concerning the crowd scene analysis, in this chapter we will overview some leading techniques in the computer vision literature designed for detecting abnormal behaviors in crowd, with a focus on existing *motion* and *model* based approaches. Then, we will give a general overview on the most recent approaches. Finally, we will extensively evaluate their performance on various challenging imaging and crowding conditions.

The rest of the paper will be organized as follows: the first part is related to the analysis of the group, in the case single individuals can be captured and modeled. In the second part we will consider the approaches of crowd scene analysis.

2   Groups: Related Work

In computer vision, the analysis of groups has occurred historically in two broad contexts: video-surveillance and meeting analysis.

Within the scope of video-surveillance, the definition of a group is generally simplified to two or more people of similar velocity, spatially and temporally close to one another [15]. This simplified definition arises from the difficulty of inferring persistent social structure from short video clips. In this case, most of the vision-based approaches perform group tracking, i.e. capturing individuals in movement and maintaining their identity across video frames, understanding how they are partitioned in groups [4, 15–19].

In meeting analysis, typified by classroom behavior [1], people usually sit around a table and remain near a fixed location for most of the time, predominantly interacting through speech and gesture. In such a scenario, activities can be finely monitored using a variety of audiovisual features, captured by pervasive sensors like portable devices, microphone arrays, etc. [20–22].

From a sociological point of view, meetings are examples of social organization that employs focused interaction, which occurs when persons openly cooperate to sustain a single focus of attention [6, 7]. This broad definition covers other collaborative situated systems of activity that entail a more or less static spatial and proxemic organization such as playing a board or sport game, having dinner, doing a puzzle together, pitching a tent, or free conversation [6], whether sitting on the couch at a friends place, standing in the foyer and discussing the movie, or leaning on the balcony and smoking a cigarette during work-break.

Free-standing conversational groups (FCGs) [8] are another example of focused encounters. FCGs emerge during many and diverse social occasions, such as a party, a social dinner, a coffee break, a visit in a museum, a day at the seaside, a walk in the city plaza or at the mall; more generally, when people spontaneously decide to be in each others immediate presence to interact with one another. For these reasons, FCGs are fundamental social entities, whose automated analysis may bring to a novel level of activity and behavior analysis.

A robust FCG detector may also impact the social robotics field, where the approaches so far implemented work on few number of people, usually focusing on a single F-formation [27–29].

Efficient identification of FCGs could be of use in multimedia applications like mobile visual search [30,31], and especially in semantic tagging [32,33], where groups of people are currently inferred by the proximity of their faces in the image plane. Adopting systems for 3D pose estimation from 2D images [34] plus an FCG detector could in principle lead to more robust estimations. In this scenario, the extraction of social relationships could help in inferring personality traits [35,36] and triggering friendship invitation mechanisms [37].

In computer-supported cooperative work (CSCW), being capable of automatically detecting FCGs could be a step ahead in understanding how computer systems can support socialization and collaborative activities: e.g., [38–41]; in this case, FCGs are usually found by hand, or employing wearable sensors.

Manual detection of FCGs occurs also in human computer interaction, for the design of devices reacting to a situational change [42,43]: here the benefit of the automation of the detection process may lead to a genuine systematic study of how proxemic factors shape the usability of the device.

The last three years have seen works that automatically detect F-formations: Bazzani et al. [9] first proposed the use of positional and orientational information to capture Steady Conversational Groups (SCG); Cristani et al. [3] designed a sampling technique to seek F-formations centres by performing a greedy maximization in a Hough voting space; Hung and Kröse [10] detected F-formations by finding distinct maximal cliques in weighted graphs via graph-theoretic clustering; both the techniques were compared by Setti et al. [12]. A multi-scale extension of the Hough-based approach [3] was proposed by Setti et al. [13]. This improved on previous works, by explicitly modeling F-formations of different cardinalities. Tran et al. [14] followed the graph based approach of [10], extending it to deal with video-sequences and recognizing five kinds of activities. Vascon *et al.* [60] employed a games-theoretic approach to deal with dominant sets in order to detect stati F-formations. Lastly, in [53] Setti *et al.* proposed a graph-cut technique that outperformed all the previous methods. In the following sections we will detail this method and present experimental results comparing all the above mentioned algorithms.

3 Graph-Cuts for F-Formation

GCFF method is strongly based on the formal definition of F-formation given by Kendon [27] (*page 209*):

> An F-formation arises whenever two or more people sustain a spatial and orientational relationship in which the space between them is one to which they have equal, direct, and exclusive access.

According to this definition, an F-formation is the proper organisation of three social spaces: *o-space*, *p-space* and *r-space* (see Fig. 1a). The o-space is a

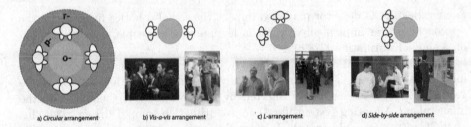

a) *Circular* arrangement   b) *Vis-a-vis* arrangement   c) *L-arrangement*   d) *Side-by-side* arrangement

Fig. 1. Structure of an F-formation and examples of F-formation arrangements. (a) Schematization of the three spaces of an F-formation: starting from the centre, *o-space*, *p-space* and *r-space*. (b–d) Three examples of F-formation arrangements: for each one of them, one picture highlights the head and shoulder pose, the other shows the lower body posture.

convex empty space surrounded by the people involved in a social interaction, where every participant is oriented inward into it, and no external people are allowed to lie. More in the detail, the o-space is determined by the overlap of those regions dubbed *transactional segments*, where as transactional segment we refer to the area in front of the body that can be reached easily, and where hearing and sight are most effective [15]. In practice, in a F-formation, the transactional segment of a person coincides with the o-space, and this fact has been exploited in our algorithm. The p-space is the belt of space enveloping the o-space, where only the bodies of the F-formation participants (as well as some of their belongings) are placed. People in the p-space participate to an F-formation using the o-space to transmit their messages. The r-space is the space enveloping o- and p-spaces, and is also monitored by the F-formation participants. People joining or leaving a given F-formation mark their arrival as well as their departure by engaging in special behaviours displayed in a special order in special portions of r-space, depending on several factors (context, culture, personality among the others); therefore, here we prefer to avoid the analysis of such complex dynamics, leaving their computational analysis as future work.

F-formations can be organised in different *arrangements*, that is, spatial and orientational layouts (see Fig. 1a–d) [16,18,27]. In F-formations of two individuals, usually we have a *vis-a-vis* arrangement, in which the two participants stand and face one another directly; another situation is the *L-arrangement*, when two people lie in a right angle to each other. As studied by Kendon [27], vis-a-vis configurations are preferred for competitive interactions, whereas L-shaped configurations are associated with cooperative interactions. In a *side-by-side* arrangement, people stand close together, both facing the same way; this situation occurs frequently when people stand at the edges of a setting against walls. *Circular* arrangements, finally, hold when F-formations are composed by more than two people; other than being circular, they can assume an approximately linear, semicircular, or rectangular shape.

Graph-Cuts for F-Formation finds the o-space of an F-formation, assigning to it those individuals whose transactional segments do overlap, without focusing on a particular arrangement. Given the position of an individual, to identify the transactional segment we exploit orientational information, which may come from the head orientation, the shoulder orientation or the feet layout, in increasing order of reliability [27]. The idea is that the feet layout of a subject indicates the mean direction along which his messages should be delivered, while he is still free to rotate his head and to some extent his shoulders through a considerable arc, before he must begin to turn his lower body as well. The problem is that feet are almost impossible to detect in an automatic fashion, due to the frequent (auto) occlusions; shoulder orientation is also complicated, since most of the approaches of body pose estimation work on 2D data and do not manage auto-occlusions. However, since any sustained head orientation in a given direction is usually associated with a reorientation of the lower body (so that the direction of the transactional segment again coincides with the direction in which the face is oriented [27]), head orientation should be considered proper for detecting transactional segments and, as a consequence, the o-space of an F-formation. In this work, we assume to have as input both positional information and head orientation; this assumption is reasonable due to the massive presence of robust tracking technologies [6] and head orientation algorithms [3,14,55].

In addition to this, we consider soft exclusion constraints: in an o-space, F-formation participants should have *equal, direct and exclusive access*. In other words, if person i stands between another person j, and an o-space centre O_g of the F-formation g, this should prevent j from focusing on the o-space, and, as a consequence, from being part of the related F-formation.

In what follows, we formally define the objective function accounting for positional, orientational and exclusion constraints aspects, and show how it can be optimised. Figure 2 gives a graphical idea of the problem formulation.

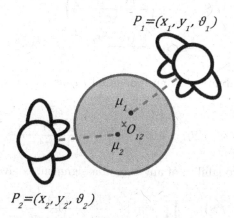

$$P_1 = (x_1, y_1, \vartheta_1)$$

$$P_2 = (x_2, y_2, \vartheta_2)$$

Fig. 2. Schematic representation of the problem formulation. Two individuals facing each other, the gray dot representing the transitional segment centre, the red cross being the o-space centre and the red area the o-space of the F-formation. (Color figure online)

3.1 Objective Function

We use $P_i = [x_i, y_i, \theta_i]$ to represent the position x_i, y_i and head orientation θ_i of the individual $i \in \{1, \ldots, n\}$ in the scene. Let TS_i be the a priori distribution which models the transactional segment of individual i. As we explained in the previous section, this segment is coherent with the position and orientation of the head, so we can assume $TS_i \sim \mathcal{N}(\mu_i, \Sigma_i)$, where $\mu_i = [x_{\mu_i}, y_{\mu_i}] = [x_i + D \cos \theta_i, y_i + D \sin \theta_i]$, $\Sigma_i = \sigma \cdot \mathbf{I}$ with $\mathbf{I}$ the 2D identity matrix, and D is the distance between the individual i and the centre of its transactional segment (hereafter called *stride*). The stride parameter D can be learned by cross-validation, or fixed a priori accounting for social facts. In practice, we assume the transactional segment of a person having a circular shape, which can be thought as superimposed to the o-space of the F-formation she may be part of.

$O_g = [u_g, v_g]$ indicates the position of a candidate o-space centre for F-formation $g \in \{1, M\}$, while we use G_i to refer to the F-formation containing individual i, considering the F-formation assignment $G_i = g$ for some g. The assignment assumes that each individual i may belong to a single F-formation g only at any given time, and this is reasonable when we are focusing one a single time, that is, an image. It follows naturally the definition of $O_{G_i} = [u_{G_i}, v_{G_i}]$, which represents the position of a candidate o-space centre for an unknown F-formation $G_i = \grave{g}$ containing i. For the sake of mathematical simplicity, we assume that each lone individual not belonging to a gathering can be considered as a *spurious* F-formation.

At this point, we define the likelihood probability of an individual i's transitional segment centre $C_i = [u_i, v_i]$ given the a priori variable TS_i.

$$\Pr(C_i | TS_i) \propto \exp\left(-\frac{\|C_i - \mu_i\|_2^2}{\sigma^2}\right) \tag{1}$$

$$= \exp\left(-\frac{(u_i - x_{\mu_i})^2 + (v_i - y_{\mu_i})^2}{\sigma^2}\right) \tag{2}$$

Hence, the probability that an individual i shares an o-space centre O_{G_i} is given by

$$\Pr(C_i = O_{G_i} | TS_i) \propto \exp\left(-\frac{(u_{G_i} - x_{\mu_i})^2 + (v_{G_i} - y_{\mu_i})^2}{\sigma^2}\right) \tag{3}$$

and the posterior probability of any overall assignment is given by

$$\Pr(C = O_G | TS) \propto \prod_{i \in [1,n]} \exp\left(-\frac{(u_{G_i} - x_{\mu_i})^2 + (v_{G_i} - y_{\mu_i})^2}{\sigma^2}\right) \tag{4}$$

with C the random variable which models a possible joint location of all the o-space centres, O_G is one instance of this joint location, and TS is the position of all the transitional segments of the individuals in the scene.

Clearly, if the number of o-space centres is unconstrained, the maximum a posteriori probability (MAP) occurs when each individual has his own separate o-space centre, generating a *spurious* F-formation formed by a single individual, that is, $O_{G_i} = TS_i$. To prevent this from happening, we associate a minimum description length prior (MDL) over the number of o-space centres used. This prior takes the same form as dictated by the Akaike Information Criterion (AIC) [10], linearly penalising the log-likelihood for the number of models used.

$$\Pr(C = O_G|TS) \propto \prod_{i \in [1,n]} \exp\left(-\frac{(u_{G_i} - x_{\mu_i})^2 + (v_{G_i} - y_{\mu_i})^2}{\sigma^2}\right) \cdot \exp(-|O_G|) \quad (5)$$

where $|O_G|$ is the number of distinct F-formations.

To find the MAP solution, we take the negative log-likelihood and discarding normalising constants, we have the following objective $J(\cdot)$ in standard form:

$$J(O_G|TS) = \sum_{i \in [1,n]} (u_{G_i} - x_{\mu_i})^2 + (v_{G_i} - y_{\mu_i})^2 + \sigma^{-2}|O_G| \quad (6)$$

As such, this can be seen as optimizing a least-squares error combined with an MDL prior. In principle this could be optimised using a standard technique such as k-means clustering combined with a brute force search over all possible choices of k to optimise the MDL cost. In practice, k-means frequently gets stuck in local optima and, in fact, using the technique described the least squares component of the error frequently increases, instead of decreasing, as k increases. Instead we make use of the graph-cut based optimisation described in [30], and widely used in computer vision [9,11,35,63].

In short, we start from an abundance of possible o-space centres, and then we use a hill-climbing optimisation that alternates between assigning individuals to o-space centres using the efficient graph-cut based optimisation [30] that directly minimises the cost (6), and then minimising the least squares component by updating o-space centres to the mean of O_g, for all the individuals $\{i\}$ currently assigned to the F-formation. The whole process is iterated until convergence. This approach is similar to the standard k-means algorithm, sharing both the assignment, and averaging step. However, as the graph-cut algorithm selects the number of clusters, we can avoid local minima by initialising with an excess of model proposals. In practice, we start from the previously mentioned trivial solution in which each individual is associated with its own o-space centre, centred on his position.

3.2 Visibility Constraints

Finally, we add the natural constraint that people can only join an F-formation if they can see the o-space centres. By allowing other people to occlude the o-space centre, we are able to capture more subtle nuances such as people being crowded out of F-formations or deliberately ostracised. Broadly speaking, an individual is excluded from an F-formation when another individual stands between him and the group centre. Taking $\theta_{i,j}^g$ as the angle between two individuals about a

---

Algorithm 1. Finding shared focal centres.

Initialise with $O_{G_i} = TS_i \quad \forall i \in [1, ..., n]$
old_cost= ∞
while $J(O_G, TS) <$old_cost **do**
  old_cost $\leftarrow J(O_G, TS)$
  run graph cuts to minimise cost (6)
  for $\forall g \in [1, ..., M]$ **do**
    if g is not empty **then**
      update $O_G \leftarrow \frac{\sum_{i \in g} TS_i}{|g|}$
    end if
  end for
end while

---

given o-space centre O_g for which is assumed $G_i = G_j = g$ and d_i^g, d_j^g as the distance of i, or j, respectively from the o-space centre O_g, the following cost captures this property:

$$R_{i,j}(g) = \begin{cases} 0 & \text{if } \theta_{i,j}^g \leq \hat{\theta}, \text{ or } d_i^g < d_j^g \\ \exp\left(K\cos(\theta_{i,j}^g)\right)\frac{d_i^g - d_j^g}{d_j^g} & \text{otherwise.} \end{cases} \tag{7}$$

and use the new cost function:

$$J'(O_G|TS) = J(O_G|TS) + \sum_{i,j \in P} R_{i,j}(G_i) \tag{8}$$

$R_{i,j}(g_i)$ acts as a visibility constraint on i regardless of the group person j is assigned to, as such it can be treated as a unary cost or data-term and included in the graph-cut based part of the optimisation. Now we turn to other half of the optimisation - updating the o-space centres. Although, given an assignment of people to a o-space centre, a local minima can be found using any off the shelf non-convex optimisation, we take a different approach. There are two points to be aware of: first, the difference between J' and J is sharply peaked and close to zero in most locations, and can generally be safely ignored; second and more importantly, we may often want to move out of a local minima. If updating an o-space centre results in a very high repulsion cost to one individual, this can often be dealt with by assigning the individual to a new group, and this will result in a lower overall cost, and more accurate labelling. As such, when optimising the o-space centres, we pass two proposals for each currently active model to graph-cuts – the previous proposal generated, and a new proposal based on the current mean of the F-formation. As the graph-cut based optimisation starts from the previous solution, and only moves to lower cost labellings, the cost always decreases and the procedure is guaranteed to converge to a local optimum.

3.3   An Explicative Example

Figure 3 gives a visual insight of our graph-cuts process. Given the position and orientation of each individual P_i, the algorithm starts by computing the transitional segments C_i. At the first iteration 0, the candidate o-space centres O_i are initialized, and are coincident with the transitional segments C_i; in this example are present 11 individuals, so 11 candidate o-space centres are generated. After iteration 1, the proposed segmentation process provides 1 singleton (P_{11}) and 5 FCGs of two individuals each. We can appreciate different configurations such as *vis-a-vis* ($O_{1,2}$), L-shape ($O_{3,4}$) and side-by-side ($O_{5,6}$). Still, the grouping in the bottom part of the image is wrong (P_7 to P_{10}), since it violates the exclusion principle. In iteration 2, the previous candidate o-space centres is considered as initialization, and a new graph is built. In this new configuration, the group $O_{7,10}$ is recognized as violating the visibility constraint and thus the related edge is penalized; a new run of graph-cuts minimization allows to correctly cluster the FCGs in a singleton (P_{10}) and a FCG formed by three individuals ($O_{7,8,9}$), which corresponds to the ground truth (visualized as the dashed circles).

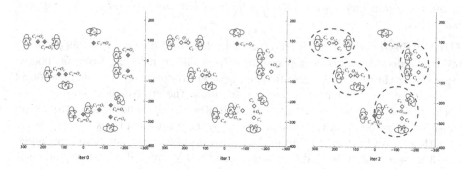

Fig. 3. An explicative example. Iteration 0: initialization with the candidate o-space centres $\{O\}$ coincident with the transitional segment of each individual $\{C\}$. Iteration 1: first graph-cuts run; easy groups are correctly clustered while the most complex still present errors (the FCG formed by P_7 and P_20 violates the visibility constraint). Iteration 2: the second graph-cuts run correctly detects the $O_{7,8,9}$ F-formation.

4   Groups: Experiments

In this section we present experimental results on five publicly available datasets employed as benchmark, and eight state of the art methods. Here we will show that GCFF definitely outperforms all the competitors, setting in all the cases new state-of-the-art scores. To conclude, we present an extended analysis of how the methods perform in terms of their ability of detecting groups of various cardinality and to test the robustness to noise, further promoting our technique.

Five publicly available datasets are used for the experiments: two from [19] (*Synthetic* and *Coffee Break*), one from [24] (*IDIAP Poster Data*), one from [52]

Table 1. Summary of the features of the datasets used for experiments.

Dataset	Data type	Detection	Detection quality	# of frames	# of people	# of people for groups' cardinality					
						1	2	3	4	5	6
Synthetic data	Synthetic	–	Perfect	100	900	300	360	240	–	–	–
IDIAP poster	Real	Manual	Very high	82	1,695	429	910	339	12	5	–
Cocktail party	Real	Automatic	High	320	1,915	174	162	246	176	275	882
Coffee break	Real	Automatic	Low	119	1,299	376	464	459	–	–	–
GDet	Real	Automatic	Very low	403	1,474	367	394	372	88	175	78

(*Cocktail Party*), and one from [5] (*GDet*). A summary of the dataset features is in Table 1, while a detailed presentation of each dataset follows.

Synthetic Data. *A psychologist generated a set of 10 diverse situations, each one repeated with minor variations for 10 times, resulting in 100 frames representing different social situations, with the aim to span as many configurations as possible for F-formations. An average of 9 individuals and 3 groups are present in the scene, together with some singletons. Proxemic information is noiseless in the sense that there is no clutter in the position and orientation state of each individual.*

IDIAP Poster Data. Over 3 h of aerial videos (resolution 654×439 px) have been recorded during a poster session of a scientific meeting. Over 50 people are walking through the scene, forming several groups over time. A total of 82 images were selected with the idea to maximise the crowdedness and variance of the scenes. Images are unrelated to each other in the sense that there are no consecutive frames, and the time lag between them prevents to exploit temporal smoothness. As for the data annotation, a total of 24 annotators were grouped into 3-person subgroups and they were asked to identify F-formations and their associates from static images. Each person's position and body orientation was manually labelled and recorded as pixel values in the image plane – one pixel represented approximately 1.5 cm.

Cocktail Party. This dataset contains about 30 min of video recordings of a cocktail party in a $30 \, m^2$ lab environment involving 7 subjects. The party was recorded using four synchronised angled-view cameras ($15 \, Hz$, 1024×768 px, jpeg) installed in the corners of the room. Subjects' positions were logged using a particle filter-based body tracker [31] while head pose estimation is computed as in [32]. Groups in one frame every 5 s were manually annotated by an expert, resulting in a total of 320 labelled frames for evaluation.

Coffee Break. The dataset focuses on a coffee-break scenario of a social event, with a maximum of 14 individuals organised in groups of 2 or 3 people each. Images are taken from a single camera with resolution of 1440×1080 px. People positions have been estimated by exploiting multi-object tracking on the heads, and head detection has been performed afterwards with the algorithm of [57], considering solely 4 possible orientations (front, back, left and right) in the image plane. The tracked positions and head orientations were then projected onto the ground plane. Considering the ground truth data, a psychologist annotated the videos indicating the groups present in the scenes, for a total of 119 frames split in two sequences. The annotations were generated by analysing each frame in combination with questionnaires that the subjects filled in.

GDet. The dataset is composed by 5 subsequences of images acquired by 2 angled-view low resolution cameras (352×328 px) with a number of frames spanning from 17 to 132, for a total of 403 annotated frames. The scenario is a vending machines area where people meet and chat while they are having coffee. This is similar to Coffee Break scenario but in this case the scenario is indoor, which makes occlusions many and severe; moreover, people in this scenario knows each other in advance. The videos were acquired with two monocular cameras, located on opposite angles of the room. To ensure the natural behaviour of people involved, they were not aware of the experiment purposes. For ground truth generation, people tracking has been carried out with the particle filter proposed in [31], while head pose estimation is performed afterwards with the method in [57] considering only 4 orientations (front, back, left and right).

We compare the GCFF algorithm with seven state of the art methods: one exploiting the concept of *view frustum* (IRPM), three based on dominant-sets (DS, IGD and GTCG), three different version of Hough Voting approaches (linear, entropic and multi-scale HVFF).

Inter-Relation Pattern Matrix (IRPM). Proposed by Bazzani *et al.* [5], uses the head direction to infer the 3D view frustum as approximation of the focus-of-attention of an individual; this is used together with proximity information to estimate interactions: the idea is that close-by people whose view frustum is intersecting are in some way interacting.

Dominant Sets (DS). Presented by Hung and Kröse [24], this algorithm considers an F-formation as a dominant-set cluster [44] of an edge-weighted graph, where each node in the graph is a person, and the edges between them measure the affinity between pairs.

Interacting Group Discovery (IGD). Presented by Tran *et al.* [58], is based on dominant sets extraction from an undirected graph where nodes are individuals and the edges have a weight proportional to how much people are interacting; the attention of an individual is modeled as an ellipse centred at a fixed offset in front of him, while the interaction between two individuals is proportional to the intersection of their attention ellipses.

Game-Theory for Conversational Groups (GTCG). In [59] the authors develop a game-theoretic framework, supported by a statistical modeling of the uncertainty associated with the position and orientation of people. Specifically, they use a representation of the affinity between candidate pairs by expressing the distance between distributions over the most plausible oriented region of attention. Additionally, they can integrate temporal information over multiple frames by using notions from multi-payoff evolutionary game theory.

Hough Voting for F-Formation (HVFF). Under this caption we consider a set of methods based on a Hough Voting strategy to build accumulation spaces and find local maxima of this function to identify F-formations. The general idea is that each individual is associated with a Gaussian probability density function which describes the position of the o-space centre he is pointing at. The pdf is approximated by a set of samples, which basically vote for a given o-space centre location. The voting space is then quantized and the votes are aggregated on squared cells, so to form a discrete accumulation space. Local maxima in this space identify o-space centres, and consequently, F-formations. Over the years, three versions of these framework have been presented: in [19] the votes are linearly accumulated by just summing up all the weights of votes belonging to the same cell, in [51] the votes are aggregated by using the weighted Boltzmann entropy function, while in [52] a multi-scale approach is used on top of the entropic version.

As accuracy measures, we adopt the metrics proposed in [19] and then extended in [53]: we consider a group as correctly estimated if at least $\lceil (T \cdot |G|) \rceil$ of their members are found by the grouping method and correctly detected by the tracker, and if no more than $1 - \lceil (T \cdot |G|) \rceil$ false subjects (of the detected tracks) are identified, where $|G|$ is the cardinality of the labelled group G, and $T \in \,]0,1]$ is an arbitrary threshold, called *tolerance threshold*. In particular, we focus on two interesting values of T: 2/3 and 1.

With this definition of *tolerant match*, we can determine for each frame the correctly detected groups (true positives – TP), the miss-detected groups (false negatives – FN) and the hallucinated groups (false positives – FP). With this, we compute the standard pattern recognition metrics precision and recall:

$$precision = \frac{TP}{TP + FP}, \qquad recall = \frac{TP}{TP + FN} \tag{9}$$

and the F_1 score defined as the harmonic mean of precision and recall:

$$F_1 = 2 \cdot \frac{precision \cdot recall}{precision + recall} \tag{10}$$

In addition to these metrics, we compute the *Global Tolerant Matching* score (GTM), which is the area under the curve (AUC) in the F_1 vs. T graph with T varying from 1/2 to 1. Since in our experiments we only have groups up to 6 individuals, without loss of generality we consider T varying with 3 equal steps in the range stated above.

Moreover, we will discuss results also in terms of group cardinality, by computing the F_1 score for each cardinality separately and then computing mean and standard deviation.

4.1   Best Results Analysis

Given the metrics explained above, the first test analyses the best performances for each method on each dataset; in practice, a tuning phase has been carried out for each method/dataset combination in order to get the best performances. Note, we did not have code for Dominant Sets [24] and thus we used results provided directly from the authors of the method for a subset of data. For this reason, average results over all the datasets are only averaged over 3 datasets, and cannot be taken into account for a fair comparison. Best parameters are found on half of one sequence by cross-validation, and kept unchanged for the remaining part of the dataset. Please note, finding the right parameters can also fixed by hand, since the stride D depends on the social context under analysis (formal meetings will have higher D, the presence of tables and similar items may also increase the diameter of the FCGs): with a given D, for example, it is assumed that circular F-formations will have diameter of $2D$. The parameter σ indicates how much we are permissive in accepting deviations from such a diameter. Moreover, D depends also on the different measure units (pixels/cm) which characterize the proxemic information associated to each individual in the scene.

Table 2 shows best results by considering the threshold $T = 2/3$, which corresponds to find at least 2/3 of the members of a group, no more than 1/3 of false subjects; while Table 3 presents results with $T = 1$, considering a group as correct if all and only its members are detected. The proposed method outperforms all the competitors, on all the datasets. With $T = 2/3$, three observations can be made: the first is that our approach GCFF improves substantially the precision (of 13% in average) and even more definitely the recall scores (of 17% in average) of the state of the art approaches. The second is that our approach produces the same score for both the precision and the recall; this is very convenient and convincing, since so far all the approaches of FCG detections have shown to be weak in the recall dimension. The third observation is that GCFF performs well both in the case where no errors in the position or orientation of the people are present (as the Synthetic dataset) and in the cases where strong noise of position and orientation is present (Coffee Break, GDet).

When moving to tolerance threshold equal to 1 (all the people in a group have to be individuated, and no false positive are allowed) the performance is reasonably lower, but the increment is even stronger w.r.t. to the state of the art, in general on all the datasets: in particular, on the Cocktail Party dataset, the results are more than twice the scores of the competitors. Finally, even in this case, GCFF produces a very similar score for precision and recall.

A performance analysis is also provided by changing the tolerance threshold T. Figure 4 shows the average F_1 scores for each method computed over all the frames and datasets. From the curves we can appreciate how the proposed method is consistently best performing for each T-value. In the legend of Fig. 4 the Global Tolerant Matching score is also reported. Again, GCFF is outperforming the state of the art, independently from the choice of T.

The reason why our approach does better than the competitors has been explained in the state of the art section, here briefly summarized: the Dominant Set-based approaches DS and IGD, even if they are based on an elegant optimization procedure, tend to find circular groups, and are weaker in individuating

Table 2. Average precision, recall and F_1 scores for all the methods and all the datasets ($T = 2/3$).

	Synthetic			IDIAP poster			Cocktail party			Coffee break			GDet			Total		
	Prec	Rec	F_1	Prec	Rec	F_1	Prec	Rec	F_1	Prec	Rec	F_1	Prec	Rec	F_1	Prec	Rec	F_1
IRPM [5]	0.85	0.80	0.82	0.82	0.74	0.78	0.56	0.43	0.49	0.68	0.50	0.57	0.77	0.47	0.58	0.70	0.49	0.56
DS [24]	0.85	0.97	0.90	0.91	0.92	0.91	–	–	–	0.69	0.65	0.67	–	–	–	0.81	0.83	0.82
IGD [58]	0.95	0.71	0.81	0.80	0.68	0.73	0.81	0.61	0.70	0.81	0.78	0.79	0.83	0.36	0.50	0.68	0.76	0.70
CTCG [59]	1.00	1.00	1.00	0.92	0.96	0.94	0.86	0.82	0.84	0.83	0.89	0.86	0.76	0.76	0.76	0.83	0.83	0.83
HVFF lin [19]	0.75	0.86	0.80	0.90	0.95	0.92	0.59	0.74	0.65	0.73	0.86	0.79	0.66	0.68	0.67	0.75	0.79	0.76
HVFF ent [51]	0.79	0.86	0.82	0.86	0.89	0.87	0.78	0.83	0.80	0.76	0.86	0.81	0.69	0.71	0.70	0.78	0.78	0.77
HVFF ms [52]	0.90	0.94	0.92	0.87	0.91	0.89	0.81	0.81	0.81	0.83	0.76	0.79	0.71	0.73	0.72	0.84	0.66	0.74
GCFF	0.97	0.98	0.97	0.94	0.96	0.95	0.84	0.86	0.85	0.85	0.91	0.88	0.92	0.88	0.90	0.89	0.89	0.89

Table 3. Average precision, recall and F_1 scores for all the methods and all the datasets ($T = 1$).

	Synthetic			IDIAP poster			Cocktail party			Coffee break			GDet			Total		
	Prec	Rec	F_1	Prec	Rec	F_1	Prec	Rec	F_1	Prec	Rec	F_1	Prec	Rec	F_1	Prec	Rec	F_1
IRPM [5]	0.53	0.47	0.50	0.71	0.64	0.67	0.28	0.17	0.21	0.27	0.23	0.25	0.59	0.29	0.39	0.46	0.29	0.35
DS [24]	0.68	0.80	0.74	0.79	0.82	0.81	–	–	–	0.40	0.38	0.39	–	–	–	0.60	0.63	0.62
IGD [58]	0.30	0.22	0.25	0.31	0.27	0.29	0.23	0.10	0.13	0.50	0.50	0.50	0.67	0.20	0.31	0.45	0.21	0.27
CTCG [59]	0.78	0.78	0.78	0.83	0.86	0.85	0.31	0.28	0.30	0.46	0.47	0.47	0.51	0.60	0.55	0.49	0.52	0.51
HVFF lin [19]	0.64	0.73	0.68	0.80	0.86	0.83	0.26	0.27	0.27	0.41	0.47	0.44	0.43	0.45	0.44	0.43	0.46	0.44
HVFF ent [51]	0.47	0.52	0.49	0.72	0.74	0.73	0.28	0.30	0.29	0.47	0.52	0.49	0.44	0.45	0.45	0.42	0.44	0.43
HVFF ms [52]	0.72	0.73	0.73	0.73	0.76	0.74	0.30	0.30	0.30	0.40	0.38	0.39	0.44	0.45	0.45	0.44	0.45	0.45
GCFF	0.91	0.91	0.91	0.85	0.87	0.86	0.63	0.65	0.64	0.61	0.64	0.63	0.73	0.68	0.71	0.71	0.70	0.71

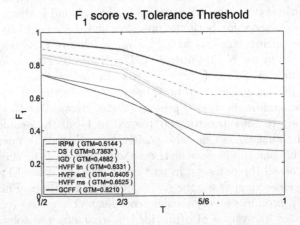

Fig. 4. Global F_1 score vs. tolerance threshold T. Between brackets in legend the Global Tolerant Matching score. Dominant Sets (DS) is averaged over 3 datasets only, because of results availability. (Best viewed in colour). (Color figure online)

other kinds of F-formations. Hough-based approaches HVFF X ($X = lin, ent, ms$) have a good modeling of the F-formation, allowing to find any shape, but rely on a greedy optimization procedure. Finally, IRPM approach has a rough modeling of the F-formation. Our approach viceversa has a rich modeling of the F-formation, and a powerful optimization strategy.

Table 4. Cocktail party – F_1 score vs. cardinality ($T = 1$).

# groups	$k = 2$	$k = 3$	$k = 4$	$k = 5$	$k = 6$	Avg	Std
	81	82	44	55	147	–	–
IRPM [5]	0.26	0.53	0.74	0.42	0.59	0.51	0.18
IGD [58]	0.06	0.52	0.66	0.73	0.85	0.56	0.30
HVFF lin [19]	0.38	0.76	0.57	0.67	0.94	0.66	0.21
HVFF ent [51]	0.45	0.75	0.69	0.73	0.96	0.71	0.18
HVFF ms [52]	0.49	0.74	0.70	0.71	0.96	0.72	0.17
GCFF	0.59	0.64	0.80	0.85	0.94	**0.76**	**0.14**

Table 5. GDet – F_1 score vs. cardinality ($T = 1$).

# groups	$k = 2$	$k = 3$	$k = 4$	$k = 5$	$k = 6$	Avg	Std
	197	124	22	35	13	–	–
IRPM [5]	0.40	0.59	0.45	0.42	0.35	0.44	**0.09**
IGD [58]	0.15	0.52	0.33	0.54	0.83	0.47	0.25
HVFF lin [19]	0.51	0.76	0.03	0.16	0.13	0.32	0.31
HVFF ent [51]	0.57	0.73	0.24	0.23	0.13	0.38	0.26
HVFF ms [52]	0.56	0.78	0.17	0.41	0.67	0.52	0.23
GCFF	0.74	0.87	0.53	0.77	0.88	**0.76**	0.14

Cardinality Analysis

As stated in [52], some methods are shown to work better with some group cardinalities. In this experiment, we systematically check this aspect, evaluating the performance of all the considered methods in individuating groups with a particular number of individuals. Since Synthetic, Coffee Break and IDIAP Poster Session datasets only have groups of cardinality 2 and 3, we only focus on the remaining 2 datasets, which have a more uniform distribution of groups cardinalities. Tables 4 and 5 show F_1 scores for each method and each group cardinality respectively for Cocktail Party and GDet datasets. In both cases the proposed method outperforms the other state of the art methods in terms of higher average F_1 score, with very low standard deviation. In particular, only IRPM gives in GDet dataset results which are more stable than ours, but they are definitely poorer.

Noise Analysis

In this experiment, we show how the methods behave against different degrees of clutter. For this sake, we consider the Synthetic dataset as starting point and we add to the proxemic state of each individual of each frame some random values based on a known noise distribution. We assume that the noise follows a Gaussian distribution with mean 0, and noise on each dimension (position, orientation) is uncorrelated. For our experiments we used $\sigma_x = \sigma_y = 20\,\mathrm{cm}$ and $\sigma_\theta = 0.1\,\mathrm{rad}$. In our experiments, we consider 11 levels of noise $L_n = 0, \ldots, 10$, where

$$\begin{cases} x_n(L_n) = x + \mathrm{randsample}(\mathcal{N}(0, L_n * \sigma_x)) \\ y_n(L_n) = y + \mathrm{randsample}(\mathcal{N}(0, L_n * \sigma_y)) \\ \theta_n(L_n) = \theta + \mathrm{randsample}(\mathcal{N}(0, L_n * \sigma_\theta)) \end{cases} \tag{11}$$

In particular, we produce results by adding noise on position only (leaving the orientation at its exact value), on orientation only (leaving the position of each individual at its exact value) and on both position and orientation. Figure 5 shows F_1 scores for each method while increasing the noise level. In this case we can appreciate that with high orientation and combined noise IGD performs comparably or better than GCFF; this is a confirmation of the fact that methods based on Dominant Sets are performing very well when the orientation information is not reliable, as already stated in [51].

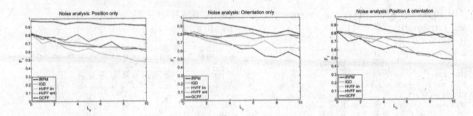

Fig. 5. Noise analysis. F_1 score vs. noise level on position (left), orientation (centre) and combined (right). (Best viewed in colour). (Color figure online)

5   Crowd: Related Work

Abnormal crowd behavior analysis has become an active research topic in recent years, so that several techniques have been proposed to automatically detecting abnormal behavior in crowds. We broadly divide the existing approach in two categories. There are two main categories named (i) *motion-based*, and (ii) *model-based*. In the following, we will give a brief overview of these two categories.

5.1   Motion Based Approaches

Motion-based referred to as methods in which motion cues such as optical flow and trajectories are used as a main source information [1,12,22,29,34,38–41,62]. Particularly, initial works on crowd behavior analysis consider crowd as a number of objects interacting in a common space, and the crowd behaviors are inferred according to the motion patterns captured from tracked objects [1,46,47]. However, detecting and tracking individuals in a crowd scenes is very challenging, due to imaging condition and density of crowds. In order to tackle this problem, proposed approaches consider crowd as a single entity instead of individual objects. For example, [1] used Finite Time Lyapunov Exponent (FTLE) field to detect and localize anomaly in a frame level from crowd videos by measuring flow instability in the segmented regions of high density crowd flows. However, this approach is limited to the structural scene where the boundary of crowd flow is well determined. To tackle this problem, a chaotic invariant approach is proposed for structural and nonstructural crowd scenes [61]. Particularly, the chaotic dynamics of crowd are computed from the crowd flow trajectory. The main limitation of this method is its computational complexity, since it use all the previous frames. Moreover, there exist also a significant number of research focus on identifying specific type of abnormality (e.g., violence, panic). The first work for detecting violence in videos was proposed in [20], which mainly focused on two person fight episodes and employed motion trajectory information of individual limbs for fight modeling and classification. This approach required limbs segmentation and tracking, which are very challenging tasks in a presence of occlusion and clutters. In [22] authors exploit the statistics magnitude of optical flow varying along the time for detecting violence behavior from dense crowded scenes. However, the effectiveness of this approach is also limited on highly dense crowd. In [40] authors quantified abnormality exploiting the statistic of motion estimated from magnitude and orientation of tracklet (short portion of trajectory) in terms of two dimensional histogram. Although, their method show promising performance on a wider ranger of crowd anomaly, choice of efficient number of bins across the scenes is under the question. In [38] authors used substantial derivative, a well-known concept in fluid dynamic, for extracting meaningful motion patterns to identify violence behaviors from video sequences. Moreover, authors demonstrate the effectiveness of the method on panic scenes. In the following we will provide brief details on [20, 22, 38] related to the motion based approaches.

Acceleration Measure Vector (Jerk). This approach can be considered as one of the first attempt in detecting human violence in video [20]. To detect violent behavior authors rely on motion trajectory information and on orientation information of a persons limbs. Then an Acceleration Measure Vector (AMV) composed of direction and magnitude of motion and jerk is defined to be the temporal derivative of AMV.

Violence Flows (ViF). It is specifically designed for classifying violent behaviors in crowds from video outbreaks [22]. The main strength of ViF lies in its ability in encoding meaningful temporal motion patterns. Initially, the magnitude of motions are estimated. Next, for each video frame a binary map is created by discarding motion values less than a predefined threshold. Then, final motion map is created by computing average along all video frames in a query video.

Substantial Derivative. Substantial Derivative (SD) is an important concept in fluid mechanics which describes the change of fluid elements by physical properties such as temperature, density, and velocity components of flowing fluid along its trajectory [4]. Unlike aforementioned approaches that only use temporal motion patterns as a main source of information [22, 36, 40], it has a capability of capturing spatial and temporal information of motion changes in a single framework [4]. Specifically, SD consists of two terms of accelerations, namely *local* and *convective* accelerations. The local acceleration captures the change rate of velocity of a certain particle with respect to time and vanishes if the flow is steady. On the other hands, the convective acceleration captures the change of velocity flow in the spatial space, therefore, it increases when particles move through the region of spatially varying velocity field in the temporal domain (see Fig. 6 for an example). Therefore, the local acceleration, which is the derivative of velocity with respect to the time, captures the instant change of flow. Whilst, the convective acceleration captures the spatial evolution of a particle moving along its trajectory. This is, in particular, useful in the abnormal crowd scenarios. Considering violence as an example of abnormal scene, where individual shows aggressive behavior toward other member of crowd, his/her motion is subject to sudden change of in velocity field with respect to the time (local acceleration). Moreover, at the same time, group of individual interact in a very different way compare with violence (or panic scenes), which lead to the significant difference in the motion trajectories of individuals compared with normal scenes in spatial domain (convective acceleration). For video representation random patch sampling with Bag-of-Word Paradigm applied from each force independently. Finally, the local and convective forces are concatenated to shape a single descriptor named *total force*.

Tables 6 and 7 provide some detail regarding the aforementioned motion based methods, including their general formula and force estimation from video sequences.

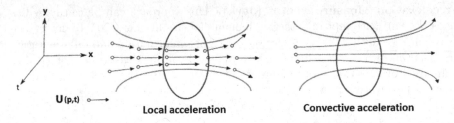

Fig. 6. An example of local and convective accelerations. The local acceleration measure instantaneous rate of change of each particle, while convective acceleration measures the rate of change of the particle moving along its trajectory. Red region indicates the particles are accelerated as it converge due to the structural change of the environment. (Color figure online)

Table 6. Summary of motion and model based forces, their general formula and estimation methods from video sequences.

Methods	Forces	General formula	Estimation from videos
Behavioral heuristics [49]	H_1	$\frac{dv_i}{dt} = \frac{v_i^{des} - v_i(t)}{\tau}$	$\frac{dv_i}{dt} \simeq a_i^{t+1} = O_i^{t+1} - O_i^t$
	$F_{ij}^{bc}(H_2)$	$F_{ij}^{bc} = n_{ji} \cdot \mathbf{g}_i(j)$	$F_i^{bc} = \frac{\sum_j a_j \cdot \mathbf{g}_i(j)}{\sum_j \mathbf{g}_i(j)}$
	$F_{ij}^{agg}(H_3)$	$F_{ij}^{agg} = n_{ji} \cdot \frac{(1 - \frac{v_i \cdot v_j}{\|v_j\| \cdot \|v_i\|})}{2} \cdot \mathbf{f}_i(j)$	$F_i^{agg} = \sum_j \left(n_{ji} \cdot \mathbf{w}_{ij} \cdot \mathbf{f}_{OO_i}^{\alpha}(j) \right)$
SFM [36]	F_{int}	$F_{int} = \frac{1}{\tau}\left(v_i^q - v_i \right) - \frac{dv_i}{dt}$	$v_i = \langle O_i \rangle$ $v_i^q = (1 - p_i)O_i + p_i \langle O_i \rangle$
Substantial derivative [38]	F^L	$F^L = m_i \frac{dv_i}{dt}$	$\frac{dv_i}{dt} \simeq a_i^{t+1} = O_i^{t+1} - O_i^t$
	F^{Cv}	$F^{Cv} = m_i\sqrt{a_x^2 + a_y^2}$	$a_x = \left(\frac{\partial\langle O_x \rangle}{\partial x} + \frac{\partial\langle O_y \rangle}{\partial y} \right) \cdot \langle O_x \rangle,$ $a_y = \left(\frac{\partial\langle O_x \rangle}{\partial x} + \frac{\partial\langle O_y \rangle}{\partial y} \right) \cdot \langle O_y \rangle$
	F^T	$F^T = F^{Cv} + F^L$	$F^T = F^{Cv}|F^L$
Jerk [20]	J	$J_i = \frac{da_i}{dt}$	$\frac{da_i}{dt} \simeq J_i = a_i^{t+1} - a_i^t$

5.2 Model Based Approaches

Model based approaches can be broadly categorized into two categories; (i) The Social Force Model (SFM), (ii) Behavioral heuristic models.

Table 7. Summary of notations.

Notations	Description formula
v_i/O_i^t	Velocity of particles/individuals i
$\langle O_i \rangle$	Average velocity of particles/individuals i
$g(i)$	Body compression subject to the individual i and neighboring individuals
$W_{i,j}$	Aggression factor between the individual i and his/her opponent j
$\theta_{i,j}$	Orientation between individual i and his/her opponents j
$d_{i,j}$	Distance between individual i and j j
R	Personal space around each examined individual
p_i	Panic factor
$f(i)$	Model the angle of view of individual i individual
a_x	Spatial acceleration in x direction
a_y	Spatial acceleration in y direction
F_{ij}^{bc}	Body contact individual i with its neighboring pedestrians j
F_{ij}^{agg}	Aggression force pedestrian i derived toward its opponent j
F^L	Local force
F^{Cv}	Convective force
F^T	Total force

The Social Force Model. The SFM originally introduced by Helbing and Molnar [23], is the historical seminal method for modeling crowd behaviors according to a set of predefined physical rules. More specifically, the SFM aimed at representing the interaction force among individuals in crowded scenes using a set of *repulsive* and *attractive* forces, which was shown to be a significant feature for analyzing crowd behaviors. Motivated by the success of SFM to reproduce crowd moving patterns, fifteen years later Mehran et al. [36] adopted the SFM and particle advection scheme to compute for detecting and localizing abnormal behavior in crowd videos. To this purpose, they considered the entire crowd as a set of moving particles whose interaction force was computed using SFM. Then, they mapped the interaction force into the image plane to obtain the force flow of each particle within frame of videos. This force map was used as the basis for extracting features which, along with the random spatial-temporal path sampling and BOW strategy, was used to assign either normal or abnormal label to each frame. Moreover, the force map was also employed to localize the anomaly region in the detected abnormal frames.

5.3 Behavioral Heuristics

Behavioral heuristic approach can be considered as a new emergent approach to disclose complex crowd dynamics [42, 43], compare to SFM [23], where single Newtonian equation used to explain crowd behaviors, it exploits the use of physics equations inspired from simple, yet effective behavioral heuristics to describe the crowd behaviors. This is, in particular useful since it is capable of capturing wider range of crowd complexities. To this end, Mohammadi et al. [49], proposed a heuristic based method inspired from [42,43], and successfully applied for violence detection in crowds [49]. Their method established based on three heuristic rules:

H1: *An individual chooses the direction that allows the most direct path to a destination point, adopting his/her moving regarding the presence of obstacles.*

H2: *In crowd situations, the movement of an individual is influenced by his/her physical body contacts with surrounding persons.*

H3: *In violent scenes, an individual mainly moves towards his/her opponents to display violent actions.*

Specifically, (H1) describe individual's internal motivation towards a goal avoiding obstacles or other individuals. While, the second heuristic rule (H2), states that individual movements are subject to the unintentional physical body contacts with his/her surrounding individuals. The third heuristic rule (H3) defines behavioral patterns within violent scenes, where there are two or more parties (e.g., police and rioters) fighting and showing violent behaviors to each other.

Then, each heuristic is formulated with physics equation(see Tables 6 and 7 for details). Next, each force is computed, independently, from video sequences. Finally, random patch sampling along with Bag-of-Word paradigm used for force representation, and all the forces are concatenated to construct the final descriptor named Visual Information Processing Signature (VIPS).

6　Crowds: Experiments

We extensively evaluated proposed approach on five benchmarks, consists of two standard datasets; Violence in Crowds [22], and Behave [8] along with two video sequences collected from web source (i.e., www.YouTube.com) which we named Panic1 and Panic2. Moreover, we assembled a new dataset, named "Violent-Cross" whose videos gathered from Violence in Crowds and CUHK [54] datasets, to show the ability of proposed approaches in cross scene recognition. Specifically, it includes 300 videos, equally divided into three classes (100 videos for each class). Figure 7 shows few samples of the benchmarks. For the Violence in Crowds we used the standard training/testing split, released along with dataset. While for Behave, panic1 and panic2, we divide the video frames into block of 10 frames with one frame overlapping among the blocks, then we labeled them into normal and abnormal blocks. For Violence-Cross, we equally divide each class into a test set of 150 videos (50 video sequences for each class) and the rest for testing. For feature representation we used Bag-of-Word Pardigm. Particularly, first we perform random 3D patch sampling from the computed force maps, then Bag-of-Word Paradigm, with fixed number of cluster centers to generate codebook. Finally, the resulting histogram of visual words are fed into the classifiers. As a choice of classifier, for Violence in Crowds and Violence-Cross, we used SVM, since we had access to negative data at the training time. While, for other datasets, we used Latent Dirichlet Allocation [7] in short LDA since we assumed that we do not have access to negative data at the training time.

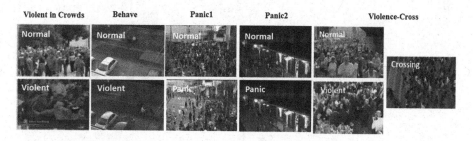

Fig. 7. Sample frames from the evaluated datasets. For first four columns; first row shows the normal crowd behaviors, while for the second row shows the abnormal behavior, specifically, violent and panic behaviors. For Violence-Cross datasets; samples from the three classes of "normal", "crossing" and "violent" behaviors.

6.1 Effect of the Random Patches

Here, we examine the effect of varying number of random patches in the range of $P \in \{100, 200, 400, 800, 1000\}$ on various datasets by fixing number of cluster center to 500. Figure 8 shows the effect of varying number of patches on optical flow, SFM [36], SD ($F^L|F^{Cv}$) [38], and VIPS [49]. As it is visible, all the descriptors show improvement in performance by increasing the number of random patches, however, we did not observe any significant improvement after reaching 1000 random patches. Moreover, we observed that VIPS shows very promising performance compare with other methods on Violence in Crowds and Behave datasets. While, SD shows its superiority in performance on panic scenes. This is understandable, since VIPS specifically designed for violence detection and aggression force plays an important role in capturing violence behavioral patterns. While, experimental results show that in panic scenarios (Painc1 and Panic2 datasets) combination of temporal and structure information in SD descriptor offers more discriminative features compare with other proposed methods.

6.2 Comparison with the State of the Art

We compared various motion and model based methods for the evaluation purpose. Particularly, for the motion based approaches, we select ViF [22], Jerk [20], total force inspired from SD [38], and for model based approach we select the SFM [36], and recently proposed heuristic based model in [49].

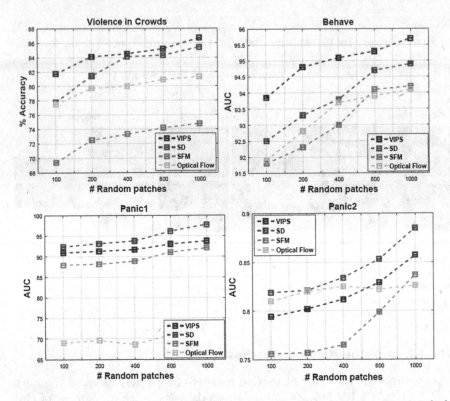

Fig. 8. Effect of varying number of random patches on the effectivness of VIPS [49], SD [38], SFM [36] and optical flow on Violence in Crowds, Behave, Panic1 and Panic2 datasets.

Table 8 show the comparison of motion and model based approaches. It is visible that among the compared methods heuristic based method outperform other competitors especially in Violence in Crowds dataset, this support the psychology studies [42,43], and highlight the strength of heuristic models in capturing wider range of crowd complexities in violence scenes which results in better performance. It is also visible that SD [38] can be considered as a close competitor to the heuristic method. Indeed, it outperforms the heuristic-based method with a high margin in Panic datasets. This is understandable since, VIPS [49] specifically designed for violence detection. In addition, it has a consist of spatial information (Local and Convective Forces, respectively), which make it capable in covering wider range of crowd complexity compared with other methods.

Moreover, we evaluate robustness of our descriptors to distinguish between acts of violence from crossing behaviors, which is a most similar approach to the act of violence. In particular, as a motion based descriptor, we select ViF [22], which is specifically design for violence detection, along with SFM [36], which is considered as one of the most well-known descriptor to detect abnormality

Table 8. Comparison of model and motion based approaches.

Descriptors	Violence in Crowds	Behave	Panic1	Panic2
ViF [22]	81.3	93.4	0.716	82.6
AMV (Jerk) [20]	74.8	94.2	90.9	83.6
SD $(F^L|F^{Cv})$ [38]	85.4	94.8	98.5	88.4
SFM [36]	74.5	94.23	91.3	84.3
VIPS [49]	86.61	95.73	94.5	85.9

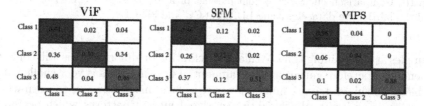

Fig. 9. Average accuracy on Violent-Cross dataset. Class1, Class2, and Class3 are referred to as violent, cross walk, and normal behaviors, respectively. ViF [22] with 57% overall accuracy; SFM [36] with 69% overall accuracy, and $VIPS$ with 92% overall accuracy.

in crowds. Figure 9 shows the confusion matrices of two state-of-the-art methods and elements of the proposed method. We observe that ViF shows a good performance on detecting acts of violence compared to the SFM, however, its overall accuracy is low since it is incapable of distinguishing violent from normal and crossing behaviors. On the other hand, we observe that VIPS shows very promising performance compared with other approaches.

Table 9. Average AUCs on Panic1, Panic2 and Behave datasets sequences with 1000 random patches and 500 cluster centers.

Generative model	Datasets		
	Panic1	Panic2	BEHAVE
LDA [7]	0.718	0.826	0.943
2D-CG [26]	0.833	0.86	0.951
3D-CG [37]	0.851	0.865	0.958

6.3   Choice of Generative Model

In general, there exist two methods for evaluating the effectiveness of descriptors for violence detection; Discriminative (e.g., SVM) and Generative (e.g., LDA [7], Counting Grids [26]) methods. Although, discriminative approaches are considered as more powerful approach for detecting abnormalities, it requires negative

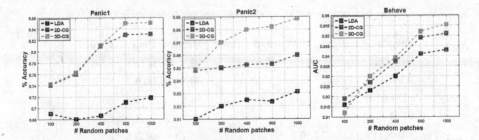

Fig. 10. Comparison effect of random patches of average AUCs (y-axis) on Panic1, Panic2, and Behave datasets, using LDA, 2D-CG and 3D-CGs.

examples at the training time. While, the definition of abnormality is still context dependent, therefore gathering huge amount of negative data is very challenging problem. To this end, one common practice to face this problem, is to learn what normality is and then abnormality considered motion patterns which deviated from learned distribution of normal behavior [36,37]. However, apart from effectiveness of descriptors in capturing crowd complexities, capability of generative models on modeling the distribution of normal data plays an important roles in detecting abnormal behaviors. To this end, we show comparison on Generative models. Particularly, we select the optical flow as a baseline, and we compared LDA with two dimensional Counting Grid (2D-CG) [26] and three dimensional Counting Grid (3D-CGs) [37] on Panic1, Panic2 and Behave datasets. Table 9 shows that 3D-CG outperform two competitors. This is understandable, since 3D-CG is able to capture spatial-temporal relationship among the bags in the feature space, while CG is only consider spatial information and LDA totally ignore intra relationship among the bags (Fig. 10).

7   Conclusions

In this chapter, we provide a broad overview along with extensive experiments of most recent state of the art methods in group detection and crowd behaviors understanding. The experimental results demonstrate that pure computer vision techniques may not be sufficient to uncover wide range of group and crowds' behaviors/dynamics, and sociological-inspired methodologies outperform other state-of-art approaches. To this end, we believe that proposing an universal approach for human behavior understanding in group and crowd levels is still can be considered as open problems, and further investigations are required to introduce new methodologies in computer vision community.

References

1. Ali, S., Shah, M.: A Lagrangian particle dynamics approach for crowd flow segmentation and stability analysis. In: 2007 IEEE Conference on Computer Vision and Pattern Recognition, pp. 1–6. IEEE (2007)

2. Ali, S., Shah, M.: Floor fields for tracking in high density crowd scenes. In: Forsyth, D., Torr, P., Zisserman, A. (eds.) ECCV 2008. LNCS, vol. 5303, pp. 1–14. Springer, Heidelberg (2008). doi:10.1007/978-3-540-88688-4_1
3. Ba, S.O., Odobez, J.M.: Multiperson visual focus of attention from head pose and meeting contextual cues. IEEE Trans. Pattern Anal. Mach. Intell. (PAMI) **33**(1), 101–116 (2011)
4. Batchelor, G.K.: An Introduction to Fluid Dynamics. Cambridge University Press, Cambridge (2000)
5. Bazzani, L., Cristani, M., Tosato, D., Farenzena, M., Paggetti, G., Menegaz, G., Murino, V.: Social interactions by visual focus of attention in a three-dimensional environment. Expert Syst. **30**(2), 115–127 (2013)
6. Benfold, B., Reid, I.: Stable multi-target tracking in real-time surveillance video. In: IEEE International Conference on Computer Vision and Pattern Recognition (CVPR), pp. 3457–3464 (2011)
7. Blei, D.M., Ng, A.Y., Jordan, M.I.: Latent Dirichlet allocation. J. Mach. Learn. Res. **3**, 993–1022 (2003)
8. Blunsden, S., Fisher, R.: The BEHAVE video dataset: ground truthed video for multi-person behavior classification. Ann. BMVA **2010**(4), 1–12 (2010)
9. Boykov, Y., Veksler, O., Zabih, R.: Fast approximate energy minimization via graph cuts. IEEE Trans. Pattern Anal. Mach. Intell. (PAMI) **23**(11), 1222–1239 (2001)
10. Bozdogan, H.: Model selection and Akaike's Information Criterion (AIC): the general theory and its analytical extensions. Psychometrika **52**(3), 345–370 (1987)
11. Campbell, N.D.F., Vogiatzis, G., Hernández, C., Cipolla, R.: Automatic 3D object segmentation in multiple views using volumetric graph-cuts. Image Vis. Comput. **28**(1), 14–25 (2010)
12. Cao, T., Wu, X., Guo, J., Yu, S., Xu, Y.: Abnormal crowd motion analysis. ROBIO **9**, 1709–1714 (2009)
13. Chan, A.B., Liang, Z.S.J., Vasconcelos, N.: Privacy preserving crowd monitoring: counting people without people models or tracking. In: IEEE Conference on Computer Vision and Pattern Recognition, CVPR 2008, pp. 1–7. IEEE (2008)
14. Chen, C., Odobez, J.M.: We are not contortionists: coupled adaptive learning for head and body orientation estimation in surveillance video. In: IEEE International Conference on Computer Vision and Pattern Recognition (CVPR), pp. 1544–1551 (2012)
15. Ciolek, T.M.: The proxemics lexicon: a first approximation. J. Nonverbal Behav. **8**(1), 55–79 (1983)
16. Ciolek, T.M., Kendon, A.: Environment and the spatial arrangement of conversational encounters. Sociol. Inq. **50**(3–4), 237–271 (1980)
17. Conte, D., Foggia, P., Percannella, G., Tufano, F., Vento, M.: A method for counting people in crowded scenes. In: 2010 Seventh IEEE International Conference on Advanced Video and Signal Based Surveillance (AVSS), pp. 225–232. IEEE (2010)
18. Cook, M.: Experiments on orientation and proxemics. Hum. Relat. **23**(1), 61–76 (1970)
19. Cristani, M., Bazzani, L., Paggetti, G., Fossati, A., Tosato, D., Del Bue, A., Menegaz, G., Murino, V.: Social interaction discovery by statistical analysis of f-formations. In: British Machine Vision Conference (BMVC), pp. 23.1–23.12 (2011)
20. Datta, A., Shah, M., Lobo, N.D.V.: Person-on-person violence detection in video data. In: Proceedings 16th International Conference on Pattern Recognition 2002, vol. 1, pp. 433–438. IEEE (2002)

21. Gong, S., Loy, C.C., Xiang, T.: Security and surveillance. In: Moeslund, T.B., Hilton, A., Krüger, V., Sigal, L. (eds.) Visual Analysis of Humans. Springer, Heidelberg (2011). doi:10.1007/978-0-85729-997-0_23

22. Hassner, T., Itcher, Y., Kliper-Gross, O.: Violent flows: real-time detection of violent crowd behavior. In: 2012 IEEE Computer Society Conference on Computer Vision and Pattern Recognition Workshops, pp. 1–6. IEEE (2012)

23. Helbing, D., Molnar, P.: Social force model for pedestrian dynamics. Phys. Rev. E **51**(5), 4282 (1995)

24. Hung, H., Kröse, B.: Detecting F-formations as dominant sets. In: International Conference on Multimodal Interfaces (ICMI), pp. 231–238 (2011)

25. Jiang, F., Wu, Y., Katsaggelos, A.K.: Detecting contextual anomalies of crowd motion in surveillance video. In: 2009 16th IEEE International Conference on Image Processing (ICIP), pp. 1117–1120. IEEE (2009)

26. Jojic, N., Perina, A.: Multidimensional counting grids: inferring word order from disordered bags of words. arXiv preprint (2012). arXiv:1202.3752

27. Kendon, A.: Conducting Interaction: Patterns of Behavior in Focused Encounters. Cambridge University Press, Cambridge (1990)

28. Kok, V.J., Lim, M.K., Chan, C.S.: Crowd behavior analysis: a review where physics meets biology. Neurocomputing **177**, 342–362 (2015)

29. Kratz, L., Nishino, K.: Anomaly detection in extremely crowded scenes using spatio-temporal motion pattern models. In: 2009 IEEE Conference on Computer Vision and Pattern Recognition, CVPR, pp. 1446–1453. IEEE (2009)

30. Ladický, L., Russell, C., Kohli, P., Torr, P.H.S.: Inference methods for CRFs with co-occurrence statistics. Int. J. Comput. Vis. **103**(2), 213–225 (2013)

31. Lanz, O.: Approximate Bayesian multibody tracking. IEEE Trans. Pattern Anal. Mach. Intell. (PAMI) **28**(9), 1436–1449 (2006)

32. Lanz, O., Brunelli, R.: Joint Bayesian tracking of head location and pose from low-resolution video. In: Stiefelhagen, R., Bowers, R., Fiscus, J. (eds.) CLEAR/RT -2007. LNCS, vol. 4625, pp. 287–296. Springer, Heidelberg (2008). doi:10.1007/978-3-540-68585-2_27

33. Li, T., Chang, H., Wang, M., Ni, B., Hong, R., Yan, S.: Crowded scene analysis: a survey. IEEE Trans. Circuits Syst. Video Technol. **25**(3), 367–386 (2015)

34. Liu, Y., Li, X., Jia, L.: Abnormal crowd behavior detection based on optical flow and dynamic threshold. In: 2014 11th World Congress on Intelligent Control and Automation (WCICA), pp. 2902–2906. IEEE (2014)

35. Lombaert, H., Sun, Y., Grady, L., Xu, C.: A multilevel banded graph cuts method for fast image segmentation. In: International Conference on Computer Vision (ICCV), pp. 259–265 (2005)

36. Mehran, R., Oyama, A., Shah, M.: Abnormal crowd behavior detection using social force model. In: IEEE Conference on Computer Vision and Pattern Recognition, CVPR 2009, pp. 935–942. IEEE (2009)

37. Mohammadi, S., Kiani, H., Perina, A., Murino, V.: A comparison of crowd commotion measures from generative models. In: Proceedings of the IEEE Conference on Computer Vision and Pattern Recognition Workshops, pp. 49–55 (2015)

38. Mohammadi, S., Kiani, H., Perina, A., Murino, V.: Violence detection in crowded scenes using substantial derivative. In: 2015 12th IEEE International Conference on Advanced Video and Signal Based Surveillance (AVSS), pp. 1–6. IEEE (2015)

39. Mousavi, H., Galoogahi, H.K., Perina, A., Murino, V.: Detecting abnormal behavioral patterns in crowd scenarios. In: Esposito, A., Jain, L.C. (eds.) Toward Robotic Socially Believable Behaving Systems - Volume II. ISRL, vol. 106, pp. 185–205. Springer, Cham (2016). doi:10.1007/978-3-319-31053-4_11

40. Mousavi, H., Mohammadi, S., Perina, A., Chellali, R., Murino, V.: Analyzing track-lets for the detection of abnormal crowd behavior. In: 2015 IEEE Winter Conference on Applications of Computer Vision, pp. 148–155. IEEE (2015)

41. Mousavi, H., Nabi, M., Kiani, H., Perina, A., Murino, V.: Crowd motion monitoring using tracklet-based commotion measure. In: 2015 IEEE International Conference on Image Processing (ICIP), pp. 2354–2358. IEEE (2015)

42. Moussaïd, M., Helbing, D., Theraulaz, G.: How simple rules determine pedestrian behavior and crowd disasters. Proc. Nat. Acad. Sci. **108**(17), 6884–6888 (2011)

43. Moussaïd, M., Nelson, J.D.: Simple heuristics and the modelling of crowd behaviours. In: Weidmann, U., Kirsch, U., Schreckenberg, M. (eds.) Pedestrian and Evacuation Dynamics 2012, pp. 75–90. Springer, Cham (2014). doi:10.1007/978-3-319-02447-9_5

44. Pavan, M., Pelillo, M.: Dominant sets and pairwise clustering. IEEE Trans. Pattern Anal. Mach. Intell. (PAMI) **29**(1), 167–172 (2007)

45. Poppe, R.: A survey on vision-based human action recognition. Image Vis. Comput. **28**(6), 976–990 (2010)

46. Rabaud, V., Belongie, S.: Counting crowded moving objects. In: 2006 IEEE Computer Society Conference on Computer Vision and Pattern Recognition (CVPR 2006), vol. 1, pp. 705–711. IEEE (2006)

47. Rittscher, J., Tu, P.H., Krahnstoever, N.: Simultaneous estimation of segmentation and shape. In: 2005 IEEE Computer Society Conference on Computer Vision and Pattern Recognition (CVPR 2005), vol. 2, pp. 486–493. IEEE (2005)

48. Rodriguez, M., Ali, S., Kanade, T.: Tracking in unstructured crowded scenes. In: 2009 IEEE 12th International Conference on Computer Vision, pp. 1389–1396. IEEE (2009)

49. Mohammadi, S., Perina, A., Kiani, H., Murino, V.: Angry crowds: detecting violent events in videos. In: Leibe, B., Matas, J., Sebe, N., Welling, M. (eds.) ECCV 2016. LNCS, vol. 9911, pp. 3–18. Springer, Cham (2016). doi:10.1007/978-3-319-46478-7_1

50. Saleh, S.A.M., Suandi, S.A., Ibrahim, H.: Recent survey on crowd density estimation and counting for visual surveillance. Eng. Appl. Artif. Intell. **41**, 103–114 (2015)

51. Setti, F., Hung, H., Cristani, M.: Group detection in still images by f-formation modeling: a comparative study. In: International Workshop on Image Analysis for Multimedia Interactive Services (WIAMIS), pp. 1–4 (2013)

52. Setti, F., Lanz, O., Ferrario, R., Murino, V., Cristani, M.: Multi-scale F-formation discovery for group detection. In: IEEE International Conference on Image Processing (ICIP) (2013)

53. Setti, F., Russell, C., Bassetti, C., Cristani, M.: F-formation detection: individuating free-standing conversational groups in images. PloS ONE **10**(5), e0123783 (2015)

54. Shao, J., Loy, C.C., Wang, X.: Scene-independent group profiling in crowd. In: Proceedings of the IEEE Conference on Computer Vision and Pattern Recognition, pp. 2219–2226 (2014)

55. Smith, K., Ba, S.O., Odobez, J.M., Gatica-Perez, D.: Tracking the visual focus of attention for a varying number of wandering people. IEEE Trans. Pattern Anal. Mach. Intell. (PAMI) **30**(7), 1212–1229 (2008)

56. Tang, S., Andriluka, M., Milan, A., Schindler, K., Roth, S., Schiele, B.: Learning people detectors for tracking in crowded scenes. In: Proceedings of the IEEE International Conference on Computer Vision, pp. 1049–1056 (2013)

57. Tosato, D., Spera, M., Cristani, M., Murino, V.: Characterizing humans on Riemannian manifolds. IEEE Trans. Pattern Anal. Mach. Intell. (PAMI) **35**(8), 1972–1984 (2013)
58. Tran, K.N., Bedagkar-Gala, A., Kakadiaris, I.A., Shah, S.K.: Social cues in group formation and local interactions for collective activity analysis. In: International Conference on Computer Vision Theory and Applications (VISAPP), vol. 1, pp. 539–548 (2013)
59. Vascon, S., Mequanint, E.Z., Cristani, M., Hung, H., Pelillo, M., Murino, V.: A game-theoretic probabilistic approach for detecting conversational groups. In: Cremers, D., Reid, I., Saito, H., Yang, M.-H. (eds.) ACCV 2014. LNCS, vol. 9007, pp. 658–675. Springer, Cham (2015). doi:10.1007/978-3-319-16814-2_43
60. Vascon, S., Mequanint, E.Z., Cristani, M., Hung, H., Pelillo, M., Murino, V.: Detecting conversational groups in images and sequences: a robust game-theoretic approach. Comput. Vis. Image Underst. **143**, 11–24 (2016)
61. Wu, S., Moore, B.E., Shah, M.: Chaotic invariants of Lagrangian particle trajectories for anomaly detection in crowded scenes. In: 2010 IEEE Conference on Computer Vision and Pattern Recognition (CVPR), pp. 2054–2060. IEEE (2010)
62. Xu, L., Gong, C., Yang, J., Wu, Q., Yao, L.: Violent video detection based on MoSIFT feature and sparse coding. In: 2014 IEEE International Conference on Acoustics, Speech and Signal Processing (ICASSP), pp. 3538–3542. IEEE (2014)
63. Xu, N., Ahuja, N., Bansal, R.: Object segmentation using graph cuts based active contours. Comput. Vis. Image Underst. **107**(3), 210–224 (2007)
64. Zhan, B., Monekosso, D.N., Remagnino, P., Velastin, S.A., Xu, L.Q.: Crowd analysis: a survey. Mach. Vis. Appl. **19**(5–6), 345–357 (2008)

Computer Graphics Theory and Applications

Real-Time Contour Image Vectorization on GPU

Xiaoliang Xiong$^{(\boxtimes)}$, Jie Feng$^{(\boxtimes)}$, and Bingfeng Zhou

Institute of Computer Science and Technology, Peking University, Beijing, China
{jonny_xiong,feng_jie,cczbf}@pku.edu.cn

Abstract. In this paper, we present a novel algorithm to convert the contour in a raster image into its vector form. Different from the state-of-art methods, we explore the potential parallelism that exists in the problem and propose an algorithm suitable to be accelerated by the graphics hardware. In our algorithm, the vectorization task is decomposed into four steps: detecting the boundary pixels, pre-computing the connectivity relationship of detected pixels, organizing detected pixels into boundary loops and vectorizing each loop into line segments. The boundary detection and connectivity pre-computing are parallelized owing to the independence between scanlines. After a sequential boundary pixels organizing, all loops are vectorized concurrently. With a GPU implementation, the vectorization can be accomplished in real-time. Then, the image can be represented by the vectorized contour. This real-time vectorization algorithm can be used on images with multiple silhouettes and multi-view videos. We demonstrate the efficiency of our algorithm with several applications including cartoon and document vectorization.

Keywords: Vectorization · Real-time rendering · GPU acceleration

1 Introduction

Vector image is a compact form to represent image with a set of geometry primitives (like points, curves or polygons). It is independent with displaying resolution so that it can be rendered at any scale without aliasing. A raster image, in contrast, uses a large pixel matrix to store the image information, which requires much more space and conveys less semantics. It can be directly mapped onto display device and rendered with high efficiency, but suffers seriously from aliasing or loss of details when the image is scaled. The advantages of vector image over raster image, make it widely used in situations such as computer-aided design, on the Internet and plenty of practical applications.

Shape-from-Silhouette (SFS) is a specific application which adopts vector form as silhouette representation. It retrieves the 3D shape of the target object from multiple silhouette images taking at different viewpoints. In SFS, silhouette boundaries are approximated by line segments to simplify the computation and achieve the real-time rendering performance. Thus, an efficient algorithm to convert the silhouettes from pixels to vectors is essential. This is the motivation

© Springer International Publishing AG 2017
J. Braz et al. (Eds.): VISIGRAPP 2016, CCIS 693, pp. 35–50, 2017.
DOI: 10.1007/978-3-319-64870-5_2

of our work. Also, it is necessary to do the raster-to-vector conversion with high efficiency in applications like high-speed document scanning and cartoon animation.

Existing vectorization methods mainly focus on the accuracy during the conversion and ideally expect to approximate both the sharp and smooth features in the raster image with less geometry primitives. Triangular mesh [17], gradient mesh [12] and diffusion curves [10] are three commonly used geometry representatives. There are some researches adopt GPU to improve the rendering speed of constructed vector image [14], but the efficiency of the vector image construction is not high enough.

In contrast, we focus primarily on the silhouettes in the raster image and explore the potential parallelism in the problem to vectorize their contours as fast as possible. Both accuracy and efficiency are concerned to satisfy practical applications. Inspired by the scanline algorithm in polygon filling, we first detect the boundary pixels line by line in parallel, resulting in a set of unorganised pixels on each line. So secondly, the relationships of these pixels are computed. We note that only adjacent lines are directly related and each two lines can be processed simultaneously. Thirdly, all the boundary pixels are organized into loops based on pre-computed relationship. Fourthly, these loops which consist of boundary pixels can be vectorized into line segments concurrently. Hence, the problem is naturally decomposed into four steps and three steps can be parallelized. With this decomposition, our algorithm becomes not so sensitive to the image resolution.

Our key contribution is a novel algorithm that vectorizes the silhouettes in a raster image with high efficiency. We make a decomposition on the problem and take advantage of the potential parallelism to get an acceleration. We also apply the algorithm into several practical situations.

2   Related Work

Comparing to raster images, vector images has the advantages of more compact in presentation, requiring less space to store, convenient to transmit and edit, artifact-free in display etc. Image vectorization techniques aim at doing the raster-to-vector conversion accurately and efficiently. It includes crude vectorization on binary images and advanced vectorization on color images.

2.1   Image Vectorization

Crude Vectorization. Crude vectorization concerns grouping the pixels in the raster image into raw line fragments and representing the original image with primary geometry like skeleton and contour polygon. It is a fundamental process in the interpretation of image elements (like curves, lines) and can be used as preprocessing of applications like cartoon animation, topographic map reconstruction, SFS, etc.

Crude vectorization is often divided into two classes: *Thinning based methods* [11] and *Non-thinning based methods* [3]. The former first thin the rastered object into a one-pixel-wide skeleton with iterative erosion, then these pixels are tracked into chain and approximated with line segments. The latter first extract the contour of the image, compute the medial axis between the contour pixels and then do the line segment approximation. *Thinning based methods* lose line width information during erosion and is time consuming. These disadvantages are compensated by *non-thinning based methods* that may have gaps at junctions. And both of these methods are sequential and need a long process time. [2] present a new medial axis pixel tracking strategy, which can preserve the width information and avoid distortion at junctions.

Advanced Vectorization. Advanced vectorization approaches concentrate on accurate approximation for all features in the raster image and take accuracy as their first consideration. *Triangle mesh based methods* [17] first sample important points in the image, then decompose this image into a set of triangles and store the corresponding pixel color on the triangle vertices. Inside each triangle, the color of each pixel can be recalculated through interpolation. [14] converts the image plane into triangular patches with curved boundaries instead of simple triangles and make the color distribution inside each patch more smooth. *Diffusion curve based methods* [10] first detect the edges in the original image, based on which it is converted into diffusion curve representation. Then a Poisson Equation is solved to calculate the final image. After vectorization by these methods, image can be effectively compressed, features are maintained or enhanced in different extent.

2.2   Image Vectorization in Applications

Cartoon Animation. In automatic cartoon animation, the artists only need to draw the key frames and in-betweens are generated by shape matching and interpolation. However, these techniques cannot be directly used in raster images, but are more suitable for vector-based graphics. Thus, a vectorization process is required to convert a raster key frame into its vector form. [18] subdivide the cartoon character into non-overlapping triangles based on which skeleton is extracted. Then artifacts are removed at the junction points and intersection areas by optimizing the triangles. There are also researches [16] on converting raster cartoon film into its vector form because the vector version is more easy to store, transmit, edit, display and so on. They take temporal coherence into consideration to alleviate flicker between cartoon frames.

Shape-from-Silhouette. Shape-from-Silhouette (SFS) is a method of estimating 3D shape of an object from its silhouette images. One famous SFS technique is the *visual hull* [6,8]. VH is defined as the maximal shape that reproduces the silhouettes of a 3D object from any viewpoint. It can be computed by intersecting the visual cones created by the viewing rays emanating from the camera center and passing through the silhouette contours, which is originally a chain of pixels. Most existing works adopt line segments as an approximation of the silhouette contour to reduce large amount of redundant computation. The conversion from silhouette contour to line segments is originally a vectorization problem and efficient algorithm is needed to decrease time consumption in VH pre-computing.

Complex hardware like multi-processors [8] and distributed system [7] are adopted to do this step to guarantee the VH computation in real-time. There are many GPU-based methods [5,13,15] to accelerate the visual hull computation, for the VH algorithm is highly parallel. Thus, it is natural to think if the preprocessing can be parallelized, too. This is the motivation of our work and draws our attention mainly on the parallelization of contour vectorization.

Document Image Processing. Document processing is a complex procedure which evolves converting the text on paper or electronic documents into features the computer can recognize. [1] present a thinning algorithm based on line sweep operation, resulting in a representation with skeletons and intersection sets, that provides extra features for subsequent character recognition. It is efficient in computation comparing to pixel-based thinning algorithm [11] which outputs skeletons only.

GPU-Acceleration in Image Vectorization. Existing GPU related work is on the vectorized image rendering. [9] introduce a novel representation for random-access rendering of antialiased vector graphics. It has the ability to map vector graphics onto arbitrary surfaces, or under arbitrary deformations. [14] develop a real-time GPU-accelerated parallel algorithm based on recursive patch subdivision for rasterizing their vectorized results. [10] also propose a GPU implementation for rendering their vectorized images described by diffusion curves.

3   Our Algorithm

Our goal is to convert the silhouettes in an input image from raster to vector form with high efficiency and accuracy. The input image is preprocessed and converted into silhouette images by thresholding or background subtraction in advance. Intuitively, the boundary pixels are detected by scanning each line in these images. Since all scanlines are independent, the detection can be done concurrently. The resulting pixels on each line are then organized into *loops* based on their *connectivity relationship* with previous line, which can be precomputed in parallel. Finally, all organized loops are vectorized into line segments independently. Figure 1 shows the process of vectorizing a cartoon color image with our method. In the following, we describe each step in detail. To clarify the description, we refer *boundary* as unordered pixels, *loop* as an ordered pixel list and *contour* as all loops of a silhouette.

Fig. 1. Vectorizing a cartoon color image with our method. (a) The input raster image. (b) The binary silhouettes. (c) The main procedure of our algorithm: ① parallel boundary detection and ② precontouring, ③ sequential contouring and ④ parallel contour vectorization. (d) The vectorization result. The contours are represented by line segments. The whole computation is completed in 9 ms, which provides possibility for real-time applications. (Color figure online)

3.1   Boundary Pixel Detecting

To rapidly extract the boundary pixels, we scan all lines in the silhouette images in parallel. A *scanline* $\hat{s}_i$ is a one-pixel-wide horizontal line that crosses the silhouette image from left to right. It is used to find the pairwise boundary pixels (I_k, O_k) of a foreground area. The collection of all scanlines are denoted as S,

$$S = \{\hat{s}_i | i = 1, \ldots, h\},$$

where h is the height of silhouette image. During scanning, when the scanline enters the foreground from background, the corresponding boundary pixel is recorded as I_k and when it leaves foreground into background, the boundary pixel is recorded as O_k. The point pair (I_k, O_k) is called an *interval* $R_k^{(i)}$ on $\hat{s}_i$, and the pixels between I_k and O_k belong to the foreground. All such pixel pairs on $\hat{s}_i$ consist its interval collection s_i,

$$s_i = \{R_k^{(i)} | R_k^{(i)} = (I_k, O_k), I_k < O_k, 1 \leq k \leq N_i\},$$

where line $\hat{s}_i$ has N_i intervals. Figure 2 shows an example of two scanlines $\hat{s}_{i_0}$ and $\hat{s}_{i_1}$. In each line, pixels are illustrated in different colors, where black indicates background, cyan for boundary pixels and gray for foreground. In the example, line $\hat{s}_{i_0}$ has 3 intervals and $\hat{s}_{i_1}$ has 4 intervals respectively.

As the independence of boundary pixel detection on each line $\hat{s}_i$, the scanning task of all lines S in the silhouette images can be allocated to multiple parallel threads, each for one scanline. This parallelization has an advantage: when the height of the image or the image number increases, we only need to add more threads and the running time is not affected too much. And it provides possibility for multiple images vectorization. The parallel scanning results in a group of foreground pixel intervals s_i on each line and the *connectivity relationship* between the lines should be computed in next step.

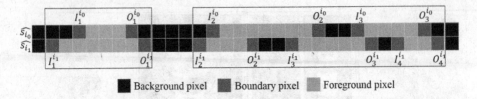

Fig. 2. Example of scanlines, intervals and segments. In this example, intervals on line $\hat{s}_{i_0}$ and $\hat{s}_{i_1}$ are divided into 2 segments, each marked with a red box. (Color figure online)

3.2 Pre-contouring

The detected boundary pixels are represented as foreground intervals s_i on each line $\hat{s}_i$. They should be organized into *loops* that enclose the object in the silhouette images. The target contour loops are denoted as B:

$$B = \{L_j | j = 1, \dots, l\},$$

where l is the loop number and each loop L_j is a ordered list of boundary pixels:

$$L_j = \{p_m | m = 1, \dots, M\},$$

That is, the loop L_j starts from p_1, goes along the silhouette and ends at p_M. If contour loops B are tracked directly on S, it is an up-down strategy that each loop stretches to pixels on next line if corresponding intervals are connected with current loop. The *connectivity relationship* between intervals on adjacent lines is needed during contour tracking and should be computed first.

For arbitrary two adjacent lines $\hat{s}_{i_0}$ and $\hat{s}_{i_1}$, their connectivity depends on the overlapping of their foreground intervals. If intervals $R_j^{(i_0)}$ in s_{i_0} and $R_k^{(i_1)}$ in s_{i_1} overlap, they consist a *segment*. In Fig. 2, $R_1^{(i_0)}$ and $R_1^{(i_1)}$ overlap, so they consist a segment, based on which we can infer these four boundary pixels are in the same loop. In this example, the rest of intervals on line $\hat{s}_{i_0}$ and $\hat{s}_{i_1}$ are divided into another segment and it has 3 intervals on i_1 and 2 on i_0 (3:2).

Theoretically, in the same segment the ratio of interval numbers on two adjacent lines can be classified into six cases:(1) *1:0* (2) *0:1* (3) *1:1* (4) *1:n* (5) *n:1* (6) *n:n* (Fig. 3). Case (1) and case (2) means interval only existing in one of the lines; Case (3) means that current loop does not change obviously from previous line to current line; case (4) and case (5) indicate loops merged or closed and new loops generated respectively; case (6) is a combination of case (4) and case (5). Each case indicates different change of loops in these lines and the boundary pixels of the included intervals are related.

Because this relationship computing depends only on the adjacent lines, it can be performed in parallel and separately accomplished as a pre-processing before contour organizing. Each parallel thread is responsible for dividing intervals on two lines into segments. With the connectivity relationship, we can organize each loop in order more efficiently.

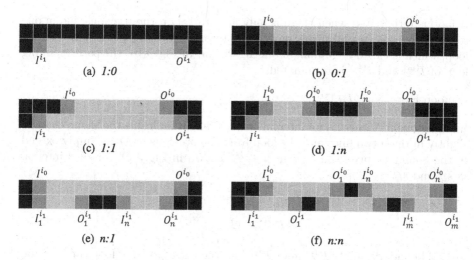

Fig. 3. Six cases of the connectivity relationship between intervals on two adjacent lines.

3.3 Contouring

Up to now, the boundary pixels are detected and pre-contoured in parallel, resulting in the foreground *intervals* and their *connectivity relationship* between adjacent lines. With these information, we can organize the boundary pixels into loops more easily, which is accomplished in each *segment*, according to the interval numbers in the two lines. During organizing, new loops may be generated, existing loops may be extended, merged, closed or branched from top to bottom in the image. The *connectivity relationship* between the two adjacent lines determines how the loop develops from the previous line to the current line, which can be directly represented by the interval numbers on each line $(|\{R_k^{(i_1)}\}| : |\{R_j^{(i_0)}\}|))$.

As described in Pre-contouring, in each individual segment, the *connectivity relationship* of adjacent lines can be classified into 6 cases, and each case means loop changes differently in these lines. Next, we will consider each case separately and show how the loops develop from previous line to current line as illustrated in Fig. 3.

– Loop Initialization (*1:0*)

During Contouring, a new loop is generated when new interval appears on current line, which does not overlap with any intervals on previous line. This loop records the boundary pixels of a presently separate region in the input image and will be complemented by the following pixels. As shown in Fig. 3(a), a loop starting from $I^{(i_1)}$ and ends at $O^{(i_1)}$ is generated.

– Loop Termination (*0:1*)

A loop is terminated when there is only an interval on the previous line in one segment. It indicates all pixels on a separate region are organised into a closed loop, which is called a *contour* in our algorithm. In Fig. 3(b), the corresponding loop of $I^{(i_0)}$ and $I^{(i_0)}$ is terminated.

– Loop Extension (*1:1*)

In one segment, if there is an interval on each line, it indicates the shape changes slightly in these two lines and the loop from the previous line can simply extend to the boundary pixels on current line. As shown in Fig. 3(c), for each interval in s_{i_0} and s_{i_1} :

$$s_{i_0} = \{R^{(i_0)} = (I^{(i_0)}, O^{(i_0)})\},$$
$$s_{i_1} = \{R^{(i_1)} = (I^{(i_1)}, O^{(i_1)})\},$$

we add boundary points $I^{(i_1)}$ and $O^{(i_1)}$ into the corresponding loops of $I^{(i_0)}$ and $O^{(i_0)}$, respectively.

– Loop Merging or Closing (*1:n*)

In this case, n intervals on the previous line change into one on current line. It means the loop number decreases and there are loops merged or closed. As shown in Fig. 3(d), there are n intervals in s_{i_0} and 1 interval in s_{i_1}:

$$s_{i_0} = \{R_j^{(i_0)} | R_j^{(i_0)} = (I_j^{(i_0)}, O_j^{(i_0)}), 1 \leq j \leq n\},$$
$$s_{i_1} = \{R^{(i_1)} = (I^{(i_1)}, O^{(i_1)})\}.$$

Hence, we add $I^{(i_1)}, O^{(i_1)}$ into the corresponding loops of $I_1^{(i_0)}, O_n^{(i_0)}$, respectively. For the rest of points in s_{i_0}, new pairs are formed as $(O_w^{(i_0)}, I_{w+1}^{(i_0)}), w = 0, \ldots,$ $n-1$. If the points of one pair belongs to the same loop, this loop will be closed, or else the different loops will be merged.

– Loop Branching (*n:1*)

On the contrary to the previous case, if 1 interval on previous line branches into n intervals on current line, new loops are generated to record the boundary pixels on the following line. In Fig. 3(e), there are n intervals in s_{i_1} and 1 interval in s_{i_0}:

$$s_{i_1} = \{R_k^{(i_1)} | R_k^{(i_1)} = (I_k^{(i_1)}, O_k^{(i_1)}), 1 \leq k \leq n\},$$
$$s_{i_0} = \{R^{(i_0)} = (I^{(i_0)}, O^{(i_0)})\}.$$

We add $I_1^{(i_1)}, O_n^{(i_1)}$ into the corresponding loop of $I^{(i_0)}, O^{(i_0)}$, respectively. For the left points in s_{i_1}, new pairs are formed as $(O_w^{(i_1)}, I_{w+1}^{(i_1)}), w = 0, \ldots, n-1$. Each pair is used for generating a new loop.

– Loop Merging(Closing) and Branching (*n:n*)

If there are more than 1 intervals on both lines in a segment, we can treat it as a combination of the case of loop merging(closing) and branching. In Fig. 3(f), there are n intervals in s_{i_0} and m intervals in s_{i_1}:

$$s_{i_0} = \{R_j^{(i_0)} | R_j^{(i_0)} = (I_j^{(i_0)}, O_j^{(i_0)}), 1 \le j \le n\},$$
$$s_{i_1} = \{R_k^{(i_1)} | R_k^{(i_1)} = (I_k^{(i_1)}, O_k^{(i_1)}), 1 \le k \le m\}.$$

We add $I_1^{(i_1)}, O_m^{(i_1)}$ into the corresponding loop of $I_1^{(i_0)}, O_n^{(i_0)}$, respectively. Loops are merged or closed for the rest of points in s_{i_0} and generated for the rest of points in s_{i_1}.

These six cases provide the rule for how to deal with boundary pixels on current line according to the *connectivity relationship* with previous line during loop organizing. This step must be in sequential manner because the boundary pixels on current line must be connected to the loops produced by previous boundary pixels. Furthermore, the computation need large memory to store the edge pixels and requires frequent memory access, which is the weakness of GPU. And this is the only step that has to be performed on CPUs. When all lines of silhouette images are processed, target loops B is generated.

3.4   Contour Vectorization

Using the method given above, the contour of the foreground can be described with a group of pixel loops B. Subsequently, we need to simplify each loop and approximate them with a set of line segments. Our approximation method is similar to the Active Contour Modeling [4]. Each loop $L_j = \{p_1 p_2 \cdots p_i \cdots p_{M-1} p_M\}$ is processed with a divide-and-conquer strategy. Let d be the maximum distance between the point p_i and line $p_1 p_M$:

$$d = max\{dist(p_i, \overrightarrow{p_1 p_M})\}.$$

if d is smaller than t (a constant threshold, we set t=1 in our experiment), $p_1 p_M$ is an approximate line segment and the discretization terminates. If not, loop L_j is divided into two sub-loops L_{j_0} and L_{j_1}:

$$L_{j_0} = \{p_1 p_2 p_i p_{\frac{M+1}{2}}\},$$
$$L_{j_1} = \{p_{\frac{M+1}{2}} \cdots p_j \cdots p_{M-1} p_M\}.$$

Then each sub-loop is tested iteratively until L_{j_0} or L_{j_1} satisfies the terminal condition or is small enough.

The vectorization of each loop is independent and we can process it with a GPU thread. When the loop number is small, the parallelism is limited and it has little improvement in performance comparing to processing each loop sequentially. The parallelizing of this step become more and more important as the increasing of the loop number.

When the four steps are completed, our algorithm can output a vector image with contour represented by line segments.

4   Experiment and Result

We implement our algorithm using CUDA on a common PC with Quad CPU 2.5 GHz, 2.75 GB RAM, and a GeForce GTX260+ graphic card. The vectorization task is decomposed into four steps, in which Boundary Detection and Pre-contouring are performed on GPU with multiple threads, each processing for different lines. Pre-contouring results are copied back to CPU for sequential computation of Contouring, and the organized contour loops are copied into the GPU for the final Vectorization.

Figures 4 and 5 show the vectorization results of some simple figures and characters. The former is used in silhouette-based applications(e.g. SFS) and the latter is inevitably used in document processing. The running time and the number of primitives used for vector representation are listed in Table 1.

4.1   Comparison

We compare our algorithm with a Floodfill-based method on time efficiency. The difference between them is the strategy of boundary pixel detection and ordering, and we use the same way to vectorize the contour loops. Floodfill based method iteratively searches the boundary pixels of the silhouette in neighborhood until all pixels are processed. Hence the running time increases exponentially with the image resolution and it depends heavily on the complexity of the scene. In contrast, our method detects and pre-contours the boundary pixels in parallel. The grouped pixels organizing depends a little on the image complexity, but not so sensitive thanks to the pre-contouring. Figure 6 shows the speed up ratio between Foodfill Based Method and our method on the three images with different resolution and gives the corresponding running time of each method. We can

Fig. 4. Vectorization results of some figures. The first row shows the input raster images of some familiar figures(Skater, Pigeon, Bird.), and the second row lists corresponding vectorization results with contour represented by line segments.

Fig. 5. Vectorization results of different characters. The first, third and fifth rows are input raster images(English, Digit and Chinese characters respectively), the second, fourth and sixth rows are corresponding vectorization results with contour represented by line segments.

Table 1. Statistics of vectorization results and running time.

Image	Resolution	Points	Edges	Loops	Time(ms)
CharP	600×600	360	53	2	6.93
CharK	600×600	837	75	1	9.04
CharU	600×600	442	64	1	7.61
Digit4	600×600	414	24	2	5.93
Digit7	600×600	573	25	1	6.43
Digit8	600×600	1036	94	3	8.18
Chinese1	600×600	848	120	2	8.83
Chinese2	600×600	1080	223	7	10.73
Chinese3	600×600	1211	171	2	11.41
Pegion	400×400	439	86	2	5.89
Skater	400×400	596	104	4	6.56
Bird	600×480	833	83	1	8.92
Winnie	500×500	2761	432	26	8.54

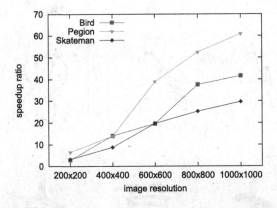

		200^2	400^2	600^2	800^2	1000^2
Pegion	FBM	29.91	105.39	306.41	510.11	799.78
	Ours	4.58	7.62	7.81	9.73	13.14
Skater	FBM	18.79	67.55	168.0	293.45	420.8
	Ours	5.62	7.73	8.52	11.61	14.17
Bird	FBM	14.51	90.53	158.26	419.94	583.28
	Ours	4.87	6.46	8.06	11.18	14.01

Fig. 6. Comparison between Floodfilled based method (FBM) and our method. The figure above shows the speedup ratio between two methods on different image resolution and the corresponding running time(ms) of each method is listed below.

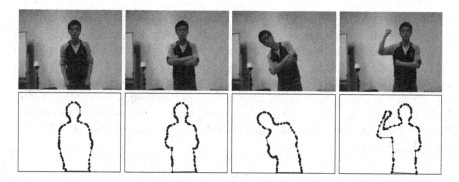

Fig. 7. Video vectorization. First row: four frames in the video. Second row: the corresponding vectorized results.

see that our method is not so sensitive to image resolution due to its parallelism and has a significant speed up especially under high image resolution.

4.2 Video Vectorization

Taking advantage of the fast speed, we apply our algorithm in video vectorization. Each frame is vectorized individually and we can achieve an average frame rate of 48 fps, which we believe will be even faster if the temporal coherence is considered. Figure 7 demonstrates the result.

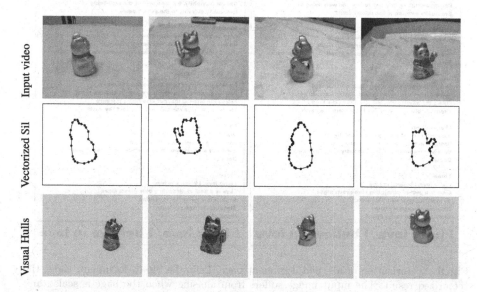

Fig. 8. Silhouettes vectorization in eight video streams and Visual Hull rendering based on the silhouettes. The first row shows 4 channel video images(eight in total), the second row is the corresponding vectorized silhouettes. Visual hulls are rendered from different viewpoints based on the silhouettes(the third row).

To further demonstrate the efficiency of our method, we perform the contour vectorization among 8-channel multi-view video streams simultaneously (which is a requirement of reconstructing dynamic visual hull) with image resolution of 600×480. Owing to the parallelization of our algorithm, the boundary pixels can be detected and pre-contoured in parallel among all video image lines at one time point. Then the pixels contouring can be parallelized between each video. Finally, each loop in the contour can be discretized into line segments in parallel. With the vectorization result, we can reconstruct and render dynamic VHs over 20 fps (Fig. 8).

4.3   Cartoon Image Vectorization

Figure 1 shows the vectorization of a cartoon image Winnie. We first binarize the input image according to the different colors and obtain a series of silhouettes. Then these silhouette images are vectorized simultaneously and result in a vector representation of the color image. Total computation can be accomplished in 9 ms.

Fig. 9. Vectorization of a typed document page. Left: the input document, right: the vectorized result. The input image suffers from aliasing when the page is scaled and our result can keep the shape of each character well.

4.4   Document Image Vectorization

Document image vectorization is challenging for complex situations may appear in scanned document or handwritten pages and the number of contours may be large enough to bring difficulties in data storing and transferring on GPUs. To demonstrate the efficiency of our algorithm, we input a typed page at the resolution of 2500×1800, and the vectorization of the characters in this page can be done in 40 ms (Fig. 9). After vectorization, each character is represented with several line segments, which can be scaled without aliasing.

5   Conclusion

We propose a hardware-accelerated algorithm to vectorize the silhouettes in the raster image with high efficiency. The problem is decomposed into four steps and three of them can be parallelized significantly. We show the efficiency of our algorithm on some challenge applications including multiple videos and document image vectorization.

The limitation of our work lies in that the contouring step is still in sequential. One feasible way to alleviate the problem is to partition the silhouette image into several parts and the contouring among them can be parallelized. However, a merge step is needed if loops between two parts are connected, which will introduce extra computation cost. And we are exploring an ideal solution for this problem.

Acknowledgements. This work is partially supported by NSFC grants #61170206, #61370112, and Specialized Research Fund for the Doctoral Program of Higher Education #20110001110077.

References

1. Chang, F., Lu, Y.-C., Pavlidis, T.: Feature analysis using line sweep thinning algorithm. IEEE Trans. Pattern Anal. Mach. Intell. **21**(2), 145–158 (1999)
2. Dori, D., Liu, W.: Sparse pixel vectorization: an algorithm and its performance evaluation. IEEE Trans. Pattern Anal. Mach. Intell. **21**(3), 202–215 (1999)
3. Jimenez, J., Navalon, J.L.: Some experiments in image vectorization. IBM J. Res. Dev. **26**(6), 724–734 (1982)
4. Kass, M., Witkin, A., Terzopoulos, D.: Snakes: active contour models. Int. J. Comput. Vis. **1**(4), 321–331 (1988)
5. Ladikos, A., Benhimane, S., Navab, N.: Efficient visual hull computation for real-time 3d reconstruction using CUDA, pp. 1–8 (2008)
6. Laurentini, A.: The visual hull concept for silhouette-based image understanding. IEEE Trans. Pattern Anal. Mach. Intell. **16**(2), 150–162 (1994)
7. Li, M., Magnor, M., Seidel, H.-P.: A hybrid hardware-accelerated algorithm for high quality rendering of visual hulls. In: Proceedings of Graphics Interface 2004, pp. 41–48. Canadian Human-Computer Communications Society (2004)
8. Matusik, W., Buehler, C., Raskar, R., Gortler, S.J., McMillan, L.: Image-based visual hulls. In: SIGGRAPH 2000, pp. 369–374. ACM (2000)

9. Nehab, D., Hoppe, H.: Random-access rendering of general vector graphics. In: ACM Transactions on Graphics (TOG), vol. 27, p. 135. ACM(2008)

10. Orzan, A., Bousseau, A., Barla, P., Winnemöller, H., Thollot, J., Salesin, D.: Diffusion curves: a vector representation for smooth-shaded images. ACM Trans. Graph. **56**(7), 101–108 (2013)

11. Smith, R.W.: Computer processing of line images: a survey. Pattern Recogn. **20**(1), 7–15 (1987)

12. Sun, J., Liang, L., Wen, F., Shum, H.-Y.: Image vectorization using optimized gradient meshes. In: ACM Transactions on Graphics (TOG), vol. 26, p. 11. ACM (2007)

13. Waizenegger, W., Feldmann, I., Eisert, P., Kauff, P.: Parallel high resolution real-time visual hull on GPU. In: 2009 16th IEEE International Conference on Image Processing (ICIP), pp. 4301–4304 (2009)

14. Xia, T., Liao, B., Yu, Y.: Patch-based image vectorization with automatic curvilinear feature alignment. In: ACM Transactions on Graphics (TOG), vol. 28, p. 115. ACM (2009)

15. Yous, S., Laga, H., Kidode, M., Chihara, K.: GPU-based shape from silhouettes. In: Proceedings of the 5th International Conference on Computer Graphics and Interactive Techniques in Australia and Southeast Asia, pp. 71–77. ACM (2007)

16. Zhang, S.-H., Chen, T., Zhang, Y.-F., Hu, S.-M., Martin, R.R.: Vectorizing cartoon animations. IEEE Trans. Vis. Comput. Graph. **15**(4), 618–629 (2009)

17. Zhao, J., Feng, J., and Zhou, B.: Image vectorization using blue-noise sampling. In: IS&T/SPIE Electronic Imaging, p. 86640H. International Society for Optics and Photonics (2013)

18. Zou, J.J., Yan, H.: Cartoon image vectorization based on shape subdivision. In: Proceedings of the Computer Graphics International 2001, pp. 225–231. IEEE (2001)

Screen Space Curvature and Ambient Occlusion

Martin Prantl[1](✉), Libor Váša[1,2], and Ivana Kolingerová[1,2]

[1] Department of Computer Science and Engineering, Faculty of Applied Sciences,
University of West Bohemia, Plzen, Czech Republic
{perry,lvasa,kolinger}@kiv.zcu.cz
[2] New Technologies for the Information Society (NTIS), University of West Bohemia,
Plzen, Czech Republic

Abstract. Curvature plays an important role in computer graphics. It helps us to better understand surfaces of various objects. We often deal with a discrete geometry. However, the exact curvature can only be evaluated for analytical surfaces and not for discrete ones. Existing algorithms estimate curvature for discrete geometry with a certain precision. Most of the time, the performance of those algorithms is low and they are not intended to be used in real time applications. Our target is to have a real time curvature estimation that can be used during interactive geometry changes, like in a virtual sculpting. This paper proposes a screen space technique which estimates two principal curvatures and their directions at interactive rates. Final curvature can be used for an ambient occlusion estimation. The proposed solution is created to fit directly into existing rendering pipelines.

Keywords: Curvature · Screen space · GPU · Ambient occlusion · Visualisation · Discrete differential geometry · Computer graphics

1 Introduction

Visualization of curvature plays an important role in computer graphics. It can help to better understand properties of surfaces, their convex and concave areas. Distinguishing them can be important during interactive modeling or sculpting. In computer graphics, the basic representation of geometry are triangle meshes. They represent only an approximation of the original geometry and the same triangle mesh can be obtained for different geometries. This causes a problem with curvature since we are not able to compute only an estimation. On the plane inside a triangle the curvature is zero. On edges of a triangle it can be zero or infinity and in triangle vertices, the curvature is infinity. To estimate curvature for the triangle mesh, there are different algorithms that are using triangle neighbors and approximations to estimate the curvature. The more triangles the model has, the better approximation we can get, but the calculation becomes slower. An exact curvature computation cannot be done for volumetric data sets, height fields, point clouds and other discrete representations, either.

© Springer International Publishing AG 2017
J. Braz et al. (Eds.): VISIGRAPP 2016, CCIS 693, pp. 51–71, 2017.
DOI: 10.1007/978-3-319-64870-5_3

The curvature approximation can be computationally expensive, especially if the input data are of a high quality (many triangles, high volume resolution, large point clouds etc.). A recalculation at each frame during interactive data changes can substantially slow down the processing. Existing methods are mostly used for static geometry and their real-time variants mostly rely on parallelization using GPU.

In the proposed approach, to mitigate this problem, the curvature is not estimated directly from the mesh, but rather from the final rendered image in the screen space. In the screen space, only data that are currently visible and interesting for the viewer are processed. Calculations are independent of triangle count of the original geometry, the only limitation is the screen resolution. There is also an advantage that the curvature can be calculated from any possible model representation with the same algorithm. There is no limitation to triangle meshes, the final scene can contain volumetric models, implicit surfaces, procedural generated geometry and other screen space generated effects, such as a water surface.

The screen space techniques have a major advantage for an existing rendering software, as they can be easily added as post-process methods or replace an existing rendering output. Nowadays, those methods are quite popular for many problems, such as water rendering, lighting, ambient occlusion and reflections. In the screen space, however, some problems may occur, usually on the object edges, where pixel flickering may occur. Another disadvantage comes directly from the screen space itself, where the geometry outside the visible area cannot contribute to the results.

The proposed screen space algorithm is designed to be used as the first and fast estimation of the curvature. For a more precise solution, the curvature should be approximated directly from the underlying models, where connectivity of the triangles can be used to improve the results quality.

Contributions of the proposed solution are:

- It overcomes the problems with the interactivity for a high number of triangles by approximating the curvature directly in the screen space.
- It can be used in the screen space as well as in the object space with little or no modification.
- It fits directly into existing rendering pipelines and uses only the usual outputs from deferred renderers (positions, normals).

The rest of this paper is organized as follows. Section 2 covers background, Sect. 3 covers some of the related work in the curvature approximation for the discrete meshes. Section 4 explains the proposed solution in object and screen space. Section 5 presents the algorithm results. Section 6 concludes the paper. Notation used in the article is as follows: symbol "·" denotes the dot product, "×" denotes the cross product, $|x|$ is the vector length and $det(X)$ is the determinant of a matrix X.

2   Basic Theory

2.1   Curvature

Only a brief introduction to curvature will be presented. For more details and proofs of theorems, the reader can study [1].

The important surface descriptors are the fundamental forms. They describe the first and second order derivatives at a given point of the surface.

The first fundamental form (I) is constructed from the first order derivatives at a surface point, which give us two tangent vectors (T_u, T_v), see Fig. 1.

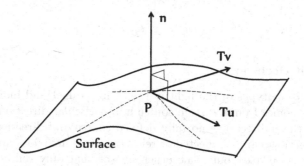

Fig. 1. Tangent vectors of surface at point P.

Vectors T_u, T_v are in general not mutually orthogonal. They are, however, orthogonal to the normal vector n to the surface at the given point. Elements of the matrix I are computed as

$$I = \begin{bmatrix} E & F \\ F & G \end{bmatrix},$$

$$E = T_u \cdot T_u, \ F = T_u \cdot T_v, \ G = T_v \cdot T_v.$$

(1)

The second fundamental form (II) is calculated from the second partial derivatives (T_{uu}, T_{vv}, T_{uv}) and the normal vector (n). The elements of the matrix II are computed as

$$II = \begin{bmatrix} L & M \\ M & N \end{bmatrix},$$

$$n = \frac{T_u \times T_v}{|T_u \times T_v|},$$

(2)

$$L = T_{uu} \cdot n, M = T_{uv} \cdot n, N = T_{vv} \cdot n.$$

Combining the fundamental forms gives the shape operator W (also known as the Weingarten operator). For every point of the surface, it tells us the change of the normalized normal vector in the direction of the tangent vector at this point. W is a 2×2 symmetric matrix that can be obtained from the first (I) and second (II) fundamental form:

$$W = I^{-1}II. \tag{3}$$

The matrix W has two real eigenvalues that correspond to the first (λ_1) and second (λ_2) principal curvatures, and the eigenvectors that correspond to the directions of the principal curvature. Mean (H) and Gaussian (K) curvature are computed from principal curvatures as:

$$H = \tfrac{1}{2}(\lambda_1 + \lambda_2),$$
$$K = \lambda_1\lambda_2. \tag{4}$$

2.2   Ambient Occlusion

There are two basic types of the light in the scene - direct and indirect. While the direct light comes from a certain source in a particular direction (it can be, e.g., a lamp), the indirect has no fully defined source and direction. It comes from every direction, can be reflected or refracted. Its source does not even have to be a light source itself but some other surface that only reflects the light. Therefore it is a very complex task, to calculate the ambient light for a given point, since all objects in the scene can be possible light sources.

Computing the full ambient lighting is virtually impossible, so simplifications have to be done. An ambient occlusion (AO) is a shading technique used to calculate the exposition of a point to ambient light. It is a global method, unlike the well-known local Phong shading, and must be computed as a function of the geometry of the entire scene. In this article, we use the labeling with $AO = 0$ for the darkest areas and $AO = 1$ for the brightest areas.

The best approximation can be computed using ray-tracing. From every point, rays are traced in all directions. If a ray intersects other geometry in the scene, the ray origin is occluded in some way, usually based on the distance of the origin and an intersection. The quality of the effect depends on the number of rays. The problem is the speed. This solution can be very slow even for a small scene, based on the number of rays from every point.

3   Related Work

As already mentioned, curvature cannot be exactly computed for discrete geometry. Instead of that, only an approximation (or estimation) can be calculated. A curvature estimation can be computed by two main categories of approaches - discrete and surface fitting. The discrete methods calculate the curvature directly from the data, while surface fitting tries to find a local approximation of the surface and calculates the curvature of this approximation. Usually, the discrete methods are faster but less accurate.

There are many algorithms for the curvature estimation. Many of them vary only in details and use the same basic ideas. Comparison of several of these algorithms can be found in [2]. Algorithms related to the design of the proposed algorithm are summarized in the next subsections.

3.1 Discrete Methods

The discrete method from [3] uses the second fundamental form to calculate the curvature estimate per triangle. These curvature estimates are then distributed to triangle vertices. Average or Voronoi area weights can be used to express the vertex curvature. This step is similar to normal vector calculations for a triangle mesh. The algorithm [4] is used for curvature estimation of volumetric datasets. This algorithm is a variation of [3], running entirely on GPU, and it is optimized for deformable meshes. However, the interactive nature of this algorithm is achieved by the GPU parallelization rather than by the algorithm itself.

An algorithm similar to [3] has been presented in [5]. It calculates the curvature for every triangle based on triangle vertices positions and unnormalized normals. By a linear interpolation, a single point and a normal is calculated for each triangle. All those values are used for curvature estimation. The estimated curvature depends on the length and quality of the normals.

The algorithm to compute curvature estimation directly from the triangles can be found in [6]. The mean curvature is computed using a discretization of the Laplace operator. Voronoi areas of triangles shared by a vertex are used as weight functions. The Gaussian curvature is calculated from the sum of vertex incidence angles, weighted by same Voronoi areas as for the mean curvature.

3.2 Surface Fitting Methods

Surface fitting methods approximate the surface by the least squares techniques. A curvature estimation algorithm for point clouds based on this approach was presented in [7].

Surface fitting can also be done by finding a local surface approximation. A Bézier patch is often used for this task. In [8], the biquadratic Bézier patch is used. The curvature is computed at a single vertex directly from the patch approximation. The computational cost is very low, but if the selected neighborhood occupies a small area, the results can be incorrect.

Another approach based on a Bézier patch was presented in [9]. First, a local Bézier surface is calculated for every triangle. From this surface, the curvature can be directly calculated. The Bézier surface patch, however, has no G^1 continuity between neighbor surfaces. On the edges, there could be a steep change in the curvature. The improvement has been done by [10], where a G^1 continuous patch is computed as a blend of Bézier surfaces over neighboring triangles. From this patch, the curvature can be directly calculated using analytic solution as proposed in [11]. A drawback of this method is that the second derivatives are much more complex to compute than for a simple Bézier patch.

The curvature computations based on surface fitting in the screen space are not very common. The only algorithm dealing with this problem known to us is [12]. It proposes the screen space curvature calculation by a sphere fitting. A point cloud is created from the screen space pixels and for each pixel, the best fitting sphere is searched. With this approach, however, only the mean curvature is calculated. The Gaussian curvature cannot be computed this way and, therefore, principal curvatures cannot be calculated either.

3.3   Ambient Occlusion

Traditional approaches to estimate AO often use screen space techniques ([13–15] etc.). They reconstruct the information for occlusion estimation from the depth and normal buffers directly in the pixel shader. This information is used for local variants of ray-tracing in some limited neighborhood. These methods are fast and offer sufficient results for real-time rendering, usually in the game industry. A common disadvantage of those methods is the noise present in the result. To overcome this problem, the final image is either down-sampled, blurred, or both approaches are used together.

Besides the traditional approaches with some version of visibility ray-tracing, a solution based on the curvature has been presented by Hattori et al. [16]. Their solution is using an approximation of the neighborhood of a point by principal curvatures (used as terms of the Taylor series). A sphere is circumscribed around the same point. The AO is calculated from the volume of the intersection of the sphere and the local approximation. This leads to an integral, with a simplified solution

$$AO(\lambda_1, \lambda_2) = \frac{2}{\pi} acos \left(\frac{-1 + \sqrt{1 + (\lambda_1 + \lambda_2)^2 r^2}}{(\lambda_1 + \lambda_2)^2} \right), \tag{5}$$

where r is the effect radius (i.e. radius of circumscribed sphere around the point).

An estimation of AO from the curvature has also been used by Griffin et al. [4]. Their solution is partially based on [16], but instead of a full computation in the pixel shader, they precomputed coefficients and used a mapping function

$$AO(\lambda_1, \lambda_2) = 0.0022(K_1 + \lambda_2)^2 + 0.0776(\lambda_1 + \lambda_2) + 0.7369. \tag{6}$$

To increase the effect of occlusion and obtain darker corners, the principal curvatures can be scaled up.

4   The Proposed Algorithm

The proposed algorithm works in the screen space and it can also be used for classic triangle meshes. The core of the algorithm is similar to the one used in [3] and uses fundamental forms as well.

First, a description of the proposed algorithm for a triangle mesh is presented. The screen space version is discussed next..

4.1   Basic Algorithm

The main idea is to describe every triangle independently by the shape operator W, recall Eq. (3). Elements of the shape operator must be calculated in order to find eigenvalues of the matrix and calculate the final curvatures.

The proposed method uses an orthonormal basis. In such a case, the first fundamental form (I) is the identity matrix which means that only the second fundamental form (II) is equivalent to the shape operator, i.e. $W = \text{II}$.

To eliminate one dimension, every triangle is transformed to a local coordinate system, also known as the tangent space. To speed up calculations, axes of the system are chosen to be orthonormal with one of the triangle vertices set as system origin. The original triangle with vertices V_1, V_2, V_3 and normal vectors Vn_1, Vn_2, Vn_3, is expressed in this local system, resulting in vertices V_{L1}, V_{L2}, V_{L3} and normals $Vn_{L1}, Vn_{L2}, Vn_{L3}$. See Fig. 2.

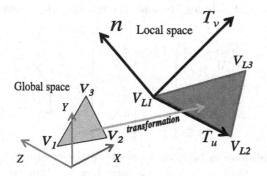

Fig. 2. Local triangle transformation.

Once the triangle is in the local space, one of the dimensions is constant and represents the plane of the triangle. In the following calculations, this dimension is not used and the problem is reduced from 3D to 2D.

The Curvature Calculation. The triangle in the local space is used to build the shape operator, as can be seen in Eq. (2), where variables L, M, N are unknown.

The shape operator describes the change of the normal over the edge of the triangle. The triangle is in the local space and one of the coordinates is constant. This coordinate is left out, which leads to 2D vectors instead of 3D. The edges of the triangle are expressed as 2D vectors

$$(u_i, v_i)^T = V_{Li} - V_{L(i+1) \bmod 3}, \tag{7}$$

and changes of the triangle normals are again 2D vectors

$$(dNu_i, dNv_i)^T = Vn_{Li} - Vn_{L(i+1) \bmod 3}, \tag{8}$$

where $i = 1, 2, 3$. Index i denotes the triangle edge index.

Changes of normals along the edges of the triangle are known. These changes together with edge vectors are used to create a system of equations to find the unknown variables L, M, N. For one edge of the triangle, we get the underdetermined system

$$\begin{bmatrix} L & M \\ M & N \end{bmatrix} \begin{bmatrix} u_1 \\ v_1 \end{bmatrix} = \begin{bmatrix} dNu_1 \\ dNv_1 \end{bmatrix}. \tag{9}$$

However, by constructing the same system for every edge of the local space triangle, an overdetermined system is obtained. The system is in the form $Ax = b$, the least squares method is used to obtain unknown variables:

$$x = (A^T A)^{-1} A^T b. \tag{10}$$

In this particular case, the matrix A is built from the triangle edge vectors $(u_i, v_i)^T, i = 1, 2, 3$ and b is the vector of changes of the triangle normals $(dNu_i, dNv_i)^T, i = 1, 2, 3$. Index i denotes the triangle edge index. Final matrices are as follows:

$$A = \begin{bmatrix} u_1 & v_1 & 0 \\ 0 & u_1 & v_1 \\ u_2 & v_2 & 0 \\ 0 & u_2 & v_2 \\ u_3 & v_3 & 0 \\ 0 & u_3 & v_3 \end{bmatrix}, b = \begin{bmatrix} dNu_1 \\ dNv_1 \\ dNu_2 \\ dNv_2 \\ dNu_3 \\ dNv_3 \end{bmatrix}, x = \begin{bmatrix} L \\ M \\ N \end{bmatrix}. \tag{11}$$

Some optimizations can be done to decrease the total number of numerical operations. The substitution $B = A^T A$ is introduced. The matrix B is symmetric and its elements can be represented by variables p, q, r:

$$B = A^T A = \begin{bmatrix} p & q & 0 \\ q & p+r & q \\ 0 & q & r \end{bmatrix}, \tag{12}$$

where $p = u_1^2 + u_2^2 + u_3^2$, $q = u_1 v_1 + u_2 v_2 + u_3 v_3$ and $r = v_1^2 + v_2^2 + v_3^2$. The inverse of the matrix B can be computed using Eq. (13). Since B is symmetric, the computation is fast and easy.

$$B^{-1} = det(B) \begin{bmatrix} p(r+p) - q^2 & -qr & q^2 \\ -qr & pr & -pq \\ q^2 & -pq & p(r+p) - q^2 \end{bmatrix} \tag{13}$$

The final step of the calculation is to calculate values for the unknown vector x. A part of this step can be simplified, because the inverse of the matrix B is symmetric (see the symmetry pattern in Eq. (14)) and the matrix A has many zero elements. A simplified multiplication can be seen in Eq. (15).

$$B^{-1} = det(B) \begin{bmatrix} b_1 & b_2 & b_3 \\ b_2 & b_4 & b_5 \\ b_3 & b_5 & b_6 \end{bmatrix}, \tag{14}$$

$$B^{-1}A^T = det(B)*$$

$$\left(\begin{bmatrix} u_1b_1 & u_1b_2 & u_2b_1 & u_2b_2 & u_3b_1 & u_3b_2 \\ u_1b_2 & u_1b_4 & u_2b_2 & u_2b_4 & u_3b_2 & u_3b_4 \\ u_1b_3 & u_1b_5 & u_2b_3 & u_2b_5 & u_3b_3 & u_3b_5 \end{bmatrix} + \begin{bmatrix} v_1b_2 & v_1b_3 & v_2b_2 & v_2b_3 & v_3b_2 & v_3b_3 \\ v_1b_4 & v_1b_5 & v_2b_4 & v_2b_5 & v_3b_4 & v_3b_5 \\ v_1b_5 & v_1b_6 & v_2b_5 & v_2b_6 & v_3b_5 & v_3b_6 \end{bmatrix} \right) \tag{15}$$

Having obtained the final vector x, we can construct the desired shape operator. From this matrix, the eigenvalues λ_1, λ_2 are computed by solving the characteristic polynomial. These values correspond to the principal curvature estimated to the triangle. The principal curvatures can be used to evaluate the mean and Gaussian curvature (see Eq. (4)).

The presented algorithm computes the curvature for each triangle. To obtain the curvature at the vertices, we have to use all adjacent triangles at the given point. The final curvature can be estimated as a simple average from all adjacent triangles or the curvature can be further weighted by the triangle area.

In the above calculations, an overdetermined system was constructed from all three edges of the triangle. To solve the system, only two edges are sufficient (values for $i = 3$ will be zero).

The curvature error of this solution can be found in Prantl et al. [17], where results of the described solution are compared with the exact curvature obtained from analytic functions.

4.2   Screen Space Version

The screen space version of the proposed algorithm was designed to fit directly into an existing deferred rendering pipeline. Only the normal vector and the depth value (from which the position is reconstructed) is required for every pixel. There could be probably some quality improvements if additional information (id of the triangle to which the current pixel belongs, the triangle size in the screen space etc.) were available, but this is not the current target.

The screen space depth buffer can be interpreted as a 2.5D function with an underlying regular grid and function values of the depth. In the screen space, there is a constant step size between neighboring pixels. These pixels are triangulated and each pixel center is taken as a triangle vertex. This screen space triangulation is converted to the world or camera space by reconstruction of the position and the normal for each pixel. This creates a simple triangulated mesh and the curvature is estimated on this mesh using the technique described in Sect. 4.1.

The algorithm from Sect. 4.1 can be used directly in the screen space. It can run entirely on the GPU, using a pixel shader. As shown in Sect. 4.1 (see Eq. (13)), the inverse matrix can be computed very fast and only six values have to be stored due to the matrix symmetry. Additionally, only two edges are needed for construction of unknown variables L, M, N to form the shape operator matrix.

$$x = det(B) \begin{bmatrix} u_1b_1 + v_1b_2 & u_1b_2 + v_1b_3 & u_2b_1 + v_2b_2 & u_2b_2 + v_2b_3 \\ u_1b_2 + v_1b_4 & u_1b_4 + v_1b_5 & u_2b_2 + v_2b_4 & u_2b_4 + v_2b_5 \\ u_1b_3 + v_1b_5 & u_1b_5 + v_1b_6 & u_2g + v_2b_5 & u_2b_5 + v_2b_6 \end{bmatrix} \begin{bmatrix} dNu_1 \\ dNv_1 \\ dNu_2 \\ dNv_2 \end{bmatrix}. \qquad (16)$$

All calculations are based on triangles that need to be reconstructed in the screen space. They are obtained directly from the currently rendered pixel and its neighbors. Based on our tests, the 4-neighborhood of the center pixel is sufficient and leads to 4 triangles (shaped like diamond around the center). However, if the neighborhood width is only one pixel, all of those triangles are not needed to compute the curvature estimation and based on our testing, the use of only one of them is sufficient.

4.3   Level of Detail

In the screen space, visible details often depend on the camera distance from the scene object. Small triangles in the world space can occupy almost all the pixels of the rendered image if the camera is very close to the surface. On the other hand, if the camera is far away, the same triangle can take only one pixel of the final image. Taking this into consideration, the level of detail can be used to improve the visual quality of the estimated curvature.

If the neighborhood with one pixel width is used, triangles of the original mesh can be seen in the estimated curvature (see Fig. 3(a)). The estimated curvature within every triangle is the same. GPU interpolates normals and positions during rendering, leading to a smooth Phong shading, but the proposed method uses differences in the positions and normals. These differences are constant (except for the numerical errors) for a flat geometry, leading to the same curvature at every inner point of each triangle.

To solve this problem, level of detail (LOD) sampling can be used. For points closer to the camera, triangles are constructed from a wider neighborhood. In [17], LOD was based on the logarithmic function and texture mipmapping. The new approach, we have used in this publication, has the same quality but with better performance. It is based on a power function and the final equation is:

$$size = size_{max} \left(\frac{1}{f^2}\right)^d + 1, \qquad (17)$$

where $size_{max}$ is the maximal size of the neighborhood, f is the distance of the camera far clip plane (in our tests, this value was always set to be ≥ 100, smaller values were clamped to this interval) and d is a current pixel depth in interval $\langle 0, 1 \rangle$ where 0 is for the closest points to the camera. The $size$ represents the step to the neighboring pixels. For this, the value should be converted to integer by omitting the fractional part. For this reason, there is the +1 term in Eq. (17). The value of $size_{max}$ can be achieved only for $d = 0$, but this value is very rare in the depth buffer.

Using this approach, the final curvature should be computed from more than one triangle. According to our observations, a maximal number of four triangles for one pixel, creating a triangle fan, is sufficient. The final curvature is calculated as an average value from all triangles. The result with LOD for the same model can be seen in Fig. 3. There are used two different sampling. In Figs. 3(a) and (b), sampling is accurate with exact normals computed directly from the function equation. In Figs. 3(c) and (d), the sampling contains noise and normals are computed from mesh itself using the algorithm from [18].

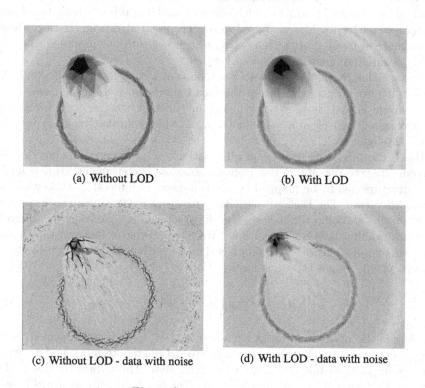

(a) Without LOD   (b) With LOD

(c) Without LOD - data with noise   (d) With LOD - data with noise

Fig. 3. Screen space curvature.

The problem with LOD are discontinuities between neighboring pixels. If the neighborhood has the size of one pixel, they are not very visible and are often not recognizable during movement. However, with an increased step size, the problem is more serious. We have used the simplest solution with the condition

$$|d - d_{neighbor}| > \frac{1}{f}, \tag{18}$$

where d is the depth of the current pixel, $d_{neighbor}$ is the depth of the neighbor with the step $size$ (see Eq. (17)) and f is the camera far clip plane. If the condition is met, e.g., there is a depth discontinuity, the LOD for the current neighbor is disabled and step is set to $size = 1$.

4.4 Ambient Occlusion

The ambient occlusion can be computed using various algorithms, as stated in Sect. 3. However, we are interested in curvature-based algorithms [4,16]. The curvature in those solutions is transferred as an additional vertex parameter. In the proposed solution, we use the proposed screen space version of the curvature estimation algorithm. Both presented curvature based AO solutions require the principal curvatures λ_1 and λ_2 which can be calculated directly from the shape operator (see Sect. 2). It is also used in our proposed algorithm. Both of these algorithms can be used in the screen space together with our proposed solution for the estimation of the principal curvatures.

The solution from [4] resembles a conversion of a color image to its greyscale version. It also uses a variation of the mean curvature (see Eqs. (4) and (6)). The problem is, that both existing algorithms are not consider the convex and concave areas. Convex areas should be dark, while concave areas are usually fully lit by light.

Based on previously mentioned information, we propose a new function to map curvature to the AO. In the proposed solution, the mean curvature is used. The curvature has to be mapped to the symmetrical interval $\langle -1, 1 \rangle$, where 0 is zero curvature. For this, we need the curvature extrema for the mapping. However, this is similar to the need of scaling factor in [4]. We know that convex areas are usually darker than concave. We have created a function that maps the mean curvature to the occlusion based on a tresshold. For values below zero (convex areas), we use the Gaussian function

$$AO = a \cdot exp\left(-\frac{(m - \mu)^2}{2\delta^2} \right), \tag{19}$$

where m is the normalized mean curvature to $\langle -1, 1 \rangle$. μ is the expected value and δ^2 is the variance. We have set those two parameters to $\mu = 0$ and $\delta^2 = 0.2$ to get a normalized function centered around zero. The parameter a sets the maximal occlusion value. We use $a = 0.9$. Values above zero (concave areas) should be lit with a maximal amount of light and therefore $AO = 1$ can be used. However, in some cases, we want a slight occlusion even in these areas to create a smoother transition. For that reason, we use a linear mapping of the interval $\langle 0, 1 \rangle$ to $\langle a, 1 \rangle$.

The results of the approxiamted AO can be seen in Sect. 5.

4.5 Limitations

Similarly to other screen space techniques, the proposed algorithm has its disadvantages. When two neighboring pixels do not come from the same part of the surface, there appears a surface discontinuity between those pixels and an artifact in the computed curvature may appear. We have proposed one possible solution in Sect. 4.3, but it is not guaranteed to work in every situation. If the depth difference is small and pixels belong to different surfaces, the problem will persist.

Another problem is related to LOD. The estimated curvature depends on the distance of the mesh from the camera, where small details are smoothed if the camera is far away from the surface. The setting of the correct LOD can improve the curvature estimation quality.

In the proposed solution, the LOD comes with a performance lost. Usually, LOD is included to increase the performance by using less samples or to simplify computations. In the proposed solution, the LOD version is less efficient due to the need of sampling more pixels than for a simple neighborhood of size 1. This can be seen in Sect. 5 in Fig. 14.

5   Experiments and Results

To test the proposed method, a PC with the following configuration was used: Intel Core i7 CPU running at 4 GHz, 32 GB of RAM memory, NVidia Geforce 960GTX graphics card with 2 GB of video memory. The algorithm was implemented in C++ and OpenGL 4.4 with GLSL shaders.

A comparison of the method described in this article in the object space can be found in Prantl et al. [17] where the curvature calculated directly from triangles in the object space is compared with the exact curvature obtained from analytical functions, tested on a sphere, the function $f_1 = 10\frac{sin(\sqrt{x^2+y^2})}{\sqrt{x^2+y^2}}$, $(x \neq 0), (y \neq 0)$ and the function $f_2 = sin(x)cos(y) + 0.1(x^2 - y^2)$, $x, y \in < -10; 10 >$. Comparison with a surface fitting algorithm based on Bézier triangles [9] is also presented. The mean squared error (MSE) is roughly twice lower for the mean than for the Gaussian curvature. The MSE values are usually low, based on the tested function and constant for sphere ($8.2 * 10^{-16}$ for the Gaussian and $7.8 * 10^{-17}$ for the mean curvature).

In the experiments presented in this article only the screen space is tested, since it is our main area of the use. The implementation of the algorithm from [12], based on [19], has been done using GLSL instead of CUDA used in the original paper.

The color gradient used for all visualisations goes from the blue for negative values to the red color for positive values. The green color in the middle represents zero. See Fig. 4.

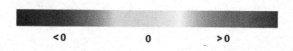

<0   0   >0

Fig. 4. The color gradient used in all presented visualizations. (Color figure online)

5.1   Screen Space Comparison

The comparison of the screen space method is done against the curvature calculated by the proposed method directly on the triangle mesh with and without the LOD active. The proposed algorithm was also compared with [12], the only other screen space technique known to us.

The tested models are: Stanford Dragon (300 000 vertices), MaxPlanck (152 403 vertices) and the function f_1 (15 000 vertices).

In the screen space, the quality of the computed curvature depends on the camera distance from the model. If we compute the curvature for the triangle mesh and render the result, with the camera moving away from the model, the triangles become smaller and more triangles can be rendered in the same pixel. This can cause an incorrect curvature to be visualized. In the proposed screen space method the problem associated with rasterization cannot happen because only visible parts are used to calculate the result and only one value is used for the final pixel. In every test, the model was tested as fully visible on the screen and the camera was moving away from the model. The dependency of MSE on the distance between the viewer and the model is shown in the following graphs.

From the graphs in Fig. 5 it can be seen that the quality of both screen space algorithms is comparable for the mean curvature. For the dragon model, using

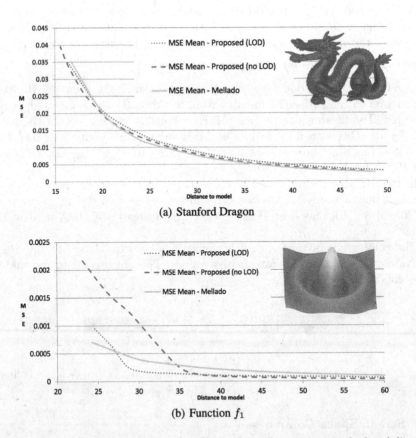

(a) Stanford Dragon

(b) Function f_1

Fig. 5. Comparison of the screen space MSE for the mean curvature calculated directly from the triangle mesh. The proposed method with and without LOD and algorithm from [12] (Mellado) were tested.

LOD has a little or no effect at all. The original model has a dense tessellation and LOD can skip fine details. On the other hand, for the model of the function, the proposed method with LOD achieves better quality.

The Gaussian curvature comparison was done only with and without LOD, since there is no other screen space method known to us that calculates the Gaussian curvature. See results in Fig. 6. The behavior is similar to Fig. 5, with a roughly doubled amount of the MSE error. This is caused by the curvature calculation, where the mean curvature is only a sum of the principal ones, while the Gaussian is computed by multiplying principal curvatures. In that case, the errors of both values are multiplied as well.

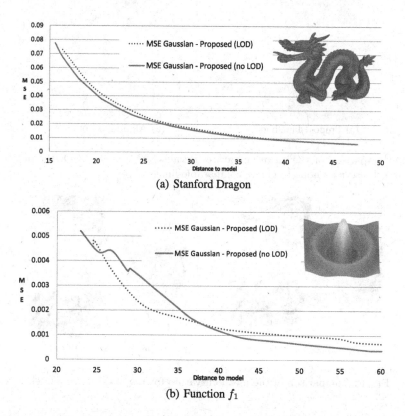

(a) Stanford Dragon

(b) Function f_1

Fig. 6. Comparison of the proposed screen space MSE for the Gauss curvature calculated directly from the triangle mesh.

The visual comparison of the proposed method with [12] can be seen in Fig. 7. Both algorithms have a comparable visual quality. The proposed method results look sharper, [12] is more blurry. See Fig. 8 for comparison of the quality of the proposed method in the screen space against the same method in the object space.

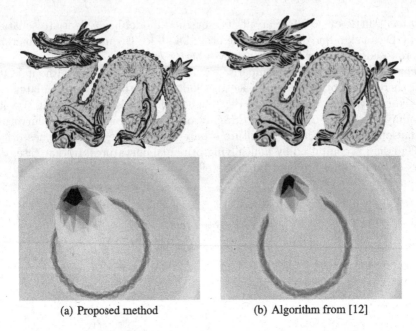

(a) Proposed method   (b) Algorithm from [12]

Fig. 7. Comparison of the mean curvature. Because [12] has no LOD, the presented comparison also uses none to create comparable images.

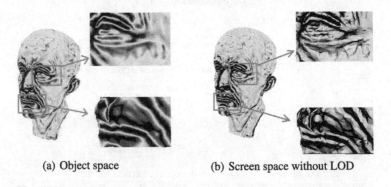

(a) Object space   (b) Screen space without LOD

Fig. 8. Comparison of the mean curvature for the MaxPlanck model.

No LOD is used to show real differences based on the camera distance. For the camera at a greater distance (full model), there is almost no visible difference. With the camera closer to the surface (detailed parts of the image), the triangles of the mesh begin to appear in the screen space curvature.

The effect of the used LOD can be seen in Figs. 9, 10 and 11. If the camera is moving away from the mesh, there is a distance, from which further there is a small or no difference between using and not using LOD. In some cases, using LOD can bring worse results as it smooths out fine details (see Fig. 9). On the other hand, in the example of the Gaussian curvature in Fig. 10, the use of

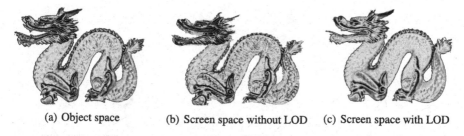

(a) Object space   (b) Screen space without LOD   (c) Screen space with LOD

Fig. 9. Comparison of the mean curvature.

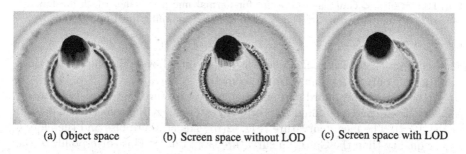

(a) Object space   (b) Screen space without LOD   (c) Screen space with LOD

Fig. 10. Comparison of Gaussian curvature using function f_1.

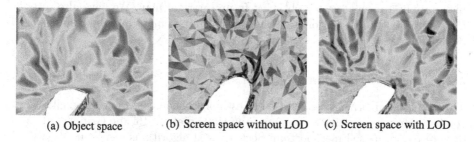

(a) Object space   (b) Screen space without LOD   (c) Screen space with LOD

Fig. 11. Detail of the mean curvature.

LOD improved the result considerably. Another comparison can be seen in the closeup in Fig. 11. If the camera moves very close to the surface, LOD is required to obtain a smooth result. Without LOD, the computed curvature appears as random colors. In some cases, e.g., in wireframe view, this visualization can be sometimes enough to see the shape. To set a suitable distance for LOD is, however, difficult - the same value does not work for all models.

The advantage of curvature estimated directly in the screen space is the possibility to reduce geometry and use normal mapping to add missing details. Our proposed algorithm can be used together with this approach. We have used a plane with details added from a normal map. Since we used fine details, LOD should be disabled in this case. The results can be seen in Figs. 12(b) and (c).

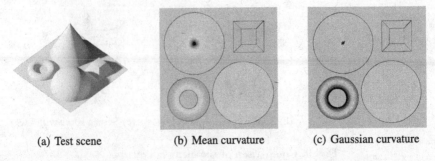

(a) Test scene   (b) Mean curvature   (c) Gaussian curvature

Fig. 12. Mean and Gaussian curvature for normal mapped plane. Black borders are caused by a high curvature, that is outside the used mapping function.

They are screenshots of a flat plane as seen from above. The view from sides would result only in a plane with no geometry.

We have used a standard normal mapping test texture that consist of a torus, a sphere, a cone and a pyramid. The original test scene with geometry can be seen in Fig. 12(a). The mean curvature (Fig. 12(b)) has a lower noise and therefore a higher quality than the Gaussian curvature (Fig. 12(c)). This corresponds to the visual quality of tests in Figs. 9 and 10. The resulting curvatures corresponds to the curvatures of real objects even if there is no position (we have a flat plane), only a normal vector obtained from a normal map. We have also tested the version with an additional displacement (bump) map, but the results were almost identical.

5.2 Ambient Occlusion

The proposed algorithm can be used to estimate the ambient occlusion either in combination with existing techniques from Hattori et al. [16] and Griffin et al. [4], or with our proposed method from Sect. 4.4. All algorithms can be used with a precomputed curvature, but in the proposed solution, the curvature is estimated directly in the screen space. The comparison of the two methods against occlusion calculated using ray-casting can be seen in Fig. 13. Both results were computed without LOD because the underlying mesh is of a high quality.

Hattori's solution uses the radius 0.25. We have tested different radii, but the results were too dark or too bright and the occlusion effect was hard to perceive. Griffin's solution offers a better visual quality. The curvatures are scaled up with factor of 5. Different scales result again in a darker or lighter effect, which is similar to Hattori's solution. The problem with the Griffin algorithm are non-white areas with zero occlusion.

Our proposed solution can be seen in Fig. 13(d). The result in our proposed solution is not as dark as [4] and looks more like ray-casted result. On the other hand, solution from [16] keeps correct white areas, but the rest of the model is too dark. We are using the tresshold a from Eq. (19) set to $a = 0.9$. If we set this tresshold to $a = 1$, we can obtain white areas as well, however, some details

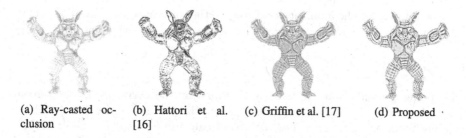

(a) Ray-casted oc-   (b) Hattori   et   al.   (c) Griffin et al. [17]   (d) Proposed
clusion   [16]

Fig. 13. Ambient occlusion estimated using the proposed curvature algorithm.

are lost. From our point of view, the configuration, we have used, offers the best
visual appereance.

5.3   Performance

The proposed method runs at interactive frame rates. Due to the independence
of the geometry, all tested models brought nearly the same results. In the tests,
the method was computed for pixel coverage of 2–100% of the screen. The depth
value of the remaining pixels was set to infinity to discard these pixels. The
comparison was done against the screen space method from [12].

The resulting performance can be seen in Fig. 14. In all tests, a decrease of
performance is partially caused by LOD computation itself but mostly by the
need of branches in the pixel shader to decide if the triangle can be used or will
be rejected as described in Sect. 4.3. With a comparable visual quality as [12]
(no LOD used), the proposed algorithm is much faster. The noisy peaks around
80–100% of the screen coverage in Fig. 14 are caused by camera movements and
some possible context switching during measurements due to the graphics driver
because the timing itself is very low.

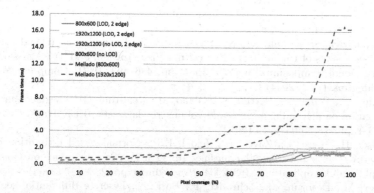

Fig. 14. Frame time based on screen pixel coverage. The proposed algorithm in versions
with only two edges (with and without LOD) against [12] (Mellado) was tested.

The timings for the ambient occlusion are the same as without it. The occlusion computations itself are very fast due to the simplicity of the used Eq. (19).

6   Conclusions

This paper presented an algorithm for estimating the curvature in the screen space, is easy to implement and runs in real-time. It can be easily added to existing rendering pipelines. The algorithm can also be used for a low polygonal geometry, where details are added from a normal map. We have tested the usability of the screen space curvature estimation for the ambient occlusion. The algorithms for the ambient occlusion estimation from curvature already exist [4,16], so we have used them directly. We have also created our new variation of the ambient occlusion approximation from the curvature.

The limitations of this technique are similar to the other screen-space algorithms. There are possible problems with surface discontinuities. Another problem is related to LOD. The estimated curvature depends on the distance of the mesh from the camera, where small details are smoothed if camera is far away from the surface. To set the correct LOD can improve the curvature estimation quality.

A reference implementation of the proposed method (the shader source code and the test application) is available at http://graphics.zcu.cz/sscurvature.html.

Acknowledgement. This work was supported by the Czech Ministry of Education, Youth and Sports – the project LO1506 and University spec. research – 1311; and by the UWB grant SGS-2016-013 Advanced Graphical and Computing Systems.

References

1. Gray, A.: Surfaces in 3-dimensional space via mathematica. In: Modern Differential Geometry of Curves and Surfaces with Mathematica, 2nd edn, pp. 394–401. CRC Press Inc., Boca Raton (1997)
2. Magid, E., Soldea, O., Rivlin, E.: A comparison of gaussian and mean curvature estimation methods on triangular meshes of range image data. Comput. Vis. Image Underst. **107**, 139–159 (2007)
3. Rusinkiewicz, S.: Estimating curvatures and their derivatives on triangle meshes. In: Proceedings of the 3D Data Processing, Visualization, and Transmission, 2nd International Symposium, 3DPVT 3004, pp. 486–493. IEEE Computer Society, Washington, D.C. (2004)
4. Griffin, W., Wang, Y., Berrios, D., Olano, M.: Real-time GPU surface curvature estimation on deforming meshes and volumetric data sets. IEEE Trans. Vis. Comput. Graph. **18**, 1603–1613 (2012)
5. Theisel, H., Rossi, C., Zayer, R., Seidel, H.P.: Normal based estimation of the curvature tensor for triangular meshes. In: 12th Pacific Conference on Computer Graphics and Applications, PG 2004, Proceedings, pp. 288–297 (2004)
6. Meyer, M., Desbrun, M., Schröder, P., Barr, A.: Discrete differential-geometry operators for triangulated 2-manifolds. In: Hege, H.C., Polthier, K. (eds.) Visualization and Mathematics III. Mathematics and Visualization, pp. 35–57. Springer, Heidelberg (2003). doi:10.1007/978-3-662-05105-4_2

7. Yang, P., Qian, X.: Direct computing of surface curvatures for point-set surfaces. In: Botsch, M., Pajarola, R., Chen, B., Zwicker, M. (eds.) Eurographics Symposium on Point-Based Graphics. The Eurographics Association (2007)

8. Razdan, A., Bae, M.: Curvature estimation scheme for triangle meshes using biquadratic Bézier patches. Comput.-Aided Des. **37**, 1481–1491 (2005)

9. Zhihong, M., Guo, C., Yanzhao, M., Lee, K.: Curvature estimation for meshes based on vertex normal triangles. Comput.-Aided Des. **43**, 1561–1566 (2011)

10. Fünfzig, C., Müller, K., Hansford, D., Farin, G.: PNG1 triangles for tangent plane continuous surfaces on the GPU. In: Proceedings of Graphics Interface 2008, GI 2008, pp. 219–226. Canadian Information Processing Society, Toronto (2008)

11. Boschiroli, M., Fünfzig, C., Romani, L., Albrecht, G.: G1 rational blend interpolatory schemes: a comparative study. Graph. Models **74**, 29–49 (2012)

12. Mellado, N., Barla, P., Guennebaud, G., Reuter, P., Duquesne, G.: Screen-space curvature for production-quality rendering and compositing. In: ACM SIGGRAPH 2013 Talks, SIGGRAPH 2013, pp. 42:1. ACM, New York (2013)

13. Mittring, M.: Finding next gen: Cryengine 2. In: ACM SIGGRAPH 2007 Courses, SIGGRAPH 2007, pp. 97–121. ACM, New York (2007)

14. McGuire, M., Osman, B., Bukowski, M., Hennessy, P.: The alchemy screen-space ambient obscurance algorithm. In: High-Performance Graphics 2011 (2011)

15. McGuire, M., Mara, M., Luebke, D.: Scalable ambient obscurance. In: High-Performance Graphics 2012 (2012)

16. Hattori, T., Kubo, H., Morishima, S.: Real time ambient occlusion by curvature dependent occlusion function. In: SIGGRAPH Asia 2011 Posters, SA 2011, pp. 48:1. ACM, New York (2011)

17. Prantl, M., Váša, L., Kolingerová, I.: Fast screen space curvature estimation on GPU. In: Proceedings of the 11th Joint Conference on Computer Vision, Imaging and Computer Graphics Theory and Applications, pp. 149–158 (2016)

18. Max, N.: Weights for computing vertex normals from facet normals. J. Graph. Tools **4**, 1–6 (1999)

19. Mellado, N.: Screen space curvature using CUDA/C++ (algorithm implementation from patate library) (2015)

Multi-Class Error-Diffusion with Blue-Noise Property and Its Application

Xiaoliang Xiong$^{(\boxtimes)}$, Haoli Fan, Jie Feng, Zhihong Liu, and Bingfeng Zhou

Institute of Computer Science and Technology, Peking University, Beijing, China
{jonny_xiong,fanhaoli,feng_jie,liuzhihong,cczbf}@pku.edu.cn

Abstract. Error-diffusion is commonly used as a sampling algorithm over a single channel of input signal in existing researches. But there are cases where multiple channels of signal need to be sampled simultaneously while keeping their blue-noise property for each individual channel as well as their superimposition. To solve this problem, we propose a novel discrete sampling algorithm called *Multi-Class Error-Diffusion* (MCED). The algorithm couples multiple processes of error-diffusion to maintain a sampling output with blue-noise distribution. The correlation among the classes are considered and a *threshold displacement* is introduced into each process of error-diffusion for solving the sampling conflicts. To minimize the destruction to the blue-noise property, an optimization method is used to find a set of optimal key threshold displacements. Experiments demonstrate that our MCED algorithm is able to generate satisfactory multi-class sampling output. Several application cases including color image halftoning and vectorization are also explored.

Keywords: Sampling · Error-diffusion · Halftoning · Image vectorization · Blue-noise

1 Introduction

Error-diffusion (ED) is originally a halftoning technique that quantizes a multi-level image to a binary one while preserving its visual appearance through diffusing the quantization error of one pixel to its neighborhood [7]. It is widely used in the industry of printing and displaying, and also an important sampling algorithm working on discrete domain. Moreover, some researchers extend its usage into digital geometry processing [1].

Previous research mainly focused on the behavior of error-diffusion sampling over a single channel of input signal [4,14,18,22]. However, there are cases where multiple channels of input signal need to be sampled simultaneously, while certain ideal properties such as blue-noise are also required for all the sampling output of these channels.

Simply overlapping the output of blue-noise sampling for multiple individual channels can not guarantee the blue-noise property of their superimposition.

© Springer International Publishing AG 2017
J. Braz et al. (Eds.): VISIGRAPP 2016, CCIS 693, pp. 72–94, 2017.
DOI: 10.1007/978-3-319-64870-5_4

Hence, we are aiming to propose a novel *Multi-Class Error-diffusion (MCED)* algorithm to solve this problem. Here, *a class* refers to the sampling process for a single channel of input signal as well as its sampling output. For an ideal multi-class error-diffusion with blue-noise property, the following requirements must be satisfied:

1. The sampling point distribution of each individual class should possess blue-noise property.
2. When the sampling output of all the classes are superimposed, no two sampling points from different classes can occupy the same position.
3. When all the sampling points from all the classes are superimposed and considered as a whole, their distribution should possess blue-noise property.

The first requirement is naturally guaranteed by the *standard ED* algorithm. For the second, we remain only one class with the highest priority and disable the others when conflict occurs. However, the selection of a certain class may disrupt the point distribution of other classes which violates the first requirement. To solve this problem, we introduce a *threshold displacement* into each process of ED and they are optimized to minimize the destruction to the blue-noise property. Since the frequency spectrum property of each class and the final output is considered during the threshold displacement optimization, the blue-noise property is promised after all classes are superimposed (the third requirement). After meeting these requirements, our MCED algorithm generates satisfactory multi-class sampling output.

The contributions of our work includes:

1. Proposing a multi-class error-diffusion framework, the validity of which can be explained by the commonly used Fourier transform [12];
2. Giving a parameter optimization method to ensure the blue-noise property of the output. Experiment results using these optimal parameters show the effectiveness of the method;
3. Several applications of the MCED are explored, showing that our algorithm is generic and applicable in many areas in computer graphics.

2 Related Work

The *original ED* is an algorithm invented for gray-scale image displaying and printing [7]. It is also frequently used in many other areas in computer graphics as a sampling algorithm [1,3,11]. There are a lot of research which aims at improving its behavior for sampling a single channel of input signal [4,13,23]. Ulichney first proposed the concept of blue-noise [18] and used it as a tool to measure the quality of ED output. Some of the work also aims at giving a solid mathematical analysis of the behavior of error diffusion algorithm. These analyses can explain or predict the results of many techniques originated from the original error diffusion algorithm [12,21].

Using ED to sample multiple channels of input signals in a coordinated way is not a novel problem. It traditionally exists in the area of color printing, where a limited number of colorants are used to reproduce a continuous-tone color image [2]. Many studies focus on solving this problem using ED technique, which is called *vector error-diffusion* in some literature, because they quantized the input signals simultaneously by treating them as a vector [6,9,10]. For example, the vector ED algorithm proposed in [5] uses an optimum matrix-valued error filter to take into account the correlation among color planes. It can generate sampling output with blue-noise property for color images, but cannot guarantee this property for each individual color channel.

Wei extends the traditional Poisson disk sampling for a single channel of signal into a *multi-class blue-noise sampling* algorithm [19]. The algorithm is able to sample a set of input signals in a correlated way while keeping the blue-noise property of the whole output. It can also precisely control the number or density of the generated sampling points. Unlike the ED which works directly in a discrete domain, this algorithm is originally designed in a continuous domain. Hence it is not suitable to be applied in certain application areas that deal with discrete domain, such as color image halftoning [20].

3   Multi-Class Error-Diffusion

Our multi-class error diffusion algorithm is built using the state-of-art standard error diffusion. In this section, we first describe the standard error diffusion utilized in our algorithm and then give our MCED framework.

3.1   Standard Error Diffusion

The *original error diffusion* algorithm given by Floyd et al. in [7] is shown inside the dashed line of Fig. 1. In this algorithm, each pixel $p(x, y) \in [0, 1]$ in the input image p is parsed with a serpentine scan line order and quantized by a quantizer:

$$Q(p', u) = \begin{cases} 1, & p' > u \\ 0, & \text{otherwise} \end{cases} \tag{1}$$

After that, the quantization error $e(x, y)$ is calculated and distributed into multiple unparsed pixels by accumulating to an error buffer $b(\cdot, \cdot)$, which is used to compensate the error. Therefore, for pixel $p(x, y)$, the actual input p' to the quantizer $Q(\cdot, \cdot)$ is $p' = p(x, y) + b(x, y)$. Here, the error filter a_{jk} is a set of constant coefficients, and the quantization threshold u in $Q(\cdot, \cdot)$ is also a fixed value, e.g. $u = 0.5$.

Some important improvements to the original ED algorithm include the introduction of a variable error filter [14] and a variable threshold value [22] to ensure the blue-noise property of the sampling output. In this paper, we use Zhou and Fang's *threshold modulated ED* algorithm to build our MCED framework. Similar as in [4], we refer to that algorithm as the *standard ED*, and its diagram is given by Fig. 1 as a whole. Unlike the original ED, the threshold u and the error filter a_{jk} here are not constant, but functions of the input pixel $p = p(x, y)$,

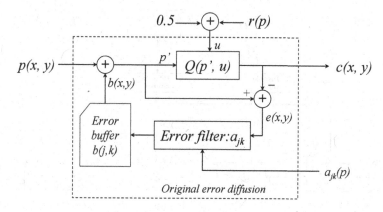

Fig. 1. The principle of the standard error-diffusion.

that is, $u = 0.5 + r(p)$, and $a_{jk} = a_{jk}(p)$. Here, $r(p) = \delta(p) \cdot \lambda(x, y)$ is a modulated white noise. The modulation strength $\delta(p)$ for the white noise $\lambda(x, y)$, and the error filter $a_{jk}(p)$ are pre-optimized so that the output of the standard ED possesses blue-noise property.

3.2   MCED Framework

Our MCED algorithm mainly concerns about simultaneously sampling on multiple channels of input signals and maintaining the blue-noise property for all the classes as well as their superimposition. If simply performing the standard ED independently on each channel of input signal, there may be sampling points from different classes situated at the same sampling position when they are superimposed, which is called *sampling conflict*. Hence, the blue-noise property of the superimposed output cannot be guaranteed.

To solve this problem, we first perform the quantization on each channel of signal, and produce a set of initial sampling outputs with sampling conflicts. Then, the conflicts are removed by disabling the outputs of certain classes based on the inter-class correlation. In this way, the initial outputs are modified to generate the final sampling points. During the process of quantization and error-diffusion, a *threshold displacement* is introduced to decrease the occurrence of sampling conflict and maintain the blue-noise property.

The framework of our MCED algorithm is illustrated in Fig. 2. It takes n channels of signals $\{p_i | i = 1, \cdots, n\}$ as input, where $p_i = p_i(x, y)$ is a 2-D discrete function that satisfies $p_i(x, y) \in [0, 1]$ and $\sum_{i=1}^{n} p_i(x, y) \leq 1$. In fact, it defines the density of sampling points to be generated at the spatial position (x, y). Specially, when p_i represents an image, element $p_i(x, y)$ is the intensity of the pixel at position (x, y).

Our framework concerns the processing of individual channels of signals as well as their correlations, and produces corresponding sampling point sets $\{c_i | i = 1, \cdots, n\}$, where $c_i(x, y) = 1$ indicates a sampling point generated at (x, y) for

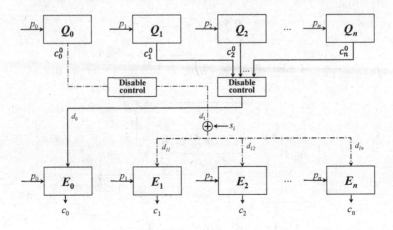

Fig. 2. The framework of our MCED algorithm, where $\{p_i | i = 1, \cdots, n\}$ are the input signals, and $p_0 = \sum_{i=1}^{n} p_i$ is an internal reference signal. After two processing steps of modified standard ED, Q_i and E_i, the framework generates blue-noise sampling outputs $\{c_i | i = 0, \cdots, n\}$. The pseudo code for the framework can be found in the Appendix A.

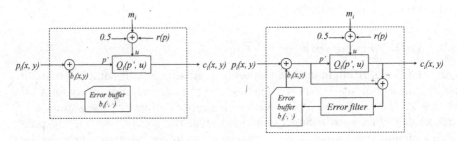

Fig. 3. Two processing steps of modified ED: Q_i (left) performs only the quantization and E_i (right) complete the final error-diffusion ($i = 0, \ldots, n$).

signal p_i, while 0 means no point generated. Therefore, the sampling process from the input p_i to the output c_i is referred as a *class* C_i.

To facilitate the inter-class correlation, we define a special internal signal p_0, whose sampling density is $p_0(x, y) = \sum_{i=1}^{n} p_i(x, y)$. Sampling to p_0 with the standard ED, we can also obtain a blue-noise output, which is used as a reference for the superimposition of the sampling output of all the classes. That means the corresponding output of p_0, denoted as c_0, will be identical to the superimposition of $\{c_i | i = 1, \cdots, n\}$. Hence, we name C_0 as a *reference class*.

Our framework parses the elements $\{p_i(x, y) | i = 0, \cdots, n\}$ in a serpentine scan line order. All $p_i(x, y)$ at the same position (x, y) are processed simultaneously, and then the processing moves to the next position. For each group of $\{p_i(x, y) | i = 0, \cdots, n\}$ at (x, y), the processing of each class includes two steps: Q_i and E_i. Q_i performs independent sampling and produces initial output $c_i^0(x, y)$ for each individual class C_i; while E_i modifies the initial output according to the correlation between the sampling classes and generate the final output $c_i(x, y)$.

3.3   Quantization and Error-Diffusion

The work flow of the two processing steps, Q_i and E_i, are shown in Fig. 3. Actually, E_i is a modified version of the standard ED given in Fig. 1. It simply adds one more variable m_i to the latter's quantization threshold. This simple modification plays an important role in our framework, because m_i will be used to introduce the inter-class correlation into the sampling, and that is the key to avoid sampling conflict and ensure the blue-noise property of c_i.

It is notable that Q_i produces only the initial sampling output, hence it performs only the quantization part of E_i. Although Q_i shares the error buffer $b_i(\cdot, \cdot)$ with E_i, it does not modify $b_i(\cdot, \cdot)$. This is because the output of Q_i will be modified in E_i and thus the quantization error made in the latter step is the one that is to be distributed for the further processing.

3.4   Removing Sampling Conflicts

In our framework, the input signal p_i, $i = 0, \cdots, n$, of each class is firstly processed by Q_i, and generates corresponding uncorrelated blue-noise output c_i^0. In order to eliminate the conflict after superimposition, the correlation between classes are introduced by the reference class C_0. Based on that, some classes with sampling conflicts will be *disabled*, i.e. they will be prohibited to produce a sampling point.

As shown in Fig. 4, when Q_0 of class C_0 does not generate a sampling point at current position (x, y), the output of all $\{E_i | i \neq 0\}$ should be forced to be 0. In other words, all E_i should be disabled from generating a sampling point at (x, y), for there is no correspondence in the final superimposition. Similarly, when Q_0 generates sampling point but none of $\{Q_i | i \neq 0\}$ does, E_0 should also be disabled because no E_i will provide sampling point to form this superimposition. The third case, in which $\{Q_i | i \neq 0\}$ generate sampling points without conflicts, is the ideal case that we are expecting and no modification to $\{c_i^0 | i \neq 0\}$ is needed. Finally, when sampling conflicts occur ($\sum_{j \neq 0} c_j^0 > 1$), most of the conflicting classes must be disabled, and only one of them with the highest priority is allowed to remain. Here, we give the priority to the class C_k with the highest average sampling density, i.e. $k = argmax_j \sum_{(x,y)} p_j(x, y)$, for $j \neq 0$ and $c_j^0 = 1$. Then, a binary *selecting signal* s_i ($i \neq 0$) is defined, where $s_i = 0$ means E_i should be disabled:

$$s_i = \begin{cases} 0, \sum_{j \neq 0} c_j^0 > 1 \text{ and } i \neq k; \\ 1, \sum_{j \neq 0} c_j^0 > 1 \text{ and } i = k; \\ 1, \sum_{j \neq 0} c_j^0 \leq 1. \end{cases} \quad (2)$$

To disable a class, we utilize another *disabling signal* to modify the quantization threshold. It is based on an important fact about the ED: If the threshold u in

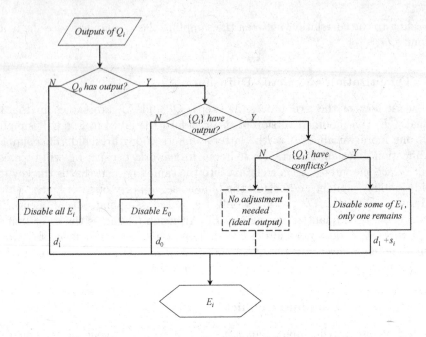

Fig. 4. Eliminating sampling conflicts by disabling some of the classes.

the quantizer $Q(p, u)$ takes a value larger than any possible value of input p, the output of Q will be forced to be 0, and the corresponding class will be disabled. As shown in Fig. 2, the disabling signals are obtained based on the initial outputs $\{c_i^0 | i = 0, \cdots, n\}$ by the *Disable control*, where $d_0 = \neg(c_1^0 \vee c_2^0 \cdots \vee c_n^0)$ and $d_1 = \neg(c_0^0)$. Then, combining with the selecting signal s_i, d_1 turns into a $d_{1i} = d_1 \wedge s_i$ for each E_i $(i \neq 0)$. Therefore, if $d_0/d_{1i} = 1$, a large value $(+\infty)$ will be added to the corresponding quantization threshold, and the class will be disabled.

In this way, d_0, d_1 and s_i facilitate the correlation described in Fig. 4. For example, when $c_0^0 = 1$ and $c_i^0 = 1$ $(i \neq 0)$, we have $d_0 = 0$ and $d_1 = 0$, and the output of E_i will be decided by its priority: if $s_i = 0$, E_i will be disabled. After the class disabling, sampling conflicts can be removed and no more than one sampling point will be generated at each position.

3.5   Maintaining Blue-Noise Property

Modifying the output of certain Q_i to remove the sampling conflicts may cause the destruction of the blue-noise property of the initial outputs. To solve this problem, we make use of another important fact for ED: For a given input signal, a constant variation of the quantization threshold u will change the distribution of output sampling points, but will not affect the average sampling density, as long as u is not constantly $+\infty$. This fact can be proofed with the analysis tool provided in [12], which will be briefly described in Sect. 7. Therefore, adding a properly chosen *threshold displacement* t_i to the threshold u will help to decrease the chance of the sampling conflict occurrence and restore the blue-noise distribution. The detail of t_i will be discussed in the next section.

Hence, the disabling signals d_0/d_{1i}, along with the threshold displacement t_i, are finally added to the thresholds of E_i via the modification term m_i in the second step. Then, E_i quantizes the input element $p_i(x, y)$ at current position (x, y), and the quantization error is distributed and accumulated to the error buffer $b(j, k)$. The above processing is repeated during the parsing of the input elements. When all the elements $p_i(x, y)$ are parsed, the final blue-noise MCED output c_i that satisfies all the requirements given in Sect. 1 can be generated.

4   Threshold Displacement

After the first step of the MCED, the output of some of the classes is disabled due to sampling conflicts. This may disrupt the blue-noise property that originally existed in the standard ED output. We solve this problem by adding a constant displacement value t_i to the quantization threshold for the input p_i, since a shift of the threshold may change the distribution of sampling points and thus may decrease the chance of sampling conflicts. Therefore, it is possible to ensure a better blue-noise output with a properly chosen t_i.

4.1   Displacement Optimization

Our goal is to find a set of optimal threshold displacements $\{t_i | i = 0, 1, \cdots, n\}$, to decrease the probability of sampling conflict occurrence. Since the sampling conflict is a interference among all the classes, the value of the displacement t_i is related to all the input signals $\{p_i | i = 1, \cdots, n\}$. For each class $C_i, i = 0, \cdots, n$, in a n-class MCED, we treat the influence from all other sampling classes as a noise, which is characterized by its strength $\sigma_i = \sum_{j \neq 0 \wedge j \neq i} p_j$. We assume that the same amount of σ_i for a given p_i will result in similar output. Since σ_i can be derived from $\sigma_i = p_0 - p_i$, then t_i can be simplified as a function of p_i and p_0. As class C_0 is designed as a reference for the superimposition of other n classes, t_0 is related with only the sum of other classes inputs, i.e. $p_0 = \sum_{i=1}^{n} p_i$.

Therefore, given an input combination $(p_0, p_i), i = 1, \cdots, n$, in a n-class MCED, we have:

$$\begin{cases} t_i = g(p_0, p_i), i = 1, \cdots, n, \\ t_0 = f(p_0). \end{cases} \quad (3)$$

Then, optimal t_i are calculated by solving an optimization problem. The optimization target function is defined according to the Fourier power spectrum of the sampling output. Based on the existing research [18,22], the blue-noise property of a sampling set can be measured by the *anisotropy* and the *lower frequency ratio* of its spectrum. In this paper, the sampling output of ED are affected by both the input signal p_i and the threshold displacement t_i. Therefore, the *anisotropy* $\alpha(p_i, t_i)$ and *lower frequency ratio* $\beta(p_i, t_i)$ for the final output c_i can be modified from their original formulations in [22]:

$$\begin{cases} \alpha(p_i, t_i) = Corre(P_0(p_i, t_i), P_{45}(p_i, t_i), P_{90}(p_i, t_i)), \\ \beta(p_i, t_i) = L(p_i, t_i), \end{cases} \tag{4}$$

where $Corre(\cdot)$ is a cross-correlation function; $P_0(\cdot)$, $P_{45}(\cdot)$ and $P_{90}(\cdot)$ are *segmented radially averaged power spectrums*; $L(\cdot)$ is the *lower frequency ratio*.

Therefore, our target function T to be minimized in the searching of optimal threshold displacements t_i is defined as:

$$T = \omega_1 \cdot \left(\omega_0 \cdot \sum_{i=1}^{N} \alpha(p_i, t_i) + (1 - \omega_0) \cdot \sum_{i=1}^{N} \beta(p_i, t_i) \right)$$
$$+ (1 - \omega_1) \cdot (\omega_0 \cdot \alpha(p_0, t_0) + (1 - \omega_0) \cdot \beta(p_0, t_0)), \tag{5}$$

where the weights $\omega_0 = 0.5$ and $\omega_1 = 0.7$ are taken in our implementation. A simplex method [15] is adopted to automatically search for the optimal displacement $\{t_i | i = 0, 1, \cdots, n\}$.

Figure 5[1] demonstrates an example of the displacement optimization. Given $p_1 = \frac{32}{255}$, $p_2 = \frac{21}{255}$, $p_3 = \frac{16}{255}$, $p_4 = \frac{12}{255}$, $p_5 = \frac{8}{255}$, $p_6 = \frac{6}{255}$, $p_7 = \frac{5}{255}$, and $p_0 = \sum_{i \neq 0} p_i = \frac{100}{255}$, the optimized threshold displacements are: $t_1 = \frac{34}{255}$, $t_2 = \frac{18}{255}$, $t_3 = \frac{52}{255}$, $t_4 = \frac{23}{255}$, $t_5 = \frac{80}{255}$, $t_6 = \frac{46}{255}$, $t_7 = \frac{17}{255}$, and $t_0 = -\frac{9}{255}$. The sampling outputs of classes $\{C_i | i = 0, \cdots, 7\}$, and their corresponding Fourier power spectra are given in the figure. Figure 5(h) is the superimposition of the dots from the 7 classes, which are colored in red, green, blue, yellow, magenta, cyan and white, respectively. The optimization is performed on 256×256 patches.

4.2   Displacement Interpolation

To decrease computational costs, we perform displacement optimization only on a set of selected *key input combinations*, by minimizing the target function T in Eq. 5. Then, the optimal threshold displacements for other input combinations can be calculated by interpolation, where $t_i = g(p_0, p_i)$, $i = 1, \cdots, n$, is implemented with a bilinear interpolation, and $t_0 = f(p_0)$ with a 1-D linear interpolation.

The *key input combinations* and their corresponding optimal threshold displacement values are shown in the tables contained in Fig. 6. The key levels are selected by an interval of $\frac{16}{255}$. For convenience, all the numbers filled into the table are 255 times of their real values. The right table gives the correspondence between the threshold displacement t_0 and the input p_0 of the reference class C_0, as $t_0 = f(p_0)$. The left table is composed of two parts:

- The values of the threshold displacement $t_i = g(p_0, p_i)$ for the given key input combinations are enumerated in the lower-left triangle area. They are obtained by solving the optimization problem of Eq. 5. The indices of p_0 and p_i are given in the bottom row and the leftmost column, respectively.

---

[1] Config the PDF reader with 100% scaling ratio and the given DPI for best viewing of the details, similar for the following figures.

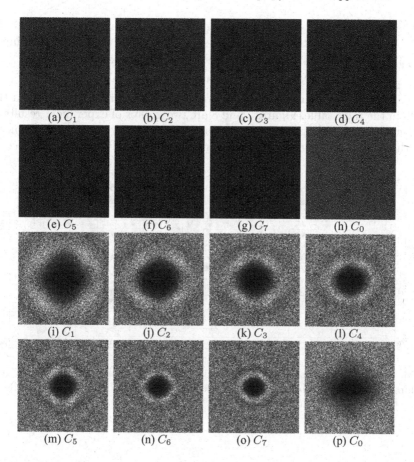

Fig. 5. Optimization result for a 7-class ED. (a)–(h) Sampling outputs of each classes C_i. (i)–(p) Corresponding Fourier power spectra. (Image best viewed at 306 DPI). (Color figure online)

- The corresponding power spectra of the sampling outputs for p_i and p_0 can be found in the upper-right triangle area, where the left image is for p_i and the right for p_0. The indices of p_0 and p_i are given in the top row and the rightmost column. The power spectra show that, for the given input combinations, our optimization successfully converges to threshold displacements that can produce outputs with ideal blue-noise property.

Hence, the threshold displacement of a key input combination can be read directly from the table. For example: given $p_0 = \frac{112}{255}$ and $p_i = \frac{16}{255}$, the displacement value $t_i = g(\frac{112}{255}, \frac{16}{255}) = \frac{49}{255}$ can be found in the cell at the 8th row and the 2nd column of the left table, and the corresponding power spectra are in the cell at the 2nd row and 8th column.

Then, for arbitrary combination (p_0, p_i) that is not included in Fig. 6, we use a bilinear interpolation to obtain their t_0 and t_i. For example, for an input $(p_0, p_i) = (\frac{122}{255}, \frac{21}{255})$, its t_i is interpolated between four key values $g(\frac{112}{255}, \frac{16}{255})$, $g(\frac{112}{255}, \frac{32}{255})$, $g(\frac{128}{255}, \frac{16}{255})$ and $g(\frac{128}{255}, \frac{32}{255})$ that exist in the table. The value of t_0 is interpolated in a similar way, but using a 1-d linear interpolation between the values in the right table. Figures 7 and 8 are two groups of experiment results of a two-class MCED using our key values and interpolation mechanism. It can be seen that the sampling results meet the requirement of MCED given in Sect. 1.

$p_0 \backslash p_i$	0	16	32	48	64	80	96	112	128	144	160	176	192	208	224	240	255	$p_0 \backslash p_i$
0	0																	0
16	0	0																16
32	0	39	0															32
48	0	49	-3	0														48
64	0	14	51	-23	0													64
80	0	28	35	3	37	0												80
96	0	56	18	43	6	-6	0											96
112	0	49	30	53	96	12	59	0										112
128	0	34	10	11	62	-26	2	93	0									128
144	0	6	26	59	5	-1	12	18	14	0								144
160	0	14	100	106	12	56	44	98	90	22	0							160
176	0	12	43	47	42	48	39	100	52	25	47	0						176
192	0	-46	28	6	0	-7	45	-36	0	25	37	1	0					192
208	0	75	54	-7	71	-33	59	23	-1	13	9	13	0	0				208
224	0	12	18	89	12	-2	75	0	0	0	12	3	0	50	0			224
240	0	16	12	9	9	12	49	-20	-2	14	50	1	9	50	46	0		240
255	0	12	12	20	12	0	29	12	44	50	18	0	50	43	50	86	0	255
$p_0 \backslash p_i$	0	16	32	48	64	80	96	112	128	144	160	176	192	208	224	240	255	

p_0	t_0
0	0
16	0
32	65
48	-35
64	-39
80	-90
96	-20
112	-15
128	-79
144	0
160	169
176	13
192	61
208	109
224	168
240	166
255	64

Fig. 6. Threshold displacement t_0 and t_i with their Fourier analysis results for different key input combinations (p_0, p_i). For convenience, the numbers in this table are all 255 times of the value by their definition.

Fig. 7. Two-class ED with density changing horizontally from $\frac{255}{255}$ to 0 in opposite directions, $p_0 = \frac{255}{255}$. (a) is the superimposition, (b) and (c) are the sampling points for the two classes, shown in red and green respectively. The image is best viewed in PDF reader with 100% scaling ratio at 150 DPI. (Color figure online)

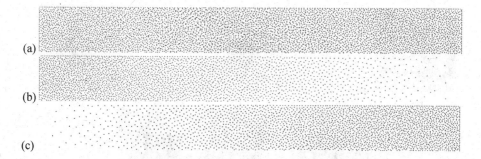

Fig. 8. Two-class ED with density changing horizontally from $\frac{51}{255}$ to 0 in opposite directions, $p_0 = \frac{51}{255}$. (a) is the superimposition, (b) and (c) are the sampling points for the two classes, shown in red and blue respectively. The image is best viewed in PDF reader with 100% scaling ratio at 150 DPI. (Color figure online)

5 Experimental Results

We proposed a multi-class ED algorithm that is able to produce blue-noise sampling points on multiple input signals as well as their superimposition. Given k channel of signals, each with n elements, the time complexity of our algorithm (Algorithm 1) is $\mathcal{O}(kn)$. Some experimental results have been shown in Sect. 4.

In this section, we compare our Multi-class ED with the per-channel standard ED [22], and our method achieves results significantly better than the latter. As illustrated in Fig. 9(b)–(g), applying our MCED to three channels of input signal, three sampling point sets with blue-noise distribution can be produced (Fig. 9(b)(d)(f), colored with red, green and blue for distinction). The corresponding Fourier power spectra in Fig. 9(c)(e)(g) demonstrate the perfect blue-noise property of each class. Figure 9(a) is a colored superimposition of the three classes. Since no sampling conflicts exist, i.e., none of the sampling point overlaps with others, there are no color other then red, green and blue in the image. Figure 9(h)–(i) show that the superimposed point sets also possesses blue-noise property.

On the contrary, if applying the standard ED to each channel of signal separately, though each set of sampling points has blue-noise distribution, a large number of sampling conflicts will occur when the three point sets are superimposed. Figure 10(a) is also a colored superimposition of the sampling point sets, where each color channel correspond to a class. Then, in Fig. 10(b)–(d) are the sampling points that do not overlap with others, and Fig. 10(e)–(h) show the conflicting sampling points, generated by the overlapping of points from different classes (The colors indicate the combination). Consequently, the superimposition set can not maintain blue-noise property (Fig. 10(i)–(j)).

The sampling conflicts are harmful in certain application areas, such as color printing. For the per-channel ED, uncontrollable overlapping of sampling points will affect the controlling of the maximum ink amount at each position, and the final printing quality. Our MCED method can help to solve this problem and hence brings an important improvement.

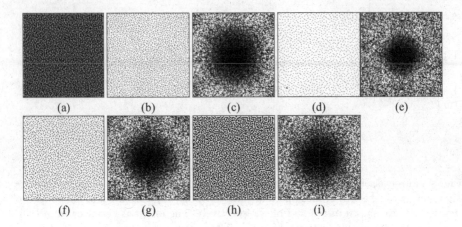

Fig. 9. A 3-class MCED sampling result. Each individual class (b–g) and their super-imposition (h–i) possess perfect blue-noise property. (Best viewed at 150 DPI.) (Color figure online)

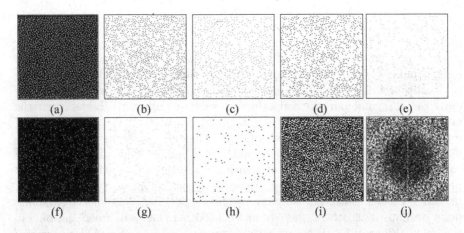

Fig. 10. A 3-class sampling using standard ED. Large number of conflicting sampling points exists (e–h), and the superimposition does not possess blue-noise property (i–j). (Best viewed at 150 DPI). (Color figure online)

6   Applications

6.1   MCED for Color Image Halftoning

ED algorithm is widely used in grayscale image halftoning due to its ideal blue-noise property [14,22]. For color image halftoning, a commonly used way is to perform standard ED independently on each color plane, and then superimpose the results to create a color halftoning. Since halftoning dots for each color plane are generated independently, the blue-noise property for their superimposition can not be guaranteed. This may result in uncontrolled color appearance in the generated halftone image (Fig. 11(p)).

In this section, we utilize our MCED to generate color halftone images that meet the requirements: The distribution of the halftone dots with the same color should possess blue-noise property; The distribution of all the dots contained in the halftone image as a whole should also possess blue-noise property.

15-Class ED for CMYK Color Halftoning

CMYK Color Images. In the state-of-the-art color printing systems, the device-independent colors are converted to the densities of the ink dots in four primary colors: Cyan, Magenta, Yellow and Black. We denote them as well as their densities as $\mathcal{D}_1$, $\mathcal{D}_2$, $\mathcal{D}_3$, $\mathcal{D}_4$, respectively, and they are the input of the halftoning algorithms. When the output halftoning image is printed using the four color inks, the dots generated on paper will appear in totally 15 colors, corresponding to all possible ink overprints.

If $\sum_{i=1}^{4} \mathcal{D}_i > 1$ within a local area in a CMYK halftone image, there must be dots with different primary colors placed at the same position, and new colors will be created. The colors generated by 2 primary colors, namely the *2nd order colors*, are denoted as $\mathcal{C}_{12}$, $\mathcal{C}_{13}$, $\mathcal{C}_{14}$, $\mathcal{C}_{23}$, $\mathcal{C}_{24}$, $\mathcal{C}_{34}$. The subscript indicate the overprinted primary colors, e.g., $\mathcal{C}_{12}$ is an overprinting of $\mathcal{D}_1$ and $\mathcal{D}_2$. In the six colors, $\mathcal{C}_{12}$, $\mathcal{C}_{13}$ and $\mathcal{C}_{23}$ correspond to blue, green and red respectively, and the remaining are very dark colors because they contain black. Similarly, the *3rd* and the *4th order colors* generated by 3 or 4 primary colors are denoted as $\mathcal{C}_{123}$, $\mathcal{C}_{124}$, $\mathcal{C}_{134}$, $\mathcal{C}_{234}$, $\mathcal{C}_{1234}$. Specially, the subsets of halftone dots generated by only one primary color are denote as $\mathcal{C}_1$, $\mathcal{C}_2$, $\mathcal{C}_3$, $\mathcal{C}_4$.

Without ambiguity, we use the same notations for the dot density and the sampling class for each color. Hence, there is a total of 15 classes to be sampled in MCED. Before applying our MCED in CMYK color halftoning, the densities of the 15 classes need to decided.

Linear Programming for Dot Densities. For a color image pixel with $\sum_{i=1}^{4} \mathcal{D}_i \leq 1$, the dot densities are simply $\mathcal{C}_1 = \mathcal{D}_1$, $\mathcal{C}_2 = \mathcal{D}_2$, $\mathcal{C}_3 = \mathcal{D}_3$, $\mathcal{C}_4 = \mathcal{D}_4$, and the densities for higher order colors are 0. While for a pixel with $\sum_{i=1}^{4} \mathcal{D}_i > 1$, the dot densities must satisfy:

$$
\begin{cases}
\mathcal{C}_1 + \mathcal{C}_2 + \mathcal{C}_3 + \mathcal{C}_4 \\
+\mathcal{C}_{12} + \mathcal{C}_{13} + \mathcal{C}_{14} + \mathcal{C}_{23} + \mathcal{C}_{24} + \mathcal{C}_{34} \\
+\mathcal{C}_{123} + \mathcal{C}_{124} + \mathcal{C}_{134} + \mathcal{C}_{234} + \mathcal{C}_{1234} = 1, \\
\mathcal{C}_1 + \mathcal{C}_{12} + \mathcal{C}_{13} + \mathcal{C}_{14} + \mathcal{C}_{123} + \mathcal{C}_{124} + \mathcal{C}_{134} + \mathcal{C}_{1234} = \mathcal{D}_1, \\
\mathcal{C}_2 + \mathcal{C}_{12} + \mathcal{C}_{23} + \mathcal{C}_{24} + \mathcal{C}_{123} + \mathcal{C}_{124} + \mathcal{C}_{234} + \mathcal{C}_{1234} = \mathcal{D}_2, \\
\mathcal{C}_3 + \mathcal{C}_{13} + \mathcal{C}_{23} + \mathcal{C}_{34} + \mathcal{C}_{123} + \mathcal{C}_{134} + \mathcal{C}_{234} + \mathcal{C}_{1234} = \mathcal{D}_3, \\
\mathcal{C}_4 + \mathcal{C}_{14} + \mathcal{C}_{24} + \mathcal{C}_{34} + \mathcal{C}_{124} + \mathcal{C}_{134} + \mathcal{C}_{234} + \mathcal{C}_{1234} = \mathcal{D}_4.
\end{cases}
\tag{6}
$$

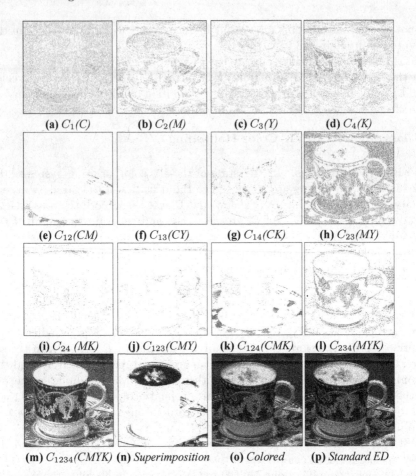

(a) $C_1(C)$ (b) $C_2(M)$ (c) $C_3(Y)$ (d) $C_4(K)$

(e) $C_{12}(CM)$ (f) $C_{13}(CY)$ (g) $C_{14}(CK)$ (h) $C_{23}(MY)$

(i) C_{24} (MK) (j) $C_{123}(CMY)$ (k) $C_{124}(CMK)$ (l) $C_{234}(MYK)$

(m) $C_{1234}(CMYK)$ (n) *Superimposition*   (o) *Colored*   (p) *Standard ED*

Fig. 11. CMYK color image halftoning using our 15-class MCED. Blue-noise dot distributions are generated for all classes (a)–(m), as well as their superimposition (n). (o): Colored superimposition. (p): Halftoning result by the standard ED. (Best viewed at 100% scaling ratio, 600 DPI). (Color figure online)

Eq. 6 is a linear programming problem, which can be solved by specifying one or more optimization target to be maximized [15]. In our experiments, we define the optimization target h as:

$$h = C_1 + C_2 + C_3 + C_{12} + C_{13} + C_{23} + C_{123}. \qquad (7)$$

This target function maximizes the colors created by C, M and Y, and minimizes those mixed with black, because black always tends to cover the appearance of other colors. When the densities of each class at each pixel are obtained, they are sent into our 15-class ED and produce an output halftone image.

Figure 11 shows an example of our 15-class ED color halftoning. It is able to create blue-noise dot distributions for all of the classes (Fig. 11(a)–(m), some

Fig. 12. More CMYK color image halftoning results using MCED. (Best viewed at 100% scaling ratio, 300DPI). (Color figure online)

empty classes are omitted), as well as their superimposition (Fig. 11(n)). Comparing our result (Fig. 11(o)) with that of the standard ED (Fig. 11(p)), it can be seen that our result demonstrates better blue-noise property. More CMYK image halftoning results using our MCED are given in Fig. 12.

Comparison with Vector Error-Diffusion. We also compared the performance of our MCED with the *vector error diffusion* (VED) [5] on color image halftoning. Both algorithms process multiple channels of signals using ED in a coordinated way. The VED treats the signals as a vector, and uses an optimum matrix-valued error filter to introduce the correlation among the color planes, hence it can generate good color halftoning results. However, it does not evaluate the dot distribution on each color planes and the conflict between them. Thus, the blue-noise property on each individual color plane cannot be promised. On the contrary, our MCED can generate blue-noise outputs on each individual channel of input signal, as well as their superimposition. Figure 13 demonstrates a per-channel comparison of the two algorithms on RGB color image halftoning. It can be seen that our MCED produce obviously better sampling results with smoother distribution and less artificial textures.

6.2 Multi-tone Error-Diffusion

Multi-toning, also known as *multi-level halftoning* [10,16], aims to reproduce a continuous tone image with dots of a limited number of intensities $\{k_i | i = 1, \cdots, n\}$ ($k_i < k_j$, if $i < j$). It is useful in printing with multiple types of inks or dot sizes. Blue-noise property is also required in multi-tone images for visually pleasant result, hence our MCED method is also a solution for multi-tone image generation.

Given a pixel (x, y) in a continuous tone image with intensity $p(x, y)$, if $k_i < p(x, y) < k_{i+1}$, then $p(x, y)$ can be simulated with a linear combination of the halftone patterns with intensity k_i and k_{i+1}:

$$p(x, y) = p_i(x, y) \cdot k_i + p_{i+1}(x, y) \cdot k_{i+1}, \tag{8}$$

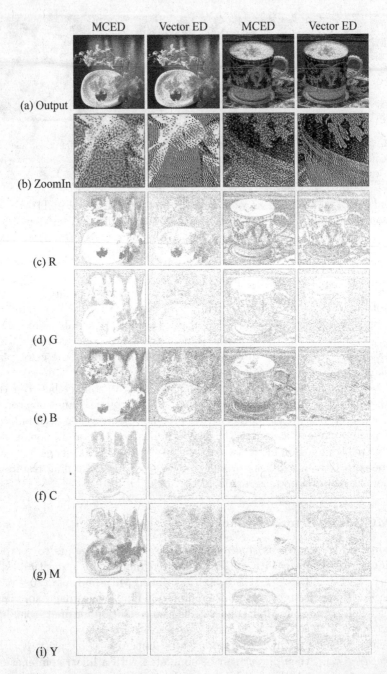

Fig. 13. Color image halftoning using our MCED method and the Vector ED [5]. (a) The superimposed color halftoning image; (b) Zoom-in viewing of the dot distribution in the red box in (a); (c)–(h) The halftoning output of the corresponding color plane. (Images best viewed at 100% scaling ratio, 600 DPI). (Color figure online)

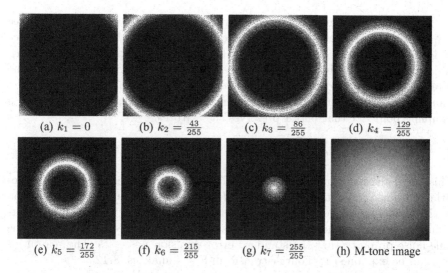

(a) $k_1 = 0$　　(b) $k_2 = \frac{43}{255}$　　(c) $k_3 = \frac{86}{255}$　　(d) $k_4 = \frac{129}{255}$

(e) $k_5 = \frac{172}{255}$　　(f) $k_6 = \frac{215}{255}$　　(g) $k_7 = \frac{255}{255}$　　(h) M-tone image

Fig. 14. An example of 7-tone ED. For an image with intensity ranges from 1 to 0 (from center to border), a multi-tone rendering of the image (h) is generated by our MCED using 7 intensities (tones) given in the figure. (a)–(g) show the distribution of the pixels with these intensities respectively. (Images best viewed at 300 DPI). (Color figure online)

where $p_i(x, y)$ and $p_{i+1}(x, y)$ are respectively the densities of the halftone pattern of pixels with intensity k_i and k_{i+1} at (x, y), and $p_i(x, y) + p_{i+1}(x, y) = 1$. For $j \neq i$ and $j \neq i + 1$, we let $p_j(x, y) = 0$.

Therefore, considering $p_i(x, y)$ as the input of class C_i in a n-class ED, a n-tone image can be generated by our MCED algorithm. Figure 14 shows an example of 7-tone image generated with our method. The halftone patterns for the given intensities $\{k_i | i = 1, \ldots, 7\}$ can be found in Fig. 14(a)–(g), and all of them possess ideal blue-noise property.

6.3 Color Image Vectorization

A typical catalog of color image vectorization methods [17] build polygon meshes on the image plane based on a set of sampling points. Then, by assigning each polygon node the image color at the same position, the original image can be converted to a vector form. The colors inside a polygon is calculated by interpolating between the colors of its nodes. Hence, the quality of the sampling point distribution is crucial for the quality of final vectorization results.

Our MCED method can provide ideal sampling point distribution for such a vectorization task. Here, the input image is in RGB, and the sampling points can be generated in the three color planes using our MCED in a similar way as in Sect. 6.1. After superimposing the sampling points, a planar triangle mesh is obtained by Delaunay triangulation, which can be used as the foundation of the vector image.

·To preserve the image features during vectorization, we extract a salience map from the gradient of the original color image. At each pixel, the salience is defined as the sum of the absolute value of the gradient components, which is calculated by a Sobel operator. The salience is separately computed in the three color planes, and the resulting salience map is also a RGB image. Hence, sampling on the salience map instead of the original image, we can obtain a sampling point set that better preserves the features.

7   Analysis and Conclusion

This paper gives an algorithm for multi-class ED. The key technique of this algorithm is to use the optimized *threshold displacement* to minimize the distortion to the blue-noise property caused by inter-class correlation in multi-class error-diffusion. Our experiment shows that this technique can effectively maintain the blue-noise property that the standard error-diffusion possesses. The reason for this can be explained by Fourier transform-based analysis [8, 12].

7.1   Analysis

According to [12], for the original ED, the power spectrum $B(u, v)$ of the Fourier transform of the output image can be written as:

$$B(u,v) = I(u,v) + F(u,v)E(u,v), \tag{9}$$

where $I(\cdot)$ and $E(\cdot)$ are the Fourier transform of the input image and the error map $e(x, y)$ generated during error-diffusion; $F(\cdot)$ is a high-pass filter defined solely by the diffusion filter.

For each class C_i, our algorithm is based on [22] by adding an extra modulation m_i to its threshold, and m_i includes the displacement t_i, which is in nature a noise from other ED classes. Also according to [12], threshold modulation is equivalent to sending to the original ED an equivalent image that is the sum of the original image and a filtered modulation, where the filter $F(\cdot)$ is exactly the one in Eq. 9. Therefore we have:

$$B(u,v) = I(u,v) + F(u,v)(D_i(u,v) + M(u,v)) + F(u,v)E'(u,v), \tag{10}$$

where $D_i(u, v)$ is the Fourier transform of d_{1i}, $M(u, v)$ is the Fourier transform of the threshold modulation $r(p)$ defined in Fig. 1, $E'(u, v)$ is the Fourier transform of the error map $e(x, y)$ for the equivalent image. Note that t_i does not appear in Eq. 10 because it is a DC component and is filtered out by $F(\cdot)$. Hence, threshold displacements do not have influence on the average density of the output image of any class.

It is also noted that in Eq. 10, only $D_i(\cdot)$ and $E'(\cdot)$ are decided by the threshold displacement $\{t_i\}$. Considering the fact that t_i actually has the effect of decreasing or increasing the amount of slow response phenomenon at the beginning of the ED, so properly chosen $\{t_i\}$ are able to minimize the amount of sampling conflict, which in turn can improve the anisotropy and lower frequency ratio defined in Eq. 4.

7.2   Limitation and Future Work

In the experiment results of our paper, smear artifacts may appear in the sampling classes with low average sampling density (Fig. 14(g)). This is because we use s_i to choose the class with the highest average intensity when sampling conflict occurs and this may cause classes with less average intensity to generate output with lower quality. Hence, the selection of s_i is a topic to be investigated.

The optimal threshold displacement $t_i = g(p_0, p_i)$ has the effect of reducing *slow response* [9], which is also called *transient effect* in some literature [22]. That effect in our MCED is shown obliviously in Fig. 15. At the top of the image, our sampling result (Fig. 15(d)) has very weak slow response than that generated by the standard ED (Fig. 15(a)). In fact, the amount of slow response directly affects the lower frequency ratio $\beta(p_i, t_i)$ in Eq. 4. Our displacement optimization automatically guides t_i to a proper value to decrease the anisotropy and lower frequency ratio, and consequently, reduces the slow response. Hence, introducing threshold displacement into the single-class standard ED to further reduce its slow response is also a future research topic to be explored.

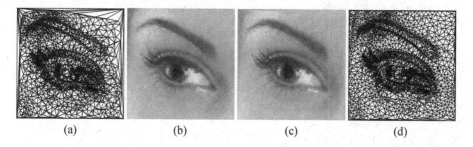

(a)   (b)   (c)   (d)

Fig. 15. Color image vectorization using our MCED vs. the standard ED. (a) and (d) are triangulation on points sampled by Standard ED and MCED, (b) and (c) are corresponding rendering result. The input image is in RGB. (Color figure online)

Acknowledgements. This work is partially supported by NSFC grants #61170206, #61370112.

A   Pseudo Code of MCED

Algorithm 1 is the pseudo-code for the framework of MCED, and the main functions are explained as follows:

GetDisplacement() evaluates t_i by accessing the lookup table (Fig. 6) we described in Sect. 4.2.

GetCoefficient() and *GetStandardThreshold*() are functions for finding appropriate diffusion coefficients and threshold for the standard ED.

Algorithm 1. Multi-Class Error-Diffusion.

1: **for** each spatial position (x,y) **do**
2:     $p_0(x,y) \leftarrow \sum_{i=1}^{n} p_i(x,y)$

3:
4: **for** each spatial position (x,y) **do**
5:   ▷ **The first step** Q_i
6:     **for** each class $i := 0$ to n **do**
7:         $t_i(x,y) \leftarrow GetDisplacement(p_0(x,y), p_i(x,y))$
8:         $a_{jk}^{(i)} \leftarrow GetCoefficient(p_i(x,y))$
9:         $u_i(x,y) \leftarrow GetStandardThreshold(p_i(x,y))$
10:         $u_i(x,y) \leftarrow u_i(x,y) + t_i(x,y)$
11:     **for** each class $i := 0$ to n **do**
12:         $c_i^0 \leftarrow Q(p_i(x,y) + b_i(x,y), u_i(x,y))$ ▷ **Eq. 1**

13:
14:   ▷ **The second step** E_i
15:     **if** $c_0^0(x,y) = 1$ **then**
16:         **if** $HaveConflict()$ **then**   ▷ i.e.:$\sum_{i \neq 0} c_i^0 > 1$
17:             $c_0(x,y) \leftarrow 1$
18:             $e_0(x,y) \leftarrow p_0(x,y) - c_0(x,y)$
19:             $minclass \leftarrow FindMaxClass()$
20:             **for** each class $i := 1$ to n **do**
21:                 **if** $i = minclass$ **then**   ▷ i.e.:$s_i = TRUE$
22:                     $c_i(x,y) \leftarrow 1$
23:                 **else**
24:                     $c_i(x,y) \leftarrow 0$
25:                 $e_i(x,y) \leftarrow p_i(x,y) - c_i(x,y)$
26:         **else if** $NoConflict()$ **then**   ▷ i.e.:$\sum_{i \neq 0} c_i^0 = 1$
27:             **for** each class $i := 0$ to n **do**
28:                 $c_i(x,y) \leftarrow c_i^0$
29:                 $e_i(x,y) \leftarrow p_i(x,y) - c_i(x,y)$
30:         **else**   ▷ **No class sampled:**$\sum_{i \neq 0} c_i^0 = 0$
31:             **for** each class $i := 0$ to n **do**
32:                 $c_i(x,y) \leftarrow 0$
33:                 $e_i(x,y) \leftarrow p_i(x,y) - c_i(x,y)$
34:     **else**   ▷ **When** $c_0^0 = 0$
35:         **for** each class $i := 0$ to n **do**
36:             $c_i(x,y) \leftarrow 0$
37:             $e_i(x,y) \leftarrow p_i(x,y) - c_i(x,y)$

38:
39:     **for** each class $i := 0$ to n **do**
40:         $DistributeError(i, x, y, e_i(x,y), a_{jk}^{(i)})$

Quantize() compares pixel value to the threshold and returns 0 if below, 1 otherwise (Eq. 1).

HaveConflict() returns TRUE if more than one sampling points from different classes situate at the current position, and *NoConflict*() indicates only one class sampled at the position.

FindMaxClass() finds the class whose sum of densities at all the spatial positions is the maximum.

DistributeError() distributes the quantization errors to neighboring pixels according to the error filter.

References

1. Alliez, P., Meyer, M., Desbrun, M.: Interactive geometry remeshing. ACM Trans. Graph. **21**, 347–354 (2002)
2. Baqai, F.A., Lee, J.-H., Agar, A.U., Allebach, J.P.: Digital color halftoning. IEEE Signal Process. Mag. **22**, 87–96 (2005)
3. Bourguignon, D., Chaine, R., Cani, M.P., Drettakis, G.: Relief: a modeling by drawing tool. In: Eurographics Workshop on Sketch-Based Interfaces and Modeling (SBM), Grenoble, France, pp. 151–160. Eurographics, Eurographics Association (2004)
4. Chang, J., Alain, B., Ostromoukhov, V.: Structure-aware error diffusion. ACM Trans. Graph. **28**, 162:1–162:8 (2009)
5. Damera-Venkata, N., Evans, B.L.: Design and analysis of vector color error diffusion halftoning systems. IEEE Trans. Image Process. **10**(10), 1552–1565 (2001)
6. Damera-Venkata, N., Evans, B.L., Monga, V.: Color error-diffusion halftoning what differentiates it from grayscale error diffusion? IEEE Signal Process. Mag. **20**, 51–58 (2003)
7. Floyd, R.W., Steinberg, L.: An adaptive algorithm for spatial greyscale. Proc. Soc. Inf. Disp. **17**(2), 75–77 (1976)
8. Gonzalez, R.C., Woods, R.E.: Digital Image Processing, 2nd edn. Addison-Wesley Longman Publishing Co. Inc., Boston (2001)
9. Haneishi, H., Suzuki, T., Shimoyama, N., Miyake, Y.: Color digital halftoning taking colorimetric color reproduction into account. J. Electron. Imaging **5**(1), 97–106 (1996)
10. Kang, H.R.: Digital Color Halftoning, 1st edn. Society of Photo-Optical Instrumentation Engineers (SPIE), Bellingham (1999)
11. Kim, S.Y., Maciejewski, R., Isenberg, T., Andrews, W.M., Chen, W., Sousa, M.C., Ebert, D.S.: Stippling by example. In: NPAR09, pp. 41–50. ACM (2009)
12. Knox, K.T., Eschbach, R.: Threshold modulation in error diffusion. J. Electron. Imaging **2**(3), 185–192 (1993)
13. Li, P., Allebach, J.P.: Tone-dependent error diffusion. In: Society of Photo-Optical Instrumentation Engineers (SPIE) Conference Series, vol. 4663, pp. 310–321 (2001)
14. Ostromoukhov, V.: A simple and efficient error-diffusion algorithm. In: SIGGRAPH 2001, pp. 567–572. ACM (2001)
15. Press, W.H., Teukolsky, S.A., Vetterling, W.T., Flannery, B.P.: Numerical Recipes in: The Art of Scientific Computing, 2nd edn. Cambridge University Press, New York (1992)
16. Rodríguez, J.B., Arce, G.R., Lau, D.L.: Blue-noise multitone dithering. IEEE Trans. Image Process. **17**(8), 1368–1382 (2008)

17. Swaminarayan, S., Prasad, L.: Rapid automated polygonal image decomposition. In: Applied Imagery and Pattern Recognition Workshop, pp. 28–28. IEEE (2006)
18. Ulichney, R.A.: Dithering with blue noise. Proc. IEEE **76**, 56–79 (1988)
19. Wei, L.-Y.: Multi-class blue noise sampling. ACM Trans. Graph. (TOG) **29**(4), 79 (2010)
20. Wei, L.-Y.: Private Communication (2012)
21. Weissbach, S., Wyrowski, F.: Error diffusion procedure: theory and applications in optical signal processing. Appl. Opt. **31**, 2518–2534 (1992)
22. Zhou, B., Fang, X.: Improving mid-tone quality of variable-coefficient error diffusion using threshold modulation. ACM Trans. Graph. **22**(3), 437–444 (2003)
23. Li, H., Mould, D.: Contrast-aware Halftoning. Computer Graphics Forum. **29**(2), 273–280 (2010). EISSN 1467-8659

Copula Eigenfaces with Attributes: Semiparametric Principal Component Analysis for a Combined Color, Shape and Attribute Model

Bernhard Egger[✉], Dinu Kaufmann, Sandro Schönborn, Volker Roth, and Thomas Vetter

Department of Mathematics and Computer Science,
University of Basel, Basel, Switzerland
{bernhard.egger,dinu.kaufmann,sandro.schoenborn,volker.roth,
thomas.vetter}@unibas.ch

Abstract. Principal component analysis is a ubiquitous method in parametric appearance modeling for describing dependency and variance in datasets. The method requires the observed data to be Gaussian-distributed. We show that this requirement is not fulfilled in the context of analysis and synthesis of facial appearance. The model mismatch leads to unnatural artifacts which are severe to human perception. As a remedy, we use a semiparametric Gaussian copula model, where dependency and variance are modeled separately. This model enables us to use arbitrary Gaussian and non-Gaussian marginal distributions. Moreover, facial color, shape and continuous or categorical attributes can be analyzed in an unified way. Accounting for the joint dependency between all modalities leads to a more specific face model. In practice, the proposed model can enhance performance of principal component analysis in existing pipelines: The steps for analysis and synthesis can be implemented as convenient pre- and post-processing steps.

Keywords: Copula Component Analysis · Gaussian copula · Principal component analysis · Parametric Appearance Models · 3D Morphable Model · Face modeling · Face synthesis · Attributes

1   Introduction

Parametric Appearance Models (PAM) describe objects in an image in terms of pixel intensities. In the context of faces, Active Appearance Models [1] and 3D Morphable Models [2] are established PAMs to model appearance and shape. The dominant method for learning the parameters of a PAM is principal component analysis (PCA) [3]. PCA is used to describe the variance and dependency in the data. Due to the sensitivity of PCA to space and scaling, seperate models are

---

B. Egger and D. Kaufmann—These authors contributed equally to this work.

© Springer International Publishing AG 2017
J. Braz et al. (Eds.): VISIGRAPP 2016, CCIS 693, pp. 95–112, 2017.
DOI: 10.1007/978-3-319-64870-5_5

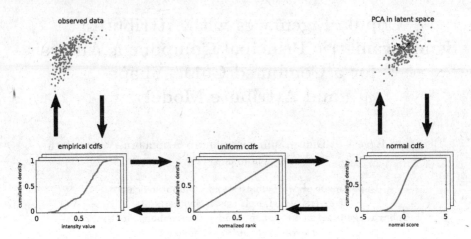

Fig. 1. This figure shows the pre- and post-processing steps necessary to use a Gaussian copula before calculating PCA.

learned for shape and appearance. Usually, PAMs are generative models that can synthesize new random instances.

Using PCA to model facial appearance leads to models which are able to synthesize instances which appear unnaturally. This is due to the assumption that the color intensities or, in other words, the marginals at a pixel are Gaussian-distributed. We show that this is a severe simplification: The pixel intensities of new samples will follow a joint Gaussian distribution. This approximation is far from the actual observed distribution of the training data and leads to unnatural artifacts in appearance.

The ability to synthesize random *and* natural instances is important when generating new face instances [4] and in face manipulation [5]. This is because human perception is very sensitive to unnatural variability in a face. On the other hand, PCA face models are used as a strong prior in probabilistic facial image interpretation algorithms [6]. Hence, such applications require a prior which follows the underlying distribution as closely as possible and, which is therefore, highly specific to faces.

In order to enhance the specificity of a PCA-based model, an obvious improvement would be the extension to a Gaussian mixture model [7]. Here, each color channel at a pixel is modeled with an (infinite) mixture of Gaussians. However, we skip this step and propose to use a semiparametric copula model directly.

A copula model provides the decomposition of the dependency and the marginal distributions such that the copula contains the dependency structure only. This separate modeling allows us to drop the parametric Gaussian assumption on the color channels and to replace them with nonparametric empirical distributions. In general, seperating all marginals from the dependency structure leads to a scale invariant description of the underlying dependency. This is desired when working with data from different modalities, living in different spaces. Scale invariance enables us to learn a combined dependency structure of shape, color and attributes. We keep the parametric dependency structure; in particular, we use a Gaussian copula because of its inherent Gaussian latent space. PCA can then be applied in the latent Gaussian space and is used to learn the dependencies of the data independently from the marginal distribution. The method is analytically analyzed in [8] and is called Copula Component Analysis (COCA). Samples drawn from a COCA model follow the empirical marginal distribution of the training data and are, therefore, more specific to the modeled object.

The additional steps for using COCA can be implemented as simple pre- and post-processing before applying PCA. The data is mapped into a space where it is Gaussian-distributed. This mapping is obtained by first ranking the data and then transforming it by the standard normal distribution. We perform PCA on the transformed data to learn its underlying dependency structure. All necessary steps are visualized in Fig. 1.

A semiparametric Gaussian copula model also provides additional benefits: First, learning is invariant to monotonic transformations of all marginals, including invariance to scaling. We explore this advantage by learning a combined color and shape model, even including attributes. Second, the implementation can be done as simple pre- and post-processing steps. Third, the model also allows changing the color space. For facial-appearance modeling, the HSV color space is more appropriate than RGB. The HSV color space is motivated by the separation of the hue and saturation components and brightness value. On the other hand, without adaptions, PCA is not applicable to facial appearance in the HSV color space because of its sensitivity to differently-scaled color channels.

In summary, methods building on PCA can easily benefit from these advantages to improve their learned model. By learning a combined shape, color and attribute model we explore scale invariance and therefore the possibility to include diverse modalities of the data in a common model.

1.1 Related Work

The Eigenfaces approach [9,10] uses PCA on aligned facial images to analyze and synthesize faces. Active Appearance Models [1] add a shape component which allows to model the shape independently from the appearance. The 3D Morphable Model [2] uses a dense registration, extends the shape model to 3D and adds camera and illumination parameters. The 3D Morphable Model allows handling appearance independently from pose, illumination and shape. These methods have a common core: They focus on analysis and synthesis of faces and all of them use a PCA model for color representation and can, therefore, benefit from COCA.

Photo-realistic face synthesis methods like Visio-lization [4] use PCA as a basis for example-based photo-realistic appearance modeling.

1.2 Outline

The remainder of the paper is organized as follows: The methods section explains the copula extension for PCA and presents the theoretical background for learning and inference. We also indicate, how to include discrete distributed data in the copula framework. Additionally, practical information for an implementation is provided. In the experiments and results section we demonstrate that facial appearance should be modeled using the copula extension. We qualitatively and quantitatively show that the proposed model leads to a facial appearance model which is more specific to faces.

2   Methods

2.1   PCA for Facial Appearance Modeling

Let $x \in \mathbb{R}^{3n}$ describe a zero-mean vector representing 3 color channels of an image with n pixels. In an RGB image, the color channels and the pixels are stacked such that $x = (r_1, g_1, b_1, r_2, b_2, b_3, \ldots, r_n, g_n, b_n)^T$. We assume that the mean of every dimension is already subtracted. The training set of m images is arranged as the data matrix $X \in \mathbb{R}^{3n \times m}$.

PCA [3] aims at diagonalizing the sample covariance $\Sigma = \frac{1}{m} X X^T$, such that

$$\Sigma = \frac{1}{m} U S^2 U^T \tag{1}$$

where S is a diagonal matrix and U contains the transformation to the new basis. The columns of matrix U are the eigenvectors of Σ and the corresponding eigenvalues are on the diagonal of S.

PCA is usually computed by a singular value decomposition (SVD). In case of a rank-deficient sample covariance with rank $m < n$ we cannot calculate U^{-1}. Therefore, SVD leads to a compressed representation with a maximum of m dimensions. The eigenvectors in the transformation matrix U are ordered by the magnitude of the corresponding eigenvalues.

When computing PCA, the principal components are guided by the variance as well as the covariance in the data. While the variance captures the scattering of the intensity value of a pixel, the covariance describes which regions contain similar color. This mingling of factors leads to results which are sensitive to different scales and to outliers in the training set. Regions with large variance and outliers could influence the direction of the resulting principal components in an undesired manner.

We uncouple variance and dependency structure such that PCA is only influenced by the dependency in the data. Our approach for uncoupling is a copula model which provides an analytical decomposition of the aforementioned factors.

2.2   Copula Extension

Copulas [11, 12] allow a detached analysis of the marginals and the dependency pattern for facial appearance models. We consider a relaxation to a semiparametric Gaussian copula model [13, 14]. We keep the Gaussian copula for describing the dependency pattern, but we allow nonparametric marginals.

Let $x \in \mathbb{R}^{3n}$ describe the same zero-mean vector as used for PCA, representing 3 color channels of an image with n pixels. Sklar's theorem allows the decomposition of every continuous and multivariate cumulative probability distribution (cdf) into its marginals $F_i(X_i), i = 1, \ldots, 3n$ and a copula C. The copula comprises the dependency structure, such that

$$F(X_1, \cdots, X_{3n}) = C(W_1, \ldots, W_{3n}) \qquad (2)$$

where $W_i = F_i(X_i)$. W_i are uniformly distributed and generated by the probability integral transformation[1].

For our application, we consider the Gaussian copula because of its inherently implied latent space

$$\tilde{X}_i = \Phi^{-1}(W_i), \quad i = 1, \ldots, 3n \qquad (3)$$

where Φ is the standard normal cdf. The multivariate latent space is standard normal-distributed and fully parametrized by the sample correlation matrix $\tilde{\Sigma} = \frac{1}{m}\tilde{X}\tilde{X}^T$ only. PCA is then applied on the sample correlation in the latent space $\tilde{X}$.

The separation of dependency pattern and marginals provides multiple benefits: First, the Gaussian copula captures the dependency pattern invariant to the variance of the color space[2]. Second, whilst PCA is distorted by outliers and is generally inconsistent in high dimensions, the semiparametric copula extension solves this problem [8]. Third, the nonparametric marginals maintain the non-Gaussian nature of the color distribution. Especially when generating new samples from the trained distribution, the samples do not exceed the color space of the training set.

2.3   Inference

We learn the latent sample correlation matrix $\tilde{\Sigma} = \frac{1}{m}\tilde{X}\tilde{X}^T$ in a semiparametric fashion using nonparametric marginals and a parametric Gaussian copula. We compute $\hat{w}_{ij} = \hat{F}_{\text{emp},i}(x_{ij}) = \frac{r_{ij}(x_{ij})}{m+1}$ using empirical marginals $\hat{F}_{\text{emp},i}$, where $r_{ij}(x_{ij})$ is the rank of the data x_{ij} within the set $\{x_{i\bullet}\}$. Then, $\tilde{\Sigma}$ is simply the sample covariance of the normal scores

$$\tilde{x}_{ij} = \Phi^{-1}\left(\frac{r_{ij}(x_{ij})}{m+1}\right), \quad i = 1, \ldots, 3n, \quad j = 1, \ldots, m. \qquad (4)$$

Equation (4) contains the nonparametric part, since $\tilde{\Sigma}$ is computed from the ranks $r_{ij}(x_{ij})$ solely and contains no information about the marginal distribution of the x's. Note, $\tilde{x} \sim \mathcal{N}(0, \tilde{\Sigma})$ is standard normal distributed with correlation matrix $\tilde{\Sigma}$. Subsequently, an eigendecomposition is applied on the latent correlation matrix $\tilde{\Sigma}$.

---

[1] The copula literature uses U instead of W. We changed this convention due to the singular value decomposition which uses $X = USV^T$.

[2] More general, a copula model is invariant against all monotonic transformations of the marginals.

Generating a sample using PCA then simply requires a sample from the model parameters

$$h \sim \mathcal{N}(0, I) \tag{5}$$

which is projected to the latent space

$$\tilde{x} = \tilde{U} \frac{\tilde{S}}{\sqrt{m}} h \tag{6}$$

and further projected component-wise to

$$w_i = \Phi(\tilde{x}_i), \quad i = 1, \ldots, 3n. \tag{7}$$

Finally, the projection to the color space requires the empirical marginals

$$x_i = \hat{F}_{\mathrm{emp},i}(w_i), \quad i = 1, \ldots, 3n. \tag{8}$$

All necessary steps are summarized in Algorithms 1 and 2 and visualized in Fig. 1.

It is possible to smoothen the empirical marginals with a kernel k and replace (8) by $x_i = k(w_i, X_{i\bullet}), \quad i = 1, \ldots, 3n.$

Algorithm 1. Learning.

Input: Training set $\{X\}$
Output: Projection matrices U, S
for *all dimensions* **do**
    for *all samples* **do**
        $\tilde{x}_{ij} = \Phi^{-1}\left(\frac{r_{ij}(x_{ij})}{m+1}\right)$
find $\tilde{U}, \tilde{S}$ such that $\tilde{\Sigma} = \frac{1}{m}\tilde{U}\tilde{S}^2\tilde{U}^T$ (via SVD)

Algorithm 2. Sampling.

Output: Random sample x
$h \sim \mathcal{N}(0, I)$
$\tilde{x} = \tilde{U}\frac{\tilde{S}}{\sqrt{m}}h$
for *all dimensions i* **do**
    $w_i = \Phi(\tilde{x}_i)$
    $x_i = \hat{F}_{\mathrm{emp},i}(w_i)$

2.4 Implementation

The additional steps for using COCA can be implemented as simple pre- and post-processing before applying PCA. Basically the data is mapped into a latent space where it is Gaussian-distributed. The mapping is performed in two steps.

First, the data is transformed to an uniform distribution by ranking the intensity values. Then it is transformed to a standard normal distribution. On the transformed data, we perform PCA to learn the dependency structure in the data.

To generate new instances from the model, all steps have to be reversed. Figure 1 gives an overview of all necessary transformations. The following steps have to be performed, e.g. in MATLAB, to calculate COCA:

```
% calculate empirical cdf
[empCDFs, indexX] = sort(X, 2);

% transform emp. cdf to uniform
[~, rank] = sort(indexX, 2);
uniformCDFs = rank / (size(rank, 2)+1);

% transform uni. cdf to std. normal cdf
normCDFs = norminv(uniformCDFs',0,1)';

% calculate PCA
[U,S,V] = svd(normCDFs, 'econ');
```

Listing 1.1. Learning.

To generate an image from model parameters, the following steps are necessary:

```
% random sample
m = size(normCDFs, 2);
h = random('norm' ,0 ,1 ,m ,1);
sample = U * S / sqrt(m) * h;

% std. normal to uniform
uniformSample = normcdf(sample, 0, 1) * (m - 1) + 1;

% uniform to emp. cdf
empSample = empCDFs(sub2ind(size(empCDFs), 1:size(data, 1), ...
    round(uniformSample')))';
```

Listing 1.2. Sampling.

These are the additional steps which have to be performed as pre- and post-processing for the analysis of the data and the synthesis of new random samples. In terms of computing resources we have to consider the following: The empirical marginal distributions F_{emp} are now part of the model and have to be kept in memory. In the learning part, the complexity of sorting the input data is added. In the sampling part, we have to transform the data back by looking up their values in the empirical distribution.

The copula extension comes with low additional effort: it is easy to implement and has only slightly higher computing costs. We encourage the reader to implement these few steps since the increased flexibility in the modeling provides a valuable extension.

2.5 Discrete Ordinal Marginals

The formulation of the coupla framework as above works with arbitrary continuous marginals. We extend the copula model for attributes, which follow discrete ordinal marginals. With this extension, we can even augment our model with attributes following binary distribution, such as gender. The underlying generative model assumes a continuous latent space, which is identified with the latent space $\tilde{X}$ of the copula. From this space, we observe the measurements via a discretization, which is related to the marginal distribution containing discontinuities. Using the cdfs of these marginals, for infering the latent space as in the previous sections, causes problems. This is because the cdf transformations $\Phi^{-1} \circ \hat{F}_{\text{emp},i} : X_i \to \tilde{X}_i$ do not change the marginal data distribution to uniform and hence do not recover the continuous latent space. Instead, these cdf transformations only change the sample space. This leads to an invalid distribution of the copula and subsequently also of the latent space.

In order to resolve this problem, we follow the approach of the extended rank likelihood [15]. This provides us with an association-preserving mapping between measurement x_{ij} and latent observation $\tilde{x}_{ij}$. The essential idea behind this approach is, that the rank relation from the observations are preserved in the latent space. The latent variables are then recovered by a Gibbs sampler, which obeys these rank relations while respecting the Gaussian copula. From this sampler, we are able to generate (continuous) latent pseudo observations $\tilde{x}$, which subsequently can be included in our model. Using this Gibbs sampler, we are able to include discrete ordinal distributed attributes with arbitrary many categories.

However, the above described Gibbs sampler causes problems in our setting, since sampling in such high dimensions is just infeasible. In our case, we want to include a binary variable (sex). Note, that a binary variable can always be considered as an ordinal variable, since the ordering of the encoding does not matter. Instead of resampling from the conditional posterior distribution $p(\tilde{x}_{\text{sex}}|\tilde{x}_{-\text{sex}}, x_{\text{sex}})$ in the latent space, we replace the label x_{sex} with logistic regression in a preprocessing step. Specifically, logistic regression provides us a (continuous) score $x'_{\text{sex}} = E(x_{\text{sex}}|x_{-\text{sex}})$, which is the conditional expectation over (a low rank approximation of) the remaining variables $x_{-\text{sex}}$. Since the score constitutes of the conditional expectation, it relates to an approximation of the conditional posterior distribution in the latent space. The variable can then be treated as a continuous variable.

3   Experiments and Results

For all our experiments, we used the texture of 200 face scans used for building the Basel Face Model (BFM) [16]. The scans are in dense correspondence and were captured under an identical illumination setting. We work on texture images and use a resolution of 1024×512 pixels. Our experiments are based on the appearance information only, the last experiment merging the appearance and shape to a combined model. We used the empirical data directly as marginal distribution. The results are rendered with an ambient illumination on the mean face shape of the BFM.

3.1   Facial Appearance Distribution

In a first experiment we investigate if the color intensities in our face data set are Gaussian-distributed. We followed the protocol of the Kolmogorov-Smirnov Test [17]. We estimate a Gaussian distribution for every color channel per pixel and compare it to the observed data. The null hypothesis of the test is that the observed data is drawn by the estimated Gaussian distribution. The test measures the maximum distance of the cumulative density function of the estimated Gaussian $\Phi_{\hat{\mu},\hat{\sigma}^2}$ and the empirical marginal distribution F_{emp} of the observed data:

$$d = \sup_x \left\| F_{\mathrm{emp}}(x) - \Phi_{\hat{\mu},\hat{\sigma}^2}(x) \right\| \tag{9}$$

Here, $\hat{\mu}$ and $\hat{\sigma}^2$ are maximum-likelihood estimates for the mean and variance of a Gaussian distribution respectively. In Fig. 2(a) we visualize the maximal distance value over all color channels per point on the surface.

We assume a significance level of $1 - \alpha = 0.05$. The critical value d_α is approximated using the following formula [18]:

$$d_\alpha = \frac{\sqrt{\ln(\frac{2}{\alpha})}}{\sqrt{2n}} \tag{10}$$

With $n = 200$ training samples we get a critical value of 0.096. Non-Gaussian marginal distributions of color intensities are present in the region of the eyebrows, eyes, chin and hair, where multi-modal appearance is present. In total for 49% of the pixels over all color channels, the null hypothesis has to be rejected. In simple monotonic regions, like the cheek, the marginal distributions are close to a Gaussian distribution. In more structured regions like the eye, eyebrow or the temple region, the appearance is highly non-Gaussian. This leads to strong artifacts when modeling facial color appearance using PCA (see Figs. 3 and 4). Since those more structured regions are fundamental components of a face, it is important to model them properly.

We also applied the Kolmogorov-Smirnov Test to the shape coordinates of our training data (see Fig. 2(b)). We observe that the observed marginal distributions

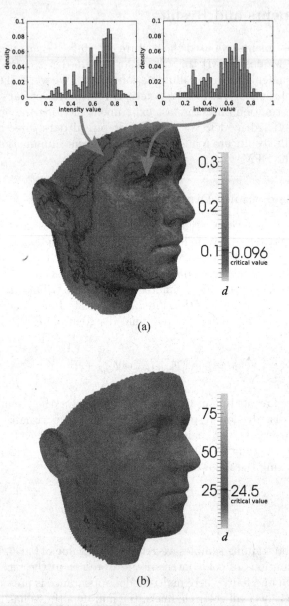

Fig. 2. The result of the Kolmogorov-Smirnov Test to compare the empirical marginal distributions of color values (a) and shape coordinates (b) from our 200 face scans with a Gaussian-reference probability distribution. We plot the highest value of the three color channels respectively dimensions per pixel, because the values for the individual components are very similar. Whilst the marginals for the shape coordinates are similar to a Gaussian distribution, the Gaussian assumption does not hold for the color marginals. We show two exemplary marginal distributions in the eye and temple region. They are not only non-Gaussian but also not similar. (Color figure online)

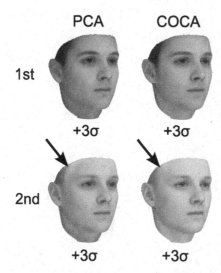

Fig. 3. PCA and COCA are compared by visualizing the first two eigenvectors with 3 standard deviations on the mean. The components look very similar, except that the PCA artifacts on the temple (arrows) in the second eigenvector do not appear using COCA. (Color figure online)

in the data are close to a Gaussian distribution. The registration of the data was performed using a nonrigid ICP algorithm by Amberg et al. [19]. The algorithm uses strong regularization techniques, therefore the Gaussian property of the shape coordinates can also be a registration artifact.

3.2   Appearance Modeling

We evaluate our facial appearance model by its capability to synthesize new instances. We measured this capability by comparing the major eigenmodes, random model instances, the sample marginal distributions and the specificity of both models. The specificity is measured qualitatively by visual examples and quantitatively by a model metric.

Model Parameters. The first few principal components store the strongest dependencies. We visualize the first two components by setting their value h_i to $\sigma = 3$ standard deviations and show the result in Fig. 3. The first parameters of PCA and COCA appear very similar in the variation of the data they model. The second principal component of PCA causes artifacts in the temple region. These artifacts are caused by the linearity of PCA. COCA is a nonlinear method and therefore, the artifacts are not present.

Random Samples. The ability to generate new instances is a key feature for generative models. A model which can produce more realistic samples is desirable

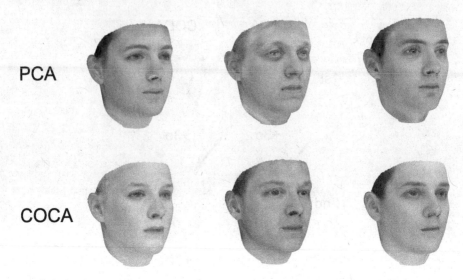

Fig. 4. The first and second row show random samples projected by PCA and COCA respectively. Using PCA, we can observe strong artifacts in the regions where the marginal distribution is not Gaussian (see Fig. 2). The improvement of COCA can be observed in the temple region, on the eyebrows, around the nostrils, the eyelids and at the border of the pupil. We chose representative samples for both methods. (Color figure online)

for various applications. For example, the Visio-lization method to generate high resolution appearances is based on a prototype generated with PCA [4].

Another field of application for the generative part of models are Analysis-by-Synthesis methods based on Active Appearance Models (AAM) or 3D Morphable Models (3DMM). They can profit from a stronger prior which is more specific to faces and reduces the search space [6].

Generating a random parameter vector leads to a random face from our PCA or COCA model. We sample h according to (5) independently for all 199 parameters and project them via PCA or COCA on the color space following (6). Random samples using COCA contain fewer artifacts and, therefore, appear much more natural (see Fig. 4). These artifacts are caused by the linearity of PCA. For non-Gaussian-distributed marginals, PCA does not only interpolate within the trained color distribution but also extrapolates to color intensities not supported by the training data.

The most obvious problem is the limited domain of the color channels: using PCA, color channels have to be clamped. The linearity constraint of PCA leads to much brighter or darker color appearance than those present in the training data in regions which are not Gaussian-distributed. In the next experiment, we show that the higher specificity is not only a qualitative result but can also be measured by a model metric.

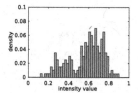
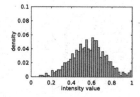
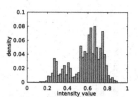

(a) Empirical marginal distribution

(b) PCA sample marginal distribution

(c) COCA sample marginal distribution

Fig. 5. The marginal distribution of the red color intensity of a single point in the eye region. (a) shows the distribution observed in the training data, (b) shows the distribution of samples drawn from a PCA model and (c) from a COCA model. (Color figure online)

Few samples od COCA contain artifacts arising from outliers in the training data which appear at the borders of the empirical cdfs. Those artifacts can be removed by slightly cropping the marginal distributions (removing the outliers) or by applying COCA in the HSV color space.

3.3 Appearance Marginal Distribution

We analyze the marginal distributions of our random faces at a single point at the border between the pupil and the sclera of the eye. In this region the Kolmogorov-Smirnov Test rejected the null hypothesis. We analyze the empirical intensity distribution of a single color channel at this point (Fig. 5(a)). The sample marginal distributions drawn from 1000 random instances generated by PCA and COCA are shown in Fig. 5(b) and (c) respectively. Whilst COCA is able to generate samples distributed similar to our input data, PCA is approximating a Gaussian distribution, which is inaccurate in a lot of facial regions.

Specificity and Generalization. To measure the quality of the PCA and COCA models, we use model metrics motivated by the shape modeling community [20]. The first metric is specificity: Instances generated by the model should be similar to instances in the training set. Therefore, we draw 1000 random samples from our model and compare each one to its nearest neighbor in the training data. We measure the distance using the mean absolute error over all pixels and color channels in the RGB-color space. The COCA model is more specific to facial appearance (see Fig. 6(a)). This corresponds to our observation of a more realistic facial appearance (Fig. 4).

Specificity should always be used in combination with the generalization model metric [20]. The generalization measures how exactly the model can represent unseen instances. We measure the generalization ability of both models using a test set and use the same distance measure as for specificity. The test data consists of 25 additional face scans not contained in the training data. We observe that both models generalize well to unseen data. PCA generalizes slightly better, see Fig. 6(b).

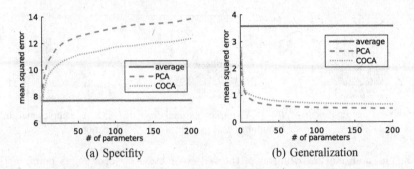

(a) Specifity   (b) Generalization

Fig. 6. (a) The specificity shows how close generated instances are to instances in the training data. The average distance of 1000 random samples to the training set (mean squared error per pixel and color channel) is shown. A model is more specific if the distance of the generated samples to the training set is smaller. We observe that COCA is more specific to faces (lower is better). (b) The generalization ability shows how exactly unseen instances can be represented by a model. The lower the error, the better a model generalizes. As a baseline, we present the generalization ability of the average face. We observe that PCA generalizes slightly better (lower is better). (Color figure online)

The third model metric is compactness - the ability to use a minimal set of parameters [20]. The compactness can be measured directly by the number of used parameters. In our experiments, the number of parameters is always the same for both models.

There is always a tradeoff between specificity and generalization. Whilst PCA performs slightly better in generalization, COCA performs better in terms of specificity. The better generalization ability of PCA comes at the price of a lower specificity and clearly visible artifacts.

3.4   Combined Shape, Color and Attribute Model

Color appearance and shape are modeled independently in AAMs and 3DMMs. Recently, it was demonstrated that facial shape and appearance are correlated [21] and those correlations were investigated using Canonical Correlation Analysis on separate shape and appearance PCA models. Attributes like age, weight, height, gender are often added to the PCA models as additional linear vectors [16] or with limitations to Gaussian marginal distributions [22].

The main reason to build separate models is a practical one – shape and color values do not live in the same space and are not scaled in the same range. Attributes are even not always continuous. Some methods approach this issue by normalization [23]. However, this approach is highly sensitive to outliers and not suitable to compare those different modalities. Since Copula Component Analysis is scale invariant and allows to include categorical data, we can directly apply it to a set of combined data.

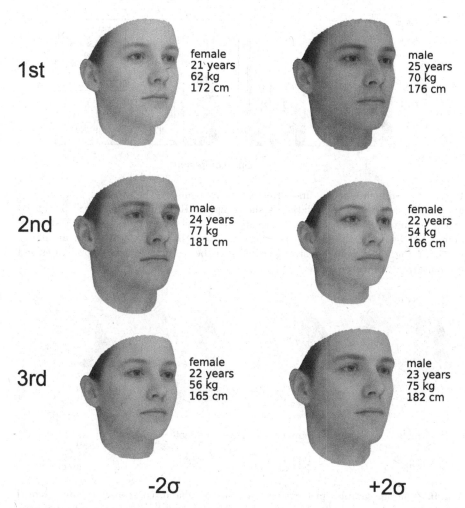

Fig. 7. We learned a common shape, color and attribute model using COCA. We visualize the first eigenvectors with 2 standard deviations, which show the strongest dependencies in our training data. Whilst the first parameter is strongly dominated by color the latter parameters are targeting shape, color and attributes (compare Fig. 8). Since the model is built from 100 females and 100 males, the first components are strongly connected to sex. The small range in age is caused by the training data which mainly consists of people with similar age. (Color figure online)

We learned a COCA model combining the color, shape and attributes information (see Figs. 7 and 9). Shape, color and attributes are combined by simply concatenating them. Age weight and height are continuous attributes and can therefore directly by integrated in the COCA model. We added gender as a binary attribute and used the strategy presented in Sect. 2.5, where we replaced the binary labels with scores, which were learned with logistic regression on

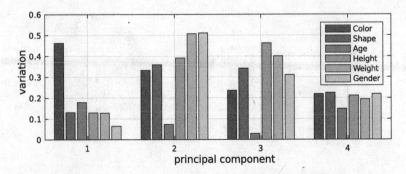

Fig. 8. The influence of the first principal components on the different modalities of our model is shown. The variation is shown as the RMS distance of the normalized attributes in the latent space (covariance matrix). Whilst the first parameter is strongly dominated by color the later parameters are targeting shape, color and attributes (compare Fig. 7). We observe strong correlations between the different modalities and attributes. (Color figure online)

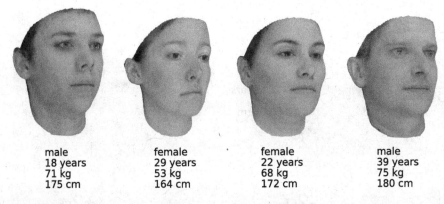

male	female	female	male
18 years	29 years	22 years	39 years
71 kg	53 kg	68 kg	75 kg
175 cm	164 cm	172 cm	180 cm

Fig. 9. Random samples projected by a common shape, color and attribute model using COCA. Our model leads to samples with consistent appearance and attributes. (Color figure online)

the covariates. The combined model allows us to generate random samples with consistent and correlated facial features. In Fig. 8 we present how the different modalities are correlated in the first parameters. By integrating this additional dependency information, the model becomes more specific [23].

4   Conclusions

We showed that the marginals of facial color are not Gaussian-distributed for large parts of the face and that PCA is not able to model facial appearance properly. In a statistical appearance model, this leads to unnatural artifacts which are easily detected by human perception. To avoid such artifacts, we propose to use PCA in a semiparametric Gaussian copula model (COCA) which allows to model the marginal color distribution separately from the dependency structure. In this model, the parametric Gaussian copula describes the dependency pattern in the data and the nonparametric marginals relax the restrictive Gaussian requirement of the data distribution.

The separation of marginals and dependency pattern enhances the model flexibility. We showed qualitatively that facial appearance is modeled better using COCA than by PCA. This finding is also supported by a quantitative evaluation using specificity as a model metric. Moreover, COCA provides scale invariance and therefore allows us to include different modalities and attributes in a unified way. We presented a combined model including shape, color, attributes like age, weight and height, and even categorical attributes like gender. The scale invariance is a key feature of COCA, it enables us interesting new applications and methods when working with statistical models.

Finally, we again want to encourage the reader to replace PCA with a COCA model, since the additional model flexibility comes with almost no implementation effort. The computer graphics and vision community is heavily modeling and working with color intensities. We believe that these intensities are most often not Gaussian-distributed and, therefore, our findings can be transferred to a lot of applications.

Acknowledgements. This work was partially supported by the Swiss National Science Foundation, project 200021_146178: Copula Distributions in Machine Learning. We would like to thank Clemens Blumer, Antonia Bertschinger and Anna Engler for their valuable inputs and proofreading.

References

1. Cootes, T.F., Edwards, G.J., Taylor, C.J.: Active appearance models. In: Burkhardt, H., Neumann, B. (eds.) ECCV 1998. LNCS, vol. 1407, pp. 484–498. Springer, Heidelberg (1998). doi:10.1007/BFb0054760
2. Blanz, V., Vetter, T.: A morphable model for the synthesis of 3D faces. In: Proceedings of the 26th Annual Conference on Computer Graphics and Interactive Techniques. ACM Press/Addison-Wesley Publishing Co., pp. 187–194 (1999)
3. Jolliffe, I.: Principal Component Analysis. Wiley, Hoboken (2002)
4. Mohammed, U., Prince, S.J., Kautz, J.: Visio-lization: generating novel facial images. ACM Trans. Graph. (TOG) **28**, 57 (2009)
5. Walker, M., Vetter, T.: Portraits made to measure: manipulating social judgments about individuals with a statistical face model. J. Vis. **9**, 12 (2009)
6. Schönborn, S., Forster, A., Egger, B., Vetter, T.: A monte carlo strategy to integrate detection and model-based face analysis. In: Weickert, J., Hein, M., Schiele, B. (eds.) GCPR 2013. LNCS, vol. 8142, pp. 101–110. Springer, Heidelberg (2013). doi:10.1007/978-3-642-40602-7_11
7. Rasmussen, C.E.: The infinite gaussian mixture model. NIPS **12**, 554–560 (1999)
8. Han, F., Liu, H.: Semiparametric principal component analysis. In: Advances in Neural Information Processing Systems, pp. 171–179 (2012)
9. Sirovich, L., Kirby, M.: Low-dimensional procedure for the characterization of human faces. JOSA A **4**, 519–524 (1987)

10. Turk, M., Pentland, A.P., et al.: Face recognition using eigenfaces. In: Proceedings: 1991 IEEE Computer Society Conference on Computer Vision and Pattern Recognition, CVPR 1991, pp. 586–591. IEEE (1991)
11. Nelsen, R.B.: An Introduction to Copulas, vol. 139. Springer Science & Business Media, Heidelberg (2013)
12. Joe, H.: Multivariate Models and Multivariate Dependence Concepts. CRC Press, Boca Raton (1997)
13. Genest, C., Ghoudi, K., Rivest, L.P.: A semiparametric estimation procedure of dependence parameters in multivariate families of distributions. Biometrika **82**, 543–552 (1995)
14. Tsukahara, H.: Semiparametric estimation in copula models. Can. J. Stat. **33**, 357–375 (2005)
15. Hoff, P.D.: Extending the rank likelihood for semiparametric copula estimation. Ann. Appl. Stat. **1**(1), 265–283 (2007)
16. Paysan, P., Knothe, R., Amberg, B., Romdhani, S., Vetter, T.: A 3D face model for pose and illumination invariant face recognition. In: Sixth IEEE International Conference On Advanced Video and Signal Based Surveillance AVSS 2009, pp. 96–301. IEEE (2009)
17. Massey Jr., F.J.: The Kolmogorov-Smirnov test for goodness of fit. J. Am. Stat. Assoc. **46**, 68–78 (1951)
18. Lothar Sachs, J.H.: Angewandte Statistik, 7th edn. Springer, Heidelberg (2006)
19. Amberg, B., Romdhani, S., Vetter, T.: Optimal step Nonrigid ICP algorithms for surface registration. In: IEEE Conference on Computer Vision and Pattern Recognition, CVPR 2007, pp. 1–8. IEEE (2007)
20. Styner, M.A., Rajamani, K.T., Nolte, L.-P., Zsemlye, G., Székely, G., Taylor, C.J., Davies, R.H.: Evaluation of 3D correspondence methods for model building. In: Taylor, C., Noble, J.A. (eds.) IPMI 2003. LNCS, vol. 2732, pp. 63–75. Springer, Heidelberg (2003). doi:10.1007/978-3-540-45087-0_6
21. Schumacher, M., Blanz, V.: Exploration of the correlations of attributes and features in faces. In: 11th IEEE International Conference and Workshops on Automatic Face and Gesture Recognition (FG), pp. 1–8. IEEE (2015)
22. Blanc, R., Seiler, C., Székely, G., Nolte, L.P., Reyes, M.: Statistical model based shape prediction from a combination of direct observations and various surrogates: application to orthopaedic research. Med. Image Anal. **16**, 1156–1166 (2012)
23. Edwards, G.J., Lanitis, A., Taylor, C.J., Cootes, T.F.: Statistical models of face images improving specificity. Image Vis. Comput. **16**, 203–211 (1998)

Representing Shapes of 2D Point Sets by Straight Outlines

Dirk Feldmann$^{(\boxtimes)}$ and Melanie Pohl

Department Scene Analysis, Fraunhofer IOSB, 76275 Ettlingen, Germany
{dirk.feldmann,melanie.pohl}@iosb.fraunhofer.de

Abstract. The problem of faithfully matching the outlines of objects that are represented by finite point sets in 2D by simple polygons is challenging if the actual shape is non-convex and features long, straight edges and only few, distinct angles. A common application for this task is the geometric reconstruction of man-made structures like buildings from LIDAR data. Using algorithms for computing hulls to outline such point sets frequently yields polygons that consist of too many short line segments joining at unexpected angles with respect to the original object. Furthermore, if the outline polygons contain large regions that correspond to holes within the underlying object, it is desirable to represent such structures by polygons as well, but increases the complexity.

We present two methods for creating outline polygons that account for the characteristics of the aforementioned kind of objects given as finite 2D point sets, and that are also suited for bordering holes. The resulting polygons have fewer vertices and angles than those obtained from hulls and are able to depict long, straight edges of the underlying objects more accurately.

Keywords: Point set · Outline · Boundary · Hull · Concave · Building · Footprints · Dominant direction

1 Introduction

The problem of finding simple, planar polygons that outline a finite point set in 2D Euclidean space arises in many practical applications. In order to automatically create polygonal 3D models (*meshes*) of buildings or even entire cities, for instance, it is common practice to employ point data acquired by LIDAR devices or from aerial photographs by means of photogrammetric methods. These points are usually based in a plane section in 2D Euclidean space. The task of creating meshes from such finite point sets comprises the detection of their outlines (also called *footprints*) [1]. The most notable solutions to this kind of problem are probably *convex hulls*. In the case of constructing 3D models of buildings, convex hulls may be inappropriate, because the outlines of many buildings or building complexes are not convex. A better approach would be the usage of methods for finding non-convex (*concave*) hulls. But since man-made objects like buildings

© Springer International Publishing AG 2017
J. Braz et al. (Eds.): VISIGRAPP 2016, CCIS 693, pp. 113–134, 2017.
DOI: 10.1007/978-3-319-64870-5_6

tend to have lots of long, straight edges, which furthermore enclose few angles of discrete measures (e.g., $\pm\frac{\pi}{2}$), true hulls rarely reflect the desired outlines very well as shown in Fig. 1: The outline of the building obtained by the Concave Hull Algorithm [2] is clearly preferable over the convex hull, but it has too many vertices and is not as straight as the underlying buildings' exterior boundaries really are. For the task of building reconstruction, having nicely straight outlines with only few vertices is desirable, because it simplifies the process of 3D model generation and the results are more realistic. Straightening outlines obtained from concave hulls using methods like the popular Ramer-Douglas-Peucker Algorithm [3,4] is not a general solution due to unintentional removal of certain corners as shown in Fig. 1(b). Please note that we do not deal with buildings having strongly curved boundaries, even though these can be found in many architectural landmarks.

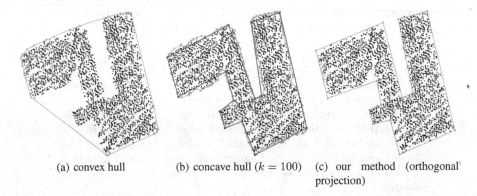

(a) convex hull   (b) concave hull ($k = 100$)   (c) our method (orthogonal projection)

Fig. 1. Using the convex hull or the Concave Hull Algorithm to generate outlines of 2D point sets obtained from man-made structures, like buildings, leads to shapes that do not necessarily reflect the actual object very well. Straightening the resulting outlines using Ramer-Douglas-Peucker Algorithm (blue line in (b)) only little improves the result. Our method based on orthogonal projection creates outlines that represent such objects more accurately. In Fig. 1(b), k denotes the number of neighbors considered (see Sect. 2.1). (Color figure online)

Furthermore, input data obtained by measurements are always noisy and the distribution of points is likely to be non-uniform. Since hulls must include every point of the set they enclose, unwanted outliers will further diminish the desired quality of the resulting outlines.

The boundaries of finite point sets located in a plane section in 2D Euclidean space may also feature larger domains where no point data are present at all (called *holes*). In the process of building reconstruction from LIDAR data, for instance, such empty domains frequently result from buildings or building complexes having inner courtyards or atria. In order to give faithful representations of the underlying objects, these holes also need to be bounded by simple polygons.

We present two methods for finding simple outline polygons of 2D point sets which may feature substantially fewer vertices and angles than hulls in order to faithfully represent shapes of structures that are typically man-made or originate from technical processes. These structures may contain holes and are characterized by straight edges and may be non-convex. Frequently, they are moreover (near-)rectangular and have only few angles and corners. Our methods are based on the *Concave Hull* Algorithm presented in [2]. Although our application aims at generating outlines for the task of building reconstruction, and we therefore demonstrate and evaluate our method in the context of that particular application, the proceedings presented in this article are general and versatile.

2   Related Work

Finding outlines of 2D point sets representing structures that have straight edges and enclose only few angles in order to obtain "good" representations of the underlying objects is an inherently ambiguous problem and strongly depends on the application. As pointed out in [5], there is no single correct outline or *footprint* of a set of points. Computing outlines by means of convex hulls, for instance, is probably reasonable if the underlying objects are (almost) convex themselves. Convex hulls are well-studied and many algorithms are known for their computation [6], but in case of non-convex objects, convex hulls naturally yield unsatisfactory outlines. Therefore, we do not further enter on methods for their generation.

α-shapes [7] are a generalization of convex hulls and allow for computing non-convex (*concave*) outlines of 2D point sets. The α-shape of a finite point set $S \subset \mathbb{R}^2$ is the boundary of the α-hull of S. The latter can be obtained by computing the Delaunay Triangulation of S and removing those k-simplexes, $k = 0, 1, 2$, whose open circumscribed circles have radii $\geq \frac{1}{\alpha}$ or that contain any other $\mathbf{p} \in S$. Using $\alpha = 0$, the α-shape of S is defined to be its convex hull.

Another approach based on Delaunay Triangulation are *characteristic shapes* (χ-shapes) as presented in [8]. The algorithm repeatedly removes the longest exterior edge whose length is greater than some value L from the initial Delaunay Triangulation of S, provided that the remaining exterior edges form a simple polygon. It is also possible to parametrize χ-shapes by a normalized length parameter $\lambda = \frac{l - l_{min}}{l_{max} - l_{min}}$, where l denotes the length of an edge and l_{min} and l_{max} are the minimal and maximal edge length from the Delaunay Triangulation, respectively.

The resulting outlines produced by α- and χ-shapes depend on the choice of the respective parameters α and l (or λ). Their choice involves a-priori knowledge about the shape of the underlying point set, or appropriate values must be found using heuristics or by trial and error.

The *Concave Hull* Algorithm in [2] is based on the idea of Jarvis' March aka. Gift Wrapping Algorithm [9] for computing convex hulls: Instead of considering all elements from the set of remaining points S, the Concave Hull Algorithm only considers the $3 \le k \le |\mathcal{S}|$ nearest neighbors of the point that has been added to the hull last. Using greater values for k results in the consideration of larger neighborhoods, and the concave hull becomes "more convex" as k increases. If $k = |\mathcal{S}|$, the algorithm computes the convex hull, because the set of points considered becomes the same as with Jarvis' March. Since our method is based on the Concave Hull Algorithm, it is briefly summarized in Sect. 2.1.

The strength of Concave Hull is its simplicity and that there is no need to compute any (Delaunay) triangulation of which most information about connectivity is not needed at all.

In [10], the *Alpha-Concave Hull* is presented, which is a simple polygon whose interior angles are $< \pi + \alpha$. If the parameter $\alpha = \pi$, the Alpha-Concave Hull is the minimal polygon of $\mathcal{S}$; in case of $\alpha = 0$, it is the convex hull.

In the field of photogrammetry, methods for generating rectangular outlines of buildings based on line fitting of contour points, like the one presented in [1], seem to be popular. The drawback of these methods is their limitation to the creation of only rectangular outlines. But even in the application of detecting building outlines, there are often boundaries that are not rectangular, like in cities that grew over centuries or that were not planed from scratch.

There are also approaches based on statistical estimates of the most appropriate outline for a given set of points [11].

Furthermore, among the methods presented so far, only α-shapes can create outlines of *holes* in 2D (see Sect. 1), but the resulting outlines are overly jagged. Representing holes, however, is essential to obtain faithful representations of atria or building complexes enclosing inner courtyards. The solutions presented in [12–14] address the detection (and also filling) of holes in point clouds that represent surfaces in 3D, so that these methods are difficult to apply to our problem.

2.1   Concave Hull Algorithm

To generate *outlines*, also called *boundaries* in the remainder of this work, we adopt the Concave Hull Algorithm in [2] which is summarized below, because it is the foundation of our approach. For a more detailed description, we refer to the original article.

Let $\mathcal{S}$ be a finite set of points in 2D Euclidean space where $\mathbf{p}_i \ne \mathbf{p}_j \ \forall \mathbf{p}_i, \mathbf{p}_j \in \mathcal{S} \wedge i \ne j$. Let furthermore $\mathbf{p}' \in \mathcal{S}$ be the point having minimal y-coordinate and $\mathcal{C}_n = (\mathbf{v}_0, \mathbf{v}_1, \dots, \mathbf{v}_n)$ the list of vertices of the outline computed so far.

Given $k \in \mathbb{N}$, $3 \le k < |\mathcal{S}|$, start with $n = 0$, $n \in \mathbb{N}$, $\mathcal{S}_{-1} = \mathcal{S}$, $\mathcal{C}_0 = (\mathbf{v}_0 = \mathbf{p}')$, $\mathbf{p} = \mathbf{p}'$ and $\mathbf{e} = -\mathbf{x}$, where $-\mathbf{x}$ denotes the negative x-axis. Remove $\mathbf{p}$ from the current set of points to obtain $\mathcal{S}_n = \mathcal{S}_{n-1} \backslash \mathbf{p}$ and compute the set $\mathcal{K}_n(\mathbf{p}) \subset \mathcal{S}_n$ of $\mathbf{p}$'s k nearest neighbors. The points $\mathbf{q}_i \in \mathcal{K}_n(\mathbf{p})$, $0 \le i < k$, are sorted according to the *clockwise* angle φ_i enclosed between the directed edge $\mathbf{e}_i = (\mathbf{q}_i - \mathbf{p})$ and edge $\mathbf{e}$ such that $j = 0$ be the index of the point $\mathbf{p}_j = \mathbf{q}_i$ where φ_i is largest, $j = 1$ the one where φ_i is second largest, etc.

Now let $\mathbf{e}_j = (\mathbf{p}_j - \mathbf{p})$ denote the directed edge from the last vertex of the outline polygon to the point $\mathbf{p}_j \in \mathcal{K}_n(\mathbf{p})$, $j = 0, 1, \dots, k-1$. Starting at $j = 0$, check if $\mathbf{e}_j$ does not intersect any of the existing edges $\overline{\mathbf{v}_{n-2}\mathbf{v}_{n-3}}, \overline{\mathbf{v}_{n-3}\mathbf{v}_{n-4}}, \ldots, \overline{\mathbf{v}_1\mathbf{v}_0}$ of the polygon made up of $\mathcal{C}_n$. If $\mathbf{e}_j$ intersects an existing edge of the hull, retry using $\mathbf{e}_{j+1}$, if $j < k$. In case no $\mathbf{e}_j$ remains, the outline computed so far is discarded, k is increased and the algorithm is restarted using $n = 0$ again. Otherwise, if no intersection of $\mathbf{e}_j$ with an existing edge is found, set $\mathbf{p}_{j'} = \mathbf{p}_j$ to be the best match in $\mathcal{K}_n(\mathbf{p})$, add $\mathbf{p}_{j'}$ to the list of vertices, i.e., $\mathcal{C}_{n+1} = \mathcal{C}_n \cup \mathbf{p}_{j'}$, and set $\mathbf{e} \leftarrow (\mathbf{v}_n - \mathbf{p}_{j'})$ as the backwards directed edge formed by the last two vertices of the hull. The process is then restarted setting $n \leftarrow n+1$ and $\mathbf{p} \leftarrow \mathbf{p}_{j'}$ until $\mathbf{p} = \mathbf{p}'$ again, or all points in $\mathcal{S}$ have been processed. In order to ensure that $\mathbf{p} = \mathbf{p}'$ a second time and that the algorithm terminates, $\mathbf{p}'$ must be added to $\mathcal{S}_n$ again after the third iteration. When $\mathbf{p} = \mathbf{p}'$ the second time, it is necessary to check if all remaining points in $\mathcal{S}_n$ are actually inside the resulting polygon formed by $\mathcal{C}_n$ so that the result is actually a hull in the proper sense. Otherwise, the outline is also discarded, k is increased and the algorithm is restarted using $n = 0$.

3   Computing Outer Boundaries

In this section we present two different approaches for computing simple polygons that outline a given finite point set $\mathcal{S} \subset \mathbb{R}^2$. The outline's edges enclose only angles of a few, distinct values, and we call them *angular outlines* in the following.

The first method relies on the concave hull and determination of dominant edge directions. Our second method modifies the Concave Hull Algorithm to directly compute an angular outline.

The advantage of the first method, presented in Sect. 3.1, is that the hull can be straightened by two predefined main angles. This approach is based on the fact that buildings are very frequently oriented along two directions to guarantee parallel walls. Furthermore, if we also take the idea of perpendicularity into account, we can force the straightened edges to form right angles. Therefore, data that are very noisy or contain holes can be bordered along the main directions or perpendicular to it.

The second approach in Sect. 3.2 is preferable if the underlying structure is not well approximated by strictly rectangular outlines or if the edges of the shape enclose more than two distinct angles. However, for this method to work, the initial orientation of the underlying shape within a Cartesian coordinate system must be provided.

3.1   Angular Outlines from Dominant Directions

Let $\mathcal{C}_m$ be the list of vertices of a concave hull of $\mathcal{S}$ and $\mathcal{E} = (\mathbf{e}_0, \ldots, \mathbf{e}_m)$ the list of edges with $\mathbf{e}_i = \overline{\mathbf{v}_i\mathbf{v}_{i+1}}$, $i = 0, 1, \ldots, (m-1)$ and $\mathbf{e}_m = \overline{\mathbf{v}_m\mathbf{v}_0}$. Since we do not consider the orientation of an edge, each difference vector $\mathbf{e}_i$ is set to $\mathbf{e}_i \leftarrow \mathrm{sgn}(e_{i,y}) \cdot \mathbf{e}_i$. For each edge $\mathbf{e}_i$ of length l_i, we compute the angle $\alpha_i \in [0, \pi]$ that is enclosed with the positive x-axis. If $\alpha_i = \pi$, we set $\alpha_i \leftarrow 0$. We analyze the directions and lengths of each edge of the concave hull polygon and compute a length-weighted angle histogram $H(\alpha_i)$. Angle α_i is assigned to the respective bin b of width w according to

$$w \cdot (b-1) \leq \alpha_i < w \cdot b, \quad b = 1, 2, \ldots, \left\lceil \frac{\pi}{w} \right\rceil$$

The value of bin b is computed as the sum of its $k(b)$ associated edges, weighted by their respective lengths, i.e., $\sum_{k(b)} l_{k(b)}$.

From this histogram, we compute the two dominant angles β and γ of the concave hull using one of the following two options:

1. Approximate the histogram $H(\alpha_i)$ by a bimodal distribution. Then β and γ correspond to the angles of the two maximums of the approximation.
2. β corresponds to the angle of the maximum value in H and γ is offset by $\pm \frac{\pi}{2}$ according to

$$\beta = \underset{0 \leq \alpha_i < \pi}{\mathrm{argmax}} \left(H\left(\alpha_i\right) \right), \quad \gamma = \begin{cases} \beta + \frac{\pi}{2}, & \beta < \frac{\pi}{2} \\ \beta - \frac{\pi}{2}, & \beta \geq \frac{\pi}{2}. \end{cases}$$

The advantage of option Eq. 1 (called "peaks" option in the following) is that the edges of the boundary obtained from straightening of the concave hull can enclose an arbitrary but fixed angle that is not necessarily a right one. Option Eq. 2 ("90°" option) ensures that these edges always enclose right angles.

From β and γ we compute the corresponding dominant directions

$$\mathbf{d}_\varphi = \begin{pmatrix} \cos\left(\varphi\right) \\ \sin\left(\varphi\right) \end{pmatrix}, \quad \varphi \in \{\beta, \gamma\}$$

and assign labels ϱ_φ to all edges of the concave hull according to the dominant angle they belong to:

$$\varrho_\varphi = \underset{\varphi}{\mathrm{argmin}} \{\angle(\mathbf{e}_i, \mathbf{d}_\varphi)\}, \quad \varphi \in \{\beta, \gamma\}.$$

Ideally, all line segments of a concave hull that represent a single straight edge of an object's outline are successively assigned the same label. We call these continuous sequences of line segments having the same label *runs*, and their number is the length of the run. In practice, due to the choice of k and the local density of the point set, long runs that would be expected to represent a single edge of the outline polygon may be "interrupted" by short ones, or labels may change alternately, especially if the two directions considered dominant differ only little from each other. In the former case, the corresponding outline polygon looks sawtooth-shaped in these areas as depicted in Fig. 2. Runs of similar length resulting from alternating directions lead to rippled shapes (see Fig. 2(c)). Such (unwanted) situations can be countered by filtering sequences of runs based on background information from the underlying application. In the area of building reconstruction, for example, we can assign thresholds for minimal edge lengths of the underlying building and accordingly eliminate short runs by reassigning labels.

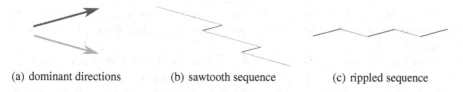

(a) dominant directions   (b) sawtooth sequence   (c) rippled sequence

Fig. 2. With two similar dominant directions like in (a), longer runs may be interrupted by shorter ones forming sawtooth-shaped edges within the outline (b), or labels may change alternately causing rippled edges (c).

To simplify the straightening procedure, it is advisable to shift the start point in the sequence of vertices of the concave hull and the entries of the label vector to ensure that a run is not split at the start or end vertex. Afterwards, we collect every run $\mathcal{T}_{\varrho_\varphi} = (\mathbf{t}_1, \mathbf{t}_2, \ldots)$ with label ϱ_φ and compute a line ℓ through the center of gravity along the corresponding main direction $\mathbf{d}_\varphi$:

$$\ell : \boldsymbol{x} = \frac{1}{|\mathcal{T}_{\varrho_\varphi}|} \sum_{\mathbf{t}_i \in \mathcal{T}_{\varrho_\varphi}} \mathbf{t}_i + \mu \, \mathbf{d}_\varphi$$

For each of these lines, we compute the intersections with the lines from preceding and following runs to obtain the final outline polygon.

Using More Than Two Dominant Directions. The "peaks" option relies on the directions of the line segments from the concave hull being distributed in a way that allows to approximate the resulting histograms by bimodal distributions. Considering more than two directions would thus only be possible, if all the different modes in a (multimodal) distribution can be identified reliably. This becomes difficult, for instance, if the actual outline of an object features edges running into more than two different, but similar directions, and which are thus not distinguishable in a histogram, even if the bin size of the histogram was small enough, since they would be considered the same mode. Besides, the number of dominant directions within an outline is generally unknown in advance or requires incorporation of information from the area of application.

In addition, trying to obtain all the directions of line segments of an outline from a single direction histogram is not necessary, and we suggest the following "divide & conquer" approach:

Once we have assigned labels to all line segments of the concave hull and identified runs based on an initial bimodal distribution, we can recursively apply the same method to the resulting runs and consider them separately. From these more localized distributions of directions, we can obtain more precise estimates for the actual direction of the corresponding edge of the outline polygon. With respect to the resulting histograms for each run, we can either accept a previously assigned direction, update/refine it, or assign new direction labels to obtain new sequences of runs. In the latter case, we may continue the recursion, or terminate based on some criterion, like the length of a run or the variance of the direction histogram. If we terminate, we may accept a previously assigned direction for the corresponding edge, or compute an average direction if we cannot find a distinct one.

With this proceeding, we would also be able to detect unwanted situations and to react accordingly: In the case of encountering a run that would result in a rippled sequence as shown in Fig. 2(c), for instance, we would find out about this if we assigned labels for the two new directions recursively to the contained line segments, because the previous run became split into many short runs with alternating labels. If outline edges associated with the run would actually represent a shallow, v-shaped crest instead, we would find out by ideally obtaining just two new runs.

However, it should be noted that with an increasing number of directions allowed, the shape of the entire outline will look less straight, and that rippled and sawtooth-shaped edges become more a problem.

3.2   Angular Outlines from Orthogonal Projection

Our second approach to computing angular boundaries modifies the Concave Hull Algorithm in Sect. 2.1 by projecting the edges of the hull onto orthogonal line segments and adjusting them. First, we explain the process of projection onto orthogonal line segments. This yields a strictly rectangular outline, which is only a "good" representation if the underlying shape is (almost) rectangular, too. The proceeding is then extended by three modifications that improve the shape of the outline in situations where strictly rectangular outlines may be perceived as unsatisfactory representations.

Orthogonal Projection. In this section, we use the same symbols from the description of the Concave Hull Algorithm in Sect. 2.1. In addition, let $\delta_i = i \cdot \frac{\pi}{2}, i = 0, 1, \ldots, 4$. When the Concave Hull has determined a candidate $\mathbf{p}_{j'}$ among the sorted k nearest neighbors $\mathcal{K}_n(\mathbf{p})$ of the current point $\mathbf{p} \in \mathcal{S}_n$ that satisfies the requirements for being a new vertex of the outline polygon, we modify the algorithm to additionally compute

$$\delta' = \underset{\delta_i}{\mathrm{argmin}} \left(|\delta_i - \varphi_{j'}| \right)$$

where φ_j is the angle associated with $\mathbf{p}_j$ such that δ' is the angle that deviates least from $\varphi_{j'}$. If δ' happens to be 0 or 2π, we discard $\mathbf{p}_{j'}$ and proceed with the next candidate in $\mathcal{K}_n(\mathbf{p})$, if there is any left. Otherwise, $\mathbf{p}_{j'}$ is discarded from the set of remaining points, just like with the Concave Hull.

Using δ', we compute the corresponding direction $\mathbf{d}$ and an orthogonal vector $\mathbf{u}$ pointing to the "right-hand side" of the ray R from $\mathbf{p}$ in direction $\mathbf{d}$, i.e., usually away from where the majority of points of $\mathcal{S}$ is located:

$$\mathbf{d} = \begin{pmatrix} \cos(\theta) \\ \sin(\theta) \end{pmatrix}, \quad \mathbf{u} = \begin{pmatrix} \sin(\theta) \\ -\cos(\theta) \end{pmatrix}$$

where $\theta = \theta_n + (\pi - \delta')$ and θ_n is the angle enclosed by the last edge e of the outline computed so far and the x-axis. Please notice that we have to use $\pi - \delta'$ instead of δ' to compute θ, because we chose clockwise angles in $[0, 2\pi]$ with the Concave Hull Algorithm. The outer boundary is thus oriented counter-clockwise. The vector $\mathbf{e}_{j'} = (\mathbf{p}_{j'} - \mathbf{p})$, which corresponds to edge $\overline{\mathbf{p}_{j'}\mathbf{p}}$, is then projected onto $\mathbf{d}$ and $\mathbf{u}$ to obtain the lengths $l_d = \mathbf{d} \cdot \mathbf{e}_{j'}$ and $l_u = \mathbf{u} \cdot \mathbf{e}_{j'}$, respectively. Next, the new vertex $\mathbf{q}$ to be added to the current outline $\mathcal{C}_n = (\mathbf{v}_0, \mathbf{v}_1, \ldots, \mathbf{v}_n)$ is computed as

$$\mathbf{q} = \mathbf{p} + |l_d| \cdot \mathbf{d} + \mathbf{w}$$

where

$$\mathbf{w} = \begin{cases} \max(l_u, 0) \cdot \mathbf{u}, & \delta' = \pi \\ l_u \cdot \mathbf{u}, & \text{otherwise} \end{cases} \tag{1}$$

Thus, if an edge is continued (i.e., $\delta' = \pi$), $\mathbf{w}$ is either zero or points towards the right-hand side. Otherwise, if the new edge introduces a turn, it may also point to the left-hand side of R. The outline polygon is then updated by considering the following two cases:

1. If $\mathcal{C}_n$ contains ≥ 2 vertices and the edge is about to be continued, we replace the last vertex by $\mathbf{q}$ and add $\mathbf{w}$ to the second last vertex:

$$\mathbf{v}_n \leftarrow \mathbf{q}, \quad \mathbf{v}_{n-1} \leftarrow (\mathbf{v}_{n-1} + \mathbf{w}) \tag{2}$$

2. Otherwise, if $\delta' = \frac{\pi}{2}$ or $\delta' = \frac{3\pi}{2}$ and the outline is about to make a left or right turn, respectively, or if it contains only one vertex so far, we append $\mathbf{q}$ to $\mathcal{C}_n$ and add $\mathbf{w}$ to $\mathbf{v}_n$:

$$\mathbf{v}_{n+1} = \mathbf{q}, \quad \mathbf{v}_n \leftarrow (\mathbf{v}_n + \mathbf{w}) \tag{3}$$

In the next iteration $n+1$, we proceed with $\mathbf{q}$ instead of $\mathbf{p}_{j'}$, after having removed $\mathbf{p}_{j'}$ from the point set $\mathcal{S}_{n+1}$. As with the Concave Hull Algorithm, we check for self-intersections of the resulting polygon. In positive cases, we also discard the polygon and start the algorithm from the beginning using $k + 1$. Likewise, we terminate if $\mathbf{p}_{j'} = \mathbf{p}'$ again, but since $\mathcal{C}_n$ is obviously not a hull, it does not necessarily include all points in $\mathcal{S}$. Therefore, we omit the final check of whether all points in $\mathcal{S}$ are located strictly inside $\mathcal{C}_n$. Figure 3 illustrates the proceeding described above.

Due to the choice of $\mathbf{w}$ in Eq. 1, we ensure that edges are dragged only to the "outside" of the shape and that more points (including $\mathbf{p}_{j'}$) are located "inside" the resulting outline if an edge is continued. In the case of left or right turns, an edge may also become dragged towards the "inside" ($l_u < 0$) in order to enclose the point set $\mathcal{S}$ more tightly. However, this may cause some points of $\mathcal{S}$ to be found again among the k nearest neighbors $\mathcal{K}_{n+1}(\mathbf{p})$ in the next iteration. Depending on the choice of k, this can lead to self-intersections and the rejection of the current k. The results of this proceeding are satisfactory in case of shapes that are (almost) rectangular everywhere. In cases where a shape's corners deviate too much from $\pm\frac{\pi}{2}$, e.g., in case of point sets derived from ancient or architecturally unusual buildings, the outlines are less appealing (see Fig. 4).

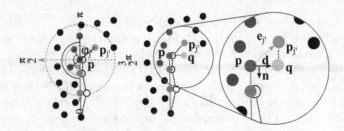

Fig. 3. Creating angular outlines by means of orthogonal projection is based on the Concave Hull Algorithm, but instead of adding $\mathbf{p}_{j'}$ to the list of vertices, its projection $\mathbf{q}$ onto $\mathbf{d}$ is added.

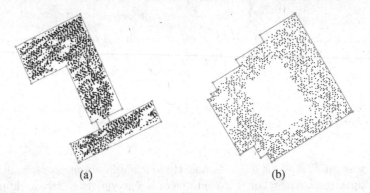

(a)   (b)

Fig. 4. Pure orthogonal projection and dragging continued edges only to the outside yields appealing results if the underlying shape is almost rectangular itself (a), but causes the appearance of unwanted steps if corners are present that are not rectangular (b).

Improvements for Non-rectangular Shapes. To make the outline match non-regular shapes more accurately, we introduce some modifications to the proceeding described in Sect. 3.2. First, we can simply omit the addition of $\mathbf{w}$ to the predecessor of the last vertex in Eq. 2 when an edge is continued. As a result, the continued edge is not necessarily perpendicular to the previous one anymore, but the outline captures a shape more accurately at corners where two edges join at angles $\varphi > \frac{\pi}{2}$. Corners where $\varphi < \frac{\pi}{2}$ remain approximated by steps, because the line segment of a continued edge is still only allowed to being dragged "outwards" due to Eq. 1. Thus, we modify Eq. 1 to obtain Eq. 4:

$$\mathbf{w} = \begin{cases} \max\left(l_u, 0\right) \cdot \mathbf{u}, & \text{if } \delta' = \pi \wedge outsideOnly = true \\ l_u \cdot \mathbf{u}, & \text{otherwise} \end{cases} \tag{4}$$

The predicate *outsideOnly* is set to *false* every time $\delta' = \frac{\pi}{2}$ or $\delta' = \frac{3\pi}{2}$ along the entire outline. It is set to *true* if $\delta' = \pi$ and if $l_u > 0$ for any point along the continued edge and remains *true* until $\delta' \neq \pi$ again. In this way, a continued edge may also become dragged towards the "inside", and it can thus enclose $\mathcal{S}$ more tightly, as long as there is no need to drag the edge outside again, i.e., $l_u < 0$ over the whole distance the edge is continued. The effect of this modification is shown in Fig. 6.

Dragging the last vertex of a continued edge towards the inside can cause more points of $\mathcal{K}_n(\mathbf{p})$ to lie on the right-hand side of the edge, because $\mathbf{p}_{j'}$ is not necessarily the first point in the descending order of enclosed, clockwise angles. Those points may be found again during the next iteration in $\mathcal{K}_{n+1}(\mathbf{p})$ and may introduce unwanted turns and corners in the outline, or disturb the algorithm by self-intersections. Therefore, we not only remove $\mathbf{p}_{j'}$ from $\mathcal{S}_n$ to obtain $\mathcal{S}_{n+1}$, but also every other point $\mathbf{p}_i \in \mathcal{K}_n(\mathbf{p})$ whose associated edge $(\mathbf{p}_i - \mathbf{p})$ encloses a clockwise angle with the outline's last edge $\mathbf{e} = (\mathbf{p} - \mathbf{v}_{n-1})$ that is larger than the one enclosed between $\mathbf{e}$ and $(\mathbf{q} - \mathbf{p})$.

A further adjustment that affects concavities of a shape is the enforcement of creating corners in the outline by modifying Eq. 2 to obtain Eq. 5:

$$\mathbf{v}_n \leftarrow \mathbf{q}, \quad \mathbf{v}_{n-1} \leftarrow \begin{cases} \mathbf{v}_{n-1} + \mathbf{w}, & \text{second last turn was right} \\ \mathbf{v}_{n-1}, & \text{otherwise} \end{cases} \tag{5}$$

In other words, we add $\mathbf{w}$ to the second last vertex of a continued edge, only if the second last turn of the outline we encountered was a right turn, and the edge is thus part of a concavity. In such concave regions, the algorithm then behaves like the version in Sect. 3.2, and problems as shown in Fig. 5(a) are prevented. The benefit of this modification, however, depends on the application and the underlying point set, and to some degree on the user's judgment or personal preferences (see Fig. 5(b)).

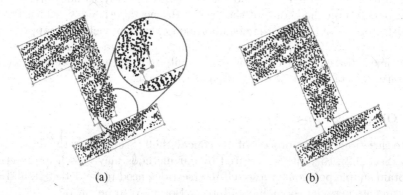

(a)   (b)

Fig. 5. (a) Due to dragging only the last vertex of outlines, edges may be moved into concavities and can unintentionally cut off too many points. (b) The modification given in Eq. 5 remedies this issue. In both cases, $k = 80$.

The Choice of Initial Direction. In the proceeding described in Sect. 3.2, the initial direction given by θ_0 is yet unspecified, but it influences the resulting

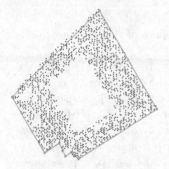

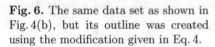

Fig. 6. The same data set as shown in Fig. 4(b), but its outline was created using the modification given in Eq. 4.

Fig. 7. The shape represented by the point set is actually tilted by $\approx 25°$ against the x-axis. Not accounting for this initial orientation yields unpleasant results.

outline as shown in Fig. 7. Using $\theta_0 = 0$ is only reasonable if the lowest edge of the underlying shape is (near) parallel to the x-axis. But since the orientation of the shape we are trying to approximate by angular outlines may be unknown in advance, we either have to let the user specify θ_0 or obtain it from analyses of the point set's orientation by statistical means, e.g., via principal component analysis (PCA) or the methods employed in Sect. 3.1.

Starting at point $\mathbf{p}'$ having lowest y-coordinate, we might also run our algorithm using $\theta_0 = 0$. The first time we encounter $\delta' \neq \pi$, we can set θ to the angle enclosed between the first edge and the x-axis. In this way, θ_0 not only depends on $\mathbf{p}'$, but also on the choice of the size of the neighborhood k, which can be unsatisfactory as outliers in the neighborhood of $\mathbf{p}'$ may yield an inappropriate outline. However, if the point set is sufficiently clean and dense in the area of the outline's first edge, this method works quite well. It can also be used to give the user a hint of the initial orientation of the shape.

3.3 Obtaining Simple Polygons

Due to the straightening process of the concave hull (Sect. 3.1) or the dragging of edges (Sect. 3.2), the outlines created by our methods may be self-intersecting. To obtain simple polygons, these self-intersections need to be detected, and the line segments between the points of intersection need to be cut off.

4 Outlines of Holes

The outlines of point sets in plane sections in 2D Euclidean space may enclose larger areas where no points are present. In the application of building reconstruction from LIDAR data, for instance, this happens if building complexes have atria or inner courtyards, because the data may only contain information about distinctly elevated structures, like roofs. We call such empty areas *holes* within the point set and define them more formally as in Definition 1.

Definition 1. *Let $S \subset \mathbb{R}^2$ be a finite point set and $\Omega(S) \subset \mathbb{R}^2$ be the domain inside a simple polygon that tightly encloses all elements of S (e.g., one of its hulls or the minimal axis-aligned bounding box). We call a connected region $\mathcal{E} \subset \Omega(S)$ of finite area a **hole (in or of S) of size** r, if $\mathcal{E}$ is large enough to fully overlap a circle of radius $r > 0$ at some location $\mathbf{c}$ such that $\forall \mathbf{p} \in S: \|\mathbf{p} - \mathbf{c}\| \geq r$. r is called the **size of hole $\mathcal{E}$**.*

Our definition of *hole* depends on the choice of polygon to border $\Omega(S)$: it is possible to choose different boundaries for S that have different numbers of holes, or holes of different sizes. In this work, we are moreover only interested in holes that do not border on the enclosing polygons of point sets.

In order to obtain faithful outlines of the objects represented by 2D points, we extend our methods by the ability to create inner boundaries. Since we are only interested in holes that are "large enough", we choose a minimum hole diameter $d_{\min}$ that shall be detected. The choice of $d_{\min}$ depends on the application and the average density of the point set. In the application of building reconstruction, for instance, $d_{\min}$ might be related to the sizes of humans or vehicles, if the holes originate from courtyards, because such structures usually serve a purpose and need to be accessible.

In contrast to outer boundaries, inner boundaries shall be oriented clockwise. This is mainly due to technical aspects and conventions, for it allows us to easily determine whether a point is located *inside* or *outside* the inner or outer boundary, respectively. Besides, we need to take into account that there might be more than one hole in a given point set.

4.1   Detecting Holes

The basic idea of detecting holes is to find points on their border and to create their outlines by applying modified versions of the methods presented in Sect. 3. In order to find points on the border of holes, we employ the following purely geometry-based heuristic (see also Fig. 8):

1. Let h be the extent of point set S along the y-axis; choose a minimum hole diameter $d_{\min}$. Let furthermore P be a simple, closed outline polygon of S, e.g., a (concave) hull or any other outline obtained by our methods.
2. Split S along the y-axis into m strips of width $w = \frac{d_{\min}}{2}$ in such a way that $m = \left\lceil \frac{2h}{d_{\min}} \right\rceil \in \mathbb{N}$.
3. Obtain disjoint point sets Y_j for each of the m strips where
$$\forall a, b \in 0, 1, \ldots, (m-1), a \neq b, \; Y_a \cap Y_{a+1} = \emptyset \wedge \bigcup_{j=0}^{m-1} Y_j = S.$$
4. For each set Y_j, sort the points in Y_j in ascending order of their x-coordinates.
5. For each two neighboring points $\mathbf{p}_i$, $\mathbf{p}_{i+1}$ in Y_j, compute the projected horizontal distance $d_x = p_{i+1,x} - p_{i,x}$.

6. If $d_x \geq 2 \cdot w$ and $\mathbf{m} = (p_{i,x} + w, p_{i,y})$ is located inside P, add the directed line segment $\overrightarrow{\mathbf{p}_i\mathbf{p}_{i+1}}$ to the list of candidates Q.

This approach will find the points on the borders of all holes at once, but we cannot tell the hole candidates from different Y_j belong to. To solve this problem, we compute an outline of the hole for the left vertex of the first line segment $\overline{s}_0 = \overline{\mathbf{a}_0\mathbf{b}_0}$ in Q as described in Sect. 4.2 and remove it from Q. For each of the remaining line segments $\overline{s}_j = \overline{\mathbf{a}_j\mathbf{b}_j} \in Q^j, Q^j = Q \setminus \{\overline{s}_k; \; k < j\}, j > 0$, we check if the midpoint $\mathbf{m}_j$ of $\overline{s}_j$ is located inside any previously computed outline of a hole, and discard $\overline{s}_j$ if $\mathbf{m}_j$ lies inside. Furthermore, we check if any edge of a new hole boundary intersects an edge of a previously computed hole outline. If two outlines intersect, we can discard one of them based on a reasonable criterion, e.g., the size of its area or its number of vertices. Another option would be to merge the two outlines in order to obtain a common outline, but we did not implement this approach due to its complexity.

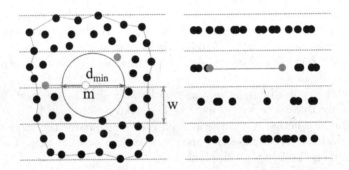

Fig. 8. To find locations on the border of a hole from where to start the generation of inner boundaries, the point set is cut into strips perpendicular to the y-axis (left). Neighboring points whose projections onto the x-axis have distances $\geq d_{min}$ are candidates (red), if $\mathbf{m}$ (red circle) is inside the point set's boundary (green). (Color figure online)

Step 6 in the heuristic sketched above reduces the chance that we erroneously take points on the border of concavities of an outer outline for candidates on hole boundaries, e.g., in the case of an u-shaped building: For such points, the corresponding point $\mathbf{m}$ would be located outside the outer outline.

Alternatively, we may suspend the search for further candidates after the first one is found. After generating the hole boundary, we can fill the hole with phantom data points based on a point density estimation. The next candidate is then detected by restarting the search procedure from the beginning.

With the approach described above, we cannot yet guarantee that the resulting outline borders a hole overlapping a circle of radius $\geq w$ and that it is therefore consistent with Definition 1: Due to "cutting" the point set perpendicular the y-axis into strips of width $w = \frac{d_{min}}{2}$, the hole has at least a radius of half the size of $\frac{w}{2}$. Cutting the point set into strips of width $d_{\min}$ might cause holes of about this size to be missed, because the strips are projected onto the x-axis and gaps might be concealed from projections of points in the top- or bottom-most strip covering a potential hole. This is a sampling problem and the strip width w should thus be smaller than $\frac{d_{\min}}{2}$ in a practical implementation.

Verifying the size of the hole found by the heuristic above turns out to be difficult, since we would need to find the center of an empty circle with radius $d_{\min}$ within the hole, but there are potentially infinite many locations. By placing a circle of radius $d_{\min}$ at the centroid of the hole's outline polygon, for example, the hole can be accepted if the circle is empty, but it cannot be safely rejected, if the circle is not empty. In practical applications that do not rely on the exact size of a hole, it may still be sufficient to find holes that are known to overlap circles with radii in the range of $\frac{w}{2}$ and w.

4.2   Bordering Holes

Creating the outlines of holes is done using slightly modified versions of the procedures described in Sect. 3 and the Concave Hull Algorithm in Sect. 2.1. The modifications basically account for changing the orientation of the resulting polygon to clockwise and the different starting location.

Our modification to the Concave Hull Algorithm to obtain a clockwise oriented polygon for outlining holes is as follows: Among all k neighbors $\mathcal{K}_n$ of the current point $\mathbf{p} = \mathbf{v}_n$, we search for the point $\mathbf{q}_i$ with the *smallest* counterclockwise angle $\varphi_i = \angle(\xi_n, \mathbf{e}_i = \overline{\mathbf{pq}_i})$ where ξ_n is the previous edge directed *backwards*, i.e., $\xi_n = \overline{\mathbf{v}_{n-1}\mathbf{v}_n}$ instead of $\overline{\mathbf{v}_n\mathbf{v}_{n-1}}$. Initially, we use the positive x-axis for ξ_0.

By choosing $\mathbf{p}'$ as the first vertex $\mathbf{a}_j$ of the line segments $\overline{s_j} \in Q^j$ obtained by the heuristic given in Sect. 4.1, we start at a point that is part of the left border of the hole. The point may furthermore be located on a sharp crease (like a $<$-shaped corner) such that the smallest φ_i may be associated with a point at the opposite border of that crease. Thus, the outline generation might fail right away if k was too large. Replacing ξ_n by an edge that is rotated counter-clockwise by $\frac{\pi}{2}$ to point inside the hole yields better results, but if the neighborhood gets even larger, the algorithm can still fail for the same reason.

Due to its position on the left border of the hole, the point $\mathbf{p}' = \mathbf{a}_j$ may not necessarily be found in a k nearest neighborhood a second time. Therefore, we also have to modify the condition for terminating the hole's concave hull computation: For each new line segment added to the hull, we additionally check if it intersects the line segment $\overline{s_j} = \overline{\mathbf{a}_j\mathbf{b}_j} \in Q^j$. Starting from point $\mathbf{a}_j$, the edges of the outline to be created will enter the "upper" part of the hole, i.e., the region above $\overline{s_j}$. It can only enter the "lower" part by crossing $\overline{s_j}$ on the right border on the side of $\mathbf{b}_j$. Once a line segment enters the "upper" part from the hole's "lower" part again, $\overline{s_j}$ is crossed a second time, and we terminate the hull computation, because the polygon has encircled the hole.

Using our method to generate the outline of a hole by means of dominant directions as presented in Sect. 3.1, we first create a clockwise oriented concave hull as described above, and apply the same proceeding to straighten the concave hull of a hole afterwards. Our method described in Sect. 3.2 merely needs to be adjusted in the same way as the Concave Hull Algorithm in order to create outlines of holes.

5   Results and Discussion

We demonstrate the usability of our methods in the area of building reconstruction by means of four different data sets D_1–D_4: the point sets already depicted in Figs. 4(a) (D1), 1 (D2) and 4(b) (D3), originate from airborne laser scans[1]. The point set depicted in Fig. 13(a) (D_4) is a synthetic data set featuring a star-shaped hole. Data sets D_1–D_3 represent non-convex building complexes, but D_1 is moreover near-rectangular in contrast to D_2 and D_3. For the purpose of generating Figures, we reduced the number of points depicted, but to create outlines, we always employed the original, denser point sets.

5.1   Results

The outline of D_1 generated by means of dominant directions (see Sect. 3.1) is shown in Fig. 9(a). With this data set, both options for determining the dominant directions ("peaks" and "90°") lead to the same rectangular outline that is significantly "straighter" than the concave hull for $k = 60$, and it satisfactorily represents the underlying shape. The straightening process causes changes in areas as highlighted in Fig. 9(b). These changes are small but help to approximate the building's true outline and compensate for noise and measurement errors.

Like with the original Concave Hull Algorithm, we can influence the shape of the resulting outlines by altering parameter k. Using larger k yields boundaries that look "more convex" than the ones created by using smaller k. This is illustrated in Fig. 10 by means of outlines of D_1 obtained from our orthogonal projection method (Sect. 3.2).

Boundaries of D_2 using dominant directions are given in Fig. 11. With this data set, the first dominant direction is almost independent of k, but increasing k also increases the uncertainty in the determination of the second main direction. Since the latter is also influenced by the point distribution and data noise, at larger k, the resulting outline looks somewhat skewed against the point set's shape. In this case, the outline created by using our "90°" option encloses the points more accurately. The remaining deviations result from the fact that the shape of D_2 is not truly rectangular. Figure 1 depicts an outline of this data set generated by means of orthogonal projection. Since the edges of that outline do not necessarily join orthogonally due to the dragging operations performed, it captures the point set's shape very well.

---

[1] D_3 is actually a copy of the original data set shown in Figs. 12 and 12(c), but we mirrored its points along the y-axis for developing and testing our methods.

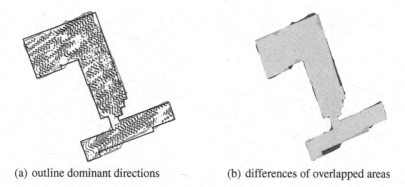

(a) outline dominant directions   (b) differences of overlapped areas

Fig. 9. The concave hulls (red) of D_1 using $k = 60$ compared to outlines generated by means of our method using dominant directions. (a) The two options "90°" (black dotted) and "peaks" (blue) lead to the same outline with substantially fewer vertices than the concave hull. (b) Areas overlapped by the concave hull (yellow + blue) and the straightened outline (yellow + green). (Color figure online)

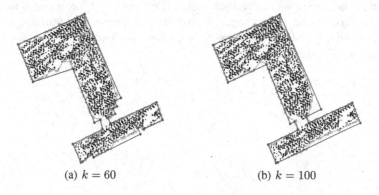

(a) $k = 60$ (b) $k = 100$

Fig. 10. Outlines of D_1 generated by using our method based on orthogonal projection (blue) compared to the concave hull (red) at two different values of k. (Color figure online)

D_3 is a real-world example of a point set acquired from a non-rectangular, non-convex building that has an atrium. Outlines of the outer and inner boundaries created by means of our methods are shown in Fig. 12. Figures 12(a) and (b) show the concave hull and outlines of holes of minimum diameter d_{min} of 100 units and 30 units, respectively. These boundaries were created by means of our dominant directions method, and the holes were found using our heuristic in Sec. 4.1. Using a smaller hole size, we are able to find and outline the additional hole in the upper right corner, whereas a larger hole size only allows for handling the atrium. However, the inner and outer boundaries produced by the straightening procedure do not faithfully represent the shape of the underlying building: The lower part of the polygon is approximated quite well when the concave hull is straightened by using the "90°" option, whereas the upper edge is matched unsatisfactorily. Using the "peaks" option, we have the opposite situation, and

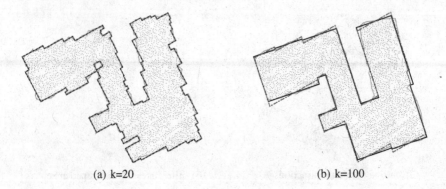

(a) k=20   (b) k=100

Fig. 11. Straightened building outlines from the concave hull (red) for $k = 20$ (a) and $k = 100$ (b) neighbors: larger k lead to smoother outlines but increase the uncertainty for computing the dominant direction. (Color figure online)

the upper edge is matched quite well. With both options, the first dominant direction is the same, but we cannot match the upper and lower edges at once.

Our method based on orthogonal projection captures the building's shape more accurately, including the upper edge, which is tilted by $\approx 12°$ towards the lower one (see Fig. 12(c)).

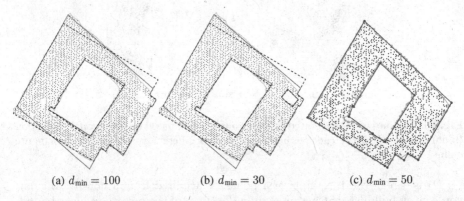

(a) $d_{min} = 100$ (b) $d_{min} = 30$ (c) $d_{min} = 50$

Fig. 12. The concave hull (red) and straightened outlines of our test data set D_3 obtained from dominant directions using "90°" (black dashed) and "peaks" (blue) options are shown in (a) and (b). (c) Shows the result obtained by orthogonal projection (blue) at $k = 20$. (Color figure online)

To put our methods for detecting and outlining holes to the test, we created the synthetic data set D_4 depicted in Fig. 13. The star-shaped hole was manually cropped and is especially challenging, because it is very sensitive to the choice of k: The edges of the spires are v-shaped and using large values of k causes them to become cut off. Using smaller values of k increases the accuracy, but also increases the uncertainty for determining the dominant directions when straightening the outline (see Figs. 13(b) and (c)). Since we only consider two dominant directions with this approach, the outline of the star loses its initial shape almost completely. By means of orthogonal projection, we get a more faithful outline of the star-shaped hole, which is not perfect, though (see Fig. 14).

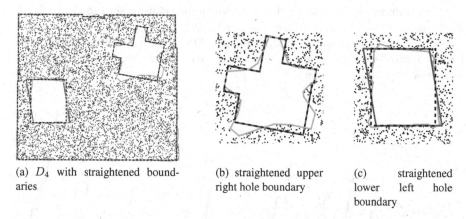

(a) D_4 with straightened boundaries

(b) straightened upper right hole boundary

(c) straightened lower left hole boundary

Fig. 13. Outlines created from our approach based on dominant directions of data set D_4 compared to the concave hull (red) at $k = 20$ and $k = 40$ for the inner and outer boundaries, respectively. With the star-shaped hole and the outer boundary, the two dominant directions obtained from the peaks of the angle histogram (blue) are the same as the ones obtained by the "90° angle" option (dashed black). In case of the other hole, the two options yield different dominant directions. (Color figure online)

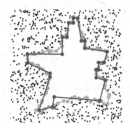

Fig. 14. Outline of the star-shaped hole in data set D_4 created by orthogonal projection (blue) in comparison to the concave hull (red) at $k = 10$. (Color figure online)

5.2 Discussion

The two methods presented in this work for generating straight outlines have different advantages over each other. Straightening concave hulls by means of dominant directions can accurately capture certain shapes of objects represented by point sets without prior knowledge about the orientation of their edges. However, it relies on the prior computation of concave hulls and is likely to yield unsatisfactory results if the shape's edges run in more than two distinct directions (cf. Figs. 12 and 13(b)).

Depending on the quality of the concave hull (i.e., mainly on the choice of k) and the point distribution, changing the histogram's bin width in our dominant directions approach can tilt the generated outline in an undesirable way (see Fig. 15). Especially by determining the second dominant direction from the histogram peaks, the choice of the bin width may cause a shift in the position of the peaks corresponding to the dominant directions. In this way, the second direction is tilted by about the same amount. The first direction can suffer from the same problem if the point set is of rather rotational symmetric shape (e.g., square or circular), of course.

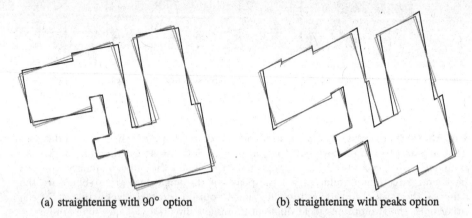

(a) straightening with 90° option   (b) straightening with peaks option

Fig. 15. As the bin width in the histograms for determining dominant directions increases (blue: 1°, red: 5°, black: 10°), the approximation of the actual directions of the outline edges becomes less distinct, which may cause the outline to become tilted. (Color figure online)

Our approach based on orthogonal projection and the dragging of edges can create faithful outlines of shapes whose edges run in more than two distinct directions, but it requires hints about the initial orientation of the shape and possibly contained holes. These hints may be obtained automatically (see Sect. 3.2) in some situations, but otherwise they have to be provided by the user. With this method it is not necessary to compute concave hulls or other intermediate polygons in advance, though.

The choice of the size k of the neighborhood considered is crucial for both our methods, but it is also subject to the distribution of the underlying points. While the Concave Hull Algorithm starts at the smallest $k = 3$ and increases the value if the outline has self-intersections or does not include all points of the set, this proceeding is not guaranteed to succeed with our methods, because our angular outlines may deliberately exclude points. In our experiments, we found that using k of about 5–10% of the number of points in the set is a reasonable initial choice, if the point distribution (excluding holes to be detected) is approximately uniform.

The decision which outline matches the underlying object "best" is up to the user who can incorporate background information about its shape. In case of the building represented by D_1 (see Figs. 9 and 10), for instance, we cannot decide whether the object is a single building complex or if there are two separate buildings, possibly linked by a corridor, if we solely rely on the information provided by the point set.

6   Conclusion and Future Work

In this article, we presented two methods based on the Concave Hull Algorithm to create straight, angular and non-convex outlines of 2D point sets and possibly contained *holes*. The outlines of point sets representing man-made structures, like buildings, created by means of our methods are "better" than concave hulls or boundaries obtained from α-shapes to the effect that our methods capture straight edges more accurately and feature only few, distinct angles.

The employment of straightening concave hulls by means of dominant directions to generate outlines is preferable if the orientation of the underlying shape is unknown or cannot be provided, or if the shape is known to have edges running in only two distinct directions, e.g., in case of rectangular buildings. By using our dominant directions approach in a recursive way as described in Sect. 3.1, we might be able to incorporate additional directions and obtain even better results.

If information about the orientation of the shape is available, our approach based on orthogonal projection can produce more faithful boundary polygons, especially if the underlying shape features edges running in more than two distinct directions. Moreover, this approach forgoes the computation of concave hulls or any other intermediate polygon.

The most challenging task is the reliable detection of holes. We encountered this problem by means of a heuristic that is subject to sampling problems and may have several theoretical limitations. Although we did not encounter major problems due to a rather specific, practical application, we would like to improve on these potential issues. For example, holes might be located close to the outer boundary of the associated point set, and they might thus share boundaries with the exterior outline. This could cause our current algorithms for bordering holes to fail, because the outline might "leave" the hole and intersect the exterior boundary.

An alternative approach for detecting holes based on the local density of point sets might be the employment of wavelet transforms. By using Haar transforms and techniques from the area of computer vision, for instance, we might also be able to locate points on the border of holes from where to start the generation of their outlines. Its application to our problem and irregular point sets might become challenging, though, because to our knowledge, efficient implementations that are currently available usually rely on point sets given on regular grids (e.g., image pixels). This problem could be overcome by rasterizing the point sets, of course, but would also introduce additional complexity.

Moreover, we only considered "isolated" point sets so far. In the application of building reconstruction, for example, this requires prior segmentation of the source data. We would like to forgo this step and determine spacings or gaps between distinct clusters of points in larger point sets to segment them while creating the corresponding outlines. This may be accomplished by a combination of modifying and constraining the neighborhoods employed by our approaches or the Concave Hull Algorithm and adjustments of our heuristic for hole detection to this problem, but remains future work to do.

References

1. Vosselman, G.: Building reconstruction using planar faces in very high density height data. In: International Archives of Photogrammetry, Remote Sensing and Spatial Information Sciences, pp. 87–92 (1999)
2. Moreira, A., Santos, M.Y.: Concave hull: a k-nearest neighbours approach for the computation of the region occupied by a set of points. In: Proceedings of 2nd International Conference on Computer Graphics Theory and Applications (GRAPP), pp. 61–68 (2007)
3. Douglas, D., Peucker, T.: Algorithms for the reduction of the number of points required to represent a digitized line or its caricature. Can. Cartogr. 10, 112–122 (1973)
4. Ramer, U.: An iterative procedure for the polygonal approximation of plane curves. Comput. Graph. Image Process. 1, 244–256 (1972)
5. Galton, A., Duckham, M.: What is the region occupied by a set of points? In: Raubal, M., Miller, H.J., Frank, A.U., Goodchild, M.F. (eds.) GIScience 2006. LNCS, vol. 4197, pp. 81–98. Springer, Heidelberg (2006). doi:10.1007/11863939_6
6. de Berg, M., Cheong, O., van Krevald, M., Overmars, M.: Computation Geometry, 3rd edn. Springer, Heidelberg (2008)
7. Edelsbrunner, H., Kirkpatrick, D., Seidel, R.: On the shape of a set of points in the plane. IEEE Trans. Inf. Theory 29, 551–559 (1983)
8. Duckham, M., Kulik, L., Worboys, M., Galton, A.: Efficient generation of simple polygons for characterizing the shape of a set of points in the plane. Pattern Recogn. 41, 3224–3236 (2008)
9. Jarvis, R.A.: On the identification of the convex hull of a finite set of points in the plane. Inf. Process. Lett. 2, 18–21 (1973)
10. Asaeedi, S., Didehvar, F., Mohades, A.: Alpha convex hull, a generalization of convex hull. The Computing Research Repository (CoRR) abs/1309.7829 (2013)
11. Wang, O., Lodha, S.K., Helmbold, D.P.: A Bayesian approach to building footprint extraction from aerial LIDAR data. In: Proceedings of 3rd International Symposium on 3D Data Processing, Visualization, and Transmission (3DPVT 2006). pp. 192–199. IEEE Computer Society, Washington, DC (2006)
12. Bendels, G.H., Schnabel, R., Klein, R.: Detecting holes in point set surfaces. In: Proceedings of 14th International Conference in Central Europe on Computer Graphics, Visualization and Computer Vision, vol. 14 (2006)
13. Wang, J., Oliveira, M.M.: Filling holes on locally smooth surfaces reconstructed from point clouds. Image Vis. Comput. 25, 103–113 (2007)
14. Wu, X., Chen, W.: A scattered point set hole-filling method based on boundary extension and convergence. In: 11th World Congress on Intelligent Control and Automation, pp. 5329–5334. IEEE (2014)

Sketching 2D Character Animation Using a Data-Assisted Interface

Priyanka Patel, Heena Gupta, and Parag Chaudhuri[✉]

Department of Computer Science and Engineering,
Indian Institute of Technology Bombay, Mumbai, India
{priyapatel,heenagupta,paragc}@cse.iitb.ac.in
https://www.cse.iitb.ac.in

Abstract. This paper presents a system to assist novice animators in creating 2D, hand-drawn, character animation. This system, called TraceMove, helps the user to sketch the character and to animate it. Frames from recorded videos of human performers are stored in a database. This is subsequently used to provide a static pose hint to the users, in the form of silhouette suggestions, as they sketch the character. The desired pose of the character is thus easy to sketch for the user as they can trace and draw over the generated pose hint. The system then predicts the next frame of the animation as a moving pose hint. This is done with the help of a user marked skeleton on a single sketched pose and a motion capture database. The sketch predicted by the system can be edited by the animator as desired. The moving and static pose hints used together, let novice artists and animators easily generate hand-drawn, 2D animated characters.

Keywords: Hand-drawn character animation · Sketch-based system · Data-assisted

1 Introduction

Hand-drawn 2D character animation is a difficult skill to master. It requires years of practice to acquire the necessary expertise required to capture fluent motion in a sequence of 2D sketches. Novice artists struggle to master the subtle concepts required to convey a certain mood or idea through their animation. The significant amount of effort required to successfully create an illusion of life [16] from sketches, often detracts and frustrates novice animators.

Our system, called TraceMove, provides hints to the animator as they draw the sketch of the character for every frame of the animation, thereby making the creation of the animation easier. Two kinds of hints are provided. The first hint is

Electronic supplementary material The online version of this chapter (doi:10.1007/978-3-319-64870-5_7) contains supplementary material, which is available to authorized users.

© Springer International Publishing AG 2017
J. Braz et al. (Eds.): VISIGRAPP 2016, CCIS 693, pp. 135–152, 2017.
DOI: 10.1007/978-3-319-64870-5_7

background silhouette image of the pose that the animator is trying to sketch at the current frame. This hint is a prediction generated by the system based on the sketch strokes that the animator has drawn on the frame so far and a database of images containing humans in various poses. The choice of following the hints provided by the system is entirely up to the animator. They can disregard the hint and sketch freely if so desired. The second hint is a prediction of the sketched pose in the next frame of the animation, once the sketch of the current frame is finished. Sketching characters in motion requires a sense of timing and rhythm of the movement [19], which is very hard to get for novice artists. We take the help of motion capture data to predict the next sketched frame for the animator. They can draw over this prediction, modify it as they wish, and proceed (see Fig. 1). These two hints can be interleaved and used as desired on the same or different frames of the animation, thereby giving the animator a lot of flexibility and complete control during the creation process.

We discuss the current literature available in this area in the next section. We follow this up with an overview of our system in Sect. 3. Subsequently, we present details of the static pose and moving pose hints, in Sects. 3 and 4. Our system was used by numerous novice animators. Their experience and feedback is discussed in the Sect. 5. We present some sketch sequences generated using our system in Sect. 6. Finally, we conclude with a discussion of the limitations of the system and directions for future work in Sect. 7.

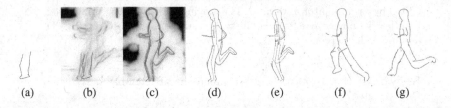

| (a) | (b) | (c) | (d) | (e) | (f) | (g) |

Fig. 1. (a) The animator starts a sketch, (b) the static pose hint updates the silhouette depending on the sketch, (c) the animator can trace over it to complete the sketch, (d)–(e) the moving pose hint predicts the sketch for the next frame from the current drawn sketch, (f)–(g) the animator can continue the process and easily complete the animation.

2   Related Work

Sketching is an art form that is ubiquitous in animation, painting and design. There have been many systems developed to make sketching accessible to novice users. The iCanDraw interface [6] helps the user by providing step by step guidance to draw the human face using a reference image and written instruction. Other systems try to match sketch strokes with images [3,9] and then these can be used to guide the sketching. ShadowDraw [11] is a sketching system that presents a shadow image that is a blend of matching object images from a database. The user can use the shadow as a draw-over guide to create a sketch of that object. A part of our system is based on ShadowDraw, in which we generate the static pose hint using methods from that paper. A gesture based rapid 3D sketching system is presented in [1] which allows novice users to sketch designs for complex objects.

Sketch-based interfaces for modelling and animation are also an active topic of research in computer graphics. Sketches have been used for modelling 3D objects from sketches by novice users [8], from design sketches [20] or for creating layered additions to exiting 3D models [5]. Sketches have also been used to pose 3D character models [14].

Sketches have also been used to drive 3D animation. The input to algorithm described in [10] is a set of hand-drawn frames. The method uses motion capture data to transfer the 2D animation to 3D, while maintaining the unique style of the input animation. In an earlier work, [4] takes user drawn stick figures as input, extracts best matched 3D skeleton poses for the input figures and generates a 3D animation. Inspired by traditional, hand-drawn animation, silhouette curves are used to stylize existing 3D animations by [13]. Motion Doodles [17] is a system that takes a stick figure of character and a motion path as input, and finally animates the figure on the path. The LifeSketch system [21] also outputs 3D animation of 2D sketch given as an input by user. However, this work makes an assumption that object is blobby and all the parts of objects are visible. More recent work by [12] creates a 3D model of an articulated character from multiple sketches and then allows animation of the character in 3D.

In other prior work [15] describe a skeleton driven 2d animation technique. The system is provided with one image and then the user sketches the skeleton for the subsequent frame. The system deforms the character according to the new position of the skeleton and creates animations automatically. However, this is very cumbersome because the user is required to draw the skeleton for all the frames and no help is provided for sketching the actual poses. Thus, we find that all prior work to our knowledge, either requires the sketch as input for the animation or provides no feedback assistance to the novice animator in creating the animation.

In contrast, our TraceMove system tries to help novice animators in sketching 2D animations. For this we not only help in the static pose sketches at individual

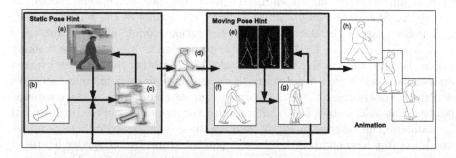

Fig. 2. Overview of the TraceMove system.

frames but also predict sketched poses for subsequent frames. These hints allow the animator to create sketched 2D character animations very quickly.

3   System Overview

An overview of the TraceMove system is shown in Fig. 2. We start by pre-processing an image database of human poses at various stages of multiple motions to edge descriptors (Fig. 2(a)). This is done only once for the entire database. Then the animator can start sketching and the static pose hint is updated, as required, based on the current sketch available to the system and the processed image database. The static pose hint is generated by blending the top ten edge figures from database that best match the sketched pose, in the edge descriptor space [11]. This is shown in Fig. 2(b)–(c). Once the animator is satisfied with the sketch (Fig. 2(d)), it is passed on to the moving pose hint generation module.

We have also pre-processed motion capture data to obtain 2D projected motion capture data (Fig. 2(e)). Now the animator draws a skeleton on the sketch by clicking joint positions on the sketch (Fig. 2(f)). This has to be done only for one pose for the entire animation. The order of clicking is shown to the animator and automatically establishes joint correspondences to the skeleton hierarchy used in the motion capture. The skeleton on the sketch is used to identify a best matching pose from the motion capture data, and the subsequent poses to the best matching pose, are used to find corresponding subsequent poses for the skeleton on the sketch, and by consequence of the sketch itself (Fig. 2(g)). At this point, the animator can choose to manually edit the predicted sketched pose, again with the help of the static pose hint or without it. This process is repeated to get sketches for all the frames of the animation.

It should be noted that the animator can choose to ignore the static and moving pose hints completely at any stage, or use them at any stage in the creation process. So the system does not stifle the freedom of the animator, but provides enough help to the novice animator to be able to create convincing sketched character animations.

The static pose hint generation module of our system is based on Shadow-Draw [11]. Our static pose hint is like the shadow image generated in that work. We have implemented our system from scratch and have made some changes to the original ShadowDraw idea which improve the quality of the generated hint.

The first part of the static pose hint generation module involves processing a database of figures of human in various poses during a motion. For walking people, we used the CASIA Gait Database [18]. For other motions, we created our own database by recording videos of various motions on 6 different users. We used the frames of these videos as figures in our database, In total the combined database has 3052 frames for 6 different kinds of motion. Example images from the database can be seen in Fig. 3. The database is processed offline, in a pre-processing step to generate a database of patch-features from the edge figures of the figures in the original database. These descriptors are then used to generate the static pose hint while the user sketches.

Fig. 3. Example images from the image database.

3.1 Generating the Database of Patch-Features

The original figures in the database are converted to edge figures, post cropping and size normalization. We use [7] to extract long edges from the figures. This is important because it is found that while sketching it is natural to draw the long edges first. So we need an algorithm that can prioritize long edges. ShadowDraw [11] uses the work presented in [2] for extracting edges. Our implementation of the same gave either faint or very thick edges, so we used the different method mentioned above.

This is followed by dividing the edge image into overlapping patches and computing a BICE descriptor for each patch [22]. We want to match the user's sketch to the figures in the database, in descriptor space. However, computing a match directly on the descriptor is expensive so it is converted to a sequence of k values, each generated by applying k different min-hash functions to the descriptor of the patch. Each sequence of these k values is a patch-feature. This is repeated n times, using a different set k hash functions each time, to get n patch-features for each patch descriptor. Therefore while matching, a potential input patch has to match multiple instances of the same descriptor to be considered a good match. This reduces both false positives and false negatives. We have used $k = 3$ and $n = 20$. We store the patch-features with a reference to the original image to which they belong, and the patch location coordinates in the original image in another database.

3.2 Generating the Static Pose Hint

As soon as the animator finishes sketching a stroke, an image of the canvas is converted to patch-features and only patches containing the strokes are matched to the database created in the previous section. Top 10 figures from which maximum number of patch-features match the patch-features from the input sketch are aligned and blended. This blended image is then multiplied with its own blurred version to strengthen edges that match in position between them and weaken others. This forms the static pose hint image. It is displayed on the drawing area, underlying the animator's sketch, and can be updated in real-time as the animator sketches. We have, however, found this to be distracting during

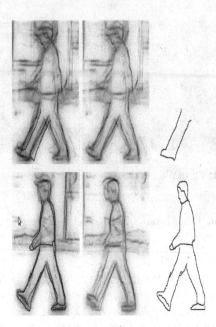

Fig. 4. First column shows the drawn sketch overlaid on the static pose hint, second and third columns show the static pose hint and the drawn sketch separately.

use. So we give the animator an option of updating and displaying the static pose hint on the canvas at the push of a button, instead of updating it continuously on sketching. The last updated static pose hint is displayed on the side in a smaller window so that the animator still has a reference for the pose being sketched but the drawing area is not obstructed by it. An example of the static pose hint is shown in Fig. 4

4   Moving Pose Hint

After successfully drawing the character in a particular pose, the animator now wants to sketch the pose in the next frame of the animation. The moving pose hint is meant to help with this. We start with a database of motion capture clips. This database currently has 6 different kinds of motions and a total of 625 frames. We project the 3D motion capture data to 2D, using a camera that projects the root node of the motion capture skeleton to the origin of the image coordinate system. We fix other camera parameters to give us desired projections of the motions being processed. It should be noted that we can only generate moving pose hints if the sketch of the character is from the a viewpoint that is close to the camera viewpoint used to generate the 2D projections of the motion capture data. The creation of the 2D projected motion capture database is a pre-processing step and has to be performed only once.

4.1   Skeleton Matching

The animator marks the skeleton on the sketch of the current pose by clicking the joint positions on the sketch. The joints have to be clicked in a particular order that is shown in the interface during the clicking (as shown in Fig. 5(a)). This has to be done only once for a single sketched pose of the entire animation and is very simple to do. The ordered clicking automatically sets up correspondence between the user marked skeleton and the motion capture skeleton.

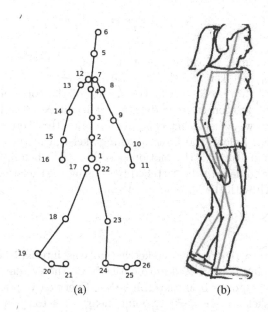

(a)   (b)

Fig. 5. (a) Order in which the skeleton nodes have to marked by the user, (b) Joint nodes for left arm have to marked even when it is occluded.

The animator has to mark the entire skeleton even if a part of the body is occluded in the current sketch. For example, as shown in Fig. 5(b), one arm of the character may be occluded but all the skeleton nodes for that limb have to marked approximately.

After the skeleton is marked on the sketch, its bones are re-scaled to match the bone length of the skeleton in the motion capture database. This is necessary because the bone lengths of the motion capture skeleton are fixed, while bone lengths of the sketch skeleton can vary with the sketch. Therefore, we determine scale factors needed to scale the sketched skeleton bones appropriately. If S_i is the scale factor for i^{th} bone, L_i^{sketch} and L_i^{mocap} are bone lengths of ith bones of the skeleton on the sketch and in the motion capture database, respectively.

$$S_i = \frac{L_i^{sketch}}{L_i^{mocap}}$$

We will also calculate the inverse scale factor $IS_i = 1/S_i$ that is used later in our calculations. This scale factor is applied to each bone of skeleton on the sketch.

After scaling the skeleton on the sketch, the system searches for the best matching frame in the motion capture data such that the pose of the skeleton in that frame best matches the sketched pose. This is done by minimizing, over all frames, a distance metric that sums the Euclidean distance between the corresponding root-centred joint coordinates of the sketch and motion capture skeleton joints.

$$D_t = \min_t \{ \sum_k dist(C_k^{sketch}, C^{mocap}, t_k) \}$$

Here D_t is the minimum value of distance metric, C_k^{sketch} is the coordinate of the kth joint of the sketched skeleton with the root of the skeleton as the origin, and the C_k^{mocap} is the similar coordinate of the corresponding joint of the skeleton in the motion capture data and t iterates over all the frames of the database. Now we can predict the next pose for the sketch from the pose of the skeleton that follows the best matching skeleton in the motion capture data. This predicted sketch is the motion pose hint. But before we can do that we need to be able to deform the sketched pose using the skeleton. This requires us to rig the sketch with the sketched skeleton.

4.2 Rigging

In order to facilitate rigging, every sketch stroke is internally converted to a Bézier curve. Rigging is computed automatically by the system on the basis of distance of the curve points from skeleton bones. Every curve point is associated with at least one skeleton bone. Curve points near a skeleton joint are associated to both bones at the joint. Weights are assigned to the curve points by inverse weighting them by their distance from the bone.

Due to ambiguity of 2D projection, there are cases where automatic rigging incorrectly associates curves with skeleton bones. This causes erroneous deformation of the sketch when the skeleton moves. In such cases, animator can correct this simply by going to manual rig mode and selecting the curve that need to be re-associated and the bone with which it needs to be associated by clicking on it. This will detach the curve from its initial bone and re-associate it to the new bone.

Figure 6(a) shows an incorrect automatic rigging output. Curves associated with different skeleton bones are of different colour. The curves of the torso get wrongly associated to an arm and move backwards as the arm swings back in a subsequent frame in Fig. 6(b), as indicated by the red arrows. Figure 6(c)–(d) shows the corrected rigging in the initial frame and how it stays in position correctly in the generated sketch for the following frame, as indicated by the green arrows.

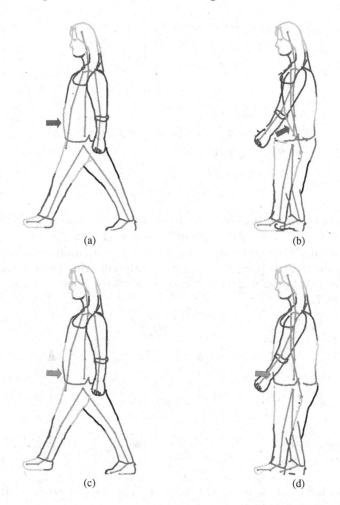

(a)   (b)

(c)   (d)

Fig. 6. (a), (b) Automatic Rigging, (c), (d) Corrected Rigging. (Color figure online)

4.3   Binding Matrix Calculation

The curves are defined in screen space and skeleton joints are defined in their own local frame. The curve associated with a particular bone need to be defined in the same frame as that of the bone so that all the transformation that are applied to joint, when applied to curve will move the curve along with the bone. To define the curve points in the joint space with which they are associated we need to find the binding matrix for all the skeleton joints. This binding matrix, B when multiplied to the curve points, transfer them to the joint local coordinate frame, with Y-axis along the bone, X- axis perpendicular to the bone and parent node as the origin. The binding matrix for the kth joint is calculated as

$$B_k = \begin{bmatrix} cos\theta & -sin\theta & 0 \\ sin\theta & cos\theta & 0 \\ 0 & 0 & 1 \end{bmatrix} \begin{bmatrix} 1 & 0 & -J_kx, \\ 0 & 1 & -J_ky \\ 0 & 0 & 1 \end{bmatrix}$$

Here

$$L_{k+1} = J_{k+1} - J_k$$
$$D = sqrt(L_{k+1}x * L_{k+1}x + L_{k+1}y * L_{k+1}y)$$
$$cos\theta = L_{k+1}y/D$$
$$sin\theta = L_{k+1}x/D$$

$$(1)$$

J_{k+1} is the coordinate of $k+1^{th}$ joint and L_{k+1} is its coordinates with respect to its parent, i.e., k^{th} joint. Now, for curve associated with k^{th} joint, the binding matrix is the product of a rotation matrix and a translation matrix. First, the translation matrix is applied to the curve which will bring the kth joint to the origin and then rotation matrix is applied to align the bone with the Y-axis as shown in Fig. 7.

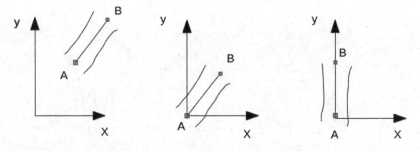

Fig. 7. (a) A bone and the associated curve, (b) Translated to origin (c) Rotated so that bone lies along Y-axis

4.4   Generating the Moving Pose Hint

We have found the frame from the motion capture database that best matches the sketched skeleton. We also know the pose of the skeleton in the frame that follows the best frame in the motion capture data. The system now finds the translation difference in coordinates for these two frames in the motion capture database and applies that difference to the current sketched skeleton, after inverse scaling it to drawn skeleton. This is done using Algorithm 1.

Here T_k is the translation difference for k^{th} joint. J_k^{t+1} and J_k^t are the 2D coordinates of kth joint for next frame and current frame of the motion capture skeleton respectively. Note that the translation difference is calculated by taking parent joint node as origin. Similarly in the subsequent step, when applying this difference to sketched skeleton joint J_{k+1}, its parent joint J_k is shifted to origin.

Algorithm 1. Generate Next Sketch Skeleton.

1: **for** every joint k of the motion capture skeleton in frames t and $t+1$ **do**
2:    $T_k = (G_{k+1}^{t+1} - G_k^{t+1}) - (G_{k+1}^t - G_k^t)$
3: **end for**
4: **for** every bone i of the sketched skeleton between joints J_{k+1} and J_k **do**
5:    $J_{k+1}^{t+1} = (J_{k+1}^t - J_k^t) + IS_i \cdot T_k + J_k^{t+1}$
6: **end for**

After applying the difference to the $(k+1)^{th}$ joint, the kth joint is shifted back to its new position that is calculated after applying the translation difference to it. The coordinates given by the J^{t+1}'s are the new predicted position of the joints of the sketched skeleton in the next frame. An example of the skeleton generated by our algorithm can be seen in Fig. 8.

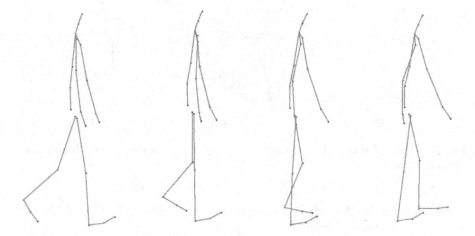

Fig. 8. First frame is the drawn skeleton, rest of the frames are generated using Algorithm 1.

We find the transformation matrix from the current sketch skeleton to the next generated sketched skeleton for every skeleton joint. We apply this transformation matrix and the binding matrix to every curve associated with a particular joint, to generate the moving pose hint. This is illustrated in Fig. 9.

4.5 Depth Adjustment

Since a sketch is 2D, some of the curves that were visible before may go behind the body and they will still be visible. For correct occlusion handling, these have to be erased via manual editing. While some new curves that were occluded before may be visible in the new frame. The animator will have to draw them. For this purpose, she can take help of the static pose hint again, if required. This

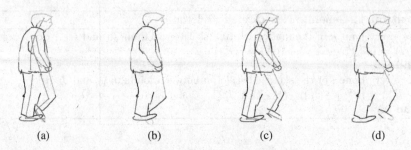

(a)   (b)   (c)   (d)

Fig. 9. (a) Current frame with skeleton, (b) Current frame without skeleton, (c) Next Frame with skeleton, (d) Next Frame without skeleton.

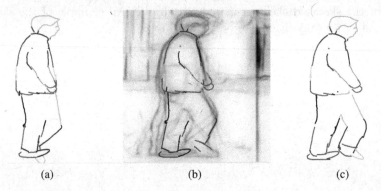

(a)   (b)   (c)

Fig. 10. (a) First frame, (b) Next frame without depth adjustment, (c) Next frame with depth adjustment.

is shown in Fig. 10. The newly drawn curves get automatically attached to the skeleton via automatic rigging.

5   User Survey

A group of 11 novice animators used the TraceMove system to create an animation of a running character. Out of these 6 users were female and 5 were male. When asked whether they like to draw or not, 4 of these users answered in the negative. Only 2 of these users had ever made a 2D animation before. In order to give the users an idea of how much effort was needed in creating a 2D animation, they were asked to sketch a running character on paper without any other assistance. Then they were asked to use TraceMove to create the animation of a running character. Their animations and the time they required to create them were recorded. After using our system, all 11 users felt that the TraceMove system made 2D animation process easier. We asked the users to rate the TraceMove system on a scale of 1 to 10, with 1 being the least helpful in the 2D animation process and 10 being the most. The user ratings are presented in Fig. 11.

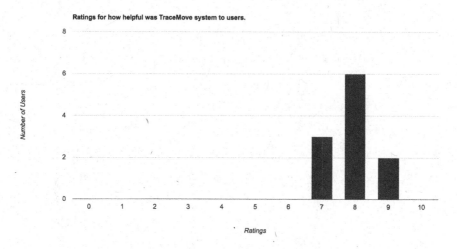

Fig. 11. User ratings for the TraceMove system.

Table 1. Time taken (in minutes) by the users to sketch a 2D run animation using the TraceMove system in 3 attempts and using pencil and paper. The users sketched 17 frames on when using the TraceMove system and 6 frames on paper.

User no.	Timing (TraceMove)			Timing (paper)
	Attempt 1 (min)	Attempt 2 (min)	Attempt 3 (min)	
1	27	22	18	
2	22	14	13	
3	26	25	20	13
4	20	14	13	6
5	12	12	12	21
6	12	14		
7	31			
8	35			
9	16			16
10	23			16
11	15			18

Table 1 presents the time taken by each user to create an animation on a computer using our system. Some users also worked with our system multiple times, each time creating a 17 frame run animation sequence. Users who were comfortable drawing on paper sketched frames of a similar animation on paper using a pencil, without any assistance. On paper, all users sketched 6 frames. It can be seen that with more practice on our TraceMove system, the time taken by a user to create an animation consistently went down. Also, the number of frames sketched to complete the animation stayed the same or went down with

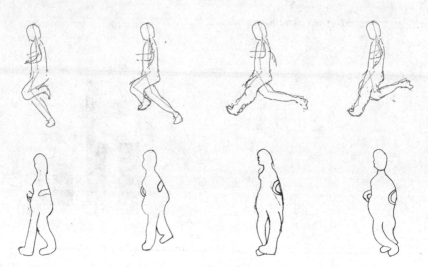

Fig. 12. Frames from attempt 3 of user 3 are in the top row. Frames drawn on paper by the same user are in the bottom row.

each attempt. The average time to sketch one frame can be seen to vary from 1 to 2 min. The animations produced on paper took much longer. The time to sketch one frame on paper varies from 2 to 3 min for most users. Frames from two of the animations produced using our system and the sequence of sketches drawn by those users on paper are shown in Figs. 12 and 13.

Fig. 13. Frames from attempt 2 of user 5 are in the top row. Frames drawn on paper by the same user are in the bottom row.

Though the response from the novice users about the system is in general positive, they did give us feedback about ways in which we can improve the system. One of the features requested was an easy undo option. The system already allows the user to delete sketched strokes but perhaps the interface to this feature is not very intuitive. Also, the automatic rigging should be improved to decrease user interaction required to correct a sketch. Some of the users commented that this was the first time they attempted hand drawn animation and the system helped them in understanding how it is done.

6   Results

We present examples of seven sketched animations generated using our Trace-Move system for various kinds of motion (Figs. 14, 15, 16, 17, 18, 19 and 20). These were all created by two novice animators who had no prior experience in hand-drawn figure animation. The actual animations can be seen in the supplementary video submitted with this paper.

Fig. 14. Frames from walk animation.

Fig. 15. Frames from another walk animation with a different character.

Fig. 16. Frames from a run animation.

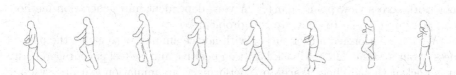

Fig. 17. Frames from a skipping animation.

Fig. 18. Frames from a jump animation.

Fig. 19. Frames from a different jumping animation.

Fig. 20. Frames from a backward walk animation.

7   Conclusion

This paper has presented a system to assist novice animators in sketching 2D character animations. The system generates a static pose hint to help in the sketching of a particular pose of a character in a frame of the animation and also generates a moving pose hint that helps sketch the subsequent frame of the animation, given the current frame. These hints are generated with the help of pre-processed, stored databases of images and motion capture data.

The current system has certain limitations. The sketches for which the moving hint can be generated must be from a viewpoint that is close to the one used in generating the 2D projected motion capture database. This can be overcome by automatic viewpoint detection on the sketch and then using the corresponding view of the 3D motion capture data at runtime. Also, during the entire sketch animation that can be generated by the system, the camera orientation relative to the character cannot change much. The hint generation modules cannot work across view-point changes. A view-dependent hint generation method can possibly be used to alleviate this problem.

We have performed a user survey with novice animators to gauge the usefulness of our system. The feedback has been positive and we feel encouraged by it. It is unclear though, how TraceMove performs as an animation learning system. We would like to understand whether using TraceMove also improves the ability of the user to animate on paper. It would be instructive to test the system with expert animators, in order to understand the efficacy of our interaction paradigms. We currently have no way to quantitatively measure the aesthetic quality of the generated animation, but the novice animators who have used our system agree that it made the task of creating the animation easier for them and gave them a handle on a skill that they would have otherwise struggled to master.

References

1. Bae, S., Balakrishnan, R., Singh, K.: Everybodylovessketch: 3D sketching for a broader audience. In: Proceedings of ACM Symposium on User Interface Software and Technology, pp. 59–68, October 2009
2. Bhat, P., Zitnick, C.L., Cohen, M., Curless, B.: Gradientshop: a gradient-domain optimization framework for image and video filtering. ACM Trans. Grap. **29**(2), 10:1–10:14 (2010)
3. Chen, T., Cheng, M.M., Tan, P., Shamir, A., Hu, S.M.: Sketch2photo: internet image montage. ACM Trans. Graph. **28**(5), 10:1–10:14 (2009)
4. Davis, J., Agrawala, M., Chuang, E., Popović, Z., Salesin, D.: A sketching interface for articulated figure animation. In: Proceedings of the 2003 ACM SIGGRAPH/Eurographics Symposium on Computer Animation, pp. 320–328 (2003)
5. De Paoli, C., Singh, K.: Secondskin: sketch-based construction of layered 3D models. ACM Trans. Graph. **34**(4), 126:1–126:10 (2015)
6. Dixon, D., Prasad, M., Hammond, T.: iCanDraw: using sketch recognition and corrective feedback to assist a user in drawing human faces. In: Proceedings of the SIGCHI Conference on Human Factors in Computing Systems, pp. 897–906 (2010)
7. Dollr, P., Zitnick, C. L.: Structured forests for fast edge detection. In: Proceedings of the 2013 IEEE International Conference on Computer Vision, ICCV 2013, pp. 1841–1848. IEEE Computer Society (2013)
8. Igarashi, T., Matsuoka, S., Tanaka, H.: Teddy: a sketching interface for 3D freeform design. In: Proceedings of the 26th Annual Conference on Computer Graphics and Interactive Techniques, SIGGRAPH 1999, pp. 409–416 (1999)
9. Jacobs, C.E., Finkelstein, A., Salesin, D.H.: Fast multiresolution image querying. In: Proceedings of the 22nd Annual Conference on Computer Graphics and Interactive Techniques, SIGGRAPH 1995, pp. 277–286 (1995)
10. Jain, E., Sheikh, Y., Hodgins, J.: Leveraging the talent of hand animators to create three-dimensional animation. In: Proceedings of the 2009 ACM SIGGRAPH/Eurographics Symposium on Computer Animation, pp. 93–102 (2009)
11. Lee, Y.J., Zitnick, C.L., Cohen, M.F.: Shadowdraw: Real-time user guidance for freehand drawing. ACM Trans. Graph. **30**(4), 27:1–27:10 (2011)
12. Levi, Z., Gotsman, C.: ArtiSketch: a system for articulated sketch modeling. Comput. Graph. Forum **32**(2), 235–244 (2013)
13. Li, Y., Gleicher, M., Xu, Y.Q., Shum, H.Y.: Stylizing motion with drawings. In: Proceedings of the 2003 ACM SIGGRAPH/Eurographics Symposium on Computer Animation, SCA 2003, pp. 309–319 (2003)
14. Öztireli, A.C., Baran, I., Popa, T., Dalstein, B., Sumner, R.W., Gross, M.: Differential blending for expressive sketch-based posing. In: Proceedings of the 12th ACM SIGGRAPH/Eurographics Symposium on Computer Animation, pp. 155–164 (2013)
15. Pan, J., Zhang, J.J.: Sketch-based skeleton-driven 2D animation and motion capture. In: Pan, Z., Cheok, A.D., Müller, W. (eds.) Transactions on Edutainment VI. LNCS, vol. 6758, pp. 164–181. Springer, Heidelberg (2011). doi:10.1007/978-3-642-22639-7_17

16. Thomas, F., Johnston, O.: The Illusion of Life: Disney Animation. Hyperion, New York (1995)
17. Thorne, M., Burke, D., van de Panne, M.: Motion doodles: an interface for sketching character motion. In: Proceedings of ACM SIGGRAPH 2004, pp. 424–431 (2004)
18. Wang, L., Tan, T., Ning, H., Hu, W.: Silhoutte analysis based gait recognition for human identification. IEEE Trans. Pattern Anal. Mach. Intell. (PAMI) **25**(12), 1505–1518 (2003)
19. Williams, R.: The Animator's Survival Kit-Revised Edition: A Manual of Methods, Principles and Formulas for Classical, Computer, Games, Stop Motion and Internet Animators. Faber & Faber Inc., London (2009)
20. Xu, B., Chang, W., Sheffer, A., Bousseau, A., McCrae, J., Singh, K.: True2Form: 3D curve networks from 2D sketches via selective regularization. ACM Trans. Graph. **33**(4), 131:1–131:13 (2014)
21. Yang, R., Wnsche, B.C.: Life-sketch: a framework for sketch-based modelling and animation of 3D objects. In: Proceedings of the Eleventh Australasian Conference on User Interface, vol. 106, pp. 61–70 (2010)
22. Zitnick, C.L.: Binary coherent edge descriptors. In: Daniilidis, K., Maragos, P., Paragios, N. (eds.) ECCV 2010. LNCS, vol. 6312, pp. 170–182. Springer, Heidelberg (2010). doi:10.1007/978-3-642-15552-9_13

Skin Deformation Methods for Interactive Character Animation

Nadine Abu Rumman[1][(✉)] and Marco Fratarcangeli[2]

[1] Department of Computer, Control, and Management Engineering,
Sapienza University of Rome, Via Ariosto 25, 00185 Rome, Italy
aburumman@dis.uniroma1.it
[2] Department of Applied Information Technology,
Chalmers University of Technology, Lindholmsplatsen 1,
41296 Gothenburg, Sweden
marcof@chalmers.se
http://www.dis.uniroma1.it/~aburumman/
http://www.fratarcangeli.net/

Abstract. Character animation is a vital component of contemporary computer games, animated feature films and virtual reality applications. The problem of creating appealing character animation can best be described by the title of the animation bible: *"The Illusion of Life"*. The focus is not on completing a given motion task, but more importantly on how this motion task is performed by the character. This does not necessarily require realistic behavior, but behavior that is believable. This of course includes the skin deformations when the character is moving. In this paper, we focus on the existing research in the area of skin deformation, ranging from skeleton-based deformation and volume preserving techniques to physically based skinning methods. We also summarize the recent contributions in deformable and soft body simulations for articulated characters, and discuss various geometric and example-based approaches.

Keywords: Character animation · Deformation · Skinning · Skeleton-based animation · Volume-preserving · Physics-based animation

1 Introduction

Producing believable and compelling skin deformations for articulated characters is a multi-disciplinary problem, which can be divided into three main problems: generating high-quality skin deformations, simulating skin contact in response to collisions, and producing secondary motion effects such as flesh jiggling when a character moves. Traditionally, bone transformations describe the position and orientation of the joints, and the skin deformation is computed by linearly blending bone transformations to the skin. However, such simple and linear blending to the bone transformations cannot be expected to capture complex deformations. In contrast, by employing a physically based method into the skinning process,

© Springer International Publishing AG 2017
J. Braz et al. (Eds.): VISIGRAPP 2016, CCIS 693, pp. 153–174, 2017.
DOI: 10.1007/978-3-319-64870-5_8

the believability of character motions is highly enhanced. Physics-based simulations manage to bring skeleton-based animation beyond the purely kinematic approach by simulating secondary motions like jiggling, volume preservation and contact deformation effects. Despite offering such realistic effects, physically based simulation is computationally demanding and complex, thus it is usually avoided in interactive applications.

The key challenge of producing believable deformations is to satisfy the conflicting requirements of real-time interactivity and believability. Believability requires achieving sufficient deformation detail, which means capturing the full range of desired effects. Producing these deformations demand at least an order of magnitude more computation time than current interactive systems. The aim of this paper is to provide a comprehensive survey on the existing skinning techniques in literature. These techniques can be classified into three categories: skeleton-based deformation (geometry and example-based skinning), volume preserving skinning and physics-based skinning methods. We discuss the existing skinning methods and how they address the above-mentioned problems, and we also highlight the advantages and disadvantages of each method.

2   Skeleton-Based Deformations

The most common approach for deforming articulated character's skin is to define the surface geometry as a function of an underlying skeletal structure. Due to the simplicity, intuitive manipulation, and the ability to quickly solve inverse-kinematics on a small subspace (i.e., the skeleton), skeleton-based methods are very popular and widely used in the animation industry. When modeling a skeleton-based deformation, the challenge is to obtain high-quality skin deformations. The current skeleton-based deformation techniques can be divided into two sections: geometry-based methods (also called smooth skinning, Sect. 2.1) and example-based methods (Sect. 2.2).

2.1   Geometric Skinning Techniques

In geometric skinning techniques, skeleton-to-skin binding is defined in a direct, geometrical way. Geometric approaches to deform articulated characters have shown reasonable results at interactive rates. We start by discussing the standard real-time method *"skeletal subspace deformation"*, also known as linear blend skinning (LBS). This method has been widely adopted in real-time applications such as games, for its computational efficiency and straightforward GPU implementation.

Linear Skinning Methods. The fundamental technique to drive the deformation of a character skin is via an underlying skeleton. Among the many proposed skeleton-based deformation techniques, linear blend skinning (LBS) is the most popular technique due to its effectiveness, simplicity, and efficiency

[57]. It has been given many different names over the years, including "*skeletal subspace deformation*", matrix palette skinning, enveloping, vertex blending, smooth skinning (Autodesk Maya), bones skinning (Autodesk 3D Studio Max), or linear blend skinning (the open-source Blender). What follows is an explanation of both the concepts and mathematics necessary to understand linear blend skinning. We will also review the algorithm and its limitations. In linear blend skinning, the basic operation is to deform the skin according to a given list of bone transformations (Fig. 1). The formulation of the LBS model requires the following input data:

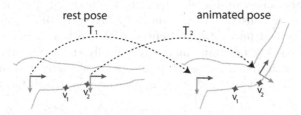

Fig. 1. An example illustrates the main concept of LBS. There are two transformations $\mathbf{T}_1$ and $\mathbf{T}_2$, corresponding to the transformations of shoulder and elbow joints from the rest pose to an animated posture.

- **Surface mesh**, a 3D model represented as a polygon mesh, where only vertex positions will change during deformations. We denote the rest pose vertices as $\mathbf{v}_i, \ldots, \mathbf{v}_n$, and $\mathbf{v}_i \in \mathbb{R}^4$ is given in homogeneous coordinates.
- **Bone transformations**, representing the current deformation using a list of matrices $\mathbf{T}_1, \ldots, \mathbf{T}_m \in \mathbb{R}^{4 \times 4}$. The matrices can be conveniently defined using an animated skeleton. $\mathbf{T}_j$ is the transformation matrix associated with bone j in its current (animated) pose, $\mathbf{R}_j^{-1}$ is the inverse transformation of the same bone in the rest pose $\mathbf{R}$. Additionally, the matrix $\mathbf{R}$ remains constant, so its inverse is also constant and can safely be pre-computed. Each bone transformation influences part of the mesh.
- **Skinning weights**, For each vertex $\mathbf{v}_i$, we have weights $w_{i,1} + \ldots + w_{i,m} \in \mathbb{R}$, in which each $w_{i,j}$ describes the amount of influence of bone j on vertex $\mathbf{v}_i$. The binding weights are normally assumed to be convex, so $w_{i,1} + \ldots + w_{i,m} = 1$ and $w_{i,j} \geq 0$.

The surface mesh is driven by a set of bones. Every vertex is associated with the bones via a bone-vertex bind weight, which quantifies the influence of each bone to the vertices. Therefore, the basic idea behind LBS is to linearly blend the transformation matrices. The skin is deformed by transforming each vertex through a weighted combination of bone transformations from the rest pose. Thus, the final transformed vertex position $\mathbf{v}_i'$ is a weighted average of its initial position transformed by each of the attached bones. The whole process can be expressed with the following equation:

$$\mathbf{v}_i' = \sum_{j=1}^{m} w_{i,j} \mathbf{T}_j \mathbf{R}_j^{-1} \mathbf{v}_i \tag{1}$$

In practice, it is rarely makes sense to attach a vertex to more than four joints. However, some old games systems used a variant dubbed rigid binding, which corresponds to allowing only one influencing bone per vertex. *Rigid binding* leads to unrealistic non-smooth deformations near joints and suffers from self-intersections. With increasing polygon budgets, linear blend skinning quickly replaced rigid binding because it allowed for smooth transitions between individual transformations (see Fig. 2). Linear blend skinning begins by assigning weights for every vertex on the skin mesh to the underlying bones. The binding weights can be constructed automatically based on the distance between the skin vertices and line segments representing the bones, but it is hard to reliably create good weights automatically. Usually the artists must paint the weights on the mesh directly, using their knowledge of anatomy. Technically, the implementation of a linear blend skinning algorithm is straightforward.

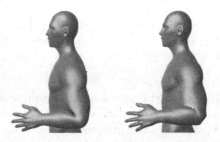

Fig. 2. *Left.* Linear blend skinning. Note the loss of volume at the elbow joint. *Right.* Rigid binding. Note the self-intersections and unnatural deformations in the areas around the elbow joint.

Linear blend skinning works very well when the blended transformations $\mathbf{T}_j$ are not very different. Issues arise if we need to blend transformations, which differ significantly in their rotation components. Despite its fast and straightforward implementation, linear blend skinning suffers from some visible artefacts when we rotate joint more than 90°. In a rotating joint, we expect the skin to rotate around the joint too, maintaining the volume. But the linear blend model instead interpolates skin vertex positions linearly between where the bones expect them, which shrinks the volume of the skin. Therefore, this linearly blending rotation leads to the well-known *"candy-wrapper"* artefact (as we can see in Fig. 3).

Figure 4 illustrates the problems of LBS, which are loss of volume when bending and the *"candy-wrapper"* artefact when twisting. The limitations of LBS have been extensively studied, where many techniques have been proposed to avoid its artefacts. One possibility is to enrich the space of skinning weights, leading to methods which are still linear but offer a wider range of deformations. These methods are called multi-linear skinning techniques [62,86], in which the

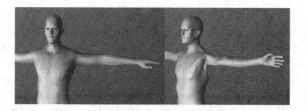

Fig. 3. The well-known *"candy-wrapper"* artefact of linear blend skinning. *Left.* The character model in its rest pose. *Right.* The model deformed with linear blend skinning, where the areas around the shoulder joint suffer from the *"candy-wrapper"* artefact and volume loss when twisting.

extra weights are learned from input examples and regularization is used to prevent overfitting. Merry et al. propose a multi-linear skinning model called *Animation Space* [59], which uses 4 weights per vertex-bone pair. However, this increase in the number of weights carries an additional cost in time and space, as well as parameter passing. While linear skinning techniques are popular due to their efficient implementations, they cannot entirely remove the *"candy-wrapper"* artefact, which is in all cases noticeable under large joint rotations. For a comprehensive survey on linear skinning techniques, we refer the interested reader to [34]. Selecting good skinning weights is critical to avoid the artefacts and generate more natural deformations. Recently, an automatic computation of skinning weights was presented in [16]. In their method, the influence weights are determined using geodesic distances from each bone, which makes the inverse-distance weights shape-aware and can even work with production meshes that may contain non-manifold geometry. Despite that associating skinning weights with the mesh vertices can be done automatically, this method tends to either increase or decrease the volume around joints.

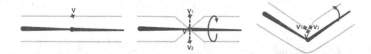

Fig. 4. *Left to right:* The skin in its rest pose. Rigid transformations (express rotation and translation). While twisting, the weighted combination of vertices v_1 and v_2 is not guaranteed to be a rigid transformation, which result in the *"candy-wrapper"* artefact. When bending, the linear interpolation of LBS between the vertices v_1 and v_2 produces v at an inadequate location, which result in a loss of volume.

Nonlinear Skinning Methods. By replacing linear blending with nonlinear blending [29,42,45], the *"candy-wrapper"* artefact can be completely avoided. Nonlinear skinning methods convert the affine rigid transformation matrices to $<quaternion, translation>$ pairs, which are then easier to blend than their matrix equivalents. In [29] some constraints are imposed on the character's rigging, whereas *spherical skinning* [45] uses a computationally expensive Singular Value Decomposition (SVD) scheme. The practical impact of these two methods is limited because of their dealing with the translational component of the skinning transformations. In contrast, dual quaternion skinning (DQS [42]) uses an approximate blending technique based on dual quaternions (essentially, two regular quaternions). DQS delivers a rigid transformation in all cases and it is almost as fast as LBS. While the underlying mathematics may not be trivial, an actual implementation of dual quaternion skinning is quite straightforward. Instead of using matrices to express the rigid transformations of the bones, dual quaternion skinning uses the geometric algebra of quaternions. The formulation of DQS is based on dual quaternions [13], which is a generalization of regular quaternions that can be used to express both the translation and the rotation. We are blending rigid transformations, however, the blended matrices $\mathbf{M}$ are no longer rigid transformations under rotations, but general affine transformations (potentially containing scale and shear factors). Therefore, instead of blending matrices $\mathbf{M} = \sum_{j=1}^{m} w_{i,j} \mathbf{T}_j \mathbf{R}_j^{-1}$. In DQS, we blend dual quaternions $\hat{\mathbf{q}}_{\mathbf{j}}$. The figure below illustrates the intuition of why DQS is better than LBS.

To compute the deformed position of a vertex with DQS. First, the transformation matrices are converted to unit dual quaternions. Then, these unit dual quaternions are blended linearly, similarly to linear blend skinning. In other words, we blend dual quaternion transformation $\hat{\mathbf{q}}_{\mathbf{j}}$, weighted by $w_{i,j}$. The result is normalized with $\| \sum_{j=1}^{m} w_{i,j} \hat{\mathbf{q}}_{\mathbf{j}} \|$ to produce the final dual quaternion used to transform a vertex from its rest pose to the deformed position. This guarantees to represent only a rotation and translation. While linearity is lost (due to the projection on unit dual quaternions), the resulting algorithm can be implemented very efficiently. Dual quaternion skinning computes deformed vertex positions according to the following formula (a more detailed explanation on dual quaternion skinning can be found in [43]):

$$\mathbf{v}_i' = \hat{\mathbf{q}}_{\mathbf{j}} \mathbf{v}_i \hat{\mathbf{q}}_{\mathbf{j}}^* \qquad (2)$$

where $\hat{\mathbf{q}}_{\mathbf{j}} = \frac{\sum_{j=1}^{m} w_{i,j} \hat{\mathbf{q}}_{\mathbf{j}}}{\| \sum_{j=1}^{m} w_{i,j} \hat{\mathbf{q}}_{\mathbf{j}} \|}$ is a unit dual quaternion and $\hat{\mathbf{q}}_{\mathbf{j}}^*$ is the conjugate of $\hat{\mathbf{q}}_{\mathbf{j}}$. Unlike matrices, by using dual quaternions there will not be any scale factor that shrinks the mesh around the joints. Dual quaternion skinning thus successfully eliminates the *"candy-wrapper"* effect, but it reveals its own artefact, called joint-bulging artefact (as we can see in Fig. 5). Moreover, dual quaternion skinning has a number of limitations especially in a production environment, as dual quaternions are unable to represent non-rigid transformations, such as shearing. The undesired *joint-bulging* artefact of DQS requires artistic manual work to be fixed. Because fixing these artefacts manually is a tedious process, automatic skinning techniques are becoming increasingly popular [5,12,35].

An interesting extension of linear blend skinning called *spline-skinning* comes from [18,19], which often produces better skinning deformations and suppresses (but not completely eliminates) the *"candy-wrapper"* artefact. Instead of using

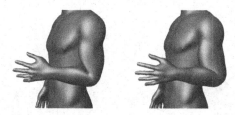

Fig. 5. A demonstration of the artefacts of linear blend skinning (LBS) and Dual quaternion skinning (DQS). *Left.* LBS suffers from volume loss while bending. *Right.* DQS successfully eliminates the *"candy-wrapper"* effect and preserve the volume of the skin, but produces the joint-bulging artefact while bending.

conventional matrix rotation, spline-skinning represents each bone of the skeleton by a spline. Furthermore, an appealing extension of DQS that is successfully applied in a production setting (Disney's *Frozen*), can be seen in [53]. For a extensive discussion on nonlinear skinning methods, we refer the reader to [41]. Whilst all the above-mentioned methods fully define the surface positions based on skeletal configuration, they cannot capture secondary motion effects and skin contact behavior in response to collision. Recently, more advanced geometric skinning methods were introduced to limit the artefacts of LBS, while keeping their simplicity. Kavan and Sorkine [44] developed a new skinning method based on the concept of joint-based deformers, which avoids the artefacts of linear blend skinning as well as the bulging artefact of dual quaternion skinning. More recently technique proposed by Jacobson and Sorkine [36], where they expanded skinning to support bending, stretching and twisting by using a slight variation on the standard skinning equations. Another impressive skinning results can be obtained using the technique presented in [83], which generates visually plausible skin deformations in real-time (see Fig. 6). Their method automatically captures contact surfaces between skin parts, without requiring any collision detection step. Moreover, they extended their framework to handle local skin contacts and produce the effect of skin elasticity (sliding effect) [84]. More recently, [48] proposed a post-processing method for dual quaternion skinning, which eliminates the joint-bulging artefacts and its suitable for real-time character animation.

In spite of improvements, skinning using geometric skinning techniques remains purely kinematic, lacking of secondary motions effects like passive jiggling motion of the fatty tissues or muscle bulging. In the next section, we present the most influential example-based skinning methods, which are able to alleviate the limitations of geometric skinning method, and add dynamic effects to the skin.

2.2   Example-Based Skinning Methods

In contrast to geometric approaches, example-based skinning methods (also known as data-driven techniques) have permitted more complex skinning effects such as muscle bulges and wrinkles, while also addressing the artefacts of linear

| linear blend skinning | dual quaternion skinning | implicit skinning |

Fig. 6. Dana model in a break-dance pose. From *left* to *right*, the model is deformed with linear blend skinning, dual quaternion skinning and implicit skinning. Note the visible loss of volume produced by LBS (*left*). Implicit skinning (*right*), however, generates visually plausible skin deformations, which avoids the artefacts of linear blend skinning, as well as the bulging artefacts of dual quaternion skinning [83].

skinning techniques. These methods take as input a series of sculpted example poses and interpolate them to obtain the desired deformation. One of the first example-based methods is pose space deformation (PSD [54]), which uses a radial basis function to interpolate correction vectors among the example poses. In pose space deformation method, *pose space* is a set of degrees of freedom of a character's model, which vary between the example poses. A particular *pose* is a set of particular values of these degrees of freedom. Pose space deformation comprises one family of approaches, in which example poses (or *local frame corrections*) are interpolated as a function of a character pose. A more sophisticated extension of PSD was presented in [79]. Their method interpolates an articulated character using example poses scattered in an abstract space. This abstract space consists of dimensions describing global properties of the 3D character, such as age and gender, in addition to dimensions that are used to describe the configuration, such as the amount of bend at the elbow joint. Moreover, PSD was generalized to support weight (weighted pose space deformation WPSD [50, 70]), which largely reduces the number of required example poses. Although WPSD can handle large-scale deformations well, it cannot provide detailed deformation and it requires more computation than the original pose space deformation (PSD). In these methods, the amount of memory grows with the number of training examples, thus they are more popular in animated feature film (DreamWorks Animation's Shrek 2) than in real-time application. To tackle this problem, [49] proposed a method similar in spirit to PSD called EigenSkin. Instead of using all the displacements for example poses, they used precomputed principal components of deformation influences on individual joints. The resulting algorithm leads to considerable memory savings and enables to transfer the computations to the GPU. Despite the fact that pose space deformation methods are simple to implement, they require tremendous effort from artists, as they have to create different poses by hand for a wide variety of examples. Another class of example-based methods, which is a direct generalization of LBS, but does not require data interpolation, is formed by methods such as single-weight enveloping (SWE [62]) and multi weight enveloping (MWE [86]). Single-weight enveloping estimated single weight per vertex with rigid character bones, with provisions made for adding additional bones. Multi-weight enveloping, however, is based on a linear framework supporting multiple weights per vertex-bone, where it provides better approximations than SWE, but at the cost of 12 weights per vertex-bone, instead of 1 weight per vertex-bone in SWE. This class of methods allows a smaller number of poses to be used to generate a larger number of deformations, while introducing more weight parameters. Thus, these numerous parameters come at a cost of complicated computation of the weights.

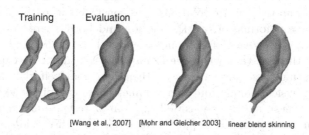

Training   Evaluation

[Wang et al., 2007]   [Mohr and Gleicher 2003]   linear blend skinning

Fig. 7. A set of example poses from an anatomically motivated arm model with both bending and twisting at the elbow. The twisting and muscle bulges are enough to prevent LBS from approximating the examples well. The technique of [62] does better, but still differ from the given example poses. The model from [85] well-approximate the examples poses. (Image taken from [85].)

As an alternative to using sculpted example poses; several example-based approaches use scanned or photographed data. Early work that uses 3D scanned poses of a human body in character skinning has been presented in [61]. Additionally, the method in [2] creates a high quality posable upper body model from range scan data and markers. In their method, to learn the skinning model, they obtain deformations corresponding to different poses by matching a subdivision surface template to the range data. Recently, more advanced example-based techniques have been effectively integrated with mesh deformation algorithms to further improve the quality of skinning [32, 77]. A rotational regression model was proposed in [85], which captures common skinning deformation such as muscle bulging (as we can see in Fig. 7) and twisting, specifically in challenging regions such as the shoulders. Park and Hodgins also introduced an interesting technique that captures and synthesizes detailed skin deformations such as bulging and jiggling [66, 67], when a character performs dynamic activities. They use a very dense and large set of markers to capture the dynamic motions. Then, they employ a second-order skinning scheme followed by a radial basis function of the residual errors to provide detailed skin deformations. While high-quality skin deformations can be captured accurately using scanned data, marker-based motion capture systems typically have a time-consuming calibration process and high hardware cost.

Example-based skinning methods are attractive since they can provide rich details from physical measurements and add realistic secondary deformation to the skeleton-based animations. Shi et al. presented an appealing method that is able to provide the jiggling of the fatty tissues in real-time by taking a surface mesh and a few sample sequences of its physical behavior [77]. Moreover, the method proposed in [32] is capable of synthesizing high resolution hand mesh deformation with rich and varying details, from only 14 examples poses. However, these approaches do not capture detailed soft-tissue deformations on a wide variety of body shapes. This limitation has been addressed by (Dyna [68]), which learns a model of soft-tissue deformations from examples using a high-resolution 4D capture system. Dyna captures surface deformations of the body at high spatial and temporal resolutions, and constructs a mathematical model for relating these deformations to the motion and body shapes of novel characters. The major drawback of example-based methods is the need for example poses. Besides the fact that when the example poses cannot be captured on a real actor, creating these poses requires either tremendous effort from an artist, or a complex physical simulation on a volumetric version of the skin mesh. In both cases, the mesh and its associated skeleton at rest are not sufficient, and further human intervention is required. An interesting discussion on example-based deformation methods, can be found in [17]. In the next section, we discuss several volume preservation methods for skinned characters, which have been proposed to tackle the loss of volume artefact of linear skinning techniques.

3   Volume Preserving Skinning Methods

Volume preservation is an important aspect in the context of skin deformation that has been addressed in a variety of research papers over the last years. Volume preservation methods allow artists to correct the volume changes through the generation of extra bulges and/or wrinkles. The method that has been proposed in [15] is one of the first methods to introduce volume preserving deformation, where they use *local volume controllers* to guarantee volume conservation of implicitly described soft substances. Moreover, *multi-resolution methods* [7] can preserve surface details by decomposing a mesh into several frequency bands. Furthermore, Funck et al. presented an appealing approach that deforms the mesh vertices based on vector field integration [20,21]. However, these two methods are either computationally expensive or do not fit into the standard animation pipeline. Angelidis and Singh developed a skinning algorithm based on a powerful embedding into the volumetric space, which enables to preserve volume locally and globally [3]. In their method, a degree of freedom is left to the artist to control the final shape, although its combination with skinning weights variation along the mesh makes this control somewhat indirect. Recently, [71] presented an automatic volume correction method to model the constant volume behavior of soft tissues. It corrects the resultant deformations of LBS using a set of local deformations. In their work, they used an automatic way to segment an organic shape into a set of regions corresponding to the main muscle and fatty tissue areas, in which volume is computed and locally corrected. Huang et al. employed a nonlinear version of the volumetric graph Laplacian, which features nonlinear volume preservation constraints [33]. Lipman et al. introduce a shape and volume preserving mesh editing technique [55], where meshes are represented by moving frames. These frames are scaled during deformation such that the volumetric shape properties are preserved.

Several impressive works that create an inner scaffolding of spring, which resist compression to maintain volume are described in [30,88]. The method in [88] provides an excellent introduction to these *interior lattice methods*. Lattice-Based freeform deformation (FFD) are widely-used in commercial software (such as Autodesk 3D Studio Max and Maya) for providing smooth deformations and preserving the volume of the skin [60]. For example, Autodesk Maya 2007 supports the notion of *flexors*. This lattice flexor uses a local FFD lattice, which can then be driven by joint transformations. However, the flexors do not support skinning transfer and the use of flexors can require significant setup and tweaking because of the multitude of lattice points. FFD was first formally proposed in [76] both as a representation for free-form solids and as a method for sculpturing solid models. Using FFD, a complex character can be deformed by positioning the control vertices of the coarse control grid. A more general extension of FFD (EFFD) was later presented by [14]. Moreover, Hsu et al. provided a method that directly manipulates the FFDs [31] and the method in [4] uses an independent deformation function to provide a more flexible FFD. Although lattice-based methods give the artist the flexibility of creating the desired deformation, they require additional setup work and the deformation is sometimes difficult to predict.

On the other hand, cage-based skinning techniques consider an appealing way to control the deformation of an enclosed fine-detailed mesh and help to preserve the volume of skin deformations [39,75]. Cage-based techniques can be considered as a generalization of the lattice-based freeform deformation. Instead of a regular control lattice, a cage is defined by a fixed-topology control lattice that is fitted to the character skin. The cage can be seen as a low-resolution abstraction of the character, which enables the user to deform a character using a simpler mesh. Most cage-based deformation methods are special case of linear blend skinning, where the handle (cage vertex) transformations are restricted to be translations and the focus is on choosing the weights. The method presented in [40] uses cage-based deformations to implement skinning templates, which offer a flexible design space within which to develop reusable skinning behavior. In their method, the skeleton drives the motion of the cage vertices using an example-based skinning technique, where the cage smoothly deforms the character model (see Fig. 8). Joshi et al. proposed a powerful cage-based deformation method based on harmonic coordinates [38] for use in high-end character articulation. Their technique guarantees that the influence of each cage vertex is non-negative and falls off with distance as measured within the cage. It generates a pleasing deformation, but computing the harmonic coordinates is not easy. In spite of that, cage-based techniques allow smooth deformation of skin geometry. Posing the cage requires manual manipulation of the cage vertices. For an overview discussion on volume-preserving deformation methods, we refer the reader to [22,65]. Another promising way to preserve the volume of the skin and to achieve realistic deformation is by applying physics-based simulation into the skin layer around the skeleton. The following section describes the vast literature on physics-based methods.

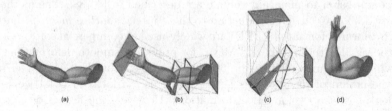

Fig. 8. Skinning with cage: (a) Input geometry with skeleton. (b) An initial cage constructed from four templates, which are associated with the hand joint, elbow joint, upper arm bone, and the shoulder joint. (c) The skeleton deforms the mesh templates. (d) The geometry is deformed by the cage, yielding a non-pinching elbow and muscle bulging. (Image taken from [40].)

4   Physics-Based Methods

In order to model dynamic phenomena, such as the vibration of fatty tissues, muscles bulging and skin contact deformations due to collisions, the animator must configure the deformation for each keyframe. While manually posing a character for each animation keyframe allows artists to create such realistic effects, this process is tedious. Therefore, an alternative method is to employ physics into the skinning process, which highly enhances the believability and realism of character motions. Accordingly, physics-based simulation manages to bring skeleton-driven animation beyond the purely kinematic approach by simulating secondary motions, such as jiggling of soft tissues when the character is moving. Those secondary motions enrich the visual experience of the animation and are essential for creating appealing characters for movie productions and virtual reality applications. After the pioneering work of Terzopoulos et al. [81] and concurrently Lasseter's animation principle *"squash and stretch"* [52]. Physical simulation has taken an important role in the animated feature film industry and computer games, where many physically based methods encouraged to simulate soft bodies or add dynamic effects to the skin. In the following subsections, we first discuss soft body simulations in (Sect. 4.1), and then physically based skinning methods in (Sect. 4.2).

4.1 Deformable and Soft Bodies Simulations

Simulating soft bodies can be achieved in different ways, and the design choice often has to balance the required accuracy and performance. The most popular techniques for simulating soft bodies in computer animation are force-based methods. In particular, most of the techniques used to simulate dynamics rely on mass spring systems, because of the simplicity and efficiency. The general idea is to represent the vertices of the mesh as mass points, governed by Newton's second law of motion, and the edges as elastic massless links (spring). Hence, the mesh is deformed when the lengths of the elastic links change. This happens when the relative position of the mass points changes due to external forces. Mass-spring systems are based on a local description of the material, in which the physics of such systems is straightforward and the simulator is easy to implement. However, to simulate a particular material, it is important to select carefully the parameters of the springs, such as the stiffness and damping. Despite that these systems are fairly easy to implement, they suffer from instability and overshooting problems under large time steps. Moreover, mass-spring systems are often not accurate, since they are strongly topology dependent and are not built based on elasticity theory. On the other hand, finite element methods (FEM) allows to model elastic materials, in which both the masses and the internal and external forces are lumped to the vertices. The vertices in the mesh are treated like mass points in a mass spring system while each element acts like a generalized spring connecting all adjacent mass points. The methods presented in [11, 81] are the first to demonstrate the effectiveness of comparatively simple mass-spring based approaches. In their methods, they applied the Lagrangian equations of motion using a finite difference scheme to simulate elastic objects with regular parametrizations. Here, the physical material properties can be described using only few parameters that are used to model soft bodies in an accurate manner. Unlike mass-spring systems, finite element methods are easy to simulate for any particular material. This makes things easier for artists in charge of modelling different types of soft bodies. Unfortunately, finite element methods are avoided in real-time applications, because they are computationally expensive and complex to implement. Various methods have been proposed to address the drawbacks of mass-spring systems and finite element methods [26, 51, 69]. A comprehensive survey of Nealen et al. [64] provides the details about these techniques. The concept of employing dynamic simulations into skinning for the purpose of character animation was introduced over two decades ago [25], where many techniques were proposed to reduce the accuracy of the simulation to help improve performance and interactivity. Capell et al. [10] used a volumetric finite element mesh to represent the deformation of skin, driven by the underlying skeleton motion. They extended their method to include rigging forces, which guide the deformation to a desired shape [9]. In their method, they effectively handled the effect of skin movement by using skeletal constraints, but by using forces that can violate the conservation of momentum makes their simulation unstable under large time steps. Shinar et al. [78] presented a framework of a two-way coupling between rigid and deformable bodies, in which they use a time integration scheme for solving dynamic elastic deformations in soft bodies interacting with rigid bodies. However, their method does not facilitate the development of an interactive animation system, because of the massive computation required for the finite elements representing the deformable body. In contrast, a possible way to accelerate the simulation of soft bodies is to focus on the surface rather than the volume [8]. In particular, Galoppo et al. [23] presented a fast method to compute the skin deformation on the surface of a soft body with rigid core. Their formulation only considers the elastic energy from skin-layer deformation, and does not include the deformation inside the volume. This may lead to inaccuracies when capturing pose-dependent deformations. All the above methods are only valid for small deformations and are unsuitable for an articulated character's large deformations. On the other hand, Müller et al. [63] achieved a good real-time performances for large rotational deformations, by using a pre-computed linear stiffness matrix to generate the deformations; their method is simple and rotationally invariant. Recently, Kim and Pollard [46] proposed an approach relying on finite element method to simulate the skin deformation, able to handle both one-way and two-way simulations. Their method generates compelling dynamic effects and the deformations are obtained at near interactive rate. Jain and Liu presented a robust approach that realistically simulate characters with soft tissue at the site of contact, where they used two-way coupling between articulated rigid bodies and deformable objects [37]. Liu et al. developed a framework that simulates and controls skeleton driven soft body characters [56]. Their method couples the skeleton dynamics and the soft body dynamics to enable two way interactions between the skeleton, the skin geometry, and the environment at interactive rates (as we can see in Fig. 9).

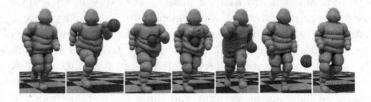

Fig. 9. Employing dynamic simulation into skinning process allows two-way interactions between the skeleton, the skin geometry, and the environment at interactive rates. (Image taken from [56].)

4.2 Physically Based Skinning

Physics-based methods are the natural choice for creating secondary motion effects such as flesh jiggling when a character is moving [24,82]. Turner and Thalmann model the elasticity of skin for character animation and simulate the fat layer by Hookian spring forces [82]. However, they treat muscles as purely controlled elements. Thus, they do not model muscles with deformable methods. Wilhelms [87] presented an approach for animated animals by simulating individual bones, muscles, soft tissues and skin. The use of muscles, soft tissues and flesh elements makes it hard to fit this approach into the skinning framework. Moreover, Hahn et al. [27,28] generated secondary skin dynamics based on the rig degrees of freedom. Their methods simulate the deformation of a character's fat and muscles in the nonlinear subspace induced by the character rig.

Fig. 10. The Method of [58] takes a skeleton and a surface mesh as input. Based on a hexahedral lattice with 106,567 cells (*center*), their method simulates the deformed surface (*right*) obeying self-collision and volumetric elasticity at 5.5 s per frame. (Image taken from [58].)

In the other direction, Kim and James [47] proposed a domain-decomposition method to simulate articulated deformable characters entirely within a subspace framework, where they combined locally rotated nonlinear subspace models to simulate the detailed deformations of the models. In order to simulate the musculotendons of the human hand and forearm, [80] add anatomic detail using the tendons and bones. While physics-based skinning methods can automatically generate secondary motion with high visual quality, they entail a significant computational burden that slows production and prohibits its use in interactive environments. McAdams et al. [58] presented a robust method using a uniform hexahedral lattice, which provides convincing deformations of the skin with contact handling. In addition, they introduce a one-point quadrature scheme and a multi-grid solver in order to improve the performance and stabilize the simulation. Although their method can capture appealing skin deformations and guarantee pinch-free geometry, it works at best at near interactive performance (as we can see in Fig. 10). Recently, Bender et al. [6] introduced a multi-layer character skinning based on shape matching with oriented particles, used to simulate the elastic behavior of a closed triangular mesh as a representation of a skin model. They make a use of position-based constraints for coupling the skeleton with the skin and handling self-collisions. In [1,72–74], a two-layered approach, the skin is first deformed with a classic linear blend skinning and then the vertex positions are adjusted using Position-based Dynamics. This allows to robustly mimic the behavior of the skin, and achieve and tune effects like volume preservation and jiggling at interactive rate.

5   Conclusion

In this paper, we have reviewed the common techniques for modelling deformations, especially those for character animation purposes. A greater attention was paid to skeleton-based methods, and physics-based methods. In skeleton-based skinning such as linear blend skinning, dual quaternion skinning and pose space deformation, surface deformation is restricted to the skeletal pose that fully defines the surface deformation. While skeleton-based methods can produce good results, the believability of the deformation using these methods is limited. Because they cannot capture secondary motions effects and skin contact deformations. Skeleton-based deformation alone is not sufficient for capturing believable skin deformations, such as skin stretching, secondary motion effects and skin contact due to collisions. In contrast, physics-based simulations bring skeleton-based animation beyond the purely kinematic approach by simulating secondary motions such as jiggling of soft tissues when the character is moving, as well as capturing skin contact deformation. Simulating such flesh-like deformations is difficult due to the coupling between the skeleton and soft skin. Moreover, the resultant deformation has a high number of independent degrees of freedom, in which it does not respect any manipulation done by the artist. Therefore, once the deformation parameters are specified, it is difficult to control the actual resulting shape of the character in every animation frame. Furthermore, physics-based methods are computationally expensive and usually avoided in interactive applications. In the future skinning research, the computational process for obtaining believable skin motion must trade-off between these requirements: it must (1) be fast enough to achieve interactive rate (i.e., >30 fps), (2) produce believable animation to minimize manual post-processing time, and (3) be controllable and stable.

Acknowledgements. The authors would like to express their deepest thanks to Dr. Bart de Keijzer for the careful proof-reading of this paper.

References

1. Abu Rumman, N., Fratarcangeli, M.: Position-based skinning for soft articulated characters. Comput. Graph. Forum **34**(6), 240–250 (2015). doi:10.1111/cgf.12533
2. Allen, B., Curless, B., Popović, Z.: Articulated body deformation from range scan data. In: Proceedings of the 29th Annual Conference on Computer Graphics and Interactive Techniques, SIGGRAPH 2002, NY, USA, pp. 612–619 (2002). http://doi.acm.org/10.1145/566570.566626
3. Angelidis, A., Singh, K.: Kinodynamic skinning using volume-preserving deformations. In: Proceedings of the 2007 ACM SIGGRAPH/Eurographics Symposium on Computer Animation, SCA 2007, San Diego, California, USA, 2–4 August 2007, pp. 129–140 (2007). http://doi.acm.org/10.1145/1272690.1272709
4. Aubert, F., Bechmann, D.: Volume-preserving space deformation. Comput. Graph. **21**(5), 625–639, September 1997. http://dx.doi.org/10.1016/S0097-8493(97)00040-X
5. Baran, I., Popović, J.: Automatic rigging and animation of 3D characters. In: ACM SIGGRAPH 2007 Papers, SIGGRAPH 2007, NY, USA (2007). http://doi.acm.org/10.1145/1275808.1276467
6. Bender, J., Müller, M., Otaduy, M.A., Teschner, M.: Position-based methods for the simulation of solid objects in computer graphics. In: EUROGRAPHICS 2013 State of the Art Reports. Eurographics Association (2013)
7. Botsch, M., Kobbelt, L.: Multiresolution surface representation based on displacement volumes. Comput. Graph. Forum **22**(3), 483–491 (2003). doi:10.1111/1467-8659.00696

8. Bro-nielsen, M., Cotin, S.: Real-time volumetric deformable models for surgery simulation using finite elements and condensation. In: Computer Graphics Forum, pp. 57–66 (1996)
9. Capell, S., Burkhart, M., Curless, B., Duchamp, T., Popović, Z.: Physically based rigging for deformable characters. In: Proceedings of the 2005 ACM SIG-GRAPH/Eurographics Symposium on Computer Animation, SCA 2005, NY, USA, pp. 301–310 (2005). http://doi.acm.org/10.1145/1073368.1073412
10. Capell, S., Green, S., Curless, B., Duchamp, T., Popović, Z.: Interactive skeleton-driven dynamic deformations. In: Proceedings of the 29th Annual Conference on Computer Graphics and Interactive Techniques, SIGGRAPH 2002, NY, USA, pp. 586–593 (2002). http://doi.acm.org/10.1145/566570.566622
11. Chadwick, J.E., Haumann, D.R., Parent, R.E.: Layered construction for deformable animated characters. SIGGRAPH Comput. Graph. **23**(3), 243–252 (1989). http://doi.acm.org/10.1145/74334.74358
12. Chen, C.H., Lin, I.C., Tsai, M.H., Lu, P.H.: Lattice-based skinning and deformation for real-time skeleton-driven animation. In: Proceedings of the 2011 12th International Conference on Computer-Aided Design and Computer Graphics, CAD-GRAPHICS 2011, pp. 306–312. IEEE Computer Society, Washington, DC (2011). http://dx.doi.org/10.1109/CAD/Graphics.2011.41
13. Clifford, W.: Mathematical papers. Technical report, June 1882
14. Coquillart, S.: Extended free-form deformation: a sculpturing tool for 3D geometric modeling. In: Proceedings of the 17th Annual Conference on Computer Graphics and Interactive Techniques, SIGGRAPH 1990, NY, USA, pp. 187–196 (1990). http://doi.acm.org/10.1145/97879.97900
15. Desbrun, M., Gascuel, M.P.: Animating soft substances with implicit surfaces. In: Proceedings of the 22nd Annual Conference on Computer Graphics and Interactive Techniques, SIGGRAPH 1995, NY, USA, pp. 287–290 (1995). http://doi.acm.org/10.1145/218380.218456
16. Dionne, O., de Lasa, M.: Geodesic voxel binding for production character meshes. In: Proceedings of the 12th ACM SIGGRAPH/Eurographics Symposium on Computer Animation, SCA 2013, NY, USA, pp. 173–180 (2013). http://doi.acm.org/10.1145/2485895.2485919
17. Feng, W.W., Kim, B.U., Yu, Y.: Real-time data driven deformation using kernel canonical correlation analysis. In: ACM SIGGRAPH 2008 Papers, SIGGRAPH 2008, NY, USA, pp. 91:1–91:9 (2008). http://doi.acm.org/10.1145/1399504.1360690
18. Forstmann, S., Ohya, J.: Fast skeletal animation by skinned arc-spline based deformation. EG 2006 Short Papers, pp. 1–4 (2006)
19. Forstmann, S., Ohya, J., Krohn-Grimberghe, A., McDougall, R.: Deformation styles for spline-based skeletal animation. In: Proceedings of the 2007 ACM SIG-GRAPH/Eurographics Symposium on Computer Animation, SCA 2007, Eurographics Association, Aire-la-Ville, Switzerland, pp. 141–150 (2007). http://dl.acm.org/citation.cfm?id=1272690.1272710
20. von Funck, W., Theisel, H., Seidel, H.P.: Vector field based shape deformations. ACM Trans. Graph. **25**(3), 1118–1125 (2006). http://doi.acm.org/10.1145/1141911.1142002
21. von Funck, W., Theisel, H., Seidel, H.P.: Volume-preserving mesh skinning. In: Proceedings of the Vision, Modeling, and Visualization Conference 2008, VMV 2008, Konstanz, Germany, 8–10 October 2008, pp. 409–414 (2008)

22. Gain, J., Bechmann, D.: A survey of spatial deformation from a user-centered perspective. ACM Trans. Graph. **27**(4), 107:1–107:21, November 2008. http://doi. acm.org/10.1145/1409625.1409629

23. Galoppo, N., Otaduy, M.A., Tekin, S., Gross, M.H., Lin, M.C.: Soft articulated characters with fast contact handling. Comput. Graph. Forum **26**(3), 243–253 (2007)

24. Gilles, B., Bousquet, G., Faure, F., Pai, D.K.: Frame-based elastic models. ACM Trans. Graph. **30**(2), 15:1–15:12 (2011). http://doi.acm.org/10.1145/1944846. 1944855

25. Girard, M., Maciejewski, A.A.: Computational modeling for the computer animation of legged figures. In: Proceedings of the 12th Annual Conference on Computer Graphics and Interactive Techniques, SIGGRAPH 1985, NY, USA, pp. 263–270 (1985). http://doi.acm.org/10.1145/325334.325244

26. Gourret, J.P., Thalmann, N.M., Thalmann, D.: Simulation of object and human skin formations in a grasping task. In: Proceedings of the 16th Annual Conference on Computer Graphics and Interactive Techniques, SIGGRAPH 1989, NY, USA, pp. 21–30 (1989). http://doi.acm.org/10.1145/74333.74335

27. Hahn, F., Martin, S., Thomaszewski, B., Sumner, R., Coros, S., Gross, M.: Rig-space physics. ACM Trans. Graph. **31**(4), 72:1–72:8 (2012). http://doi.acm.org/ 10.1145/2185520.2185568

28. Hahn, F., Thomaszewski, B., Coros, S., Sumner, R., Gross, M.: Efficient simulation of secondary motion in rig-space. In: Proceedings of the ACM SIGGRAPH/Eurographics Symposium on Computer Animation, SCA 2013 (2013)

29. Hejl, J.: Hardware skinning with quaternions. In: Kirmse, A. (ed.) Game Programming Gems, vol. 4, pp. 487–495. Charles River Media, Newton Centre (2004)

30. Hong, M., Jung, S., Choi, M.H., Welch, S.: Fast volume preservation for a mass-spring system. Comput. Graph. Appl. IEEE **26**(5), 83–91 (2006)

31. Hsu, W.M., Hughes, J.F., Kaufman, H.: Direct manipulation of free-form deformations. In: Proceedings of the 19th Annual Conference on Computer Graphics and Interactive Techniques, SIGGRAPH 1992, NY, USA, pp. 177–184 (1992). http:// doi.acm.org/10.1145/133994.134036

32. Huang, H., Zhao, L., Yin, K., Qi, Y., Yu, Y., Tong, X.: Controllable hand deformation from sparse examples with rich details. In: Proceedings of the 2011 ACM SIGGRAPH/Eurographics Symposium on Computer Animation, SCA 2011, NY, USA, pp. 73–82 (2011). http://doi.acm.org/10.1145/2019406.2019416

33. Huang, J., Shi, X., Liu, X., Zhou, K., Wei, L.Y., Teng, S.H., Bao, H., Guo, B., Shum, H.Y.: Subspace gradient domain mesh deformation. ACM Trans. Graph. **25**(3), 1126–1134 (2006). http://doi.acm.org/10.1145/1141911.1142003

34. Jacka, D., Reid, A., Merry, B., Gain, J.: A comparison of linear skinning techniques for character animation. In: Proceedings of the 5th International Conference on Computer Graphics, Virtual Reality, Visualisation and Interaction in Africa, AFRIGRAPH 2007, NY, USA, pp. 177–186 (2007). http://doi.acm.org/10.1145/ 1294685.1294715

35. Jacobson, A., Panozzo, D., Glauser, O., Pradalier, C., Hilliges, O., Sorkine-Hornung, O.: Tangible and modular input device for character articulation. ACM Trans. Graph. **33**(4), 82:1–82:12 (2014). http://doi.acm.org/10.1145/2601097. 2601112

36. Jacobson, A., Sorkine, O.: Stretchable and twistable bones for skeletal shape deformation. In: Proceedings of the 2011 SIGGRAPH Asia Conference, SA 2011, NY, USA, pp. 165:1–165:8 (2011). http://doi.acm.org/10.1145/2024156.2024199

37. Jain, S., Liu, C.K.: Controlling physics-based characters using soft contacts. ACM Trans. Graph. (SIGGRAPH Asia) **30**, 163:1–163:10 (2011). http://doi.acm.org/10.1145/2070781.2024197

38. Joshi, P., Meyer, M., DeRose, T., Green, B., Sanocki, T.: Harmonic coordinates for character articulation. ACM Trans. Graph. **26**(3) (2007). http://doi.acm.org/10.1145/1276377.1276466

39. Ju, T., Schaefer, S., Warren, J.: Mean value coordinates for closed triangular meshes. ACM Trans. Graph. **24**(3), 561–566 (2005). http://doi.acm.org/10.1145/1073204.1073229

40. Ju, T., Zhou, Q.Y., van de Panne, M., Cohen-Or, D., Neumann, U.: Reusable skinning templates using cage-based deformations. ACM Trans. Graph. **27**(5), 122:1–122:10 (2008). http://doi.acm.org/10.1145/1409060.1409075

41. Kavan, L., Collins, S., O'Sullivan, C.: Automatic linearization of nonlinear skinning. In: Proceedings of the 2009 Symposium on Interactive 3D Graphics and Games, I3D 2009, NY, USA, pp. 49–56 (2009). http://doi.acm.org/10.1145/1507149.1507157

42. Kavan, L., Collins, S., Zára, J., O'Sullivan, C.: Skinning with dual quaternions. In: Proceedings of the 2007 Symposium on Interactive 3D Graphics and Games, I3D 2007, NY, USA, pp. 39–46. ACM, New York (2007)

43. Kavan, L., Collins, S., Zára, J., O'Sullivan, C.: Geometric skinning with approximate dual quaternion blending. ACM Trans. Graph. **27**(4), 105:1–105:23 (2008). http://doi.acm.org/10.1145/1409625.1409627

44. Kavan, L., Sorkine, O.: Elasticity-inspired deformers for character articulation. ACM Trans. Graph. **31**(6), 196:1–196:8 (2012). http://doi.acm.org/10.1145/2366145.2366215

45. Kavan, L., Zára, J.: Spherical blend skinning: a real-time deformation of articulated models. In: Proceedings of the 2005 Symposium on Interactive 3D Graphics and Games, I3D 2005, NY, USA, pp. 9–16 (2005). http://doi.acm.org/10.1145/1053427.1053429

46. Kim, J., Pollard, N.S.: Fast simulation of skeleton-driven deformable body characters. ACM Trans. Graph. **30**(5), 121:1–121:19 (2011). http://doi.acm.org/10.1145/2019627.2019640

47. Kim, T., James, D.L.: Physics-based character skinning using multi-domain subspace deformations. In: Proceedings of the 2011 ACM SIGGRAPH/Eurographics Symposium on Computer Animation, SCA 2011, NY, USA, pp. 63–72 (2011). http://doi.acm.org/10.1145/2019406.2019415

48. Kim, Y., Han, J.: Bulging-free dual quaternion skinning. J. Vis. Comput. Anim. **25**(3–4), 323–331 (2014). doi:10.1002/cav.1604

49. Kry, P.G., James, D.L., Pai, D.K.: Eigenskin: Real time large deformation character skinning in hardware. In: Proceedings of the 2002 ACM SIGGRAPH/Eurographics Symposium on Computer Animation, SCA 2002, NY, USA, pp. 153–159 (2002). http://doi.acm.org/10.1145/545261.545286

50. Kurihara, T., Miyata, N.: Modeling deformable human hands from medical images. In: Proceedings of the 2004 ACM SIGGRAPH/Eurographics Symposium on Computer Animation, SCA 2004, Eurographics Association, Aire-la-Ville, Switzerland, pp. 355–363 (2004). http://dx.doi.org/10.1145/1028523.1028571

51. Larboulette, C., Cani, M.P., Arnaldi, B.: Dynamic skinning: adding real-time dynamic effects to an existing character animation. In: SCCG, pp. 87–93 (2005)

52. Lasseter, J.: Principles of traditional animation applied to 3D computer animation. SIGGRAPH Comput. Graph. **21**(4), 35–44 (1987). http://doi.acm.org/10.1145/37402.37407

53. Lee, G.S., Lin, A., Schiller, M., Peters, S., McLaughlin, M., Hanner, F.: Enhanced dual quaternion skinning for production use. In: ACM SIGGRAPH 2013 Talks, SIGGRAPH 2013, NY, USA, p. 9:1 (2013). http://doi.acm.org/10.1145/2504459.2504470

54. Lewis, J.P., Cordner, M., Fong, N.: Pose space deformation: a unified approach to shape interpolation and skeleton-driven deformation. In: Proceedings of the 27th Annual Conference on Computer Graphics and Interactive Techniques, SIGGRAPH 2000, pp. 165–172. ACM Press/Addison-Wesley Publishing Co., New York (2000). http://dx.doi.org/10.1145/344779.344862

55. Lipman, Y., Cohen-Or, D., Gal, R., Levin, D.: Volume and shape preservation via moving frame manipulation. ACM Trans. Graph. **26**(1) (2007). http://doi.acm.org/10.1145/1189762.1189767

56. Liu, L., Yin, K., Wang, B., Guo, B.: Simulation and control of skeleton-driven soft body characters. ACM Trans. Graph. **32**(6), 215:1–215:8 (2013). http://doi.acm.org/10.1145/2508363.2508427

57. Magnenat-Thalmann, N., Laperrière, R., Thalmann, D.: Joint-dependent local deformations for hand animation and object grasping. In: Proceedings on Graphics Interface 1988, Canadian Information Processing Society, Toronto, Ontario, Canada, pp. 26–33 (1988). http://dl.acm.org/citation.cfm?id=102313.102317

58. McAdams, A., Zhu, Y., Selle, A., Empey, M., Tamstorf, R., Teran, J., Sifakis, E.: Efficient elasticity for character skinning with contact and collisions. ACM Trans. Graph. **30**(4), 37:1–37:12 (2011). http://doi.acm.org/10.1145/2010324.1964932

59. Merry, B., Marais, P., Gain, J.: Animation space: a truly linear framework for character animation. ACM Trans. Graph. **25**(4), 1400–1423 (2006). http://doi.acm.org/10.1145/1183287.1183294

60. Milliron, T., Jensen, R.J., Barzel, R., Finkelstein, A.: A framework for geometric warps and deformations. ACM Trans. Graph. **21**(1), 20–51 (2002). http://doi.acm.org/10.1145/504789.504791

61. Min, K.H., Baek, S.M., Lee, G.A., Choi, H., Park, C.M.: Anatomically-based modeling and animation of human upper limbs. In: Proceedings of International Conference on Human Modeling and Animation (2000)

62. Mohr, A., Gleicher, M.: Building efficient, accurate character skins from examples. In: ACM SIGGRAPH 2003 Papers, SIGGRAPH 2003, NY, USA, pp. 562–568 (2003). http://doi.acm.org/10.1145/1201775.882308

63. Müller, M., Dorsey, J., McMillan, L., Jagnow, R., Cutler, B.: Stable real-time deformations. In: Proceedings of the 2002 ACM SIGGRAPH/Eurographics Symposium on Computer Animation, pp. 49–54. ACM, New York (2002)

64. Nealen, A., Mueller, M., Keiser, R., Boxerman, E., Carlson, M.: Physically based deformable models in computer graphics. Comput. Graph. Forum **25**(4), 809–836 (2006). doi:10.1111/j.1467-8659.2006.01000.x

65. Nieto, J., Susin, A.: Cage based deformations: a survey. In: González Hidalgo, M., Mir Torres, A., Varona Gómez, J. (eds.) Deformation Models. LNCVB, vol. 7, pp. 75–99. Springer, Netherlands (2013). http://dx.doi.org/10.1007/978-94-007-5446-1_3

66. Park, S.I., Hodgins, J.K.: Capturing and animating skin deformation in human motion. In: ACM SIGGRAPH 2006 Papers, SIGGRAPH 2006, NY, USA, pp. 881–889 (2006). http://doi.acm.org/10.1145/1179352.1141970

67. Park, S.I., Hodgins, J.K.: Data-driven modeling of skin and muscle deformation. In: ACM SIGGRAPH 2008 Papers, SIGGRAPH 2008, NY, USA, pp. 96:1–96:6 (2008). http://doi.acm.org/10.1145/1399504.1360695

68. Pons-Moll, G., Romero, J., Mahmood, N., Black, M.J.: Dyna: a model of dynamic human shape in motion. ACM Trans. Graph. **34**(4), 120:1–120:14 (2015). http://doi.acm.org/10.1145/2766993

69. Popović, J., Seitz, S.M., Erdmann, M.: Motion sketching for control of rigid-body simulations. ACM Trans. Graph. **22**(4), 1034–1054 (2003). http://doi.acm.org/10.1145/944020.944025

70. Rhee, T., Lewis, J.P., Neumann, U.: Real-time weighted pose-space deformation on the GPU. Comput. Graph. Forum **25**(3), 439–448 (2006). doi:10.1111/j.1467-8659.2006.00963.x

71. Rohmer, D., Hahmann, S., Cani, M.P.: Exact volume preserving skinning with shape control. In: Proceedings of the 2009 ACM SIGGRAPH/Eurographics Symposium on Computer Animation, SCA 2009, NY, USA, pp. 83–92 (2009). http://doi.acm.org/10.1145/1599470.1599481

72. Rumman, N.A., Fratarcangeli, M.: Position based skinning of skeleton-driven deformable characters. In: Proceedings of the 30th Spring Conference on Computer Graphics, SCCG 2014, NY, USA, pp. 83–90 (2014). http://doi.acm.org/10.1145/2643188.2643194

73. Rumman, N.A., Fratarcangeli, M.: State of the art in skinning techniques for articulated deformable characters. In: Proceedings of the 11th Joint Conference on Computer Vision, Imaging and Computer Graphics Theory and Applications, pp. 198–210 (2016)

74. Rumman, N.A., Schaerf, M., Bechmann, D.: Collision detection for articulated deformable characters. In: Proceedings of the 8th ACM SIGGRAPH Conference on Motion in Games, MIG 2015, NY, USA, pp. 215–220 (2015). http://doi.acm.org/10.1145/2822013.2822034

75. Savoye, Y., Franco, J.S.: Cageik: dual-Laplacian cage-based inverse kinematics. In: Proceedings of the 6th International Conference on Articulated Motion and Deformable Objects, AMDO 2010, pp. 280–289 (2010). http://dl.acm.org/citation.cfm?id=1875984.1876011

76. Sederberg, T.W., Parry, S.R.: Free-form deformation of solid geometric models. In: Proceedings of the 13th Annual Conference on Computer Graphics and Interactive Techniques, SIGGRAPH 1986, NY, USA, pp. 151–160 (1986). http://doi.acm.org/10.1145/15922.15903

77. Shi, X., Zhou, K., Tong, Y., Desbrun, M., Bao, H., Guo, B.: Example-based dynamic skinning in real time. ACM Trans. Graph. **27**(3), 29:1–29:8 (2008). http://doi.acm.org/10.1145/1360612.1360628

78. Shinar, T., Schroeder, C., Fedkiw, R.: Two-way coupling of rigid and deformable bodies. In: Proceedings of the 2008 ACM SIGGRAPH/Eurographics Symposium on Computer Animation, SCA 2008, Eurographics Association, Aire-la-Ville, Switzerland, pp. 95–103 (2008). http://dl.acm.org/citation.cfm?id=1632592.1632607

79. Sloan, P.P.J., Rose III., C.F., Cohen, M.F.: Shape by example. In: Proceedings of the 2001 Symposium on Interactive 3D Graphics, I3D 2001, NY, USA, pp. 135–143 (2001). http://doi.acm.org/10.1145/364338.364382

80. Sueda, S., Kaufman, A., Pai, D.K.: Musculotendon simulation for hand animation. ACM Trans. Graph. **27**(3), 83:1–83:8 (2008). http://doi.acm.org/10.1145/1360612.1360682

81. Terzopoulos, D., Platt, J., Barr, A., Fleischer, K.: Elastically deformable models. SIGGRAPH Comput. Graph. **21**(4), 205–214 (1987). http://doi.acm.org/10.1145/37402.37427

82. Turner, R., Thalmann, D.: The elastic surface layer model for animated character construction. In: Thalmann, N.M., Thalmann, D. (eds.) Proceedings of Computer Graphics International 1993. CGS CG International Series, pp. 399–412. Springer, Heidelberg (1993). doi:10.1007/978-4-431-68456-5_32

83. Vaillant, R., Barthe, L., Guennebaud, G., Cani, M.P., Rohmer, D., Wyvill, B., Gourmel, O., Paulin, M.: Implicit skinning: real-time skin deformation with contact modeling. ACM Trans. Graph. **32**(4), 125:1–125:12 (2013). http://doi.acm.org/10.1145/2461912.2461960

84. Vaillant, R., Guennebaud, G., Barthe, L., Wyvill, B., Cani, M.P.: Robust iso-surface tracking for interactive character skinning. ACM Trans. Graph. **33**(6), 189:1–189:11 (2014). http://doi.acm.org/10.1145/2661229.2661264

85. Wang, R.Y., Pulli, K., Popović, J.: Real-time enveloping with rotational regression. ACM Trans. Graph. **26**(3) (2007). http://doi.acm.org/10.1145/1276377.1276468

86. Wang, X.C., Phillips, C.: Multi-weight enveloping: least-squares approximation techniques for skin animation. In: Proceedings of the 2002 ACM SIGGRAPH/Eurographics Symposium on Computer Animation, SCA 2002, NY, USA, pp. 129–138 (2002). http://doi.acm.org/10.1145/545261.545283

87. Wilhelms, J.: Modeling animals with bones, muscles, and skin. Technical report, University of California (1994)

88. Zhou, K., Huang, J., Snyder, J., Liu, X., Bao, H., Guo, B., Shum, H.Y.: Large mesh deformation using the volumetric graph Laplacian. In: ACM SIGGRAPH 2005 Papers, SIGGRAPH 2005, NY, USA, pp. 496–503 (2005). http://doi.acm.org/10.1145/1186822.1073219

Appealing Avatars from 3D Body Scans: Perceptual Effects of Stylization

Reuben Fleming[1,2]($\boxtimes$), Betty J. Mohler[2], Javier Romero[3], Michael J. Black[3], and Martin Breidt[2]

[1] Sheffield Hallam University, Sheffield, UK
r.fleming@shu.ac.uk
[2] Max Planck Institute for Biological Cybernetics, Tübingen, Germany
[3] Max Planck Institute for Intelligent Systems, Tübingen, Germany

Abstract. Using styles derived from existing popular character designs, we present a novel automatic stylization technique for body shape and colour information based on a statistical 3D model of human bodies. We investigate whether such stylized body shapes result in increased perceived appeal with two different experiments: One focuses on body shape alone, the other investigates the additional role of surface colour and lighting. Our results consistently show that the most appealing avatar is a partially stylized one. Importantly, avatars with high stylization or no stylization at all were rated to have the least appeal. The inclusion of colour information and improvements to render quality had no significant effect on the overall perceived appeal of the avatars, and we observe that the body shape primarily drives the change in appeal ratings. For body scans with colour information, we found that a partially stylized avatar was perceived as most appealing.

Keywords: 3D body scan · Stylization · Avatar · Perception · Virtual character · Appeal

1 Introduction

Virtual avatars are frequently used in games, virtual worlds and for online communications. How an avatar is perceived by others is considered extremely important but creating a highly detailed and realistic avatar does not necessarily produce appealing results [1]. *Appeal*, one of the twelve principles of animation [8] is commonly used to describe well designed, interesting and engaging characters. In the same way animated movie and game characters require appeal in order for people to engage with them, virtual avatars also require appeal in order to engage others.

In this paper, we explore different ways in which we can increase the appeal of 3D body scans via stylization. In particular, we explore which styles make the virtual avatars most appealing, and what is the optimal amount of stylization for the most successful ones. We also examine the role of other factors in the

© Springer International Publishing AG 2017
J. Braz et al. (Eds.): VISIGRAPP 2016, CCIS 693, pp. 175–196, 2017.
DOI: 10.1007/978-3-319-64870-5_9

appeal of the final stylized avatar, namely the realism of the renders and the appearance of the original subject. For this, we acquire 3D body scans of real people using a state-of-the-art capturing system and automatically create virtual avatars with different styles and degrees of stylization. We render these avatars with and without colour information and also improve the render quality to simulate varying levels of realism. We then study, in multiple experiments, which factors affect the perceived appeal of these avatars.

2   Background

Human-like virtual characters are becoming more and more present in our lives. Since virtual environments often try to replicate the real world, a natural design choice for these virtual characters is to make them look as real as possible. This is becoming easier thanks to the improvements of 3D capturing systems in terms of speed, accuracy, quality and price. However, not all virtual characters are designed with reality in mind. Highly stylized avatars, such as the ones used by the Xbox and Wii game consoles, seem to suggest that people find the stylized look very appealing. The stylized approach to avatar creation has traditionally required the user to manually define and personalize their avatar; creating a true resemblance of that person is hard to achieve. Game studios are already attempting to cut out the manual work associated with this process by capturing gamers' features and applying them to in-game characters. For example *NBA 2K15* [18] attempts to capture the player's face while *Kinect Sports Rivals* [13] captures the entire body. Extracting information about the player is currently achieved using peripherals such as Microsoft's *Kinect 2* and Sony's *PS4* camera.

Despite the availability of both high quality, realistic body scans and increasingly convincing avatars from consumer peripherals, highly stylized characters are still extremely popular and can be seen in many animated movies, games and even web-based support services. Animation studios such as *Disney*, *Blue Sky* and *Pixar* tend to aim for a highly stylized look when designing characters. Attempts to increase realism in stylized characters have often led to negative reactions from viewers [12]. Examples of this include films such as *Polar Express* [21], *Mars Needs Moms* [19], and *The Adventures of Tintin* [17].

Furthermore, negative reactions to the almost, but not quite, photo-realistic characters seen in some film and game productions [3] and the human-like robots that inspired the "Uncanny Valley" hypothesis [14] suggest that stylizing (therefore decreasing realism), rather than attempting to improve realism, could actually produce more appealing results. For example, Inkpen and Sedlins [6] conducted a survey into peoples' comfort when communicating with avatars and found that although respondents were comfortable interacting with both realistic and cartoon looking avatars, some avatars rated highly realistic were also felt to be eerie or creepy. It is therefore possible that, even if one created a photo-real digital replica of a human, a stylized version could still be more appealing.

Experiments performed by Seyama and Nagayama [16], where they morphed between artificial and real faces, suggest that stylization should be done consistently in order to avoid viewer discomfort as abnormal features (e.g. cartoon eyes applied to a human head) produced the lowest pleasantness scores in their experiments. These conclusions match similar findings by MacDorman et al. [9] who found that a mismatch between the size and texture of the eyes and face was especially prone to making a character look eerie. Interestingly, again for faces, results by McDonnell et al. [11] suggest that both highly realistic and highly abstract render styles were both considered appealing but render styles in-between the two were considered unappealing. Recently Zell et al. [20] found that while face shape was important for portraying realism, it was the material – specifically the albedo map – that played the key role in influencing the perceived appeal of CG characters and a stylized smooth looking skin texture (obtained via blurring) was perceived as being more appealing than a realistic skin texture.

To go beyond previous work on face appeal and investigate the perceptual effects of stylization on full-bodied avatars, we conducted two separate experiments. The aim of the first experiment was to investigate the effects of stylization when applied to average male and female body shapes containing no colour information. This allowed us to focus purely on body shape. In the second experiment we stylized 3D body scans of individual people. The aim here was to assess the effect of the subjects' appearance (both in terms of body shape and colour information) and render quality on perceived appeal and realism. Overall the aim of these experiments is to better understand how to create more appealing avatars from 3D body scans.

3   Stylization of 3D Body Scans

3.1   3D Body Scanning

To represent realistic human body shape and appearance we first captured detailed 3D body scan and colour information from 4 actors (2 female).

The system is composed of 22 measurement units, each including a pair of black and white cameras observing a projected speckle light pattern, and one 5 megapixel colour camera that captures the body appearance 2 ms after the speckle image, synchronized with white flash lights. The system provides very good coverage of the entire human body and can resolve the 3D locations of a point on the body to approximately 1 mm.

Raw scans contain noise, occasional holes and can be difficult to manipulate. Consequently we registered a common 3D body template to the scans in order to obtain a coherent mesh topology for each. For this we used a statistical 3D body model that compactly parametrises body deformations automatically in terms of shape changes due to subjects' identity and body pose [4]. The model is based on approximately 2100 body scans of male and 1700 body scans of female subjects from the US and EU CAESAR dataset [15].

Through this process we obtained 3D geometry that resembles the original 3D scans but also shares a common layout. We call these meshes *registrations*. Since registrations share mesh topology, their vertex positions fully describe the body geometry. Therefore, morphing between two or more bodies can be performed as a weighted sum of the corresponding vertex positions across registrations.

Once registered, we extracted both albedo and pre-lit colour maps [2] from each scan (using the UV texture coordinate layout defined in the template). The albedo maps were illuminated using simple point lights within 3ds Max 2014 (Autodesk, USA). The pre-lit colour maps (Fig. 1 right), provided a more realistic look and, unlike the albedo maps (Fig. 1 left), required no CG lighting due to the lighting information already being embedded within the maps.

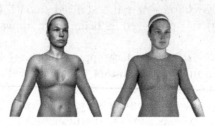

Fig. 1. Detail showing female actor (F_1) with CG lit albedo map (left), and pre-lit colour map (right).

3.2   Shape Stylization

For our initial experiment, 10 style templates (5 male and 5 female) were created in the digital sculpting software Zbrush (Pixologic, USA) (Fig. 2). The two most appealing styles for each gender were then used again for our second experiment.

The style templates were created by a trained 3D artist and were modeled to match the body proportions of existing stylized designs (see Table 1). The

Table 1. Style template references.

Gender	Style name	Description
Female	Marvel	Typical female Marvel superheroine (Marvel Entertainment, USA)
Female	Disney	"Rapunzel" from the animated movie "Tangled" (Disney, 2010)
Female	Sony	"Sam Sparks" from the animated movie "Cloudy with a Chance of Meatballs" (Sony, 2009)
Female	Pixar	"Princess Merida" from the animated movie "Brave" (Pixar, 2012)
Female	Barbie	"Barbie" doll (Mattel, USA)
Male	Marvel	Typical male Marvel superhero (Marvel Entertainment, USA)
Male	Disney	"Kristoff" from the animated movie "Frozen" (Disney, 2013)
Male	Sony	"Flint Lockwood" from the animated movie "Cloudy with a Chance of Meatballs" (Sony, 2009)
Male	Dreamworks 1	"Metro Man" from the animated movie "Megamind" (Dreamworks Animation, 2010)
Male	Dreamworks 2	"Megamind" from the animated movie "Megamind" (Dreamworks Animation, 2010)

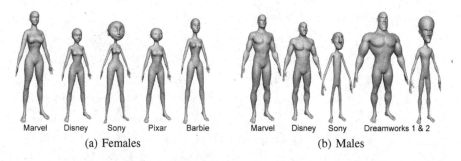

(a) Females   (b) Males

Fig. 2. Style templates created from character reference.

style templates used the same topology as the registrations, therefore direct morphing between registrations and style templates was made possible by simply interpolating vertex positions. While creating the styles is a time-consuming manual task that requires a skilled artist, it is worth noting that thanks to the shared topology, this only needs to be done once and then the templates can be reused for all future scans.

3.3   Body Shape Generation

For experiment 1 we used average male and female body shapes (Fig. 4). As the process for generating these shapes was identical, the following describes the process for generating the average female body M (Fig. 3 top left). The average

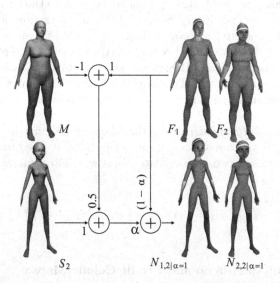

Fig. 3. Stylization process for the female Disney style S_2.

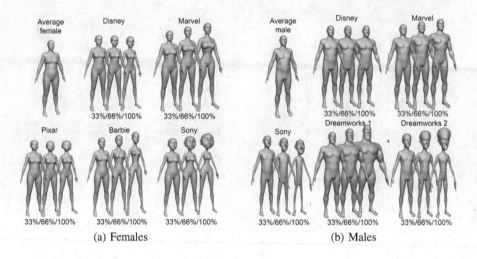

(a) Females   (b) Males

Fig. 4. Styles used in experiment 1.

shape was computed by averaging the $n \approx 2100$ female or the $n \approx 1700$ male original registrations O_i (as described in Sect. 3.1): $M = \frac{1}{n} \sum_{i=1}^{n} O_i$.

In experiment 2, rather than stylize the average body shape, we stylized registrations F obtained from 3D body scans of individual people. In order to maintain the characteristic body features (even when fully stylized) we augmented the resulting morph between registration F_i and template S_j with data derived from the registration: Half of the Euclidean difference between the mean body shape M and the registration F_i was added back to the current body style template S_j (right term of Eq. 1). See Fig. 3 for an illustration of the process.

We obtain the final stylized avatar shape $N_{i,j|\alpha}$ used for one trial (with identity i, style j and stylization level $\alpha \in [0 \dots 1]$) by linear interpolation between the original registration F_i and the current style template S_j, augmented by the individual body characteristics. This helped to maintain the more subtle characteristics of the original body shape while keeping the overall proportions of the stylized models.

Equation 1 was implemented using the *Morpher* modifier in 3ds Max, morphing the scans from their original state to the 100% stylized model (see Fig. 3 bottom for examples). Two intermediate states were also generated at 33% and 66% stylization.

$$N_{i,j|\alpha} = (1 - \alpha) F_i + \alpha \left(S_j + \frac{1}{2}(F_i - M) \right) \quad (1)$$

3.4 Stylization of Albedo and Pre-lit Colour Maps

For experiment 2, we used the albedo and pre-lit colour maps extracted from our scans. These required some manual clean-up work as they contained visual marker information (recorded for a different purpose) and had some missing data (Fig. 5). This was achieved using basic retouching tools and techniques in Photoshop (Adobe, USA). Note that normally this step would not be necessary as subjects would be scanned without markers.

Fig. 5. Original albedo map (left), cleaned up and grey body suit added (center), stylized result (right).

As all actors wore swimwear during scanning, a grey body suit template was placed over each map (Fig. 5 center) to avoid participants becoming distracted by the clothing. Due to the shared topology, and thus the shared UV texture layout, this only needed to be done once for all characters. To match the realism of the pre-lit colour maps, a realistic version of the suit was rendered using area lights and global illumination in the V-Ray 2.0 renderer (Chaos Group, Bulgaria). The albedo and pre-lit colour maps were also stylized in order to recreate a similar look to most of the style references. A typical feature film animation look is often represented through heavily saturated colours and soft surface details. To achieve this, our colour stylization technique was implemented as a Photoshop action and involved Photoshop's edge-preserving Surface Blur filter and increasing the colour vibrancy (increasing the saturation of less saturated colours), controlled by masking regions (Fig. 5 right).

Both the albedo and pre-lit colour maps had a resolution of 1024×1024 pixels. Parameter α in Eq. 1 was used to linearly blend between the normal and stylized versions of these maps.

4   Experiment 1

For our first experiment we rendered avatars with no colour information within the Unity 4 real-time game engine (Unity Technologies, Denmark). We displayed an average male and female body, 5 different style templates for each body and 2 in-between levels of stylization (33 and 66%) for each template (Fig. 4). This resulted in a total of 32 avatars (the average male and female body shapes plus 30 stylized bodies). Stimuli were presented in a random order.

4.1   Experimental Design

Male and female stimuli were shown to two separate groups of people. We tested 18 participants (10 female, mean age 26.5 yr, SD = 7.5) for the female stimuli, and 17 particpants (8 female, mean age 29.5 yr, SD = 5.9) for the male stimuli. Both were conducted in accordance with the Declaration of Helsinki. All participants signed informed consent and were financially compensated. We asked participants for their judgements of appeal on a 7-pt Likert scale when observing our stimuli. Participants were instructed to answer quickly and simply provide their first impression. Participants were also given an in-depth definition of appeal before they began:

High Appeal: "you find the person extremely appealing. This may include finding the person extremely engaging, likeable and/or generally pleasing to the eye."

Average Appeal: "you neither find the person appealing nor unappealing. They are acceptable but essentially you are indifferent, finding them neither engaging nor disengaging, likeable nor unlikeable."

Low Appeal: "you find the person extremely unappealing. This may include finding the person extremely disengaging, unlikeable and/or repellent."

4.2   Experiment 1: Results for Female Avatars

A partially stylized avatar was generally rated as being more appealing than the average female body shape while the amount of stylization for optimal appeal depends on the style (Fig. 6 and Table 2).

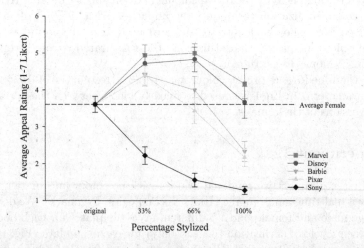

Fig. 6. Female appeal ratings on a 7-pt Likert scale (n = 18). Ratings for the average female have been included in this graph as a reference. Error bars represent one standard error of the mean in all presented figures.

Table 2. Mean female appeal ratings (standard deviations in parentheses). The average female body shape was rated as 3.61 (SD = 0.92) on appeal.

Style	33% stylized	66% stylized	100% stylized
Marvel	4.94 (1.211)	5.00 (1.085)	4.17 (1.724)
Disney	4.72 (1.127)	4.83 (1.425)	3.67 (1.847)
Barbie	4.39 (1.195)	4.00 (1.572)	2.33 (1.138)
Pixar	4.39 (1.092)	3.44 (1.504)	2.17 (1.043)
Sony	2.22 (1.003)	1.56 (0.784)	1.28 (0.461)

We ran a repeated measures ANOVA on the rating of appeal for female avatars with two within subject factors: Style (5 levels: Marvel, Disney, Sony, Dreamworks 1, Dreamworks 2) and Percent Stylized (4 levels: average female, 33, 66, 100%) and one between subject factor: Participant Gender (2 levels: male, female). There was no significant effect of participant gender on rating of appeal on female avatars, $F(1,16) = 0.027$, $p = 0.871$, $\eta_p^2 = 0.002$.

There was a significant effect of percent stylized on appeal ratings, $F(3,48) = 8.485$, $p < 0.001$, $\eta_p^2 = 0.347$. 100% stylized had significantly lower appeal ratings than both 33% and 66% while both 33% and 66% stylized female avatars were found to be more appealing than the average female body (as revealed by post-hoc comparisons using Bonferroni adjustments for multiple comparisons ($p < 0.001$)).

There was also a significant effect of style on appeal ratings, $F(4,64) = 63.03$, $p < 0.001$, $\eta_p^2 = 0.798$. The five styles did not have the same effect on appeal ratings across percent stylized (as revealed by post-hoc comparisons using Bonferroni adjustments). Most notably, the Sony style had a significantly lower rating of appeal than all of the styles ($p < 0.001$) and was always rated as less appealing than the average female body. The appeal ratings for the Marvel and Disney styles did not significantly differ from each other (p = 1.00) and appeal ratings for the Barbie and Pixar styles also did not significantly differ from each other (p = 0.907). All other comparisons between styles resulted in significantly different appeal ratings ($p < 0.05$).

These main effects were conditioned upon a significant interaction between style and percent stylized, $F(12,192) = 8.154$, $p < 0.001$, $\eta_p^2 = 0.422$. We therefore tested for an effect of percent stylized for each individual style. The effects of percent stylized were significant for all styles ($p < 0.05$) except for the Marvel style (p = 0.12). Not surprisingly, there was a simple effect of style found when testing at each percent stylized, driven by the low ratings of appeal for Sony ($p < 0.05$). In addition, at 100% stylized the Marvel and Disney appeal ratings were significantly greater than both Barbie and Pixar appeal ratings (as revealed by post-hoc comparisons using Bonferroni adjustments ($ps < 0.05$)).

These results show that for all styles except Sony, a partially stylized avatar leads to the greatest appeal ratings and depending on the style either 33% or 66% leads to optimal appeal. Excluding the Sony style, an improvement upon the original appeal ratings for the average female body of approximately 29% could be achieved by stylizing the body shape.

4.3   Experiment 1: Results for Male Avatars

A partially stylized avatar was often rated as more appealing than the average male body shape while the amount of stylization for optimal appeal depends on the style (Fig. 7 and Table 3).

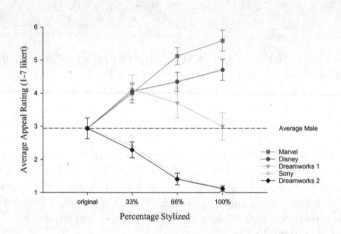

Fig. 7. Male appeal ratings on a 7-pt Likert scale (n = 17). Ratings for the average male have been included in this graph as a reference.

Table 3. Mean male appeal ratings (standard deviations in parentheses). The average male body shape was rated as 2.94 (SD = 1.30) on appeal.

Style	33% stylized	66% stylized	100% stylized
Marvel	4.00 (1.12)	5.12 (1.06)	5.59 (1.33)
Disney	4.06 (1.03)	4.35 (1.17)	4.71 (1.31)
Dreamworks 1	4.12 (1.76)	3.71 (1.86)	3.00 (1.73)
Sony	2.41 (1.33)	1.47 (0.62)	1.12 (0.49)
Dreamworks 2	2.29 (0.985)	1.41 (0.71)	1.12 (0.33)

We ran a repeated measures ANOVA on the rating of appeal for male avatars with two within subject factors: Style (5 levels: Marvel, Disney, Sony, Dreamworks 1, Dreamworks 2) and Percent Stylized (4 levels: average male, 33, 66, 100%) and one between subject factor: Participant Gender (2 levels: male, female). There was no significant effect of Participant Gender, $F(1,15) = 1.831$, $p = 0.196$, $\eta_p^2 = 0.109$.

There was no significant effect of percent stylized on appeal ratings, $F(3,45) = 2.745$, $p = 0.429$, $\eta_p^2 = 0.059$. There was however a significant effect of style on appeal ratings, $F(4,60) = 67.743$, $p < 0.001$, $\eta_p^2 = 0.819$. The five styles did not have the same effect on appeal ratings across percent stylized (as revealed by post-hoc comparisons using Bonferroni adjustments). Most notably the Sony and Dreamworks 2 styles had significantly lower appeal ratings than the other styles at all percentage stylized ($p < 0.001$). These two styles were always rated as less appealing than the average male body shape and did not significantly differ from each other ($p < 0.001$). Two other style comparisons were not significantly different in appeal ratings, specifically Marvel and Disney ($p = 0.084$) and Disney and Dreamworks I ($p = 0.434$). All other comparisons between styles resulted in significantly different appeal ratings ($p < 0.05$). Marvel clearly received the highest appeal rating (4.90), followed by Disney (4.37) and then finally Dreamworks I (3.61).

These main effects were conditioned upon a significant interaction between style and percent stylized, $F(12,180) = 21.056$, $p < 0.001$, $\eta_p^2 = 0.584$. We therefore tested for an effect of percent stylized for each individual style. The effects of percent stylized were significant for all styles ($p < 0.05$) except for the Disney style (p $= 0.238$). Not surprisingly, there was a simple effect of style found when testing at each percent stylized driven by the low ratings of appeal for Sony and Dreamworks 2 ($p < 0.05$). In addition, only at 100% stylized were the Marvel and Disney appeal ratings significantly greater than Dreamworks I (as revealed by post-hoc comparison using Bonferroni adjustments ($p < 0.05$)). For Disney and Marvel, appeal ratings were significantly greater than the average male avatar at all percent stylized (($p < 0.05$), with Bonferonni adjustments for multiple comparisons). For Dreamworks 1 only 33% stylized was rated as significantly more appealing than the average male body shape ($p = 0.01$).

These results show that for all styles except Sony and Dreamworks 2, a stylized avatar leads to the greatest appeal ratings and depending on the style, either 33% (Dreamworks 1) or 100% (Marvel and Disney) leads to optimal appeal. Excluding the Sony and Dreamworks 2 styles, an improvement upon the original appeal ratings for the average male body of approximately 63% could be achieved by stylizing the body shape.

5   Experiment 2

Both the Marvel and Disney styles provided the most significant improvements for appeal ratings in experiment 1 so were therefore used to stylize the body scans of real people in our second experiment. For this experiment, 28 images were rendered in 3ds Max at 500 × 1000 pixels using our CG lit albedo maps, 7 rendered images per character: three percentages at two styles each, plus the original body scans (2 male and 2 female). Another 28 images were rendered in 3ds Max using our pre-lit colour maps at the same resolution (Fig. 8).

5.1   Experimental Design

There were 61 Participants (35 female) with an average age of 28.08 yr (SD = 7.6) and none of the participants had been involved in experiment 1. The experiment

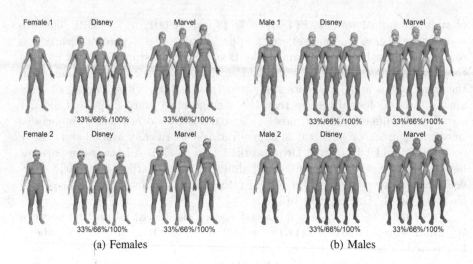

(a) Females   (b) Males

Fig. 8. Styles rendered using pre-lit colour maps.

was conducted in accordance with the Declaration of Helsinki. All participants signed informed consent and were compensated. There was no time limit assigned to the sessions but participants were instructed to answer quickly and provide their first impression. The experiment lasted approximately one hour.

There were two tasks in this experiment. First, participants provided appeal ratings as in experiment 1. Second, participants were asked to dynamically create the most appealing and repelling avatars.

Task 1: Likert Scale Ratings
Experiment 2 was a repeated measures design and involved rating male and female avatars on a 5-pt Likert scale and had three within subject factors for female avatars: Actor Identity (2 levels: F_1, F_2), Style (2 levels: Marvel, Disney), Percent Stylized (3 levels: 33, 66, 100%) and three within subject factors for male avatars: Actor Identity (2 levels: M_1, M_2), Style (2 levels: Marvel, Disney), Percent Stylized (3 levels: 33, 66, 100%) and one between subject factor of Render Quality (2 levels: albedo maps with CG lighting (42 participants), pre-lit colour maps without CG lighting (19 participants)).

In addition to appeal, we asked our participants six further questions (Table 4) to see whether our stylization technique had a significant effect on other character traits[1].

This resulted in a total of 7 questions being asked for 28 avatars (2 styles × 3 stylization levels) + 4 original body scans (2 male and 2 female). We additionally asked the participants who saw the avatars with pre-lit colour maps to rate realism. The avatars were presented in a random order. Questions were asked for each avatar in the same order as in Table 4.

---

[1] Our questions were inspired by existing research, e.g. *appeal* [10,11] and *sympathy* [5], *neuroticism* inspired by the Big Five personality traits [7].

Table 4. Avatar questionnaire.

1	Do you feel as if you can sympathise with this person?
2	Are you repelled by this person?
3	Would you accept help from this person?
4	Do you find this person engaging?
5	How appealing do you find this person?
6	How charming do you find this person?
7	Does this person come across as being neurotic?

Task 2: Interactive Creation of Appealing Characters

In a second task, 37 participants were asked to specify the exact level of stylization they found to be most appealing/repelling by interactively adjusting a slider in real-time within the Unity 4 game engine.

In addition to the stylized result controlled by the slider, the participants were also shown the original 3D body scan next to the avatar they were manipulating, providing them with a non-stylized reference. Albedo maps and CG lighting were used for this task. The dependent variable was the chosen percentage of stylization and the fixed factor was the style (Marvel and Disney) and the actor identity (F_1, F_2, M_1 and M_2). Each trial always began with the slider value set to 0% (i.e. showing the original body shape). This resulted in eight trials (two styles for each of the four actor identities). They were presented in random order and the instructions were as follows:

"Make this person as *appealing* as possible.", and "Make this person as *repelling* as possible."

For this task participants were instructed to take their time and be as accurate as possible. Participants were also instructed to adjust the slider once more after deciding on a level of stylization to verify their decision before submitting a rating.

5.2   Experiment 2: Task 1: Results

For the most part responses to our questions show a similar pattern across percent stylized for all actors: an increase in rating at 33% and a decrease in rating at 100% (as shown in Fig. 9). We decided to analyze only appeal and realism ratings as appeal seems to be representative of the majority of the questions in Table 4. Indeed the mean appeal rating did not significantly differ from all other questions ($p < 0.05$) (with the exceptions only being neurotic and repelling that were inverted in scale). Realism appears to be the only question that demonstrates a unique pattern across percent stylized so we analyzed realism ratings separately.

Appeal Rating Results

Average appeal rating for the original female body scans was 2.69. The average appeal ratings for female style templates was 3.49 for Marvel and 3.19 for

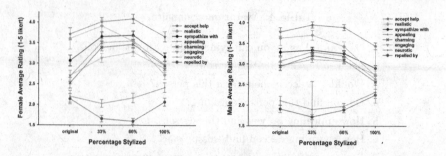

Fig. 9. Female (left) and male (right) ratings on a 5-pt Likert scale for all questions, collapsed across all styles, identities and render quality (n = 61, n = 19 for realism). Note: Repelling and neurotic ask for negative traits and therefore have inverted values.

Disney. The average appeal ratings for the original male body scans was 3.08 (slightly higher than the female original body scans). The average appeal ratings for the male style templates was slightly lower than the female styles at 3.2 for Marvel and 2.89 for Disney (see Table 5 and Fig. 10). In relative terms, our method achieves approximately 34% improvement in appeal ratings for the original female body scans by stylizing at 33% or 66% towards Marvel or Disney styles and approximately 8% improvement in appeal ratings for the original male body scans by stylizing 33% towards Disney or 66% towards Marvel styles.

Table 5. Female and male appeal ratings (standard deviations in parentheses).

Level	Actor F_1		Actor F_2		Actor M_1		Actor M_2	
	Marvel	Disney	Marvel	Disney	Marvel	Disney	Marvel	Disney
Original	2.56 (1.13)	2.56 (1.13)	2.82 (1.02)	2.82 (1.02)	3.15 (1.18)	3.15 (1.18)	3.00 (1.03)	3.00 (1.03)
33%	3.52 (0.96)	3.36 (1.11)	3.62 (0.90)	3.49 (0.92)	3.25 (1.12)	3.20 (1.14)	3.44 (1.15)	3.33 (1.00)
66%	3.57 (1.02)	2.89 (1.13)	3.87 (1.02)	3.64 (1.05)	3.36 (1.24)	2.98 (1.04)	3.43 (1.04)	3.03 (1.10)
100%	2.93 (0.89)	2.62 (1.11)	3.34 (1.14)	2.89 (1.02)	2.72 (1.11)	2.30 (0.94)	3.03 (1.13)	2.52 (1.06)

Since the style templates and actors are unique for female and males we ran separate repeated measures ANOVA on the rating of appeal for male and female avatars with three within-subject factors: Actor Identity (2 levels, F_1, F_2/M_1, M_2), Style (2 levels: Marvel, Disney) and Percent Stylized (4 levels: original, 33, 66, 100%) and two between subject factors of Participant Gender (2 levels: male, female) and of Render Quality (2 levels: albedo maps with CG lighting, pre-lit colour maps without CG lighting).

Surprisingly, for both male and female avatars we found no significant effect of render style on appeal ratings (the albedo maps with CG lighting versus the pre-lit colour maps without CG lighting stimuli), Female: $F(1,57) = 0.833$, $p = 0.365$, $\eta_p^2 = 0.014$, Male: $F(1,57) = 0.933$, $p = 0.338$, $\eta_p^2 = 0.02$. We also found no effect of participant gender on appeal ratings for both the female avatars:

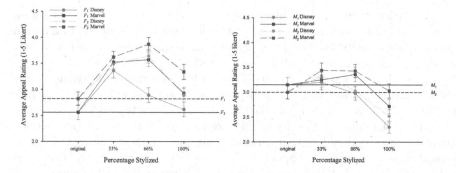

Fig. 10. Female (left) and male (right) appeal ratings on a 5-pt Likert Scale, collapsed across both render styles ($n = 61$).

$F(1,57) = 0.152$, $p = 0.698$, $\eta_p^2 = 0.003$ and the male avatars: $F(1,57) = 0.022$, $p = 0.883$, $\eta_p^2 = 0.000$.

Similar to experiment 1, we found a significant effect of percent stylized on appeal ratings for female avatars: $F(3,171) = 22.044$, $p < 0.001$, $\eta_p^2 = 0.279$ and male avatars: $F(3,171) = 15.267$, $p < 0.001$, $\eta_p^2 = 0.211$. Post-hoc comparisons using Bonferroni adjustments revealed that the original female body scans are rated significantly lower than 33% and 66% ($ps < 0.001$), but not 100% stylized female avatars ($p = 0.291$). Also 33% and 66% stylized female avatars do not significantly differ from each other on appeal ratings ($p = 1.0$), while 100% stylized female avatars are significantly lower rated on appeal than both 33% and 66% female stylzied avatars ($p < 0.001$, as revealed through post-hoc comparisons using Bonferroni adjustments). The difference in appeal can be seen as an increase in appeal from the original to 33% or 66% stylized female avatars, followed by a reduction in appeal at 100% to similar appeal ratings as the original female body scans. Post-hoc comparisons using Bonferroni adjustments revealed that 33% stylized male avatars are rated significantly higher on appeal than the original male body scans ($p = 0.041$) and 100% stylized is significantly lower rated on appeal than the original body scans, 33% and 66% stylized male avatars ($p < 0.05$). The difference in appeal can be seen as a slight increase in appeal from the original to 33% stylized male avatars, followed by a slight decline at 66% and a steeper decline in appeal with 100% stylization of male avatars.

For males and females, we also found a significant effect of style on appeal ratings, females: $F(1,57) = 15.885$, $p < 0.001$, $\eta_p^2 = 0.218$ and males: $F(1,57) = 15.184$, $p < 0.001$, $\eta_p^2 = 0.210$. For both males and females, the gender respective Marvel template is significantly higher rated on appeal than the gender specific Disney template (as revealed through post-hoc comparisons using Bonferroni adjustments, $p < 0.001$). Only for the females did we find a significant effect of actor identity, females: $F(1,57) = 8.237$, $p < 0.05$, $\eta_p^2 = 0.126$ and males: $F(1,57) = 0.253$, $p = 0.617$, $\eta_p^2 = 0.004$. A stylized female F_1 is significantly lower rated on appeal than a stylized female F_2 (as revealed through post-hoc comparisons using Bonferroni adjustments, $p < 0.001$).

These effects suggest that both percent stylized and style influence the ratings of appeal of male and female avatars and for females there is an additional influence of actor identity. For the female avatars, these main effects were conditioned upon a significant interaction between actor identity, style and percent stylized, $F(3,171) = 3.45$, $p = 0.035$, $\eta_p^2 = 0.055$ and a for both male and female avatars there was a significant interaction between style and percent stylized, Female: $F(2,118) = 5.34$, $p < 0.01$, $\eta_p^2 = 0.083$ and Male: $F(2,118) = 4.27$, $p = 0.016$, $\eta_p^2 = 0.067$. The Disney F_1 style at 66% seems to drive the interaction between actor identity, style and percent stylized. This one data point deviates from the typical pattern of improved appeal at 33%, greater improvement at 66% and then a decrease in appeal at 100% stylized, as revealed through post-hoc comparisons using Bonferroni adjustments, $p < 0.001$. The male Disney avatars stylized at 100% have significantly lower appeal ratings than all other data points, where Marvel does not always decrease in appeal at 100% stylized.

Realism Rating Results

Realism ratings for the original female body scans was on average 3.61. Average realism for the stylized female avatar at 33% was 3.91, at 66% was 3.54, and at 100% was 2.72.

Realism ratings for the original male body scans was on average 3.64. Realism ratings were quite similar for the male avatars as compared to the female avatars. Specifically, average realism for the stylized male avatars at 33% was 3.7, at 66% was 3.44, and at 100% was 2.73 (see Fig. 11 and Table 6).

Interestingly for the all avatars at 33% stylized there was no decrease in ratings of realism as compared to the original scans, at 66% only the female F_1 Disney style and the male Disney style received a decreased realism rating. All 100% stylized body scans (both male and female) received a significantly lower realism rating.

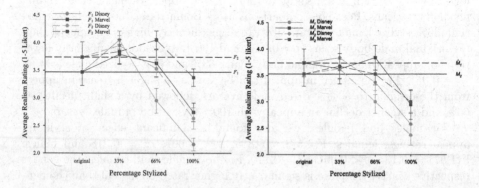

Fig. 11. Female (left) and male (right) realism ratings on a 5-pt Likert scale (n = 19,17 respectively).

We ran two separate repeated measures ANOVA on the rating of realism for both male and female avatars with three within subject factors: Actor Identity

Table 6. Female and male realism ratings (standard deviations in parentheses).

Level	Actor F_1		Actor F_2		Actor M_1		Actor M_2	
	Marvel	Disney	Marvel	Disney	Marvel	Disney	Marvel	Disney
Original	3.47 (1.07)	3.47 (1.07)	3.74(1.09)	3.74 (1.09)	3.75 (1.15)	3.75 (1.15)	3.53 (1.02)	3.53 (1.02)
33%	3.79 (0.79)	4.05 (0.71)	3.84(0.96)	3.95 (0.78)	3.63 (1.26)	3.84 (1.07)	3.68 (1.06)	3.63 (0.76)
66%	3.63 (1.07)	2.95 (0.97)	4.05(0.71)	3.53 (0.70)	3.84 (1.02)	3.37 (0.01)	3.53 (0.91)	3.00 (0.94)
100%	2.74 (0.73)	2.16 (0.50)	3.37(0.68)	2.63 (0.83)	2.95 (1.22)	2.58 (0.96)	3.00 (0.94)	2.37 (0.83)

(2 levels, F_1, F_2/M_1, M_2), Style (2 Levels: Marvel, Disney) and Percent Stylized (3 levels: 33, 66, 100) and one between subject factor: Participant Gender (2 levels: male, female). Since we had not asked for realism ratings for the albedo colour maps there is no between subject factor for ratings of realism.

We found no effect of participant gender on realism ratings for both the female avatars: $F(1,17) = 0.196$, $p = 0.664$, $\eta_p^2 = 0.011$ and the male avatars: $F(1,17) = 0.004$, $p = 0.949$, $\eta_p^2 = 0.000$.

We found a significant effect of percent stylized on realism ratings for the female avatars: $F(3,51) = 9.917$, $p < 0.001$, $\eta_p^2 = 0.368$ and the male avatars: $F(3,51) = 14.846$, $p < 0.001$, $\eta_p^2 = 0.5 = 466$. Post-hoc comparisons using Bonferroni adjustments revealed that there was a significant decrease in realism between 100% stylized and the 33% and 66% stylized female avatars ($p < 0.001$), and a significant decrease in realism between 100% stylized and the original, 33% and 66% stylized male avatars ($p < 0.01$). All other comparisons were not significantly different. For both males and females, partially stylized avatars had quite similar ratings for realism as compared to the original body scans, while for the male avatars realism was significantly lower at 100% and for the female avatars even 100% stylized did not have significantly lower average realism than the original female body scans ($p = 0.07$).

We found a significant effect of style on realism ratings for the female avatars, $F(1,17) = 35.898$, $p < 0.000$, $\eta_p^2 = 0.679$ and the male avatars, $F(1,17) = 6.045$, $p < 0.05$, $\eta_p^2 = 0.262$. For both male and female avatars the gender respective Marvel template is rated to have a significantly higher realism rating compared to the gender respective Disney template (as revealed through post-hoc comparisons using Bonferroni adjustments, $p < 0.05$).

Again similar to the appeal ratings, only for the females did we find a significant effect of actor identity, female avatars: $F(1,17) = 19.242$, $p < 0.001$, $\eta_p^2 = 0.531$, male avatars: $F(1,17) = 3.259$, $p = 0.089$, $\eta_p^2 = 0.161$. Consistent with appeal ratings, a stylized female F_1 is rated significantly lower for realism than a stylized female F_2 (as revealed through post-hoc comparisons using Bonferroni adjustments, $p < 0.001$).

For the both the female and male avatars, these main effects were conditioned upon a significant interaction between style and percent stylized, female avatars: $F(3,51) = 13.751$, $p < 0.001$, $\eta_p^2 = 0.447$; male avatars: $F(3,51) = 7.67$, $p < 0.001$, $\eta_p^2 = 0.311$ and for the female avatars a significant interaction between actor identity and percent stylized $F(2,36) = 6.610$, $p < 0.005$, $\eta_p^2 = 0.269$. For the male avatars, this can be summarized by Disney having a more negative effect on realism than Marvel at increasing percent stylized. For the female avatars, these interactions can be summarized by the styles not changing realism ratings at 33% stylized, but as stylization increases, the styles and the actors are being perceived with a variety of different ratings of realism. High levels of Marvel stylization result in decreased realism ratings for F_1 but not so for F_2, where the Marvel stylization seems to have the same influence on realism ratings across both female actors. These results are consistent with our appeal ratings and may shed some light on why 100% stylization is seen as less appealing, because it is also seen as less realistic.

5.3   Experiment 2: Task 2: Interactive Creation of Avatars Results

When asked to interactively create the most appealing female avatar using sliders it was extremely rare that participants chose either the original body scan (3%) or the fully stylized avatar (5% of the trials). The average stylization amount chosen was 40.9% across actors and styles for an appealing avatar (see Fig. 13 and Table 7).

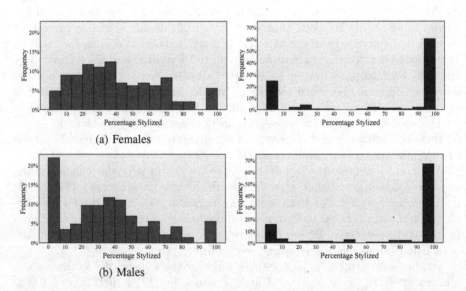

(a) Females

(b) Males

Fig. 12. Frequency of stylization level chosen interactively for the most appealing avatar (left) and most repelling avatar (right), collapsed across actor identity and style (n = 37).

When asked to interactively create the most appealing male avatar using sliders the most common response for participants was to choose the original male body scan (15.5%) while the fully stylized avatar was rarely chosen (4.7% of the trials). The average stylization amount chosen was 35.6% across actors and styles for an appealing male character.

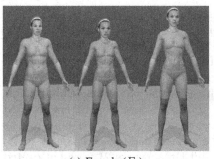 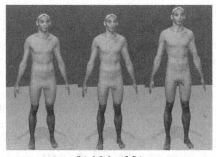

(a) Female (F_1)   (b) Male (M_2)

Fig. 13. Female avatars: Original body scan F_1 (left), 38% Disney (center), 51% Marvel (right). Male avatars: Original body scan M_2 (left), 30% Disney (center) 51% Marvel (right).

Table 7. Average stylization level in the method of adjustment task for most appealing character per actor and style (standard deviations in parentheses).

Style	F_1	F_2	M_1	M_2
Marvel	50.77% (26.3)	45.56% (24.90)	44.38% (27.30)	51.03% (36.93)
Disney	37.65% (23.3)	29.71% (22.02)	16.96% (17.60)	30.20% (23.04)

Due to bimodal distributions for the repelling avatars for both male and female, we decided to present the histograms for the percent stylized chosen (see Fig. 12). When participants were asked to create the most repelling female avatar they chose the 100% stylized version for 58% of the total trials and the original scan for 24% of the total trials. When asked to create the most repelling male avatar they chose the 100% stylized version for 64.2% of the total trials and the original scan for 13.5% of the total trials. This further confirms that the most appealing avatar is a partially, and specifically not a fully, stylized one. Overall, these results are also consistent with the repelling ratings from task 1. However, the degree to which the original body scans and 100% stylized male and female avatars were rated as repelling was at most only just above 2 on the 5-pt Likert scale (see Fig. 9). Important therefore to note is that original body scans and 100% stylized avatars were not seen as extremely repelling, but rather more repelling as compared to partially stylized avatars.

6   Discussion

For experiment 1, appeal ratings for the average male and female body shapes were highest when some level of stylization was applied (usually 33% or 66% but in some cases 100%). However, not all styles affect the resulting appeal score in a similar manner. For females, the Disney and Marvel styles were found to be consistently more appealing than the average body shapes (peaking at 66%). Styles with exaggerated proportions tended to be rated lower than styles with more human-like proportions, with the Barbie and Pixar styles producing lower appeal ratings than the Disney and Marvel styles. The Sony style (featuring the most exaggerated proportions) was rated as being consistently less appealing than the average female body shape regardless of the level of stylization applied. Male results show a slightly different pattern with the Marvel and Disney styles being considered significantly more appealing than the average male body shape and receiving the highest appeal ratings at 100%. However, the exaggerated muscular style (Dreamworks 1), while most appealing at 33%, was perceived as being only slightly more appealing than the average male body shape at 100%. The other exaggerated male styles (Sony and Dreamworks 2) were both rated to be consistently less appealing than the average male body shape. For both males and females, our results seem to suggest a preference for less exaggerated body proportions.

For our second experiment involving the stylization of body scans featuring colour information, our stylization technique proved to be less effective for males. While a partially stylized avatar was rated as being most appealing for both males and females, the increase in appeal ratings was highest for females with even the fully stylized avatars receiving similar, or in some cases higher appeal ratings than the original actors. Interestingly, the male Marvel and Disney styles originally rated as being most appealing at 100% in experiment 1 had the opposite effect when being applied to our male actors and produced the least appealing avatars at 100%. This could be due to the already muscular physique of actors M1 and M2 compared to the average body shape used in experiment 1, resulting in a higher initial appeal rating and a less obvious change in body shape when being stylized.

We found that efforts to make the avatars appear more realistic (through improvements in render quality) had no significant impact on appeal ratings. Considering our results, it could be that in contradiction to previous work [20], body shape is playing a larger role in the influence of perceived appeal than the inclusion (and modification) of surface colour, however further testing is needed to better understand this. Realism ratings displayed a similar pattern to appeal ratings and may provide some insight as to why higher levels of stylization are less desirable than partial stylization, as the highest level of stylization was rated lowest in realism. Interestingly, 33% and in some cases even 66% stylized body scans were not seen as less realistic as compared to the original body scans. This could be due to the stylization of the colour map drawing attention away from imperfections in the surface colour. The fact that there is not an immediate decrease in realism with stylization also is a result that further supports the use of stylization to create more appealing avatars.

We hope to further investigate the perception of realism in more depth to better understand the role realism plays in the appeal of our avatars and specifically further investigate why partial stylized avatars receive similar realism ratings as our original body scans.

A limitation of our findings is the small number of templates and body registrations presented in this paper and is also considered important for future research.

7   Summary

The stylization technique presented in this paper allows for the semi-automatic creation of more appealing personalized avatars from 3D body scans for use in games, virtual environments and online communications.

For stylizing 3D body scans, the main steps are the automatic registration of a statistical 3D body model to the body scan data, followed by the stylization through feature-preserving morphing with colour filtering and blending. Our technique can be automatically applied to any human body scan after registration of the scan to our 3D body template. The only manual process involved is the initial sculpting of the style templates. However, once created, these can be applied to an infinite number of body scans and based on the individual templates, can be applied at the optimal level of stylization required to maximize the appeal of these body scans. The rest of the process is fully automated. Based on the results presented in this paper, the proposed system can increase appeal ratings by approximately 34% for females (ranging from 29–39%) and 8.5% for males (ranging from 1.6% to 14.7%) depending on the actor and style.

We conducted multiple experiments in which participants were asked to rate the appeal of different avatars. Our results demonstrate that some stylization (approximately 48% for female and 44% for male) is perceived as most appealing on average across all styles and actors. This suggests that 3D body scans can be made to look more appealing with some level of stylization applied. Finally, when we asked our participants to interactively create the most appealing and repelling avatars from 3D body scans, we found again that a partially stylized avatar is the most appealing at approximately 41% for females and 36% for males. In contrast, when considering the percent of stylization people used to create repelling avatars, we found that 82% of the chosen female avatars and 78% of the chosen male avatars were made to be either 100% stylized or the original body scans themselves. For both male and female, a highly stylized avatar was most often chosen as the most repelling avatar.

Our results show that partially stylized male and female 3D body scans are perceived as most appealing as compared to the original and fully stylized body scans.

Acknowledgements. We would like to thank Chris Ferguson, Joe Smallwood, Anna Wellerdiek, Alex Holmes and Michael Geuss as well as all the participants of our experiments.

References

1. Boberg, M., Piippo, P., Ollila, E.: Designing avatars. In: Proceedings of the 3rd International Conference on Digital Interactive Media in Entertainment and Arts, DIMEA 2008, pp. 232–239. ACM, New York (2008). http://doi.acm.org/10.1145/1413634.1413679
2. Bogo, F., Romero, J., Loper, M., Black, M.J.: FAUST: dataset and evaluation for 3D mesh registration. In: Proceedings IEEE Conference on Computer Vision and Pattern Recognition (CVPR). IEEE, Piscataway, NJ, June 2014

3. Chaminade, T., Hodgins, J., Kawato, M.: Anthropomorphism influences perception of computer-animated characters actions. Soc. Cogn. Affect. Neurosci. **2**(3), 206–216 (2007). http://scan.oxfordjournals.org/content/2/3/206.abstract

4. Hirshberg, D.A., Loper, M., Rachlin, E., Black, M.J.: Coregistration: simultaneous alignment and modeling of articulated 3D shape. In: Fitzgibbon, A., Lazebnik, S., Perona, P., Sato, Y., Schmid, C. (eds.) ECCV 2012. LNCS, vol. 7577, pp. 242–255. Springer, Heidelberg (2012). doi:10.1007/978-3-642-33783-3_18

5. Hodgins, J., Jörg, S., O'Sullivan, C., Park, S.I., Mahler, M.: The saliency of anomalies in animated human characters. ACM Trans. Appl. Percept. **7**(4), 22 (2010). http://doi.acm.org/10.1145/1823738.1823740

6. Inkpen, K.M., Sedlins, M.: Me and my avatar: exploring users' comfort with avatars for workplace communication. In: Proceedings of the ACM 2011 Conference on Computer Supported Cooperative Work, CSCW 2011, pp. 383–386. ACM, New York (2011). http://doi.acm.org/10.1145/1958824.1958883

7. Johnson, J.A., Ostendorf, F., Johnson, J.A., Psycholog, D., State, P.: Clarification of the five-factor model with the abridged big five dimensional circumplex. J. Pers. Soc. Psychol. **65**(3), 563–576 (1993)

8. Johnston, O., Thomas, F.: Disney Animation: The Illusion of Life. Abbeville Press, New York (1981)

9. MacDorman, K.F., Green, R.D., Ho, C.C., Koch, C.T.: Too real for comfort? uncanny responses to computer generated faces. Comput. Hum. Behav. **25**(3), 695–710 (2009). http://www.sciencedirect.com/science/article/pii/S0747563208002379

10. McDonnell, R.: Appealing virtual humans. In: Kallmann, M., Bekris, K. (eds.) MIG 2012. LNCS, vol. 7660, pp. 102–111. Springer, Heidelberg (2012). doi:10.1007/978-3-642-34710-8_10

11. McDonnell, R., Breidt, M., Bülthoff, H.H.: Render me real: investigating the effect of render style on the perception of animated virtual humans. ACM Trans. Graph. **31**(4), 91:1–91:11 (2012). http://doi.acm.org/10.1145/2185520.2185587

12. Melina, R.: Is 'Mars Needs Moms' Too Realistic? (2011). http://www.livescience.com/33198-is-mars-needs-moms-too-realistic.html. Accessed 18 June 2014

13. Microsoft: Kinect Sports Rivals. Computer Game (2014)

14. Mori, M., MacDorman, K., Kageki, N.: The uncanny valley [from the field]. IEEE Robot. Autom. Mag. **19**(2), 98–100 (2012)

15. Robinette, K.M., Blackwell, S., Daanen, H., Boehmer, M., Fleming, S.: Civilian American and European surface anthropometry resource (CAESAR), final report. Volume 1. Summary. Technical report, US Air Force (2002)

16. Seyama, J., Nagayama, R.S.: The uncanny valley: effect of realism on the impression of artificial human faces. Presence: Teleoperators Virtual Environ. **16**(4), 337–351 (2007). http://dx.doi.org/10.1162/pres.16.4.337

17. Spielberg, S.: The Adventures of Tintin. Animated Feature Film (2011)

18. Take-Two Interactive: NBA 2K15. Computer Game (2014)

19. Wells, S.: Mars Needs Moms. Animated Feature Film (2011)

20. Zell, E., Aliaga, C., Jarabo, A., Zibrek, K., Gutierrez, D., McDonnell, R., Botsch, M.: To stylize or not to stylize?: the effect of shape and material stylization on the perception of computer-generated faces. ACM Trans. Graph. **34**(6), 184:1–184:12 (2015). http://doi.acm.org/10.1145/2816795.2818126

21. Zemeckis, R.: The Polar Express. Animated Feature Film (2004)

Information Visualization Theory and Applications

On the Visualization of Hierarchical Relations and Tree Structures with TagSpheres

Stefan Jänicke[(✉)] and Gerik Scheuermann

Image and Signal Processing Group, Leipzig University, Leipzig, Germany
{stjaenicke,scheuermann}@informatik.uni-leipzig.de

Abstract. Tag clouds are widely applied, popular visualization techniques as they illustrate summaries of textual data in an intuitive, lucid manner. Many layout algorithms for tag clouds have been developed in the recent years, but none of these approaches is designed to reflect the notion of hierarchical distance. For that purpose, we introduce a novel tag cloud layout called *TagSpheres*. By arranging tags on various hierarchy levels and applying appropriate colors, the importance of individual tags to the observed topic gets assessable. To explore relationships among various hierarchy levels, we aim to place related tags closely. Various usage scenarios from the digital humanities, sports, aviation and natural disaster management point out the benefit of TagSpheres for different domains. In addition, we highlight that TagSpheres is also a novel layout approach for tree structures.

Keywords: Tag clouds · Text visualization · Hierarchical data · Tree layout

1 Introduction

The usage of tag clouds to visualize textual data is a relatively novel technique, which was rarely applied in the past century. In 1976, Stanley Milgram was one of the first scholars who generated a tag cloud to illustrate a mental map of Paris, for which he conducted a psychological study with inhabitants of Paris, aiming to analyze their mental representation of the city [1]. In 1992, a German edition of "Mille Plateaux", written by the French philosopher Gilles Deleuze, was published with a tag cloud printed on the cover to summarize the book's content [2]. This idea to present a visual summary of textual data can be seen as the primary purpose of tag clouds [3]. But the popularity of tag clouds nowadays is attributable to a frequent usage in the social web community in the 2000s as overviews of website contents. Although there are known theoretical problems concerning the design of tag clouds [4], they are generally seen as a popular social component perceived as being fun [5]. With the simple idea to encode the frequency of terms to a given topic, tag clouds are intuitive, comprehensible visualizations, which are widely used metaphors (1) to display summaries of textual

© Springer International Publishing AG 2017
J. Braz et al. (Eds.): VISIGRAPP 2016, CCIS 693, pp. 199–219, 2017.
DOI: 10.1007/978-3-319-64870-5_10

data, (2) to support analytical tasks such as the examination of text collections, or even (3) to be used as interfaces for navigation purposes on databases.

In the recent years, various algorithms that compute effective tag cloud layouts in an informative and readable manner have been developed. One of the most popular techniques is Wordle [6], which computes compact, intuitive tag clouds and can be generated on the fly using a web-based interface.[1] Although the produced results are very aesthetic, the different used colors do not transfer information and the final arrangement of tags depends only on the scale, and not on the content of tags or potential relationships among them. Some approaches attend to the matter of visualizing more information than the frequency of terms with tag clouds – most often to compare textual summaries of different data facets.

In this paper, we present the tag cloud design *TagSpheres*, which endeavors to effectively visualize hierarchies in textual summaries. The motivation arose from research on philology. Humanities scholars wanted to analyze the clause functions of an ancient term's co-occurrences. Querying the large database, the scholars often face numerous results in the form of text passages. When only plain lists are provided to interact with the results, the discovery of significant text passages and the analysis of the contexts in which the chosen term was used becomes laborious. To support this task, we provide summaries of text passages in the form of interactive tag clouds that group terms in accordance to their distance to the search term. So, the humanities scholar gets an overview, and she is able to retrieve text passages of interest on demand.

We designed TagSpheres in a way that various types of text hierarchies can be visualized in an intuitive, comprehensible manner. To emphasize the wide applicability of TagSpheres, we list several examples from the digital humanities, sports, aviation and natural disaster management. That TagSpheres can be further used to generate layouts for tree structures is outlined in Sect. 5.

2   Related Work

Although tag clouds rather became popular in the social media, research in visualization attended to the matter of developing various layout techniques in the last years. A basic tag cloud layout is a simple list of words placed on multiple lines [7]. In such a list, tags are typically ordered by their importance to the observed issue, which is encoded by font size [8]. An alphabetical order is also often used, but a study revealed that this order is not obvious for the observer [5]. Later, more sophisticated tag cloud layout approaches that rather emphasize aesthetics than meaningful orderings were developed. A representative technique is Wordle [6,9], which produces compact aesthetic layouts with tags in different colors and orientations, but both features do not transfer any additional information. A Wordle showing the most important terms in Edgar Allan Poe's *The Raven* is given in Fig. 1(a).

Various approaches highlight relationships among tags by forming visual groups. In thematically clustered or semantic tag clouds, the detection of tags

---

[1] http://www.wordle.net/.

(a) Wordle of Edgar Allan Poe's *The Raven*.   (b) Comparing the co-occurrences of three Latin terms with TagPies.

Fig. 1. Tag cloud visualizations.

belonging to the same topic is supported by placing these tags closely [10]. Traditional, semantic word lists place clustered tags successively [11]. More sophisticated layout methods often use force directed approaches with semantically close terms attracting each other [12–14]. After force directed tag placement, tag cloud layouts can be compressed by removing occurring whitespaces [15].

Some methods generate individual tag clouds for each group of related tags, and combine the resultant multiple tag clouds to a single visual unity afterwards. An example is the Star Forest method [16], which applies a force directed method to pack multiple tag clouds. Other approaches use predefined tag cloud containers, e.g., user-defined polygonal spaces in the plane [17], polygonal shapes of countries [18], or Voronoi tesselations [19]. Newsmap uses a treemap layout [20] to group newspaper headlines of the same category in blocks [21]. Morphable Word Clouds morph the shapes of tag cloud containers in order to visualize temporal variance in text summaries [22]. For the comparison of the tags of various text documents, a ConcentriCloud divides an elliptical plane into sectors that list shared tags of several subsets of the underlying texts [23]. Due to the rather independent computation of individual tag clouds – which often leads to large whitespaces in the final composition step – the above mentioned methods can be seen as sophisticated small multiples. A rather traditional small multiples approach is Words Storms [24] that supports the visual comparison of textual summaries of documents.

Tag clouds also have been used to visualize trends. Parallel Tag Clouds generate alphabetically ordered tag lists as columns for a number of time slices and highlight the temporal evolution of a tag placed in various columns on mouse interaction [25]. In contrast, SparkClouds attach a graph showing the tag's evolution over time [26]. Other approaches overlay time graphs with tags characteristic for certain time ranges [27].

Only few approaches generate multifaceted tag cloud layouts in a single, continuous flow that includes the positioning of all tags belonging to various groups. RadCloud visualizes tags belonging to various groups within a shared elliptical area [28]. In Compare Clouds, tags of two media frames (MSM, Blogs) are comparatively visualized in a single cloud [29]. To support the comparative analysis of multiple tag groups, TagPies are arranged in a pie chart manner [30]. An example showing the comparative visualization of the co-occurrences of Latin terms is shown in Fig. 1(b).

Table 1. Characteristics of usage scenarios for TagSpheres.

Domain	Digital humanities (see Sect. 4.1)	Sports (see Sect. 4.2)	Aviation (see Sect. 4.3)	Natural disaster management (see Sect. 4.4)
Task	Analyzing the clause functions of the co-occurrences of a search term T	Comparing the performances of teams in championships	Observing all direct flights from an airport or a city	Exploring the risks of natural disasters of countries (World Risk Index 2015)
H_1	Search term T	Best performing teams	Departure airport/city	Countries with very high disaster risks
$H_2, \ldots, H_n$	Co-occurrences in dependency on the word distance to T	Teams grouped by decreasing performance	Federal (H_2), continental (H_3) and worldwide flights (H_4)	Countries with decreasing disaster risks
n	4	5, 6, 8	2..4	5
$w(t)$	Number of (co-)occurrences of t	Number of a team's appearances	Inverse distance weighting between departure and arrival airports/cities	A country's disaster risk percentage
$p(t)$	Equally labeled tag of a higher hierarchy level	Same team if already placed on a higher hierarchy level	Previously placed tag of the same country/continent	Previously placed tag of the same continent
Strong tag relations	Equally labeled tags	Same teams if placed on multiple hierarchy levels	Departure/arrival airports/cities	N/A
Weak tag relations	Spelling variants	N/A	Airports/cities of the same country/continent	Countries of the same continent

Although techniques like TagPies or Parallel Tag Clouds are capable of visualizing sequences of tag groups, none of the mentioned approaches endeavors to visually encode generic hierarchical information intuitively in a single, compact, aesthetic tag cloud. TagSpheres – presented in this paper – are designed to fill this gap.

3   Designing TagSpheres

The central idea of TagSpheres is the visualization of textual summaries that comprise hierarchical information. This paper provides various usage scenarios that exemplify the existence of hierarchies in textual data (see Sect. 4). An overview of the characteristics of these examples is given in Table 1.

Given n hierarchy levels $H_1, \ldots, H_n$, the top hierarchy level H_1 contains tags representing the focus of interest of a usage scenario. All other tags are divided into $n-1$ groups in dependency on their hierarchical distance according to the observed topic, or to the tags on H_1. Each tag t in TagSpheres has a weight $w(t)$ reflecting its importance, and an optional predecessor tag $p(t)$ representing a relationship to another tag that was placed before t and usually belongs to a higher hierarchy level. In dependency on the observed topic, it might be necessary to place the same tag on several hierarchy levels to encode the change of a tag's importance among hierarchies. In such cases, predecessor tags help to visually link these tags.

3.1   Design Decisions

When designing TagSpheres, we use the following, well-established design features for tag clouds:

- **Font Size:** Evaluated as the most powerful property [31], font size encodes the weight $w(t)$ of a tag.
- **Orientation:** As rotated tags are perceived as "unstructured, unattractive, and hardly readable" [32], we do not rotate tags to keep the layout easily explorable.
- **Color:** Being the best choice to distinguish categories [32], various colors are assigned to tags belonging to different hierarchy levels. As TagSpheres encode the distance to a given topic, the usage of a categorial color map is inappropriate. Unfortunately, suitable sequential color maps as provided by the ColorBrewer [33] produce less distinctive colors even for a small number of hierarchy levels, so that adjacent tags belonging to different hierarchy levels are hard to classify. Following the suggestions given by Ware [34], we defined a divergent cold-hot color map using red for the first hierarchy level and blue for tags belonging to the last hierarchy level n. To avoid uneven visual attraction of tags, we only chose saturated colors that are in contrast to the white background. Example color maps for up to eight hierarchy levels are shown in Fig. 2(a).

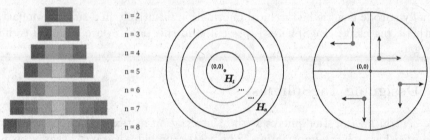

(a) Resultant color maps for $n = 2, \ldots, 8$ hierarchy levels.

(b) Using spheres for the tags of different hierarchy levels.

(c) Vectors for occlusion check to guarantee hierarchical coherence.

Fig. 2. TagSpheres layout algorithm details.

(a) Placing all tags of H_1.

(b) Placing a tag without predecessor.

(c) Placing a tag with predecessor.

Fig. 3. Determining tag positions using an Archimedean spiral.

3.2 Layout Algorithm

In preparation, the tags are sorted by increasing hierarchy level, so that all tags within the same hierarchical distance to H_1 are placed successively. The tags of each hierarchy level are ordered by decreasing weight to ensure that important tags are circularly well distributed.

To avoid large whitespaces, a problem addressed by Seifert et al. [35], our method follows the idea of the Wordle algorithm [6] – permitting overlapping tag bounding boxes if the tags' letters do not occlude – to determine the positions of tags. So, we obtain compact, uniformly looking tag clouds for the underlying hierarchical, textual data. To ensure well readable tag clouds, we use a minimal padding between letters of different tags.

As shown in Fig. 2(b), we aim to visually compose tags of the same hierarchy level in the form of spheres around the tag cloud origin at $(0,0)$. Initially, we iteratively determine positions for the tags of H_1 in the central sphere using an Archimedean spiral originating from $(0,0)$. An example is given in Fig. 3(a). For each tag t of the remaining hierarchy levels $H_2, \ldots, H_n$, we also use $(0,0)$ as spiral origin, if $p(t)$ is not provided (see Fig. 3(b)). If $p(t)$ is defined, we use the predecessor's position as spiral origin (see Fig. 3(c)). As a consequence, hierarchically related tags are placed closely and visually compose in the form of rays originating from $(0,0)$ as shown in Fig. 6(a). In contrast to other spiral based tag cloud algorithms, we avoid to cover whitespaces with tags of hierarchy level H_i within spheres of already processed hierarchy levels $H_1, \ldots, H_{i-1}$. Dependent on the quadrant in the plane, in which a tag shall be placed, we search for already placed tags intersecting two vectors originating from the dedicated position as illustrated in Fig. 2(c). If no intersections are found, we place the tag. This approach coheres all tags of a hierarchy level as a visual unity outside the inner bounds of the previously processed hierarchy levels' spheres.

3.3   Interactive Design

Implemented as an open source JavaScript library,[2] TagSpheres can be dynamically embedded into web-based applications. With mouse interaction, we enable the user to detect hierarchically related tags quickly. Thereby, we distinguish between strongly and weakly related tags, which are defined in dependency on the underlying usage scenario (see Table 1). Related tags are shown on mouseover (see Fig. 4). For strongly related tags we use a black font on transparent backgrounds having the hierarchy level's assigned color. In contrast, weakly related tags retain their saturated font color, but gray, transparent backgrounds indicate relationships.

TagSpheres provide a configurable tooltip displayed when hovering or clicking a tag to be used, e.g., to list all related tags and their weights. The mouse click function can be used for displaying additional information. e.g., to link to external sources, or to show text passages containing the chosen tag.

3.4   Limitations

The main objective of the presented layout algorithm is to combine a hierarchical information of textual data with the aesthetics of tag clouds. In contrast to the usual approach to always initialize an Archimedean spiral at the tag cloud origin $(0,0)$ when determining the position of a tag, the usage of predecessor tags as spiral origins slightly affects the uniform appearance of the result in some cases (e.g., see Fig. 7). Occasionally, little holes occur, and – at the expense of visualizing the hierarchical structure of the underlying data – the tag cloud boundaries get distorted.

The proposed hot-cold color map used to visually convey hierarchical distance generates well distinguishable colors when the number of hierarchy levels is small. For a larger number of hierarchies as displayed in Fig. 6(c) or Fig. 10, closely positioned tags of different levels may become visually indistinct, especially when only few tags belong to a certain level.

The current TagSpheres design does not take the distribution of tags throughout different hierarchy levels into account. In use cases with a steadily increasing

---

[2] http://tagspheres.vizcovery.org.

or decreasing number of tags per hierarchy level it gets possible that a considerable proportion of the color maps' bandwidth is used for a comparatively small portion of tags. An assignment of colors taking the density distribution of the tags' weights into account could overcome this issue.

4   Use Cases

TagSpheres are applicable whenever statistics of unstructured text shall be visualized in the form of a tag cloud and a decent hierarchy among the tags exists. This section illustrates usage scenarios of TagSpheres for text-based data from four different domains: digital humanities, sports, aviation and natural disaster management.

4.1   Digital Humanities Scenario

Within the digital humanities project *eXChange*,[3] historians and classical philologists work with a database containing a large amount of digitized historical texts in Latin. Usually, humanities scholars pose keyword based search queries and often receive numerous results, which are hard to revise individually. As a consequence, the generation of valuable hypotheses is a laborious, time-consuming process. To facilitate the humanities scholars' workflows, we develop visual interfaces that attempt to steer the analysis of search results into promising directions.

TagPies – also developed within the *eXChange* project – are tag clouds arranged in a pie chart manner that support the comparison of multiple search query results [30]. Using a TagPie, humanities scholars analyze contextual similarities and differences of the observed terms – an example is given in Fig. 1(b). Whereas the tags of the same groups are placed in the same circular sectors in TagPies to support their comparative analysis, the intention of TagSpheres is the visualization of hierarchical information. This supports approaching a further research interest of the humanities scholars: the analysis and classification of a term's co-occurrences according to their clause functions. For this purpose, the scholars require four-level TagSpheres displaying the following tags:

H_1: search term T,
H_2: co-occurrences of T with word distance 1,
H_3: co-occurrences of T with word distance 2, and
H_4: co-occurrences of T with word distance 3 up to word distance m.

The font size of T on level H_1 encodes how frequent the search term occurs in the underlying text corpus; the font sizes of all other terms reflect their number of co-occurrences with T in dependency on the corresponding distance. On H_4, font sizes are normalized in relation to the distance range $m - 2$. A tag on hierarchy level H_i receives a predecessor tag if the corresponding term occurs on one of the previous layers $H_{i-1}, \ldots, H_1$.

---

[3] http://exchange-projekt.de/.

Two use cases provided by the humanities scholars involved in the *eXChange* project shall illustrate the utility of TagSpheres to support the classification of a term's co-occurrences by their clause functions.

The **first use case** (see Fig. 4) outlines the analysis of the co-occurrences of the Latin term **morbo** (disease). The humanities scholar discovered and classified terms in similar relationships to the given topic. In large distances, the humanities scholar found objects in form of affected parts of the body, e.g., head (*caput*), soul (*animo*) and limbs (*membrorum*), affected persons, e.g., son (*filius*), woman (*mulier*) and king (*rex*), and related places, e.g., Rome (*romam*), church (*ecclesia*) and *villa*. Closer to *morbo* (most often with distance 1 or 2), typical attributes and predicates can be found. Whereas attributes describe the type or intensity of the disease, e.g., pestilential (*pestifero*), heavy (*gravi*), deadly (*exitiali*) and acute (*acuto*), the occurring predicates illustrate the disease's progress, e.g., seize (*correptus*), dissappear (*periit*) and worsening (*ingravescente*). Adjacent to *morbo*, specific terms for "moral" diseases, e.g., greediness (*avaritiae*), arrogance (*superbiae*) and lust (*concupiscentiae*), and actual

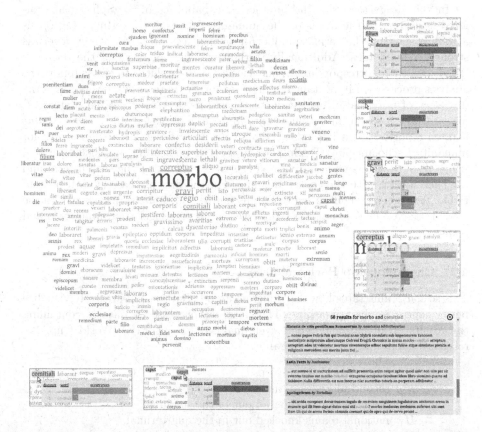

Fig. 4. The analysis of co-occurrences of the Latin term *morbo* (disease) on word distance.

diseases like jaundice ([morbo] *regio*), leprosy (*leprae*) and two common names for epilepsy (*[morbo] comitiali, [morbo] sacro*) occur.

The **second use case** (see Fig. 5) illustrates the exploration of the co-occurrences of the Latin term ***vino*** (vine). Like in the previous example, attributes of vine like precious (*pretioso*), sweet (*dulci*), new (*novo*), good (*bono*), white (*albo*) or "the best" (*optimo*) co-occur next to *vino*. Also closely positioned, usually with distance 1 or 2, are verbs describing (1) what people do with vine, e.g., drink (*postati, bibitur*), mix (*miscetur*) or swill (*lavabit*), and (2) what vine does to people, e.g., inebriate (*inebriatus, crapulatus*), rave (*furere*) or degenerate (*degenerantes*). In larger distances, subjects associated with vine can be found, e.g., people (*homines, populus*), saints (*sancti*), lord (*dominus*) or drunks (*ebrii*). Rather unexpected was the dominant usage of *vino* in Christian texts – visible through co-occurring terms like bread (*panem*), blood (*sanguis*), Body of Christ (*corpus, christi*) or sacrifice (*sacrificium*) – in contrast to a less frequent usage in classical texts. But, the humanities scholar stated that the visualization vividly reflects the classical tricolon "*vino – frumento* (grain) – *oleo* (oil)" as a list of important groceries in antiquity for soldiers to survive.

In this usage scenario, the interaction capabilities of TagSpheres are tailored according to the needs of the humanities scholars. Hovering a tag opens a tooltip showing the term's number of occurrences on all hierarchy levels as strongly related tags. Additionally, variant spellings or cases of the term are listed with their corresponding frequencies as weakly related tags to support the analysis process. An important requirement for the humanities scholars was the discovery of potentially interesting text passages, but they desired a straightforward access to the underlying texts in general. This so-called *close reading* is often reported as an important component when designing visualizations for humanities scholars [36]. TagSpheres support close reading by clicking a tag, which displays the corresponding text passages containing the search term and the clicked term with the chosen distance. For the first use case (Fig. 4, bottom right), text passages containing the terms *morbo* and *comitiali* are shown. In the second use case (Fig. 4, bottom right), we see text passages containing *vino* and *frumento*.

4.2   Championship Performances

This scenario illustrates how TagSpheres can be used to comparatively visualize performances in championships. Therefore, we processed a dataset containing the results of all national teams ever qualified for the FIFA World Cup. We receive the following six-level hierarchy:

H_1: FIFA World Champions,
H_2: second placed national teams,
H_3: national teams knocked out in the semifinal,
H_4: national teams knocked out in the quarterfinal,
H_5: national teams knocked out in the second round, and
H_6: national teams knocked out in the (first) group stage.

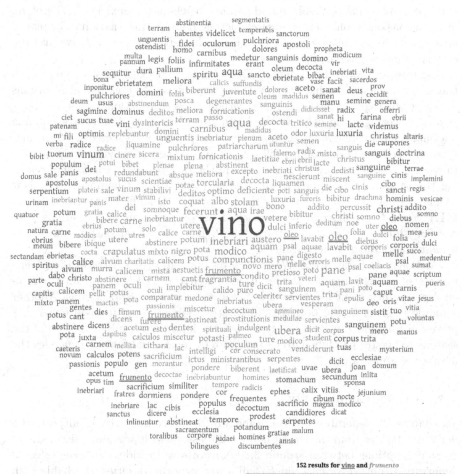

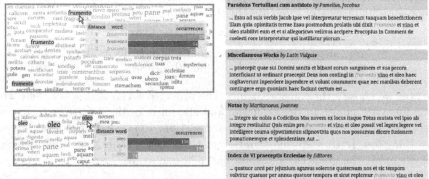

Fig. 5. The analysis of co-occurrences of the Latin term *vino* (vine) on word distance.

The nations' names are used as tags and font size encodes how often a national team partook a championship round *without* reaching the next level. Therefore, most nations occur on various hierarchy levels. If a tag t for a nation to be placed on H_i was already placed at a higher hierarchy level $H_{i-1}, \ldots, H_1$, we use the corresponding tag as predecessor $p(t)$.

Figure 6(a) shows the resultant TagSpheres. Especially this scenario illustrates the benefit of using the positions of predecessor tags as spiral origins for successor tags. In most cases, the various tags of a nation are closely positioned. Hovering a tag displays the all-time performance of the corresponding national team for all championship rounds in a tooltip. Expectedly, *Brazil* and *Germany* achieved very good results, especially in the last championship rounds. In contrast, *Italy* was often knocked out in the first round, but in case of reaching the semifinal (8x), *Italy* often became FIFA World Champion (4x). Few nations have a 100% success rate in the group stage. Qualified three times for the FIFA World Cup, *Senegal* always reached the quarterfinals. Most nations, e.g., *Sweden*, show the expected pattern "the higher the championship round, the smaller the number of appearances".

Analogously to the FIFA World Cup results, Fig. 6(b) illustrates the performances of all national teams ever participated the UEFA European Championship – pointing out Germany and Spain as most successful nations. Another example is given in Fig. 6(c) that illustrates the success of football clubs ever played in England's first league. Here, we use the average rank at the end of the seasons to cluster 68 clubs into eight hierarchy levels, and font size encodes the number of appearances.

4.3   Airport Connectivity

To analyze the federal, continental and worldwide connectivity of airports, we derived a dataset from the OpenFlights database,[4] which provides a list of direct flight connections between around 3,200 airports worldwide. With the selected departure airport d (or city) on H_1, all other airports (or cities) reachable with a non-stop flight cluster into three further hierarchy levels:

H_2: airports/cities in the same country as d,
H_3: airports/cities on the same continent as d, and
H_4: all other reachable worldwide airports/cities.

As tags we chose either airport names, the provided IATA codes,[5] or the corresponding city names. In this scenario, font size encodes the inverse geographical distance between the departure airport $d = \{lat_d, lon_d\}$ and an arrival airport $a = \{lat_a, lon_a\}$. To keep the deviation to the actual distance as small as possible, we apply the great circle distance G [37], defined as

$$G = 6378 \cdot \arccos\Big(\sin(lat_d) \cdot \sin(lat_a) + \cos(lat_d) \cdot \cos(lat_a) \cdot \cos(lon_d - lon_a)\Big).$$

---

[4] http://openflights.org/data.html.
[5] http://www.iata.org/services/pages/codes.aspx.

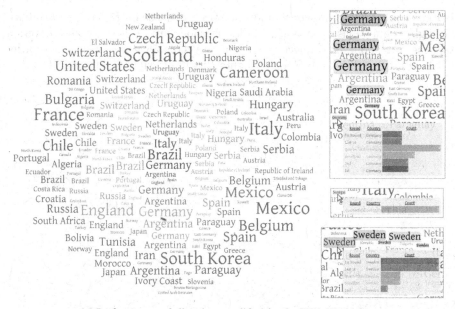

(a) Performances of all nations qualified for the FIFA World Cup.

(b) Performances of all nations qualified for the UEFA European Championship.

(c) Performances of English first league football clubs from 1888/89 – 2014/15.

Fig. 6. Visualizing championship performances with TagSpheres.

Predecessor tags are used to place airports or cities of the same country or continent closely. For a tag t to be placed on H_3, we choose the first placed tag with the same associated country as predecessor, if existent; for H_4, we choose the first placed tag with the same associated continent. Thus, a predecessor tag $p(t)$ in this scenario always belongs to the same hierarchy level as t.

Figure 7 shows TagSpheres for non-stop flights from various airports or cities. All examples show that airports/cities of the same countries/continents are placed closely in clusters. For Sydney, no tags are placed on H_3, and for Cagliari, no flight connections to airports outside Europe exist. When the user hovers a tag, the corresponding connection and the travel distance are shown in a tooltip. Clicking a tag redirects to Google Flights[6] listing possible flight connections.

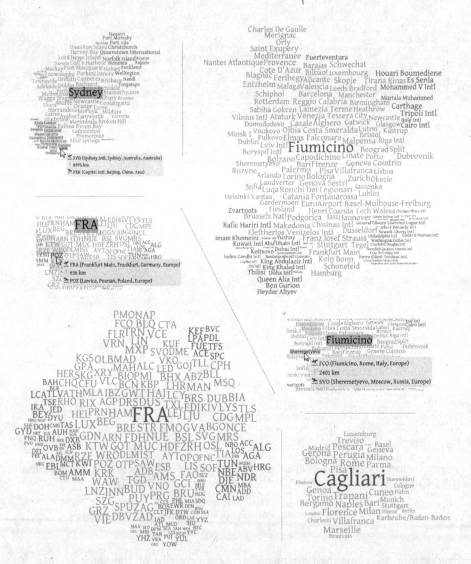

Fig. 7. Direct flight connections from airports in Sydney, Rome, Frankfurt and Cagliari.

---

[6] https://www.google.com/flights/.

Fig. 8. World Risk Index 2015 visualized with TagSpheres.

4.4   World Risk Index 2015

The World Risk Report[7] analyzes disaster risks of countries. Thereby, the exposure of a country towards natural hazards (e.g., earthquakes, tsunamis, cyclones or floods) is compared to the country's vulnerability, which depends on living conditions and economic circumstances. Each country in the database receives a disaster risk percentage – Vanatu being the country with the highest risk (36.72%) and Qatar the country with the lowest risk (0.08%). All countries are clustered into five classes from *very high* to *very low* disaster risk, which are used to generate a thematic map[8] with countries colored according to these classes. The World Risk Index 2015 visualized with TagSpheres (see Fig. 8) uses the disaster risk classes as hierarchy levels:

H_1: countries with very high disaster risks (10.40%–36.72%),
H_2: countries with high disaster risks (7.31%–10.39%),
H_3: countries with medium disaster risks (5.47%–7.30%),
H_4: countries with low disaster risks (3.47%–5.46%), and
H_5: countries with very low disaster risks (0.08%–3.46%).

---

[7] http://www.worldriskreport.org/.
[8] http://tinyurl.com/htkw8h8.

In contrast to a thematic map, we highlight the actual, individual disaster risk percentage of each country with font size. To approximate geographical relations, we use predecessor tags to place country names belonging to the same continent closely.

5 Visualizing Tree Structures with TagSpheres

Numerous algorithms have been developed to visualize large tree structures [38]. Usually, explicit tree representations in the form of node-link diagrams focus on highlighting branching patterns, e.g., [39,40]; the visualization of values associated with individual nodes plays only a minor role. On the other hand, in implicit tree representations, e.g., tree maps [20], bubble charts [41] or pie chart variants [42], links are not drawn but hierarchical relationships between nodes are illustrated with nesting techniques. But, only few implicit tree layout algorithms communicate the actual values of nodes [43,44].

Applying the TagSpheres algorithm to tree structures yields an implicit node-link diagram that visualizes the values of nodes without explicitly displaying links. But, TagSpheres indicate structural relationships by using the parent of a node in the tree as predecessor tag, by applying variable font size to illustrate the number of a node's children, and by using the interaction functionality to highlight individual paths in the tree. This way, we gain a novel tree layout that rather favors the representation of nodes than links. Two examples presenting tree layouts generated with TagSpheres are outlined below.

5.1 Airport Connectivity

Using the OpenFlights database, we can construct a (minimum spanning) tree that reflects all possible flight connections from a selected departure airport d. As in Sect. 4.3, d is the only tag on hierarchy level H_1. All other hierarchy levels compose in dependency on the number of stops it takes to reach another airport. So, H_2 contains all airports reachable with a non-stop flight, H_3 contains all airports reachable with one stop, and so on. As the maximum number of stops is six, we get eight hierarchy levels. In case of multiple possible flight connections having the same number of stops when traveling between two airports, we keep the connection with the shortest geographical distance. Thus, each airport has a clearly defined predecessor. The resultant TagSpheres with Rome-Fiumicino (FCO) as departure airport is shown in Fig. 9. As the underlying tree is well balanced and the average number of children (outdegree) is relatively high (around 5.2 children per inner node), structural relationships are only faintly visible in the outer spheres. Paths are shown on mouse selection indicating the stops between d and the selected airport as well as available connecting flights to other airports. In contrast to other node-link diagrams, the values of all 3.228 nodes and their distances to the root node are easily recognizable with TagSpheres. Thereby, the font size of a tag reflects the number of connecting flights of the corresponding airport.

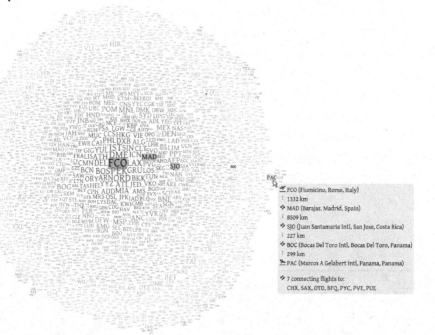

Fig. 9. Flight connections from Rome-Fiumicino (FCO).

5.2   Bible Family Tree

More than 600 verses of the Bible describe familial relationships, e.g., between husbands and wives or between fathers and children. Tying all these information together results in the Bible family tree.[9] It contains 416 nodes (persons), the maximum depth of the tree is 74, and the average number of children of inner nodes is 1.7. Using a vertical dendrogram layout[10] supports the analysis of global structural relationships, but the values of nodes are only locally visible. With TagSpheres, the values of all nodes are readable in the global view. In contrast to the previous example, the sparseness of the tree and scaling the font size according to the outdegree of a node fairly indicate present relationships, which can be further explored with mouse interaction.

---

[9] Bible genealogy data taken from BibleFamilyTree.info, Copyright © 2013 by The Psalm 119 Foundation. Used by permission. (http://www.ThePsalm119Foundation. org).

[10] http://biblefamilytree.info/.

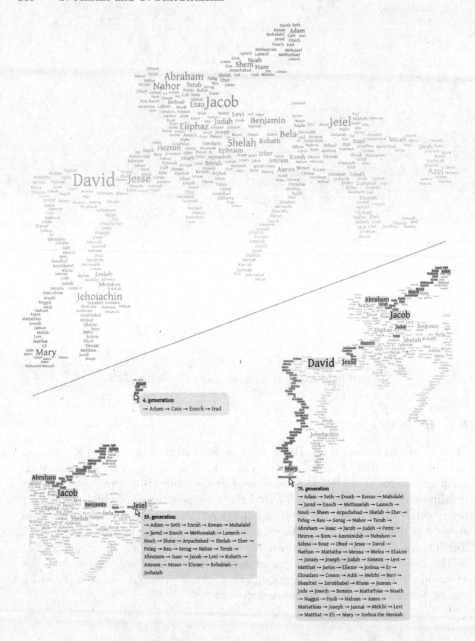

Fig. 10. Bible family tree visualized with TagSpheres.

6   Conclusion

We introduced TagSpheres that arrange tags on several hierarchy levels to transmit the notion of hierarchical distance in tag clouds. We accentuate relationships between different hierarchy levels by placing hierarchically related tags closely. The original motivation to design TagSpheres was to support humanities scholars in analyzing the clause functions of a search term's co-occurrences (see Sect. 4.1). Aspects of a positive evaluation of the TagSpheres design during the corresponding *eXChange* project are outlined in the previous version of this paper [45]. Further usage scenarios in sports, aviation and natural disaster management outline the inherence of hierarchical textual information in various domains and the usefulness of TagSpheres as they provide an interesting view on this type of data. In addition, we pointed out that the TagSpheres also serves as a novel tree layout algorithm. Although the value of this approach is yet to be evaluated, two use cases in aviation and theology indicate it's potential.

Despite few listed limitations, TagSpheres might be applicable to a multitude of further research questions from other areas. Also imaginable is the combination of TagSpheres and TagPies to support the comparative analysis of different textual summaries with hierarchical information.

Acknowledgements. The authors thank Judith Blumenstein for preparing the digital humanities use cases. This research was funded by the German Federal Ministry of Education and Research.

References

1. Milgram, S., Jodelet, D.: Psychological maps of Paris. In: Environmental Psychology, pp. 104–124 (1976)
2. Deleuze, G., Guattari, F.: Tausend Plateaus. Kapitalismus und Schizophrenie II. Merve Verlag, Berlin (1992)
3. Sinclair, J., Cardew-Hall, M.: The folksonomy tag cloud: when is it useful? J. Inf. Sci. **34**, 15–29 (2008)
4. Viégas, F.B., Wattenberg, M.: TIMELINES: tag clouds and the case for vernacular visualization. Interactions **15**, 49–52 (2008)
5. Hearst, M., Rosner, D.: Tag clouds: data analysis tool or social signaller? In: Proceedings of the 41st Annual Hawaii International Conference on System Sciences, p. 160 (2008)
6. Viégas, F., Wattenberg, M., Feinberg, J.: Participatory visualization with wordle. IEEE Trans. Vis. Comput. Graph. **15**, 1137–1144 (2009)
7. Viégas, F., Wattenberg, M., van Ham, F., Kriss, J., McKeon, M.: ManyEyes: a site for visualization at internet scale. IEEE Trans. Vis. Comput. Graph. **13**, 1121–1128 (2007)
8. Murugesan, S.: Understanding Web 2.0. IT Prof. **9**, 34–41 (2007)
9. Jo, J., Lee, B., Seo, J.: WordlePlus: expanding wordle's use through natural interaction and animation. Comput. Graph. Appl. **35**(6), 20–28 (2015). IEEE
10. Lohmann, S., Ziegler, J., Tetzlaff, L.: Comparison of Tag Cloud Layouts: Task-Related Performance and Visual Exploration. In: Gross, T., Gulliksen, J., Kotzé, P., Oestreicher, L., Palanque, P., Prates, R.O., Winckler, M. (eds.) INTERACT 2009. LNCS, vol. 5726, pp. 392–404. Springer, Heidelberg (2009). doi:10.1007/978-3-642-03655-2_43

11. Schrammel, J., Tscheligi, M.: Patterns in the clouds - the effects of clustered presentation on tag cloud interaction. In: Ebert, A., Veer, G.C., Domik, G., Gershon, N.D., Scheler, I. (eds.) HCIV -2011. LNCS, vol. 8345, pp. 124–132. Springer, Heidelberg (2014). doi:10.1007/978-3-642-54894-9_9
12. Cui, W., Wu, Y., Liu, S., Wei, F., Zhou, M., Qu, H.: Context preserving dynamic word cloud visualization. In: Pacific Visualization Symposium (PacificVis), pp. 121–128. IEEE (2010)
13. Wang, J., Zhao, J., Guo, S., North, C., Ramakrishnan, N.: ReCloud: semantics-based word cloud visualization of user reviews. In: Proceedings of the 2014 Graphics Interface Conference, GI 2014, pp. 151–158. Canadian Information Processing Society (2014)
14. Liu, X., Shen, H.W., Hu, Y.: Supporting multifaceted viewing of word clouds with focus+context display. Inf. Vis. **14**(2), 168–180 (2014)
15. Wu, Y., Provan, T., Wei, F., Liu, S., Ma, K.L.: Semantic-preserving word clouds by seam carving. In: Computer Graphics Forum, vol. 30, pp. 741–750. Wiley Online Library (2011)
16. Barth, L., Kobourov, S.G., Pupyrev, S.: Experimental comparison of semantic word clouds. In: Gudmundsson, J., Katajainen, J. (eds.) SEA 2014. LNCS, vol. 8504, pp. 247–258. Springer, Cham (2014). doi:10.1007/978-3-319-07959-2_21
17. Paulovich, F.V., Toledo, F., Telles, G.P., Minghim, R., Nonato, L.G.: Semantic wordification of document collections. In: Computer Graphics Forum, vol. 31, pp. 1145–1153. Wiley Online Library (2012)
18. Nguyen, D.Q., Tominski, C., Schumann, H., Ta, T.A.: Visualizing tags with spatiotemporal references. In: 2011 15th International Conference on Information Visualisation (IV), pp. 32–39 (2011)
19. Seifert, C., Kienreich, W., Granitzer, M.: Visualizing text classification models with voronoi word clouds. In: Proceedings of the International Conference Information Visualisation (IV), London (2011)
20. Shneiderman, B., Plaisant, C.: Treemaps for Space-Constrained Visualization of Hierarchies (1998)
21. Weskamp, M.: Newsmap (2015). http://newsmap.jp/. Accessed 15 Nov 2015
22. Chi, M., Lin, S., Chen, S., Lin, C., Lee, T.: Morphable word clouds for time-varying text data visualization. IEEE Trans. Vis. Comput. Graph. **21**, 1415–1426 (2015)
23. Lohmann, S., Heimerl, F., Bopp, F., Burch, M., Ertl, T.: ConcentriCloud: word cloud visualization for multiple text documents. In: 19th International Conference on Information Visualisation (2015)
24. Castellà, Q., Sutton, C.: Word storms: multiples of word clouds for visual comparison of documents. In: Proceedings of the 23rd International Conference on World Wide Web, WWW 2014, pp. 665–676. ACM (2014)
25. Collins, C., Viégas, F., Wattenberg, M.: Parallel tag clouds to explore and analyze faceted text corpora. In: IEEE Symposium on Visual Analytics Science and Technology, VAST 2009, pp. 91–98 (2009)
26. Lee, B., Riche, N., Karlson, A., Carpendale, S.: SparkClouds: visualizing trends in tag clouds. IEEE Trans. Vis. Comput. Graph. **16**, 1182–1189 (2010)
27. Shi, L., Wei, F., Liu, S., Tan, L., Lian, X., Zhou, M.: Understanding text corpora with multiple facets. In: 2010 IEEE Symposium on Visual Analytics Science and Technology (VAST), pp. 99–106 (2010)
28. Burch, M., Lohmann, S., Beck, F., Rodriguez, N., Di Silvestro, L., Weiskopf, D.: RadCloud: visualizing multiple texts with merged word clouds. In: 2014 18th International Conference on Information Visualisation (IV), pp. 108–113 (2014)

29. Diakopoulos, N., Elgesem, D., Salway, A., Zhang, A., Hofland, K.: Compare clouds: visualizing text corpora to compare media frames. In: Proceedings of IUI Workshop on Visual Text Analytics (2015)
30. Jänicke, S., Blumenstein, J., Rücker, M., Zeckzer, D., Scheuermann, G.: Visualizing the results of search queries on ancient text corpora with tag pies. In: Digital Humanities Quarterly (2016)
31. Bateman, S., Gutwin, C., Nacenta, M.: Seeing things in the clouds: the effect of visual features on tag cloud selections. In: Proceedings of the Nineteenth ACM Conference on Hypertext and Hypermedia, HT 2008, pp. 193–202. ACM (2008)
32. Waldner, M., Schrammel, J., Klein, M., Kristjánsdóttir, K., Unger, D., Tscheligi, M.: FacetClouds: exploring tag clouds for multi-dimensional data. In: Proceedings of Graphics Interface 2013, GI 2013, pp. 17–24. Canadian Information Processing Society (2013)
33. Harrower, M., Brewer, C.A.: ColorBrewer.org: an online tool for selecting colour schemes for maps. Cartogr. J. **40**, 27–37 (2003)
34. Ware, C.: Information Visualization: Perception for Design. Elsevier (2013)
35. Seifert, C., Kump, B., Kienreich, W., Granitzer, G., Granitzer, M.: On the beauty and usability of tag clouds. In: 12th International Conference on Information Visualisation, IV 2008, pp. 17–25 (2008)
36. Jänicke, S., Franzini, G., Cheema, M.F., Scheuermann, G.: On close and distant reading in digital humanities: a survey and future challenges. In: Borgo, R., Ganovelli, F., Viola, I. (eds.) Eurographics Conference on Visualization (EuroVis) - STARs. The Eurographics Association (2015)
37. Head, K.: Gravity for Beginners. University of British Columbia 2053 (2003)
38. Schulz, H.J.: Treevis.net: a tree visualization reference. IEEE Comput. Graphics Appl. **31**, 11–15 (2011)
39. Nguyen, Q.V., Huang, M.L.: A space-optimized tree visualization. In: IEEE Symposium on Information Visualization, INFOVIS 2002, pp. 85–92 (2002)
40. Holten, D.: Hierarchical edge bundles: visualization of adjacency relations in hierarchical data. IEEE Trans. Vis. Comput. Graph. **12**, 741–748 (2006)
41. Zhao, H., Lu, L.: Variational circular treemaps for interactive visualization of hierarchical data. In: 2015 IEEE Pacific Visualization Symposium (PacificVis), pp. 81–85 (2015)
42. Vliegen, R., van Wijk, J.J., van Der Linden, E.J.: Visualizing business data with generalized treemaps. IEEE Trans. Vis. Comput. Graph. **12**, 789–796 (2006)
43. Collins, C., Carpendale, S., Penn, G.: DocuBurst: visualizing document content using language structure. Comput. Graph. Forum **28**, 1039–1046 (2009)
44. Ghoniem, M., Cornil, M., Broeksema, B., Stefas, M., Otjacques, B.: Weighted maps: treemap visualization of geolocated quantitative data. In: IS&T/SPIE Electronic Imaging, International Society for Optics and Photonics, p. 93970G (2015)
45. Jänicke, S., Scheuermann, G.: Tagspheres: Visualizing hierarchical relations in tag clouds. In: GRAPP/IVApp (2016)

Visual Analysis of Character and Plot Information Extracted from Narrative Text

Markus John[1][(✉)], Steffen Lohmann[2], Steffen Koch[1], Michael Wörner[1], and Thomas Ertl[1]

[1] Institute for Visualization and Interactive Systems (VIS), University of Stuttgart, Universitätsstraße 38, 70569 Stuttgart, Germany
markus.john@vis.uni-stuttgart.de
[2] Fraunhofer Institute for Intelligent Analysis and Information Systems (IAIS), Schloss Birlinghoven, 53757 Sankt Augustin, Germany

Abstract. The study of novels and the analysis of their plot, characters and other information entities are complex and time-consuming tasks in literary science. The digitization of literature and the proliferation of electronic books provide new opportunities to support these tasks with visual abstractions. Methods from the field of computational linguistics can be used to automatically extract entities and their relations from digitized novels. However, these methods have known limitations, especially when applied to narrative text that does often not follow a common schema but can have various forms. Visualizations can address the limitations by providing visual clues to show the uncertainty of the extracted information, so that literary scholars get a better idea of the accuracy of the methods. In addition, interaction can be used to let users control and adapt the extraction and visualization methods according to their needs. This paper presents ViTA, a web-based approach that combines automatic analysis methods with effective visualization techniques. Different views on the extracted entities are provided and relations between them across the plot are indicated. Two usage scenarios show successful applications of the approach and demonstrate its benefits and limitations. Furthermore, the paper discusses how uncertainty might be represented in the different views and how users can be enabled to adapt the automatic methods.

Keywords: Text visualization · Visual text analytics · Digital humanities · Distant reading · Narrative text · Uncertainty visualization · NLP

1 Introduction

Common tasks in literary science are studying novels and analyzing their plot, characters and other entities. Literary scholars are interested in getting an overview of the plot and its characters, the relationships between them and their evolution during the plot [37]. The digitization of literature and the proliferation of electronic books (*ebooks*) provide new means to support these tasks with visual abstractions that are automatically generated from ebooks.

© Springer International Publishing AG 2017
J. Braz et al. (Eds.): VISIGRAPP 2016, CCIS 693, pp. 220–241, 2017.
DOI: 10.1007/978-3-319-64870-5_11

Traditionally, literary scholars read and analyze novels in a sequential way by using so-called *close reading*. In contrast to this, Moretti introduced the idea of *distant reading* [32]. Instead of carefully reading and analyzing a literary work, distant reading abstracts the text by providing visualizations, such as graphs that depict the genre change of historical novels, maps to represent geographical aspects of the plot, or trees to classify various types of detective stories [21]. These visual abstractions can convey useful information and assist in exploring and understanding complex relationships, verifying hypotheses as well as forming new research ideas.

In order to provide visual abstractions for literary works, a combination of automatic methods and interactive visualization techniques is required. When dealing with ebooks, natural language processing (NLP) methods are the first choice for automatic analysis. Using these methods, entities such as characters and places can be extracted from ebooks. This enables the development of visual abstractions that allow to explore these entities and their relationships in more detail.

However, the combination of automatic NLP methods and visual representations often involve a certain degree of uncertainty. The NLP methods are normally trained on specific text corpora, such as newspaper or journal article texts, and do not provide entirely correct results for texts that differ from the training corpus thematically, chronologically or stylistically. Since the visual abstractions are derived through NLP methods, the uncertainty, or rather the errors, are reflected in them.

However, inaccurate methods can facilitate, for example, the work of a human annotator by highlighting and providing interactive adaptation possibilities of the extracted text characteristics. Visual clues could indicate uncertain annotations, which are good candidates for manual annotation, and users can correct or confirm these classifications and trigger a retraining to improve the NLP method, thereby implementing a feedback loop as discussed by Sacha et al. [41]. Furthermore, visual clues can assist literary scholars in getting a better idea about the accuracy of the NLP methods and ease text exploration and decision making.

This paper is an extended version of a work presented at IVAPP 2016 [24]. It presents ViTA, a web-based approach that aims to provide literary scholars with visual abstractions to facilitate character analysis in novels. It utilizes automatic named entity extraction and visualizes relationships between characters and places based on their co-occurrence, also considering temporal aspects. The basic idea of ViTA is to highlight patterns, such as specific characters and places or groups of characters that interact with each other at certain places over time. By offering several views, such patterns are made easily recognizable with the approach and provide the starting point for a deeper analysis. This can result in a better understanding of the plot, in particular related to the characters and their relationships.

The main contributions of this work are: (1) A web-based approach that offers a wide range of interactive features to facilitate character analysis in narrative texts. (2) Several visual abstractions that provide aggregated and interrelated information and allow for an interactive navigation to the corresponding passages in the text. (3) A discussion about the importance of uncertainty visualizations to support the analysis and decision-making, and how users can be enabled to adapt automatic methods.

The rest of the paper is structured as follows: Sect. 2 summarizes related work, before the ViTA approach is detailed in Sect. 3. This is followed by two use cases demonstrating the applicability and usefulness of ViTA in Sect. 4. Section 5 provides a discussion of the approach and Sect. 6 concludes the paper with a summary and outlook on future work.

2   Related Work

Since our approach is concerned with the visual abstraction of text, we first summarize existing work in this area in Sect. 2.1. Next, we report on visual analytics attempts in the field of literary science and review systems that are most closely related to our approach in Sect. 2.2. Last, we discuss in Sect. 2.3 how uncertainty can be represented in visualizations and how users can steer and adapt automatic methods.

2.1   Visual Text Abstraction

Several techniques for visually summarizing and abstracting text documents have been developed over the last couple of years.

One compact visualization method related to our approach is *literature fingerprinting* [25], which uses a pixel-based technique that represents each text unit as a single pixel and visually groups them into higher level units. A similar technique is used in Seesoft [11], which has been designed as a visual fingerprint summarization of source code to graphically represent software statistics. The intention of Tilebars [17] is the visual representation of search results comparable to the fingerprint idea, while FeatureLens [9] also uses a pixel-based attempt to explore interesting text patterns and to find co-occurrences in texts.

Another popular technique to visually summarize text documents are *word clouds* [13,29]. They usually depict the most frequently used words of a text with the font size scaled according to the word frequencies. Word clouds enable literature scholars to get a first impression of the main terms and topics of a text and can provide a useful starting point for deeper analyses [19,47].

To visualize relational information of a text document, visualization techniques such as PhraseNets [46] and WordTrees [48] were suggested. They depict syntactic, lexical or hierarchical relationships that exist between the words of a text as node-link diagrams. In contrast, Oelke et al. [35] use an adjacency matrix to encode the development of relations between entities across a text document.

Inspired by Munroe's hand-drawn illustration "Movie Narrative Charts" [33], a new visual technique has emerged as so-called *storyline* or *plot view* visualization. It aims to portray the dynamic relationship between entities in a story over time. Tanahashi and Ma [45] as well as Liu et al. [28] propose design considerations and optimization approaches for generating aesthetically appealing storyline visualizations.

2.2 Visual Text Analytics

In recent years, several approaches for visual text analytics have been introduced in different domains. Examples can be found in social media [10], patent analysis [26] or opinion mining [34], among others.

There are also quite a number of visual analytics approaches in the field of literary science. Jänicke et al. [22] propose several techniques for the visualization and comparison of text that is reused in different documents in order to support literary scholars in discovering and exploring intertextual similarities. Abdul-Rahman et al. [1] present a rule-based solution for poetry visualization, allowing for high-level interactions with the end users in a closed loop. They use glyphs to encode phonetic units and visual links to show phonetic and semantic relationships. The VarifocalReader [27] supports literary scholars by combining distant and close reading and by enabling intra-document explorations through advanced navigation concepts. It integrates machine learning techniques, search mechanisms and several visual abstractions.

Oelke et al. [35] discuss the analysis of prose literature by using the aforementioned literature fingerprinting technique. Their approach visually abstracts implicit relationships between characters and encodes their development within the analyzed novel. However, it does not allow to directly work with the text resource. Vuillemot et al. [47] present the system POSvis, which extracts named entities from literary text and focuses on the exploration of networks of characters. POSvis offers multiple coordinated views, including word clouds and self-organizing graphs, equipped with filter methods to review the vocabulary of novels. While this is closely related to our work, we do not pursue the goal to review the vocabulary in the context of one or more entities filtered by part of speech. Instead, we aim to support the analysis of characters and their relationships in the storyline of a novel, based on named entity extraction and co-occurrence analyses.

Another system closely related to our work is Jigsaw [43], which has been designed to support analysts during foraging and sense-making activities in collections of textual reports and other sets of documents. It provides multiple coordinated views including lists, scatter plots, word clouds and graph visualizations that allow tracking entities and exploring their relationships across the document collections. Jigsaw has been primarily designed for *inter*-document analysis, whereas we are interested in *intra*-document analysis, i.e., we support the analysis of a single text document at a time. Apart from that, Jigsaw follows a rather generic approach that does not focus on fictional literature and the analysis of characters but provides general-purpose visualizations for different kinds of entities extracted from the documents.

2.3 Visualization of Uncertainty

The inclusion of information about the quality of data, over which people are reasoning, has been gaining attention across different domains, such as education [8], geographic and geologic mapping [14], or statistics [42]. For analysts, it is important to have access to information about the accuracy and reliability of data in order to correctly interpret the data and make informed decisions.

In particular in the domain of *geographic information systems*, a number of methods for uncertainty visualization have been proposed [30,36]. Griethe and Schumann [16] summarize different techniques to indicate uncertainty in data, including:

- *Visual variables*, such as color, size, position, angle, texture or transparency.
- *Extra objects*, such as labels, images or glyphs.
- *Animation*, such as speed, duration, motion blur, range or extent of motion.
- *Other human senses*, such as incorporation of acoustics, changes in pitch, volume, rhythm, vibration, or flashing textual messages.

As an example, Collins et al. [7] present a visualization intended to reveal the uncertainty and variability inherent in statistically-derived lattice structures. Their approach uses different visual variables, including transparency, color, and size, to expose the inherent uncertainty in statistical processing and thus helps analysts in making more informed decisions. Such visual clues can also help literary scholars in getting a better feeling of the accuracy of automatically extracted information, especially in distant reading when scholars are not deeply familiar with the texts. In addition, visual clues can assist scholars in finding and recognizing errors in visual abstractions faster.

There are a couple of attempts to combine automatic NLP and visualization methods to enhance text analysis tasks. However, only few works implement an interactive and visual approach for the user-steered adaptation of the NLP methods. Brown et al. [3] present a technique to develop similarity functions for machine learning models in a visual and interactive way. The technique does not require profound knowledge on the complex parameters of those models, but can also be controlled by users with little experience in machine learning. Similarly, Hu et al. [20] suggest a semantic interaction approach that allows users to adapt parameters by user interactions on spatializations of the underlying models. Endert et al. [12] also use such a semantic interaction approach to support sensemaking in document collections. It combines analytic interactions in a spatialization (e.g., document repositioning, text highlighting, search, annotations) with updates to the model responsible for generating the spatial layout. A related approach is presented by Heimerl et al. [18] that lets users interactively train a support vector machine for the classification of text documents.

3 Visual Text Analysis with ViTA

A large interest has grown in web-based systems for literature analysis that are easy to use and do not require any skills in computational linguistics [38]. Against this background, the ViTA approach has been implemented as a web application that is easily available to literary scholars and other user groups and does not require any installation on the user's side. The implementation is based on standard web technologies and can be run with a modern web browser supporting HTML5, SVG, CSS, and JavaScript.[1] It provides different visual abstractions representing specific characteristics of the analyzed narrative text and highlighting search results that illustrate the development of characters in a storyline.

The web application offers automatic methods for importing ebooks, extracting entity information and visualizing this information. The developed visualizations include word clouds, fingerprints of characters and places, a graph representation indicating connections between characters and a plot view that illustrates the relationships between characters and places in the story of a novel over time.

3.1   Text Processing

There are a variety of formats for the digital representation of novels. One widely used format is EPUB, which is a free and open standard that encodes structure and layout information besides the actual text of the novel. Many digital libraries, such as Project Gutenberg[2], offer ebooks in EPUB format or alternatively as plain text.

EPUB and plain text are also the two formats supported by our approach. It does not require the plain text to be structured in a specific way. However, if the structure of chapters or other metadata should be considered in the analysis, this information must be given in the text file. We therefore utilize some simple markup to structure ebooks provided in plain text, such as those by Project Gutenberg. The markup can be used to add chapter headings, line breaks and comments as well as other metadata (e.g., the title, author(s), publication date, publisher, edition or genre of a book) manually or automatically.

Once the ebook is loaded into the system, it is processed in a linguistic analysis pipeline, consisting of tokenization, sentence splitting and named-entity recognition. The ViTA implementation offers three different analysis tools that users can choose from: Stanford CoreNLP[3], OpenNLP[4] and ANNIE[5]. All three tools perform state-of-the-art NLP but use different techniques that each have their advantages and limitations. Depending on the use case and type of novel, users can select the NLP tool that is most suitable for the analysis. As this is often not clear from the start, they can also run the linguistic analysis several times with all three tools, compare the different outputs and choose the one that produces the best results. The current implementation only supports the processing of English texts; however, it can be extended to other languages if required.

---

[1] A video and demo of the web implementation are available at http://textvis. visualdataweb.org.

[2] http://www.gutenberg.org.

[3] http://nlp.stanford.edu/software/corenlp.shtml.

[4] http://opennlp.apache.org.

[5] https://gate.ac.uk/ie/annie.html.

The users can set several other parameters to configure the analysis and visualizations. Most importantly, they can control whether stop words are removed and whether person and place names starting with a lowercase letter should be considered. These configuration parameters are only shown on demand and are intended for experienced users. By default, ANNIE is used for NLP processing and unlikely character and place names are removed, as these settings produced the best results for most of the novels we tested.

3.2   Overview Page

After an ebook has been linguistically analyzed, an overview page is shown, listing metadata about the book and providing links to the visual abstractions. As an example, the overview page of the classic adventure novel "Around the World in 80 Days" by Jules Verne is shown in Fig. 1.

The metadata is listed in the middle of the screen (Fig. 1ⓑ). Some of it (e.g., the title, author, and release date) is directly taken from the text source (if provided), while the computation of other metadata (e.g., the number of words and chapters) requires some basic text analysis. Yet other metadata, such as the main characters listed on the overview page, can only be determined by using advanced text analysis, in this case named-entity recognition. This advanced text analysis is computationally expensive and can take some time depending on the size of the novel. For instance, the advanced analysis of the novel "Around the World in 80 Days" with the NLP tool ANNIE requires around 50 s on the current server that hosts the demo application (Intel Core i7-4930K with 3.4 GHz and 4 GB RAM), whereas the basic analysis is completed in less than 5 s.

As a general strategy, we therefore decided to show the results of each analysis step whenever they are available; for example, the web application already shows the results of the basic analysis even though the advanced analysis is still running. This strategy also applies to the visualizations provided by our

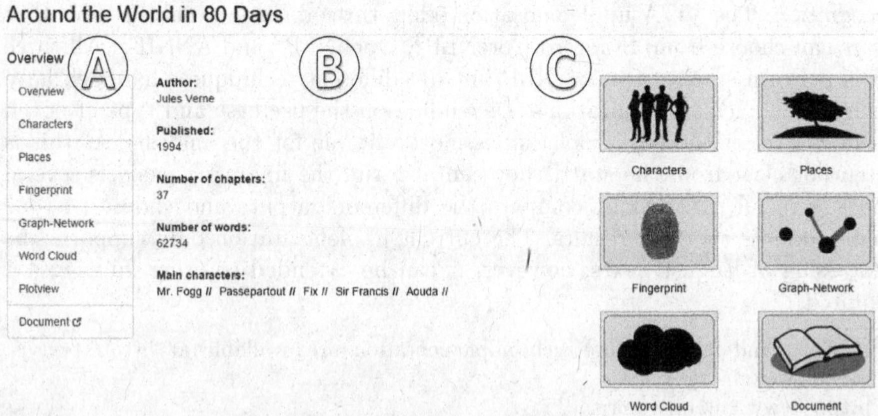

Fig. 1. Overview page listing metadata about the novel "Around the World in 80 Days".

approach. For instance, while the basic word cloud visualization can quickly be generated, the advanced word cloud as well as most other visualizations require more sophisticated NLP processing and can therefore not be shown before this processing is completed.

The interactive visualizations are available via the menu on the left (Fig. 1ⓐ), or via the icons on the right which are shown on the overview page (Fig. 1ⓒ). The fingerprint visualizations of the main characters can be directly opened from the overview page by clicking on the character links. The main characters are determined by counting the occurrences of all characters in the novel, with those that appear most often are assumed to be the protagonists. This simple measure worked surprisingly well for the novels we tested, in particular, since we also consider variations of the character names as detailed in Sect. 3.4.

The application assigns a unique color to each of the main characters. This color is shown on the overview page and constantly used for that character on all pages and in all visualizations. We created two color schemes, one for users with color vision deficiencies, consisting of four distinct colors determined by using the ColorBrewer 2.0[6], and the other for users with normal vision consisting of seven distinct colors using the categorical color scheme of D3[7]. If there are more than seven characters in a novel, the rest of them are shown in a gray color.

3.3   Characters and Places

To get an overview of the extracted characters and places, users can open either the characters or places view. Initially, the most frequently occurring entity is preselected in both views, complemented by a list of all extracted entities (Fig. 2ⓒ), where it is possible to search for and switch between entities.

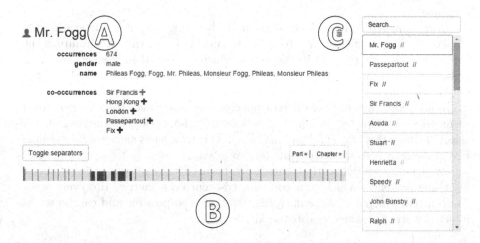

Fig. 2. Character view listing information about the character *Mr. Fogg* extracted from the novel.

[6] http://colorbrewer2.org.
[7] https://github.com/mbostock/d3/wiki/Ordinal-Scales#categorical-colors.

228   M. John et al.

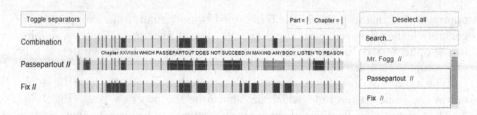

Fig. 3. Combined fingerprint visualization of two selected characters: *Passepartout* and *Fix*.

For each selected entity, a profile is provided, containing information about its occurrences, the detected gender and alternative names (for characters), and listing other entities that co-occur most frequently with that entity (Fig. 2ⓐ). In addition, the fingerprint visualization of the selected entity is shown (Fig. 2ⓑ).

3.4 Fingerprint Visualization

After the users obtained a first overview, they can further analyze the characters and places with the fingerprint visualization. It shows the temporal distribution of the entity occurrences in the novel. Blue and red bars represent parts and chapters of a book. Black blocks depict text segments where the entities occur. By clicking on the *toggle separators* button, the part and chapter bars can be hidden to get the 'plain' fingerprint.

Longer blocks indicate that entities occur often in that segment of the book, while shorter blocks appear when a character or place is only briefly mentioned. However, a longer block does not necessarily mean that entities are mentioned in every consecutive sentence, but it is sufficient if they are mentioned every few sentences.

Users can highlight a block segment by hovering over it to determine the respective chapter, shown in a tooltip. By clicking on it, they can jump to the corresponding text passage in the novel, which is opened in the text view. In that view, all occurrences are highlighted with the assigned specific color of the entity. This supports users in finding and analyzing text passages faster.

In addition, they can select multiple characters or places to get a combined fingerprint, as depicted in Fig. 3. The example shows the conjunction of two selected characters (*Passepartout* and *Fix*). This way, users can easily determine text passages where selected characters co-occur.

By default, we define that entities co-occur if they appear together in at least one sentence, which is a common co-occurrence metric. However, other co-occurrence metrics (range of words, etc.) are also possible and can be set as internal system parameters if required.

3.5 Character Network

To further investigate the character relations, users can switch to the character network view. This view contains a force-directed graph visualization that

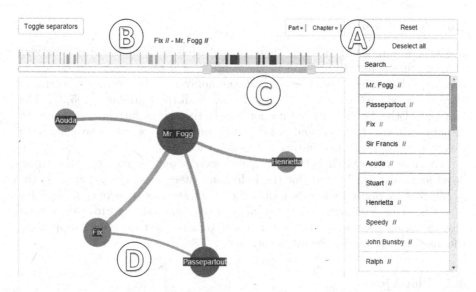

Fig. 4. Character network visualization representing the co-occurrence graph of a selected text segment. This example represents connections between the main characters of the novel "Around the World in 80 Days". The nodes of the graph represent the characters and the edges the number of sentences in which they co-occur.

represents connections between characters, as depicted in Fig. 4. The nodes of the graph represent the characters and the edges the number of sentences in which each pair of characters co-occurs. The node size and edge thickness are scaled proportionally to the characters' individual and co-occurrence frequency respectively. This helps to get a quick overview of the main characters and their connections.

The character network is once again complemented by a fingerprint visualization (Fig. 4ⓑ), indicating where the characters occur in the novel, and with a range slider (Fig. 4ⓒ) that lets users limit the view to a certain range of the novel (e.g., a single chapter). This has the advantage that users are not only enabled to analyze the overall structure of a novel but also the course of the relationships between characters, at least on the level of character co-occurrences. Users can select an edge in the graph (Fig. 4ⓓ) to display the co-occurrences of two related characters in the combined fingerprint visualization.

Initially, up to seven characters are preselected for the graph visualization, based on their occurrence frequency. The list of characters is displayed in that view again (Fig. 4ⓐ), so that users can search for characters and decide which ones are shown in the graph visualization by selecting or deselecting them in the list. That way, the graph visualization can be dynamically adapted according to the goals of the user. It can also be panned, zoomed, and rearranged to further support the analysis.

3.6   Word Cloud

If users are interested in terms that appear together with character mentions or in getting a first impression of the content of a book, they can switch to the word cloud view. Word clouds are commonly used by literary scholars, as they are considered easy to understand despite all their limitations [5, 31]. The font size of the words is scaled proportionally to their occurrence frequency to visually indicate this value and put it in relation to other words. The exact word frequencies can be viewed in tooltips on demand.

The user can switch between a global word cloud representing the entire novel and local word clouds for the individual characters. The latter show the words that co-occur most often with the characters, as depicted in Fig. 5 for *Mr. Fogg*. This gives users some flexibility in their analysis, by providing a visually appealing overview of the novel or a novel character as well as supporting the discovery of new ideas and hypotheses.

3.7   Plot View

Finally, users can switch to the plot view to get a better idea of the dynamic relationships between characters. It reuses and extends an implementation of the University of Waterloo[8], which takes annotation data and automatically generates a storyline visualization in the spirit of the aforementioned "Movie Narrative Charts" [33].

Our approach adapts the visualization and displays the ten most frequently occurring characters as lines and every chapter as a node (cf. Fig. 6ⓐ). The horizontal axis represents the plot of the novel and the vertical grouping of lines indicates which characters co-occur in the chapters. If two or more lines share a node, this means that the corresponding characters co-occur frequently in that chapter. When hovering over a node, a tooltip lists the characters and places which co-occur in that chapter. Hovering over a line highlights the whole line as well as the name of the corresponding character.

The plot view supports users in getting a rough idea of the course of the storyline. It allows to quickly identify when and where characters come together or go separate ways and whether groups of characters exist. In the plot view of Fig. 6, one can see, for example, that *Mr. Fogg* and *Fix* interact for the first time in chapter seven (Fig. 6ⓑ), while *Aouda* and *John Bunsby* enter the plot not before the middle of the novel (Fig. 6ⓒ).

3.8   Text View

To support literary scholars in their common workflow and allow for close reading, we also provide a text view where they can directly work with the text. Recognized chapters are listed as hyperlinks on the left (Fig. 7ⓐ), while the text is presented on the right (Fig. 7ⓑ). The focused chapter is emphasized with bold

---

[8] http://csclub.uwaterloo.ca/~n2iskand/?page_id=13/.

Fig. 5. Word cloud for the character *Mr. Fogg*.

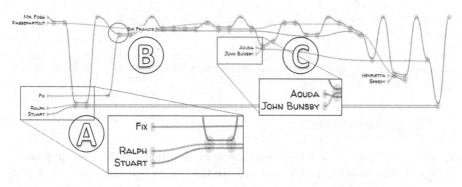

Fig. 6. Plot view of Jules Verne's novel "Around the World in 80 Days".

type. After clicking on a chapter, the text view jumps to the beginning of that chapter. If the user reached the text view from the fingerprint, graph network or word cloud view, the selected entity or entities are highlighted in the assigned color. Furthermore, there is a possibility for searching any other word or text passage as well as to reset the highlighting with a reset button.

The text view displays a vertical fingerprint next to its scrollbar. The idea is to provide both a visual representation of the distribution of entities and the possibility to inspect a text passage in detail, in order to support both distant and close reading. When hovering over the fingerprint blocks, the corresponding text passage is displayed in a tooltip (Fig. 7ⓑ), and after clicking on one, the text view jumps to the corresponding position. Additionally, the literary scholars

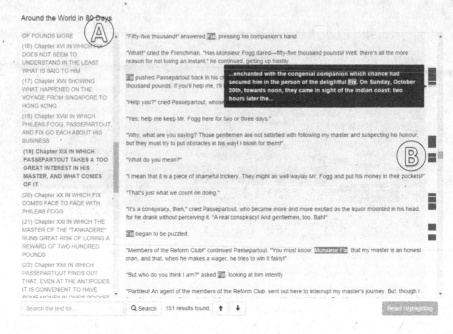

Fig. 7. Text view with selected character *Fix*.

can navigate to the next or previous occurrence of an entity by clicking on the up and down arrow buttons.

4   Usage Scenarios

In the following, we present two usage scenarios that demonstrate the applicability and usefulness of the ViTA approach. In the ePoetics project[9], we developed methods to support the analysis of German poetics – a form of early scholarly works on literature that formed the foundation of modern literature science. During our collaboration with colleagues from the literature department, we discussed the usefulness of direct access to text sources when using visual abstractions. This discussion inspired the development of an approach for analyzing narrative texts in a similar way, but taking into account the specific aspects of the plot and the social network of characters described in the text.

Although this was not the primary focus of the project, our colleagues emphasized its usefulness for their own work and for teaching. We therefore decided to set up a corresponding student project in oder to find out whether creating such visual abstractions from text mining results would be possible using off-the-shelf NLP techniques and tested the applicability on well-known novels.

For the usage scenarios, we selected a modern and an old English novel for analysis by a fictitious literary scholar. She has previous knowledge about the

---

[9] http://www.epoetics.de.

novels, since she read them some time ago, and is now trying to retrace the storyline and to detect the most important characters and events with our ViTA approach.

4.1 Analysis of the Novel "Harry Potter and the Half-Blood Prince"

In our first usage scenario, we present an analysis of the novel "Harry Potter and the Half-Blood Prince" by J. K. Rowling. It is the sixth and penultimate novel in the Harry Potter series and was published in 2009. The series chronicles the adventures of the young wizard Harry Potter and his quest to defeat the dark wizard Lord Voldemort, who strives to rid the wizarding world of Muggle (non-magical) heritage.

In a first step, the literary scholar explores and analyzes the character and network view. That way, she gets a quick overview of the main characters and their relationships. During the analysis, she encounters the name Slughorn and is surprised because she cannot remember him. To find out more about Slughorn, she selects the name in the character view (Fig. 8ⓐ) and uses the fingerprint visualization to navigate directly to its first occurrence, which is opened in the text view.

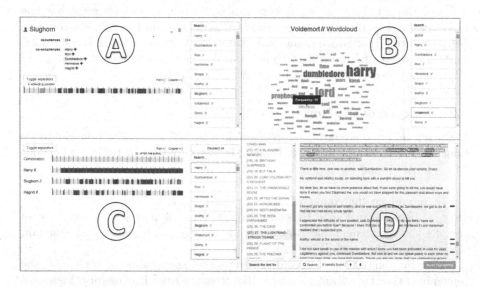

Fig. 8. Some of the visualizations used in the analysis of the novel "Harry Potter and the Half-Blood Prince": character view ⓐ, word cloud ⓑ, fingerprint visualization ⓒ and text view ⓓ.

While reading some paragraphs in the text view, she finds out that Dumbledore, the headmaster of the wizarding school Hogwarts, convinced Slughorn to return as potions teacher. Afterwards, she vaguely remembers that Dumbledore tasked Harry Potter with retrieving a memory from Slughorn that contains crucial information about Voldemort. To pursue this assumption, she switches to the word cloud view and explores the word clouds of Slughorn and Voldemort (Fig. 8ⓑ).

In the word cloud of Voldemort, she identifies the term Horcrux (an object in which a dark wizard has hidden a fragment of his soul for the purpose of attaining immortality) and remembers a conversation between Slughorn and Harry Potter at the home of Hagrid, the gamekeeper of Hogwarts.

In order to find the text passage, she uses the fingerprint view and selects the three characters Slughorn, Hagrid and Harry Potter. She determines that all three characters only co-occur at one text passage (Fig. 8ⓒ). Consequently, she jumps to that text passage and finds that Harry Potter succeeds in retrieving the memory, which shows Voldemort asking for information on creating Horcruxes.

In the following, she switches to the plot view since she is interested in examining the course of the storyline again. She still knows that Dumbledore dies at the end of the book and that Draco Malfoy, the son of one of Voldemort's followers, and Severus Snape, a professor at Hogwarts, are involved in his death. By analyzing the different chapters (nodes) and occurring characters (lines) in the plot view, she quickly recognizes the chapter of Dumbledore's death and navigates directly to the text of this chapter.

With the aid of the vertical fingerprints next to the text view's scroll bar, she can easily analyze the relevant text passages as depicted in Fig. 8ⓓ. She confirms her recollection that Draco Malfoy was chosen by Voldemort to kill Dumbledore. Furthermore, she finds out that Malfoy was unable to bring himself to do it and that Snape accomplished it with a deadly curse.

4.2   Analysis of the Novel "The Hobbit"

In the second usage scenario, our fictitious literary scholar analyzes the children's book "The Hobbit" by J. R. R. Tolkien. It was published in 1937 and is about Bilbo, a hobbit and the protagonist, and his adventures with dwarfs, elves, trolls and a dragon.

To reproduce the course of the novel, she starts her analysis on the plot view, as depicted in Fig. 9. By scanning the view, she gets a quick overview of the plot and remembers that Bilbo's adventure begins at his home with Gandalf, a wizard, and 13 dwarfs (Fig. 9ⓐ).

After jumping to the text and reading some passages in the text view, she remembers that they want to recover the treasure from Erebor (also known as the Lonely Mountain, former home to the greatest dwarf kingdom) and Bilbo is hired as their "burglar", since hobbits are small and unobtrusive.

Once she returns to the plot view, she notices the name of the creature Gollum (Fig. 9ⓑ) – originally a hobbit – and recalls that Bilbo wins a magical ring from him in a riddle war. However, she is unsure which further role Gollum plays in the plot and whether he co-occurs with other characters. To gain insights into this question, she switches to the graph view and immediately recognizes that

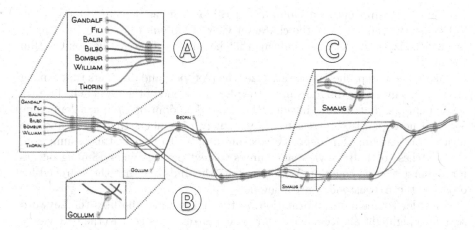

Fig. 9. Plot view showing some of the main characters of J. R. R. Tolkien's novel "The Hobbit".

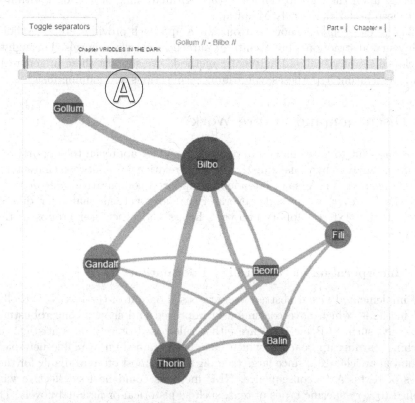

Fig. 10. Character network showing some of the main characters of "The Hobbit" and their co-occurrence frequencies. It can be observed that the character "Gollum" interacts only with "Bilbo" in the novel, while there seem to be strong ties between "Gandalf", "Thorin", "Bilbo" and the other main characters in the novel.

only a relationship between Gollum and Bilbo exist, as depicted in Fig. 10ⓐ. Subsequently, she activates the character view of Gollum to examine his fingerprint visualization, and thus confirms that he only appears at this point within the novel.

As the next step, she further analyzes the plot view and discovers that Smaug (Fig. 9ⓒ), a great fire dragon, enters the plot at the end of the novel. The literary scholar knows that Smaug invaded the dwarf kingdom of Erebor a long time ago and now guards the treasure. She vaguely remembers that Bilbo and the dwarfs are searching for a specific treasure object. To inspect this assumption, she switches to the text view and jumps to text passage where Smaug occurs for the first time in the plot. While reading, she finds out that the searched-for object is the Arkenstone, a great jewel.

In order to get more information on the Arkenstone, she uses the keyword search to highlight all occurrences as vertical fingerprints next to the text view's scroll bar. By analyzing the text passages, she determines, for example, that the Arkenstone is a heirloom of the dwarf kings and that, at the end of the novel, it is placed upon the chest of Thorin – the legitimate king of Erebor – within his tomb deep under the Lonely Mountain.

The usage scenarios show that our ViTA approach provides literary scholars with visual abstractions that facilitate character analysis in novels. Through the developed set of visual and automatic methods, we support them in generating and investigating hypotheses, confirming recollections and gaining insights.

5 Discussion and Future Work

It is important to note that we developed the ViTA approach to support *intra-document* analysis by visualizing the dynamic relationships between characters in a narrative text. ViTA is not intended to support a comparative *inter-document* analysis. However, we provide several visual abstractions that offer different views on the text to support literary scholars and other user groups in their analysis.

5.1 Incorporation of Uncertainty Information

The implemented visual abstractions are based on out-of-the-box NLP toolkits and methods, which have continuously improved and show strong robustness. Typically, such NLP methods are either rule-based or rely on statistical and machine learning approaches, which have been trained on available newspaper or journal article texts, since large training sets are most often available for these types of texts. As a consequence, NLP methods could be less effective when applied to very specific types of text, such as historical or fictional novels. This can results in uncertainties and errors in the visual abstractions. An uncritical interpretation of the provided visualizations can therefore cause confusion and misunderstandings.

Interactive visualizations can play an important role in addressing these problems. We intend to provide visual clues that communicate the quality of an automatic analysis. This could help literary scholars to interpret automatically generated results more accurately. Colors are particularly suited for the representation of such visual clues, for example, green could represent a very reliable result, whereas red could indicate high uncertainty. As mentioned before, labels are also suitable to communicate uncertainty information, for example, by adding a confidence value (e.g., given as percentage) for each extracted entity. However, labeling can be challenging as the uncertainty information must not obscure other labels or graphical elements in the visualization, and must be easy to spot by users at the same time. Therefore, labels would be particularly useful in the character and places views of ViTA, as there is sufficient space to represent uncertainty information next to the detected entities.

Another possibility to indicate the accuracy of an extraction method is to use glyph-based techniques like Chernoff faces [4]. However, these techniques have the drawback that they do not scale to a large number of entities and that users first need to learn how to read them. A more appropriate method to show the confidence value of automatic methods with the help of visual clues is color saturation. For example, light colors could indicate entities with a high uncertainty, which are good candidates for manual correction, while strong colors could indicate entities which are assumed to be accurate. Using saturation to communicate uncertainty is easy to understand and can also be easily integrated in the visualizations provided by ViTA. For example, color saturation can be added to nodes and lines to represent the accuracy of the automatically generated results in the graph network or plot view. Moreover, saturation might be used to display the accuracy of recognized terms in the word cloud view or of the extracted entities in the text view. The use of animations to represent uncertainty has to be handled with care, as animations are known to quickly confuse users with transient information and may result in perceptual overload.

5.2   Other Extensions and Improvements

We plan to provide better means to interactively control and adapt automatic NLP methods as they are used in ViTA. In this regard, visual clues can support users in finding uncertain extraction results, which are good candidates for manual correction. Users could then correct or confirm the extracted information one after the other or according to some predefined priority schemes. This could finally help to improve the automatic methods and analysis results, also when applied to similar texts.

The current implementation for the detection of co-occurrences between entities is based on named entity recognition. We determine that two entities co-occur if they both appear in the same sentence. This approach works well in many cases, however, it can be improved in various aspects. One option would be to let the literary scholars set the co-occurrences range variable, for example, increase the context to several sentences or a whole paragraph. Also, methods for coreference resolution might be integrated to detect alternative occurrences of the entities in the text and compute more precise frequency values [39, 44]. However, existing coreference techniques are again mostly trained on non-fictional texts and have to be adapted (where appropriate) to the type of text and user needs. We also plan to take the semantics of the text into account to better investigate the relationships between entities. Once again existing approaches [15, 40] could be integrated to provide the literary scholars with additional information for their analyses. In general, we aim at supporting a broader spectrum of natural language processing methods in the future.

Another topic for future work is the challenge of visual scalability when working with very long texts. Most of the visual abstractions used in ViTA scale quite well. However, the fingerprint visualization and the plot view become less useful with a growing text length. Since the available screen space remains constant, they are increasingly compressed until they finally overlap. To address this problem, some focus+context technique could be integrated, such as a fisheye distortion [2], or an overview+detail approach to present multiple views with different levels of abstractions [6].

Apparently, the presented ViTA approach can also be extended by additional visualizations. For example, views that visualize geographical information [23] might be integrated if the story of the text is based on real world places, such as the above analyzed novel "Around the World in 80 Days", or own cartographic material in the case of fictional places like in the novel "The Hobbit". Furthermore, the existing visualizations could be improved and extended. One such extension could be different graph layouts for the character network to get a better impression of the constellation of characters or to add more information, such as semantic relations between characters. Another possible extension could be to equip the plot view with automatically detected events (e.g., the "Battle of Helm's Deep" in case of the novel "The Lord of the Rings"), enabling users to better track and identify the course of the story.

6   Conclusion

In this work, we presented an easily accessible web-based approach for visualizing the relation of characters and places in a novel. The approach incorporates different natural language processing toolkits to extract named entities, and provides possibilities to set parameters for variable analyses. Furthermore, it provides several visual abstractions that support literary scholars in a flexible and comprehensive analysis of the novel characters. The approach facilitates distant reading, in particular, and provides a starting point for new ideas, hypotheses and further analyses. All views enable literary scholars to jump to the corresponding text passages and thus allow for working with the text directly. We presented an implementation of the approach and provided two usage scenarios that illustrate its applicability and usefulness.

Acknowledgements. We would like to thank our students Sanjeev Balakrishnan, Felix Do, Sebastian Frank, Paul Kuznecov, Vincent Link, Eduard Marbach, Jan Melcher, Christian Richter, Marc Weise, and Marvin Wyrich who implemented the approach in a student project. This work has partly been funded by the German Federal Ministry of Education and Research (BMBF) as part of the 'ePoetics' project and as part of the Center for Reflected Text Analysis 'CRETA' at University of Stuttgart.

References

1. Abdul-Rahman, A., Lein, J., Coles, K., Maguire, E., Meyer, M., Wynne, M., Johnson, C.R., Trefethen, A., Chen, M.: Rule-based visual mappings - with a case study on poetry visualization. Comput. Graph. Forum **32**(3pt4), 381–390 (2013)
2. Bederson, B.: Fisheye menus. In: Proceedings of the 13th Annual ACM Symposium on User Interface Software and Technology, UIST 2000, pp. 217–225. ACM (2000)
3. Brown, E., Liu, J., Brodley, C., Chang, R.: Dis-function: learning distance functions interactively. In: Proceedings of the IEEE Conference on Visual Analytics Science and Technology, VAST 2012, pp. 83–92 (2012)
4. Chernoff, H.: The use of faces to represent points in k-dimensional space graphically. J. Am. Stat. Assoc. **68**(342), 361–368 (1973)
5. Clement, T., Plaisant, C., Vuillemot, R.: The story of one: humanity scholarship with visualization and text analysis. In: Proceedings of the Digital Humanities Conference, DH 2009. HCIL-2008-33 (2009)
6. Cockburn, A., Karlson, A., Bederson, B.: A review of overview+detail, zooming, and focus+context interfaces. ACM Comput. Surv. **41**(1), 1–31 (2009)
7. Collins, C., Carpendale, S., Penn, G.: Visualization of uncertainty in lattices to support decision-making. In: Proceedings of the 9th Joint Eurographics/IEEE VGTC Conference on Visualization, EuroVis 2007, pp. 51–58. Eurographics Association (2007)
8. Epp, C.D., Bull, S.: Uncertainty representation in visualizations of learning analytics for learners: current approaches and opportunities. IEEE Trans. Learn. Technol. **8**(3), 242–260 (2015)
9. Don, A., Zheleva, E., Gregory, M., Tarkan, S., Auvil, L., Clement, T., Shneiderman, B., Plaisant, C.: Discovering interesting usage patterns in text collections: integrating text mining with visualization. In: Proceedings of the 16th ACM Conference on Conference on Information and Knowledge Management, CIKM 2007, pp. 213–222. ACM (2007)
10. Dou, W., Wang, X., Skau, D., Ribarsky, W., Zhou, M.: Leadline: interactive visual analysis of text data through event identification and exploration. In: Proceedings of the IEEE Conference on Visual Analytics Science and Technology, VAST 2012, pp. 93–102 (2012)
11. Eick, S., Steffen, J., Sumner, E.E.: Seesoft-a tool for visualizing line oriented software statistics. IEEE Trans. Softw. Eng. **18**(11), 957–968 (1992)
12. Endert, A., Fiaux, P., North, C.: Semantic interaction for sensemaking: inferring analytical reasoning for model steering. IEEE Trans. Vis. Comput. Graph. **18**(12), 2879–2888 (2012)
13. Feinberg, J.: Wordle. In: Beautiful Visualization, pp. 37–58. O'Reilly (2010)
14. Fisher, P.F.: Visualizing uncertainty in soil maps by animation. Cartogr.: Int. J. Geog. Inf. Geovisualization **30**(2–3), 20–27 (1993)
15. Gildea, D., Jurafsky, D.: Automatic labeling of semantic roles. Comput. Linguist. **28**(3), 245–288 (2002)

16. Griethe, H., Schumann, H.: The visualization of uncertain data: methods and problems. In: Proceedings of Simulation und Visualisierung 2006, SimVis 2006, pp. 143–156. SCS (2006)
17. Hearst, M.A.: Tilebars: visualization of term distribution information in full text information access. In: Proceedings of the SIGCHI Conference on Human Factors in Computing Systems, CHI 1995, pp. 59–66. ACM/Addison-Wesley (1995)
18. Heimerl, F., Koch, S., Bosch, H., Ertl, T.: Visual classifier training for text document retrieval. IEEE Trans. Vis. Comput. Graph. 18(12), 2839–2848 (2012)
19. Heimerl, F., Lohmann, S., Lange, S., Ertl, T.: Word cloud explorer: text analytics based on word clouds. In: 47th Hawaii International Conference on System Sciences, HICSS 2014, pp. 1833–1842. IEEE (2014)
20. Hu, X., Bradel, L., Maiti, D., House, L., North, C.: Semantics of directly manipulating spatializations. IEEE Trans. Vis. Comput. Graph. 19(12), 2052–2059 (2013)
21. Jänicke, S., Franzini, G., Cheema, M.F., Scheuermann, G.: On close and distant reading in digital humanities: a survey and future challenges. In: Eurographics Conference on Visualization - STARs, EuroVis 2015. The Eurographics Association (2015)
22. Jänicke, S., Geßner, A., Büchler, M., Scheuermann, G.: Visualizations for text re-use. In: Proceedings of the 5th International Conference on Information Visualization Theory and Applications, IVAPP 2014, pp. 59–70. Scitepress (2014)
23. Jänicke, S., Heine, C., Stockmann, R., Scheuermann, G.: Comparative visualization of geospatial-temporal data. In: Proceedings of the 3rd International Conference on Information Visualization Theory and Applications, IVAPP 2012, pp. 613–625. Scitepress (2012)
24. John, M., Lohmann, S., Koch, S., Wörner, M., Ertl, T.: Visual analytics for narrative text - visualizing characters and their relationships as extracted from novels. In: Proceedings of the 7th International Conference on Information Visualization Theory and Applications, IVAPP 2016, pp. 27–38. Scitepress (2016)
25. Keim, D., Oelke, D.: Literature fingerprinting: a new method for visual literary analysis. In: Proceedings of the IEEE Symposium on Visual Analytics Science and Technology, VAST 2007, pp. 115–122 (2007)
26. Koch, S., Bosch, H., Giereth, M., Ertl, T.: Iterative integration of visual insights during scalable patent search and analysis. IEEE Trans. Vis. Comput. Graph. 17(5), 557–569 (2011)
27. Koch, S., John, M., Wörner, M., Müller, A., Ertl, T.: VarifocalReader - in-depth visual analysis of large text documents. IEEE Trans. Vis. Comput. Graph. 20(12), 1723–1732 (2014)
28. Liu, S., Wu, Y., Wei, E., Liu, M., Liu, Y.: Storyflow: tracking the evolution of stories. IEEE Trans. Vis. Comput. Graph. 19(12), 2436–2445 (2013)
29. Lohmann, S., Ziegler, J., Tetzlaff, L.: Comparison of tag cloud layouts: task-related performance and visual exploration. In: Gross, T., Gulliksen, J., Kotzé, P., Oestreicher, L., Palanque, P., Prates, R.O., Winckler, M. (eds.) INTERACT 2009. LNCS, vol. 5726, pp. 392–404. Springer, Heidelberg (2009). doi:10.1007/978-3-642-03655-2_43
30. MacEachren, A.M., Robinson, A., Hopper, S., Gardner, S., Murray, R., Gahegan, M., Hetzler, E.: Visualizing geospatial information uncertainty: what we know and what we need to know. Cartogr. Geogr. Inf. Sci. 32(3), 139–160 (2005)
31. McNaught, C., Lam, P.: Using wordle as a supplementary research tool. Qual. Rep. 15(3), 630–643 (2010)
32. Moretti, F.: Graphs, Maps, Trees: Abstract Models for a Literary History. Verso (2005)

33. Munroe, R.: Movie Narrative Charts (2009). http://xkcd.com/657/
34. Oelke, D., Hao, M., Rohrdantz, C., Keim, D., Dayal, U., Haug, L., Janetzko, H.: Visual opinion analysis of customer feedback data. In: Proceedings of the IEEE Symposium on Visual Analytics Science and Technology, VAST 2009, pp. 187–194 (2009)
35. Oelke, D., Kokkinakis, D., Keim, D.A.: Fingerprint matrices: uncovering the dynamics of social networks in prose literature. Comput. Graph. Forum 32(3pt4), 371–380 (2013)
36. Pang, A.T., Wittenbrink, C.M., Lodha, S.K.: Approaches to uncertainty visualization. Vis. Comput. 13(8), 370–390 (1997)
37. Phelan, J.: Reading People, Reading Plots: Character, Progression, and the Interpretation of Narrative. University of Chicago Press, Chicago (1989)
38. Plaisant, C., Rose, J., Yu, B., Auvil, L., Kirschenbaum, M.G., Smith, M.N., Clement, T., Lord, G.: Exploring erotics in Emily Dickinson's correspondence with text mining and visual interfaces. In: Proceedings of the 6th ACM/IEEE-CS Joint Conference on Digital Libraries, JCDL 2006, pp. 141–150. ACM (2006)
39. Raghunathan, K., Lee, H., Rangarajan, S., Chambers, N., Surdeanu, M., Jurafsky, D., Manning, C.: A multi-pass sieve for coreference resolution. In: Proceedings of the 2010 Conference on Empirical Methods in Natural Language Processing, EMNLP 2010, pp. 492–501. ACL (2010)
40. Ruiz-Casado, M., Alfonseca, E., Castells, P.: Automatising the learning of lexical patterns: an application to the enrichment of wordnet by extracting semantic relationships from Wikipedia. Data Knowl. Eng. 61(3), 484–499 (2007)
41. Sacha, D., Stoffel, A., Stoffel, F., Kwon, B.C., Ellis, G., Keim, D.A.: Knowledge generation model for visual analytics. IEEE Trans. Vis. Comput. Graph. 20(12), 1604–1613 (2014)
42. Skeels, M., Lee, B., Smith, G., Robertson, G.G.: Revealing uncertainty for information visualization. Inf. Vis. 9(1), 70–81 (2010)
43. Stasko, J., Görg, C., Liu, Z.: Jigsaw: supporting investigative analysis through interactive visualization. Inf. Vis. 7(2), 118–132 (2008)
44. Stoyanov, V., Cardie, C., Gilbert, N., Riloff, E., Buttler, D., Hysom, D.: Coreference resolution with reconcile. In: Proceedings of the ACL 2010 Conference Short Papers, pp. 156–161. ACL (2010)
45. Tanahashi, Y., Ma, K.-L.: Design considerations for optimizing storyline visualizations. IEEE Trans. Vis. Comput. Graph. 18(12), 2679–2688 (2012)
46. Van Ham, F., Wattenberg, M., Viegas, F.: Mapping text with phrase nets. IEEE Trans. Vis. Comput. Graph. 15(6), 1169–1176 (2009)
47. Vuillemot, R., Clement, T., Plaisant, C., Kumar, A.: What's being said near "Martha"? Exploring name entities in literary text collections. In: Proceedings of the IEEE Symposium on Visual Analytics Science and Technology, VAST 2009, pp. 107–114 (2009)
48. Wattenberg, M., Viegas, F.: The word tree, an interactive visual concordance. IEEE Trans. Vis. Comput. Graph. 14(6), 1221–1228 (2008)

Visual Querying of Semantically Enriched Movement Data

Florian Haag[✉], Robert Krüger, and Thomas Ertl

Institute for Visualization and Interactive Systems, University of Stuttgart,
Universitätsstraße 38, 70569 Stuttgart, Germany
{florian.haag,robert.krueger,thomas.ertl}@visus.uni-stuttgart.de

Abstract. Visual data exploration is used to reveal unknown patterns that, however, need to be validated, refined, and extracted for a final presentation and reporting. We contribute VESPa, a pattern-based visual query language for event sequences. With VESPa, analysts can formulate hypotheses gained and query the data for matches. In an interative analysis loop the pattern can be altered with further restrictions to narrow down the result set. Our language allows for (1) hypothesis expression and refinement, (2) visual querying, and (3) knowledge externalization. We focus on semantically enrichend movement data, used in law enforcement, consumer, and traffic analysis. To evaluate the applicability we present two case studies as well as a user study consisting of comprehensive and composition tasks.

Keywords: Visual query language · Semantic movement analysis

1 Introduction

Rising amounts of movement data and an increasing recording precision offer new possibilities for urban and transport planning, customer analysis, and law enforcement. While visual exploration is an important task to reveal the unknown, analysts often have specific questions about the data. Regarding urban planning tasks, an analyst may be interested in checking whether a certain suburb is well connected to the inner city area, or more precisely, if there are any times of the day when roads are congested and the daily commute takes particularly long. To improve sales and quality of service, it is of interest to understand indoor behavior patterns such as the flow of customers during their stay in a mall, or visitors exploring different exhibits in a museum. Also, concerning surveillance tasks, spatio-temporal happenings need to be verified. Here, especially gatherings are of interest, that is, when several persons meet at a specific time at a specific place. More abstractly, each of these examples revolves around a sequence of consecutive spatio-temporal events that involve a single or multiple persons. Figure 1 exemplarily illustrates data of three persons' movement paths and the event sequences therein.

This publication is an extended version of our IVAPP 2016 paper [1]

© Springer International Publishing AG 2017
J. Braz et al. (Eds.): VISIGRAPP 2016, CCIS 693, pp. 242–263, 2017.
DOI: 10.1007/978-3-319-64870-5_12

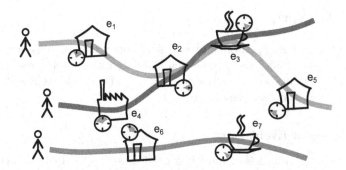

Fig. 1. A sketch of a movement dataset that stores consecutive stays and meetings of people at specific locations and time as sequences of events. For instance, the person in the middle has the event sequence $e_4 \rightarrow e_2 \rightarrow e_3$ and meets the person on top in e_2 and e_3.

Detecting such situations can be complex, as multiple conditions like varying timespans, different places, persons, and orderings of events need to be expressed. This is especially true for domain experts in the aforementioned domains who have no programming or database experience. Visual query notations can help to hide this complexity under an easier to understand set of visual items. To the best of our knowledge, none of the existing graphical query notations for event data (e.g. [2,3]) visualizes meetings of multiple persons at certain points of interest.

To address this gap, we propose a new notation to visually express such spatio-temporal event sequences of one or more persons. Once defined, the visual event sequence can be used as a graphical representation of a hypothesized sequence of events. Likewise, it can be used as a pattern to filter the data and find any instances of the presumed event sequence, thereby confirming the hypothesis. Based on the results, the event sequence pattern can also be iteratively refined in a verification loop, based on the sensemaking process by Pirolli and Card [4]. Our approach is especially helpful for domain experts of an application field (e.g. urban planning, digital forensics) who are not familiar with formal text-based query languages such as SQL. Furthermore, users who could express their queries in SQL might still benefit from a visual representation to facilitate error checking and communication.

After discussing related work in the fields of event data models, event visualization, and event filtering in Sect. 2, we present VESPa, our Visual Event Sequence Pattern notation, in Sect. 3 and justify our design choices. The general idea of using the technique is outlined in Sect. 4. We demonstrate the usefulness of VESPa with several pattern examples, two case studies (Sect. 6), and a small user study (Sect. 7), both using a prototypical implementation (Sect. 5).

2   Related Work

We have looked at various options to represent concrete event sequences and to visualize the events found therein, as well as related spatio-temporal data. The findings from that literature survey provide a basis to discuss possible query visualizations for such event sequences.

2.1   Modeling of Event Data

There are several variations of modeling event-based data. In basic event logs, events are usually associated with a timestamp and carry some structured information on the event [5,6]. In the case of multiple logging sources, that information can also include a hint about the origin of the event message [7].

For the purpose of representing action sequences described in texts, a timeline of events can be defined as a series of intervals that are defined by an action and related resources [8]. Similarly, events are defined with respect to sensor data [9–11], where a timespan can be supplied as an additional attribute [12].

Entries in spatio-temporal event logs can be associated with timespans and locations [13]. Such entries may include data that implicitly or explicitly points out a connection between related events [14,15].

All of these works do not involve any visualization or querying of events, and therefore are not comparable to our work. However, we will base our own model of event data on these works.

2.2   Visualization of Event Data

Temporal event data are frequently visualized in a timeline, such as by lines expressing states [16]. These data are commonly split up to distinguish different persons [17] or event types [10,18]. Additional information on single events or on sets of events can be encoded into the timeline [19,20]. In some cases, the lines still conform to a time axis, but are reshaped to express some information themselves [21,22].

When looking at the visualization of spatio-temporal events in particular, it becomes apparent that the spatial and the temporal dimensions are often expressed in separate, but linked views [20,23]. Alternative approaches use a space-time-cube [24,25], or they integrate aggregated spatial data into the temporal visualization [26]. Moreover, additional visual elements can indicate the temporal properties of data shown in a spatial view [27].

Rather than the exact geographical reality, a semantic view on locations (e.g., location names or categories rather than coordinates) can be extracted from geographic databases or generated from social networks [28,29]. The link between the geographical locations and the logical locations can be displayed [30], but in some cases, the primary interest lies in the semantic locations [31,32], or they are even the only data available or cleared for use [33].

While all of the approaches mentioned above show event sequences, they do not help to express and visualize queries on them.

2.3   Filtering of Event Data

A variety of approaches for visually expressing generic filter queries have been proposed in the past [34–37]. There are specialized query visualizations for spatial data, some using symbolic representations of geographical relationships [38,39] and others working directly on concrete maps [40]. The symbolic representations lend themselves to use cases that work with logical locations.

For selectively finding particular pieces of information in spatio-temporal data, the spatial and the temporal dimensions are split up in some concepts [41–43]. Also, a graphical notation that expresses exact relative relationships between locations or areas and timespans has been proposed [44]. The only other visual approach that we are aware of displays the query restrictions in a space-time-cube [45]. Some of these concepts can be further abstracted to replace the geographical association with restrictions based on logical locations [42], and time-related works provide some further ideas on how to visually represent certain features required for temporal queries [46].

So far, only few approaches specifically deal with the visual specification of queries for event sequences, which could be used for finding particular event patterns and validating hypotheses about event sequences. Those that do only visualize the event sequence, but they do not emphasize the query structure and its restrictions in a visual way other than aligning sequential states horizontally or vertically [2,16,47]. Only few concepts represent series of events to find on a timeline [48], or as nodes with temporal constraints [49], rather than just displaying results in a visual way [16]. Still, relations between distinct objects are usually not supported [3]. When they are, these are expressed by matching colors in otherwise visually disjoint elements [50].

As opposed to the aforementioned work, we focused on the graphical representation of event sequence patterns of multiple persons. Rather than just recognizing similar sequences [16], we are looking for actual overlaps that allow for an interaction of persons. Moreover, we want to visually express the overlaps of event sequences by actual connections rather than just by matching properties such as node colors [50] to make the connection explicit.

3   Query Notation

We contribute a new visual notation to represent patterns of event sequences. The notation can be used to express hypotheses and it can be executed as a data query, for example, to answer which kinds of locations people meet at before going to a restaurant together when evaluating the layout of a shopping mall. While this sequence of events can still be described relatively easily, queries can often become more complex: "Find all pairs of persons who travel together from one place to another place, where they arrive after 3 PM, after one of them has first visited an arbitrary location and then an ice cream parlor, while the other one has arrived from another location, where he or she has not met the first person." still describes a relatively simple sequence of events, yet the text is already quite long and possibly confusing. Queries of the same or a higher complexity can also be required, for instance, in digital forensics, when investigators try to find out whether people meet after having visited certain locations that are connected to a crime.

In the above example and the aforementioned literature [5,6], certain basic components of the query become apparent: There are several distinct *persons*. These persons stay at different locations, and the time at which they stayed there is relevant to determine whether the persons might possibly have met. Therefore, such a stay can be called an *event*. Furthermore, the *movements* of each person between the respective locations are considered, connecting events to *event sequences*. Finally, *restrictions* can be imposed on the elements mentioned so far, such as the type of a location, or an arrival time.

Based upon these components, we have defined a minimal set of visual elements. These elements can be connected in a node-link diagram, which we call Visual Event Sequence Pattern (VESPa). This visualization has two purposes:

1. It can be used to express a hypothesis about the event sequences of one or more known or unknown persons.
2. It can be executed as a database query, where each element from the event sequence pattern is mapped to a database item, for each result.

3.1   Query Elements

With the aforementioned goals and requirements, VESPa consists of the following elements:

Person Node: A person node as depicted in Fig. 2(a) is a placeholder for a movable subject, such as a person.

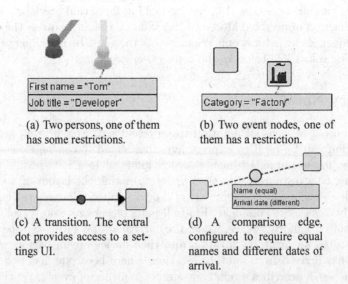

(a) Two persons, one of them has some restrictions.

(b) Two event nodes, one of them has a restriction.

(c) A transition. The central dot provides access to a settings UI.

(d) A comparison edge, configured to require equal names and different dates of arrival.

Fig. 2. The basic visual elements used in VESPa.

Event Node: An event node, shown in Fig. 2(b), is a single node in an event sequence pattern, like events found in related work [16,48,49]. In spatio-temporal contexts, it denotes a stay of one or more persons at a particular location and timespan.

Transition: A transition unidirectionally connects two event nodes (Fig. 2(c)), comparable to the TimeSpan bars from PatternFinder [48]. It represents how persons move on from one event node to another.

Restriction Tag: A restriction tag expresses an absolute condition that can be imposed on a person node or on an event node. Examples are depicted in Figs. 2(a) and (b).

Comparison Edge: A comparison edge connects two person nodes, two event nodes (as shown in Fig. 2(d)), or a person node and an event node, to enforce a relationship between any of their attributes. For instance, attribute values of two such nodes can be required to be equal or different. We have chosen to add this additional representation of restrictions, as conditions fulfilled with respect to another node have been considered in related concepts [37,51].

Based upon the aforementioned visual query elements, various compositions are possible. In the following, we will present some advanced compositions and restrictions by example.

Person nodes are always indicated next to the first event node in the sequence of the respective person and connected to it with a thick, semi-transparent line for clarification. There is only one person node for each person in the dataset. We chose this design, as restrictions on persons are globally valid throughout the whole event sequence. A simple example of a VESPa query with one person is shown in Fig. 3(a). It shows a person who moves from an arbitrary place to a place categorized as a *restaurant*.

Category = "Restaurant"

(a) A single person travels to a restaurant.

Category = "Restaurant"

(b) Two persons travel to a restaurant together.

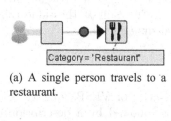

(c) Two persons meet after individual events.

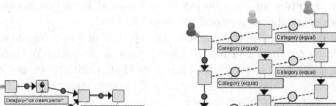

(d) Three persons visit three places of the same category.

Fig. 3. Various exemplary event sequence patterns.

Transitions are shown as colored lines. The color(s) of each transition match(es) the color(s) of the involved persons[1]. For instance, Fig. 3(b) shows how two persons meet at an arbitrary place and move on together to a restaurant from there—the example from the very beginning of this section.

The arrowhead indicates the direction of each transition. As event nodes represent locations *at a specific time* in spatio-temporal contexts, cycles of transitions and undirected transitions are disallowed. A transition may include several legs of a journey, as a specified number of events to skip can be supplied for each transition in the pattern.

Based upon these definitions, we can also represent the complicated pattern described textually in the beginning of this section, as shown in Fig. 3(c). As an example of how comparison edges can be used, Fig. 3(d) shows a query that finds three persons that subsequently visit three similar (with respect to their category attribute) places in the same order.

3.2   Interpretation of Event Nodes

As pointed out above, we chose to define an *event node* in a query as a place at a particular time. This allows us to focus on encounters between several persons, which thus need to coincide both in time and place. Furthermore, we define that no two event nodes can refer to stays of persons at the same place and at overlapping timespans.

Accordingly, Fig. 4(a) shows an event sequence pattern that represents independent event sequences of two persons who do not meet. Expressed as shown in Fig. 4(b), the persons meet at a restaurant (the central node marked with a fork-and-knife-symbol). In Fig. 4(c), both persons visit a particular restaurant (in a dataset where each restaurant has a unique name), but they do *not* meet. As both depicted restaurant event nodes refer to the same restaurant, and no two event nodes can be mapped to the same location at the same time, it is implied that the two persons stayed there at different times.

3.3   Query Semantics

In the following, we define the exact semantics of VESPa queries. First, we describe the structure of the queried data, followed by a description of how VESPa query elements are mapped to result elements.

We assume the small database schema depicted in Fig. 5, essentially a set of persons, each of which can have linear sequences of events. We have chosen this schema as its simple structure illustrates that VESPa does not require a specifically optimized dataset. Also, we expect the conversion of data from other schemas to ours to be sufficiently easy.

---

[1] For this paper, colors were chosen so as to be distinguishable both on colored and greyscale printouts. Conceptually, alternative color schemes can be chosen that are specific to user requirements, for instance, color blindness.

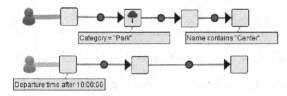

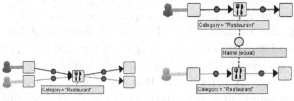

(a) Event sequences of two persons who do not meet each other.

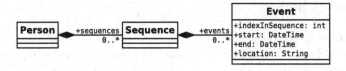

(b) Two persons meet each other in a restaurant.

(c) Two persons visit the same restaurant, but at different times (and hence do not meet).

Fig. 4. Contrast between persons who visit the same place at the same time and persons who visit the same place at different times.

Person		+sequences	Sequence		+events	**Event**

Fig. 5. UML diagram of the database schema assumed for the data that can be filtered with VESPa. Persons and events may have additional columns for domain-specific attributes that can be used in restrictions.

Formally, a dataset matching the schema from Fig. 5 can be seen as a tuple $d \subseteq \text{Persons}_d \times \text{Sequences}_d \times \text{Events}_d$ of three disjoint sets. We are interested in meetings between persons, but in the above schema, each event belongs to exactly one person. Therefore, a pre-processing step is necessary to replace Events_d with Events'_d—a set of events linked to one or more persons and part of one or more event sequences. Each element $e_d \in \text{Events}'_d$ contains a set $e_d.\text{persons} \subseteq \text{Persons}_d$ of participants of the event.

Events'_d contains copies of all events from Events_d, as well as any events describing possible meetings. Formally, any element of Events'_d is based upon a set $\tilde{E}_d \subseteq \text{Events}_d$ of events from the original dataset. Then, $\tilde{E}_d$ must satisfy the condition

$$\forall \tilde{e}_{d,\alpha}, \tilde{e}_{d,\beta} \in \tilde{E}_d, \tilde{e}_{d,\alpha} \neq \tilde{e}_{d,\beta} : (\tilde{e}_{d,\alpha}.\text{location} = \tilde{e}_{d,\beta}.\text{location})$$
$$\wedge (\tilde{e}_{d,\alpha}.\text{start} < \tilde{e}_{d,\beta}.\text{end}) \wedge (\tilde{e}_{d,\beta}.\text{start} < \tilde{e}_{d,\alpha}.\text{end})$$

A VESPa query $q \subseteq \text{Persons}_q \times \text{Events}_q$ can then be used to filter the dataset d. For each result of the query, the function

$$\text{valueOf} : \text{Persons}_q \cup \text{Events}_q \rightarrow \text{Persons}_d \cup \text{Events}'_d$$

maps abstract person nodes and event nodes from the query to concrete persons and events from the dataset, respectively, and no two nodes from the query can be mapped to the same person or event from the dataset. Furthermore, the conditions imposed by restrictions and comparison edges are adhered to.

The mapping of event nodes from Events_q to elements of Events'_d is then defined as follows: Let $\tilde{P}_q \subseteq \text{Persons}_q$ be a set of persons from the query that is bijectively mapped to the elements of a set $\tilde{P}_d \subseteq \text{Persons}_d$. Let $e_{q,1}, e_{q,2} \in \text{Events}_q$ be two events from the query that are connected by a transition t_q from $e_{q,1}$ to $e_{q,2}$, where t_q applies to the elements of $\tilde{P}_q$. The mapping can then determine $\text{valueOf}(e_{q,1}) = e_{d,\alpha}$ and $\text{valueOf}(e_{q,2}) = e_{d,\beta}$ such that:

- $e_{d,\alpha}$ and $e_{d,\beta}$ equal the first and last, respectively, in a sequence of events $\dot{e}_{d,1}, \ldots, \dot{e}_{d,\dot{n}} \in \text{Events}'_d$ connected by transitions.
- $\bigwedge_{i=1}^{\dot{n}} \dot{e}_{d,i}.\text{persons} \supseteq \tilde{P}_d$.
- $\dot{n} - 2$ equals the number of intermediate events that may be skipped by t_q.

4   Usage Schema

VESPa is most powerful when it is coupled with a visualization to explore the query results. Figure 6 shows a typical analysis process and explains how VESPa can be integrated.

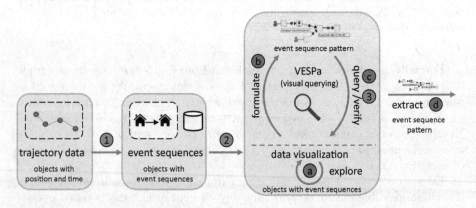

Fig. 6. Usage schema of VESPa in a visual analysis process. The blue numbers represent computational steps. The red labels highlight analysis steps of the user. (Color figure online)

The blue numbers in the schema figure indicate computational steps, while the red labels highlight analysis steps conducted by the user. As a computational preprocessing step, trajectory data (e.g., retrieved via GPS), is transformed into event sequences (① in Fig. 6). This can be done with any semantic enrichment process, e.g., mapping the destinations of trips to points or areas of interest, as proposed by Parent et al. [28]. This is, however, beyond the scope of this paper.

Subsequently, the sequence data is stored in a database with a schema like the one presented in Fig. 5. As a next step, a data visualization can be computed (②). This can include maps and timelines that allow the analyst to explore the data in time and space (a). Gained hypotheses can then be formulated as a graphical pattern by using the pattern-based visual query language VESPa (b). When the analyst executes the pattern (c), the database is queried and returns results that satisfy the query (③).

More precisely, the dataset is scanned based upon the VESPa queries— suitable instructions or textual queries to the underlying database are automatically generated. The result of this scan is a set of mappings of the elements from the VESPa query to elements from the actual dataset. The analyst can explore them in the result visualization and use the information to iteratively refine the query with VESPa or create new queries. Finally, a verified pattern serves as a compact representation of the finding and may be included in a report document.

5   Implementation

We have implemented VESPa in a system based on the Java Swing toolkit. It supports all of the features described in Sect. 3, except for comparison edges between a person and an event node. Also, there is a hard-coded limitation to six persons.

All VESPa queries depicted in this work are based on screenshots of the prototype. Therefore, circular user interface elements for accessing configuration pop-ups are visible in the center of edges.

In the prototype, we use SQLite for executing the queries, though other SQL databases could also be used. For that reason, each event sequence pattern is translated by the prototype to an equivalent SQL query. We use the aforementioned schema depicted in Fig. 5, with an additional *Person* column to directly retrieve an event owner. The extraction of $Events'_d$ from $Events_d$ as described in Sect. 3.3 happens on the fly while executing the query, as precomputing all mutual events would multiply the data volume. As we rely on the event index in a sequence, our implementation cannot recognize consecutive meetings of overlapping sets of persons at the same location.

Based on the schema, Listing 1.1 shows SQL code equivalent to the small event sequence from Fig. 3(b). Even with such a simple schema, the SQL statement is inconveniently long and not directly comprehensible. The SQL statement might be slightly shortened by avoiding composite keys in the schema, but the basic complexity of matching events to find possible meetings between persons would remain. We consider this a sign that our visual event sequence pattern constitutes a simplification over the textual specification of the same query with a formal language.

Listing 1.1. SQL statement equivalent to the event sequence pattern from Fig. 3(b), based on the schema from Fig. 5: The graphical representation in Fig. 3(b) is more concise.

```
 1  SELECT P1.id, Ea_1.sequence AS "Sequence1",
 2    Ea_1.indexInSequence, Eb_1.indexInSequence,
 3    P2.id, Ea_2.sequence AS "Sequence2",
 4    Ea_2.indexInSequence, Eb_2.indexInSequence
 5  FROM Person P1, Person P2, Event Ea_1, Event Ea_2, Event Eb_1, Event
        Eb_2
 6  WHERE (P1.id <> P2.id)
 7    AND (Ea_1.sequence = Eb_1.sequence) AND (Ea_1.indexInSequence + 1
 8    = Eb_1.indexInSequence)
 9    AND (Ea_2.sequence = Eb_2.sequence) AND (Ea_2.indexInSequence + 1
10    = Eb_2.indexInSequence)
11    AND (Ea_1.Person = P1.id) AND (Ea_2.Person = P2.id)
12    AND (Ea_1.location = Ea_2.location) AND (Ea_1.start < Ea_2.end
13      AND Ea_2.start < Ea_1.end) AND (NOT ((Ea_1.location = Eb_1.location)
14      AND (Ea_1.start < Eb_1.end) AND (Eb_1.start < Ea_1.end)))
15    AND (Eb_1.Person = P1.id) AND (Eb_2.Person = P2.id)
16    AND (Eb_1.location = Eb_2.location) AND (Eb_1.start < Eb_2.end
17      AND Eb_2.start < Eb_1.end) AND (Eb_1.ATTR_STATE_CATEGORY = "
            Restaurant")
```

6   Case Studies

In the following sections, we describe two scenarios of application for VESPa using two different datasets. Based on these, we want to show how the query notation can be applied to actual datasets to find particular event patterns and retrieve specific information. The case studies do not just present hypothetical possibilities—the presented filter operations have actually been executed in the prototype described in Sect. 5, which then also retrieved and displayed query results mentioned in the case studies.

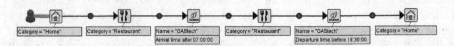

Fig. 7. The event sequence pattern finds behavior sequences that fit the expected daily working routine. Most GAStech employees start their day with a coffee, work at GAStech, have a normal lunch break before working again, and leave no later than 6:30 PM to go home.

6.1   Social Event Detection

The first dataset was released as part of the VAST Challenge 2014 [52]. Every year, a synthetic dataset is created that covers various patterns that have to be found using Visual Analytics approaches. The 2014 challenge concerns surveillance tasks. In Mini-Challenge 2, the task was to detect frequent, but also suspicious behaviors of people in the two weeks leading up to a kidnapping. In a fictitious city, people work at the company GAStech and rent cars that get tracked with GPS devices. With some additional data provided, such as credit card transactions and points of interest (POIs), we extract event sequences from the movement between logical locations as described in Fig. 1.

These sequences serve as input data for VESPa (see Sect. 4), which we embedded in an analysis system with multiple linked views (see Fig. 8). By means our filter approach, we are able to formulate and prove various hypotheses, including meetings of suspicious people, working patterns outside of business hours, or abnormal sequences of movement destinations.

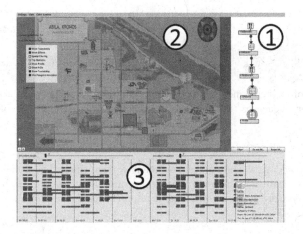

Fig. 8. VESPa ① is embedded in an interactive visual analysis system [53] that further consists of a map ② and a temporal sequence view ③ that shows the sequences of all persons (rows) along a horizontal time axis. The coloring of events is based on the annotated map regions. Here, the analyst filtered for a daily working routine similar to Fig. 7. The temporal view shows the result set.

We have adjusted the set of restrictions and attributes supported by the filter prototype to the data available in the datasets used. In this case, the available filters included *name*, *category*, *employment title* and *type*, *time* and *date* of *arrival* and *departure* at an event node, and the amount of *payments*, if any. By exploring the data, we detect fuzzy repetitive routines using multiple spatial and temporal views. Obviously, these are daily working patterns that are mostly similar, except for some minor temporal differences and some other outliers (see Fig. 9(a)). To create a hypothesis based on our suspicion, we express these routines with our filter notation. This visual query is then executed, which means that SQL query code is automatically generated from the filter pattern. It is sent to the database, whereupon any matching event sequences are retrieved. Step by step, we can refine our hypothesis by adding various restrictions. For example, we define the working day to start at 7 AM at the earliest and to end no later than 6:30 PM with a lunch break in between, according to the usual business habits of GAStech. Lastly, after an iterative refinement, our pattern shows that employees start their day with a coffee at one out of various restaurants before they go to work till noon. For lunch, they also mostly go out, and then work till the early evening. Then, some of them go home directly, while others first meet at various bars. The final pattern is depicted in Fig. 7. From this query, we get all daily routines that exactly fit this pattern (see Fig. 9(b)). Accordingly, when we invert the results, we get all sequences that do *not* match the pattern (see Fig. 9(c)). These sequences might be interesting to look into. For example, there are some people who work outside the usual working hours, who do not take a lunch break, or who do not appear at work, but go shopping.

(a) A snippet of unfiltered movement behavior sequences over three days.

(b) A snippet of filtered sequences over two days. Here, the pattern shown in Figure 7 is applied.

(c) Inversion of results reveals some anomalies.

Fig. 9. A sequence view shows event sequences over time. The events are colored according to their category: home in blue, restaurant and cafés in cyan, business places in red, and other store types in orange and pink. (Color figure online)

In our further analysis, we want to investigate the behavior of one of these persons in depth. He is a GAStech CEO (chief executive officer) who arrives late during the week. Thus, his behavior significantly differs from that of other employees. After we have explored the data with various visual tools, we wonder whether the CEO might be involved in any kinds of suspicious activities outside of work. We therefore construct a filter pattern to verify the hypothesis as follows: First, we create an event node, that we set not to match the GAStech company building. We then set the person for this event to be the CEO. Lastly, we add another person node that is not restricted any further, either. By doing so, we formulate a query that finds events during which the executive meets any other person outside of GAStech (see Fig. 10).

As a result we get three main events. Besides official meetings at the company GAstech on Friday, the CEO also meets three other executives at the golf course. Furthermore, he meets other employees for lunch on Saturday and for dinner on Sunday. While this result does not fully confirm our hypothesis, knowledge about the meetings outside of work proved helpful for a complete overview of the events leading up to the kidnapping. Our findings match the official solution of the challenge.

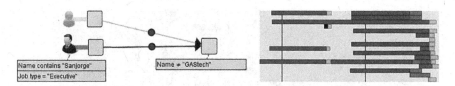

Fig. 10. The sequence pattern (top) finds events where the GAStech CEO named Sanjorge meets somebody else outside the company building of GAStech. The sequence view (bottom) visualizes results, revealing that the CEO meets people at restaurants (cyan) and at the golf course (green). The person node representing the CEO has been configured to use a custom icon to emphasize the restriction to one particular person. (Color figure online)

6.2   Transportation Analysis

The second use case is based on a large real-world dataset [54] that is retrieved from CRAWDAD, an online platform for open source data. Movements of 320 taxis in Rome were recorded in the course of 30 days at a high resolution. Overall, the data consists of more than 12,000 trips. Again, we mapped trip origins and destinations to areas of interest, such as business-related and historic city districts. As the dataset is very large, we applied a semi-automated enrichment approach [29] that extracts POIs (points of interest) from Foursquare, a social media service. We then manually reviewed and refined the results. Lastly, we enriched the trips of the taxis with this information, which resulted in a temporal sequence of POI visits for each taxi (event sequence) as described in Fig. 1. This data serves as input data for our filter approach.

Exemplarily, we showcase a query to find movements starting at the airport, which is far out of town. We suspect that taxis are often used to get to the requested destination. The result of this query reveals that taxis frequently travel to several hotels, located in the city and the upper north of Rome. We then refine our filter query for sequences from an airport to hotels only (see Fig. 11).

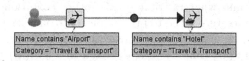

Fig. 11. The pattern finds sequences starting at any airport and finishing at any hotel.

For this scenario, we use an interactive geographic map that shows the location of the events and the corresponding transitions (routes), as can be seen in Fig. 12. This reveals that there are various taxi trips to pick up hotel guests at the International Airport outside of the city to drive them to their hotels (see Fig. 12, bottom). Continuing our analysis, we query the dataset for any tourist activities starting from these hotels. As suspected, taxis are used to visit common sights, such as Vatican City and the Colosseum. Further queries could investigate taxi usage with respect to temporal patterns. For example, one might expect more trips to historic places during weekends.

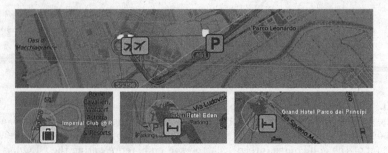

Fig. 12. In an interactive geographic map, the filter results can be investigated. Here, icons visualize location types. Movements between locations are drawn in red. The upper image shows the departure of taxis at the airport. The images on the bottom show arrivals of taxis at various hotels. (Color figure online)

7   User Experiment

To evaluate comprehension and composition understandability of VESPa, we have conducted a small experiment with five users. Our goal was to find out whether users can, after a short introduction, express event sequences with our notation, and whether they are able to recognize what event sequence patterns assembled by someone else mean. Moreover, we wanted to gain some insight on possible issues and improvements for our visual notation, based on user knowledge about visualization and impressions from reading and creating event sequence patterns based on our concept. For that purpose, we prepared six comprehension tasks and six composition tasks that make use of all visual elements.

7.1   Comprehension Tasks

In the six comprehension tasks, participants were provided with completed event sequence patterns. The task was to describe in detail what event sequence patterns are depicted, and what restrictions and other conditions were expressed. Moreover, participants were asked to briefly speculate on the nature of the events that might match each of the event sequence patterns. Task complexity ranged from very simple graphs (like the one depicted in Fig. 3(a)) to more complex ones, such as the one shown in Fig. 13.

7.2   Composition Tasks

For the composition tasks, textual descriptions of six queries for event sequences were prepared. Participants were asked to use the interactive prototype to assemble an appropriate event sequence patterns that would help to solve each of the textual queries. Again, the first few queries were quite basic (e.g., "Find all users who started at a restaurant, traveled to a factory, and traveled to another restaurant from there."), while the last ones were more complex ("Are there any users that, after starting at different locations, meet at one location, travel to another location together, and move on to different places?").

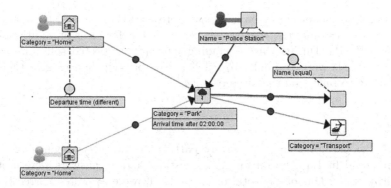

Fig. 13. One of the most complex comprehension tasks from the user study: this event sequence pattern matches two individuals convening in a park, possibly at night, while a third person arrives from a police station. One of the two initial persons joins the third individual to return to the police station, while the remaining person heads for a transportation place (for instance, a train station or an airport).

7.3   Materials and Equipment

All materials for the user study were prepared in German, the native language of the participants. We prepared a brief printed description of the general idea and the graphical elements found in the concept, similar to Sect. 3 in this work. The twelve tasks were printed on paper, as well. As we wanted to emphasize and evaluate the intuitive mental transition between the real-world environment and our event sequence patterns, we used layman's terms in the descriptions and tasks rather than the scientific terminology found in this paper—for instance, a *spatio-temporal event node* became an *event at a particular time and place*, and a *comparison edge* became a *comparison line*.

The implemented prototype was displayed on a 19" monitor. While there was no written documentation for the user interface, participants had an opportunity to get used to the interaction specifics of the implementation before starting to work on the tasks. In addition, they were allowed to ask questions concerning the interaction throughout the user study, as the evaluation focused on the visual notation, not the context menus or settings dialog boxes of the editor.

7.4   Participants and Procedure

We recruited five participants (four males, one female), all of whom are researchers from the field of visualization. Each participant conducted the study separately, while two of the authors were present to give directions and note down any responses.

After reading the description of the concept and familiarizing themselves with the prototype, participants were first given the sheets with the comprehension tasks and solved them. Subsequently, the sheet with the composition tasks was handed out to the participants, and they used the implemented prototype to create solutions. Eventually, participants were asked for general thoughts and suggestions concerning the visual concept, as well as for ideas of additional use cases for VESPa. During this concluding interview, expected solutions of the completed trials were gradually revealed to the participants, to give them an opportunity to reflect on the discrepancies.

7.5   Results

Overall, the performance of the study participants was very promising—each of them could find a solution for each question, and all answers were at least partially correct. Out of the total of 60 trials, 42 were answered correctly—23 comprehension trials and 19 composition tasks—while answers for the others usually contained only a single mistake related to the interpretation or the identity of event nodes. For the additional question about the possible nature of the events described by the patterns, the participants' understanding matched the core of our back-story. They merely interpreted minor parts differently, e.g., the movement of the third person to the police station in Fig. 13.

Comments were mostly positive; several of the participants noted that the notation was entirely clear to them once difficulties had been discussed after completing the trials. Two participants found the visualization "intuitive" and easy to learn. Another participant pointed out that achieving an overview with our event sequence pattern notation was inherently easier than with any text-based language.

While we had told participants to interpret event nodes with the same name as referring to the same location, several of the participants struggled slightly with this definition, and one explicitly wished for an *identical place* connector between event nodes. In general, participants were quick to identify each event node with one location, and thus none of the participants correctly recognized that in one of the comprehension tasks, two event nodes actually represented the same location at different (non-overlapping) times. One participant pointed out that the category icons displayed in some event nodes to express categories such as *Home* or *Factory* visually conveyed to her that each event node was equivalent to a place. In turn, some of the participants neglected the necessity to enforce that two event nodes be mapped to two different locations, when the locations were explicitly required to be different.

Minor remarks about the graphical representation, which were partially related to our particular implementation rather than the concept in general, asked for an increased thickness of transitions especially when several persons were involved and a smart positioning of restriction boxes. Moreover, a clearer distinction between transitions and the (equally colored, but thicker and partially transparent) connector lines between person nodes and event nodes (cf. for example Fig. 13) was suggested.

7.6   Discussion of Results

The most frequent issue encountered by participants was the interpretation of an event node as a place *at a specific time*. None of them had a definitive idea how to improve the notation, however, and all agreed that the definition was logical, just not intuitive at first.

A possible way to mitigate this difficulty is a time-related symbol on an event node, such as a little clock, as a visual clue that the event node also has a temporal aspect. That clock symbol might even be modified in a domain-specific way to emphasize temporal restrictions, if any, such as an absolute start or end time or a minimum or maximum duration. Another suggestion was related to placing event nodes on some kind of a timeline, which may work well when the temporal relationship between two event nodes (before, after, same time) is known. On the other hand, this can create additional problems for event nodes whose relative temporal relationship is unknown (or irrelevant), or between event nodes with absolute and relative temporal restrictions.

Two participants were unsure about the notion of persons *meeting* one another at the beginning of an event sequence. They ended up adding one additional "entry event node" for each person, rather than starting with a mutual event node (cf. Fig. 3(b)) right away. Based on the discussion with the respective participants, it seems likely that a stronger connection of person nodes with the event nodes, rather than just with connected transitions like in our prototype, might emphasize the idea that the persons "meet" when they start out at an initial mutual event node.

The answers of one participant in the user study also revealed the importance of a careful selection of category names and icons. That particular user generally interpreted event nodes restricted to locations of the category *Transport* as the process of transportation itself, rather than a location related to transportation (a train station, an airport, etc.), as the icon in question showed the vehicles rather than the building.

8   Conclusion and Future Work

We have presented VESPa, a visual notation for specifying patterns of event sequences. With this notation, ideas about possible event sequences can be visualized in an abstract and concise way. The visualization can then be used to filter databases of actual event sequences, to check hypotheses about the events in such a database, and also to express common patterns found therein. As opposed to existing approaches, overlapping event sequences of several persons can be explicitly expressed in VESPa. Rather than keeping space and time separate like related work, VESPa uses spatio-temporal event nodes that make queries for meetings of persons very straightforward.

While our event sequence pattern visualization seems adequate for expressing certain complex patterns, and was accordingly positively commented on by participants of our qualitative study, the experiment also revealed some issues that need to be addressed. The temporal aspects of event nodes needs to be clarified;

future research might also lead to domain-specific solutions. Possible extensions to the visual notation include directedness of comparison edges to support asymmetric relationships between event nodes or person nodes, and restrictions on transitions. More importantly, concepts that are currently not supported, such as repeated or alternative subsequences of events, need to be considered.

We have shown the applicability of the concept to certain spatio-temporal movement information in our case studies, but we see the potential to use the event sequence pattern notation in more generic contexts, some of which were suggested by study participants. Beside the use case of surveillance and digital forensics, event sequence patterns might also help recognize movement patterns for other purposes such as marketing analysis, website visitor navigation (where event nodes could map to single pages), wild animals in their habitat, or of fictional characters to describe or find specific scenes in novels or movies. Furthermore, the event sequence patterns might be useful for different kinds of analysis tasks, for instance gaze patterns found in eye-tracking data, data packets in network traffic, or cross-thread resource access. Lastly, our visual event sequence pattern notation can be used in a prescriptive way to specify abstract itineraries for time management assistance systems, which could then be automatically completed to a concrete itinerary based on transportation schedules, business location information, and calendars of co-workers or friends.

Acknowledgements. This work was supported by the Horizon 2020 project *CIMPLEX*, grant no. 641191.

References

1. Haag, F., Krüger, R., Ertl, T.: VESPa: a pattern-based visual query language for event sequences. In: Proceedings of the 7th International Conference on Information Visualization Theory and Applications (IVAp. 2016), vol. 7 (2016)
2. Wongsuphasawat, K., Plaisant, C., Taieb-Maimon, M., Shneiderman, B.: Querying event sequences by exact match or similarity search: design and empirical evaluation. Interact. Comput. **24**, 55–68 (2012)
3. Zgraggen, E., Drucker, S.M., Fisher, D., DeLine, R.: (s—qu)eries: Visual regular expressions for querying and exploring event sequences. In: Proceedings of CHI 2015, pp. 2683–2692. ACM (2015)
4. Pirolli, P., Card, S.: The sensemaking process and leverage points for analyst technology as identified through cognitive task analysis. In: Proceedings of International Conference on Intelligence Analysis, MITRE, pp. 2–4 (2005)
5. Makanju, A., Zincir-Heywood, A.N., Milios, E.E.: Storage and retrieval of system log events using a structured schema based on message type transformation. In: Proceedings of SAC 2011, pp. 528–533. ACM (2011)
6. Gaaloul, W., Bhiri, S., Godart, C.: Discovering workflow transactional behavior from event-based log. In: Meersman, R., Tari, Z. (eds.) OTM 2004. LNCS, vol. 3290, pp. 3–18. Springer, Heidelberg (2004). doi:10.1007/978-3-540-30468-5_3
7. Abela, J., Debeaupuis, T., Consultants, H.S.: Universal format for logger messages (1999). http://tools.ietf.org/html/draft-abela-ulm-05
8. Do, Q.X., Lu, W., Roth, D.: Joint inference for event timeline construction. In: Proceedings of EMNLP-CoNLL 2012, pp. 677–687. ACL (2012)

9. Atrey, P., Maddage, M., Kankanhalli, M.: Audio based event detection for multimedia surveillance. In: Proceedings of ICASSP 2006, vol. 5, pp. 813–816. IEEE (2006)
10. Heydekorn, J., Nitsche, M., Dachselt, R., Nürnberger, A.: On the interactive visualization of a logistics scenario: requirements and possible solutions. In: Proceedings of IWDE 2011, pp. 1–7. Technical report (Internet): Elektronische Zeitschriftenreihe der Fakultät für Informatik der OVGU Magdeburg (2011)
11. Kim, P.H., Giunchiglia, F.: Life logging practice for human behavior modeling. In: Proceedings of SMC 2012, pp. 2873–2878 (2012)
12. Atrey, P.K., Kankanhalli, M.S., Jain, R.: Timeline-based information assimilation in multimedia surveillance and monitoring systems. In: Proceedings of VSSN 2005, pp. 103–112. ACM (2005)
13. Peuquet, D.J., Duan, N.: An event-based spatiotemporal data model (ESTDM) for temporal analysis of geographical data. Int. J. Geogr. Inf. Syst. 9, 7–24 (1995)
14. Huang, Y., Zhang, L., Zhang, P.: A framework for mining sequential patterns from spatio-temporal event data sets. IEEE Trans. Knowl. Data Eng. 20, 433–448 (2008)
15. Jiang, F., Yuan, J., Tsaftaris, S.A., Katsaggelos, A.K.: Anomalous video event detection using spatiotemporal context. Comput. Vision Image Underst. 115, 323–333 (2011)
16. Plaisant, C., Milash, B., Rose, A., Widoff, S., Shneiderman, B.: LifeLines: visualizing personal histories. In: Proceedings of CHI 1996, pp. 221–227. ACM (1996)
17. Kumar, V., Furuta, R., Allen, R.B.: Metadata visualization for digital libraries: interactive timeline editing and review. In: Proceedings of DL 1998, pp. 126–133. ACM (1998)
18. Tao, C., Wongsuphasawat, K., Clark, K., Plaisant, C., Shneiderman, B., Chute, C.G.: Towards event sequence representation, reasoning and visualization for EHR data. In: Proceedings of IHI 2012, pp. 801–806. ACM (2012)
19. Krstajić, M., Bertini, E., Keim, D.: CloudLines: compact display of event episodes in multiple time-series. IEEE TVCG 17, 2432–2439 (2011)
20. Fischer, F., Mansmann, F., Keim, D.A.: Real-time visual analytics for event data streams. In: Proceedings of SAC 2012, pp. 801–806. ACM (2012)
21. Havre, S., Hetzler, B., Nowell, L.: ThemeRiver: visualizing theme changes over time. In: Proceedings of InfoVis 2000, pp. 115–123. IEEE (2000)
22. Guo, X., Li, J., Yang, R., Ma, X.: NEI: a framework for dynamic news event exploration and visualization. In: Proceedings of VINCI 2014, pp. 121–128. ACM (2014)
23. Marcus, A., Bernstein, M.S., Badar, O., Karger, D.R., Madden, S., Miller, R.C.: Twitinfo: aggregating and visualizing microblogs for event exploration. In: Proceedings of CHI 2011, pp. 227–236. ACM (2011)
24. Kapler, T., Wright, W.: GeoTime information visualization. Inf. Vis. 4, 136–146 (2005)
25. Tominski, C., Schumann, H., Andrienko, G., Andrienko, N.: Stacking-based visualization of trajectory attribute data. IEEE TVCG 18, 2565–2574 (2012)
26. Guo, H., Wang, Z., Yu, B., Zhao, H., Yuan, X.: TripVista: triple perspective visual trajectory analytics and its application on microscopic traffic data at a road intersection. In: Proceedings of PacificVis 2011, pp. 163–170. IEEE (2011)
27. Sun, G., Liu, Y., Wu, W., Liang, R., Qu, H.: Embedding temporal display into maps for occlusion-free visualization of spatio-temporal data. In: Proceedings of PacificVis 2014, pp. 185–192. IEEE (2014)

28. Parent, C., Spaccapietra, S., Renso, C., Andrienko, G., Andrienko, N., Bogorny, V., Damiani, M.L., Gkoulalas-Divanis, A., Macedo, J., Pelekis, N., Theodoridis, Y., Yan, Z.: Semantic trajectories modeling and analysis. ACM Comput. Surv. **45**, 42:1–42:32 (2013)

29. Krüger, R., Thom, D., Ertl, T.: Visual analysis of movement behavior using web data for context enrichment. In: Proceedings of PacificVis 2014, pp. 193–200. IEEE (2014)

30. Zhu, X.Y., Guo, W., Huang, L., Hu, T., Gao, W.X.: Pan-information location map. ISPRS Archives **XL–4**, 57–62 (2013)

31. Nguyen, T., Loke, S., Torabi, T.: The community stack: concept and prototype. In: Proceedings of AINAW 2007, vol. 2, pp. 52–58 (2007)

32. Westermann, U., Jain, R.: Toward a common event model for multimedia applications. IEEE Multimedia **14**, 19–29 (2007)

33. Andrienko, N., Andrienko, G., Fuchs, G.: Towards privacy-preserving semantic mobility analysis. In: EuroVis Workshop on Visual Analytics, pp. 19–23. Eurographics Association (2013)

34. Shneiderman, B.: Dynamic queries for visual information seeking. IEEE Softw. **11**, 70–77 (1994)

35. Seifert, I.: A pool of queries: interactive multidimensional query visualization for information seeking in digital libraries. Inf. Vis. **10**, 97–106 (2011)

36. Soylu, A., Giese, M., Jimenez-Ruiz, E., Kharlamov, E., Zheleznyakov, D., Horrocks, I.: OptiqueVQS: towards an ontology-based visual query system for big data. In: Proceedings of MEDES 2013, pp. 119–126. ACM (2013)

37. Russell, A., Smart, P., Braines, D., Shadbolt, N.: NITELIGHT: a graphical tool for semantic query construction. In: Proceedings of SWUI 2008, vol. 543 of CEUR-WS (2008)

38. Morris, A., Abdelmoty, A., El-Geresy, B., Jones, C.: A filter flow visual querying language and interface for spatial databases. GeoInformatica **8**, 107–141 (2004)

39. Wu, S., Otmane, S., Moreau, G., Servières, M.: Design of a visual query language for geographic information system on a touch screen. In: Kurosu, M. (ed.) HCI 2013. LNCS, vol. 8007, pp. 530–539. Springer, Heidelberg (2013). doi:10.1007/978-3-642-39330-3_57

40. Kumar, C., Heuten, W., Boll, S.: Geographical queries beyond conventional boundaries: regional search and exploration. In: Proceedings of GIR 2013, pp. 84–85. ACM (2013)

41. Boyandin, I., Bertini, E., Bak, P., Lalanne, D.: Flowstrates: an approach for visual exploration of temporal origin-destination data. Comput. Graph. Forum **30**, 971–980 (2011)

42. Certo, L., Galvão, T., Borges, J.: Time automaton: a visual mechanism for temporal querying. J. Visual Lang. Comput. **24**, 24–36 (2013)

43. Krüger, R., Thom, D., Wörner, M., Bosch, H., Ertl, T.: TrajectoryLenses - a set-based filtering and exploration technique for long-term trajectory data. Comput. Graph. Forum **2013**, 451–460 (2013)

44. Bonhomme, C., Trépied, C., Aufaure, M.A., Laurini, R.: A visual language for querying spatio-temporal databases. In: Proceedings of GIS 1999, pp. 34–39. ACM (1999)

45. D'Ulizia, A., Ferri, F., Grifoni, P.: Moving GeoPQL: a pictorial language towards spatio-temporal queries. GeoInformatica **16**, 357–389 (2012)

46. Monroe, M., Lan, R., Morales del Olmo, J., Shneiderman, B., Plaisant, C., Millstein, J.: The challenges of specifying intervals and absences in temporal queries: a graphical language approach. In: Proceedings of CHI 2013, pp. 2349–2358. ACM (2013)
47. Gotz, D., Stavropoulos, H.: DecisionFlow: visual analytics for high-dimensional temporal event sequence data. IEEE TVCG **20**, 1783–1792 (2014)
48. Fails, J., Karlson, A., Shahamat, L., Shneiderman, B.: A visual interface for multivariate temporal data: finding patterns of events across multiple histories. In: VAST 2006, pp. 167–174 (2006)
49. Dionisio, J.D., Cárdenas, A.F.: MQuery: a visual query language for multimedia, timeline and simulation data. J. Visual Lang. Comput. **7**, 377–401 (1996)
50. Jin, J., Szekely, P.: QueryMarvel: a visual query language for temporal patterns using comic strips. In: Proc. VL/HCC 2009, pp. 207–214 (2009)
51. Fegeras, L.: VOODOO: a visual object-oriented database language for ODMG OQL. In: W13. The First ECOOP Workshop on Object-Oriented Databases (1999)
52. Visual Analytics Community: VAST 2014 Challenge - the Kronos incident (2014). http://vacommunity.org/VAST+Challenge+2014
53. Krüger, R., Herr, D., Haag, F., Ertl, T.: Inspector gadget: integrating data preprocessing and orchestration in the visual analysis loop. In: EuroVis Workshop on Visual Analytics (EuroVA). The Eurographics Association (2015)
54. Bracciale, L., Bonola, M., Loreti, P., Bianchi, G., Amici, R., Rabuffi, A.: CRAWDAD data set roma/taxi (v. 2014–07-17) (2014). http://crawdad.org/roma/taxi/

Correlation Coordinate Plots: Efficient Layouts for Correlation Tasks

Hoa Nguyen[1(✉)] and Paul Rosen[2]

[1] Scientific Computing and Imaging Institute, University of Utah,
Salt Lake City, USA
hoanguyen@sci.utah.edu
[2] Department of Computer Science and Engineering, University of South Florida,
Tampa, USA

Abstract. Correlation is a powerful measure of relationships assisting in estimating trends and making forecasts. It's use is widespread, being a critical data analysis component of fields including science, engineering, and business. Unfortunately, visualization methods used to identify and estimate correlation are designed to be general, supporting many visualization tasks. Due in large part to their generality, they do not provide the most efficient interface, in terms of speed and accuracy for correlation identifying. To address this shortcoming, we first propose a new correlation *task-specific* visual design called Correlation Coordinate Plots (CCPs). CCPs transform data into a powerful coordinate system for estimating the direction and strength of correlation. To extend the functionality of this approach to multiple attribute datasets, we propose two approaches. The first design is the Snowflake Visualization, a focus+context layout for exploring all pairwise correlations. The second design enhances the CCP by using principal component analysis to project multiple attributes. We validate CCP by applying it to real-world data sets and test its performance in correlation-specific tasks through an extensive user study that showed improvement in both accuracy and speed of correlation identification.

Keywords: Correlation identification · Correlation visualization · Multidimensional data visualization

1 Introduction

Correlation is a powerful metric that provides a predictive relationship between variables used in science, engineering, and business [17,26,32]. A correlation coefficient is a measure of the strength and direction of such a relationship. While correlation is a powerful metric, visual examination is also critical. The many-to-one relationship between data and a correlation coefficient may obscure important features of the data. In Anscombe's Quartet (see Fig. 1) [1], 4 distributions (i.e. the many relationship) have identical correlation coefficients (i.e. the

© Springer International Publishing AG 2017
J. Braz et al. (Eds.): VISIGRAPP 2016, CCIS 693, pp. 264–286, 2017.
DOI: 10.1007/978-3-319-64870-5_13

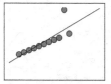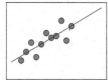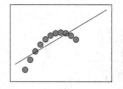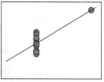

Fig. 1. Anscombe's Quartet [1] shows 4 distributions with an outlier, noise, non-linearity, and non-relationship, respectively, that all have correlation coefficients of 0.816.

one relationship). Visual examination can disambiguate the variations to outliers (case 1), noise (case 2), non-linearity (case 3), and non-relationship (case 4).

Both scatterplots (SCP) [20] and parallel coordinates plots (PCP) [19] are capable of being used to investigate correlation. However, that does not mean one should not infer that these are the *optimal* tools for performing such tasks. In situations where correlation is the most important data feature, these encodings are arguably non-optimal [12,22]. This challenge is exacerbated by the increasing desire to analyze multi-attribute data. A number visualization techniques exist for this analysis [2,4,29], with Scatterplot Matrices (SPLOMs) and PCPs remaining the most popular. SPLOMs simultaneously show all possible combinations of attribute, but the plots become small as the number of combinations grows quadratically. For PCPs, the series of axes grow linearly, but the interface relies heavily upon interaction.

The critical shortcoming to these methods is in their design goal—they are designed as general-purpose tools for performing a wide variety of analytic tasks. No special consideration has been made to any single task, meaning that while they *can be* used to identify correlation, they are *not designed* for it.

With these limitations in mind, we have developed a new, *correlation task-specific* visual design called Correlation Coordinate Plots, or CCPs (see Fig. 2(a–c)). CCPs use design attributes, such as axis shape and a simple, yet effective, point transform to enable quick and accurate determination of correlation direction and strength.

To support multi-attribute analysis we developed 2 different approaches. The first is a focus+context style circular layout for CCPs, called the Snowflake Visualization (see Fig. 2(d)). This visualization represents a compromise where the screen space needed to represent additional attributes grows linearly in the focus and quadratically in the context region. Interaction is still relied upon for full investigation. In the second approach, we have extended the visual metaphors of the CCP to support a single visual interface for multi-attribute analysis by using principal component analysis (PCA) of the data.

To validate the efficacy of our new approaches, use case examples and a user study are used. Our user study had novice and expert subjects perform correlation-related tasks in SCP, PCP, and CCP environments. Our results confirmed that CCP methods outperform SCP and PCP in accuracy and timing.

In summary, the contributions of this paper are:

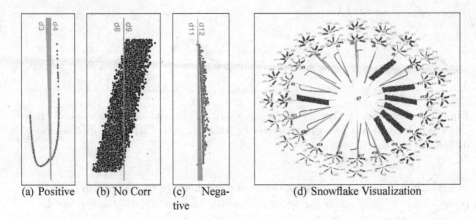

(a) Positive   (b) No Corr   (c)   Nega-   (d) Snowflake Visualization
                                        tive

Fig. 2. Correlation Coordinate Plots (CCPs) transform data into a coordinate system better suited to investigating correlation between attributes. (a–c): Example CCPs show positive, no, and negative (or anti-) correlation, respectively. (d): Snowflake Visualization is a focus+context interface that combines CCPs for 1 attribute to all others in the middle (i.e. the focus) and CCPs for all other pairings on the perimeter (i.e. the context).

- a task-specific visualization, the Correlation Coordinate Plot, designed to efficiently identify correlations;
- a circular layout, the Snowflake Visualization, that provides an efficient focus+content style visualization of all pairwise relationships in multi-attribute data;
- a single plot visualization for exploring multi-attribute correlations using PCA; and
- a use case analysis and user study confirming the superior performance of CCP with correlation-related tasks when compared to SCP and PCP.

2   Related Work

2.1   Correlation

Correlation is a metric calculated on data that can be used to model and predict relationships [17,32]. The "quality of relationship" is often measured using a correlation coefficient [6,31], with positive correlation indicating 2 attributes are increasing together, while negative or anti-correlation indicates that 1 attribute increases and the other decreases. There are several correlation coefficient measures, the most common of which is the Pearson Correlation Coefficient (PCC) [25,28]. PCC, $\rho(x,y)$, measures the linear relationship between 2 attributes x and y with means $\bar{x}$ and $\bar{y}$ and standard deviations σ_x and σ_y. It is defined as:

$$\rho(x,y) = \frac{cov(x,y)}{\sigma_x \sigma_y} = \frac{\Sigma(x_i - \bar{x})(y_i - \bar{y})}{\sigma_x \sigma_y}. \tag{1}$$

As far as correlation in visualization is concerned, there are 2 schools of thought. The first is to show metrics on data, not the data themselves. Examples include Corrgrams [10] and Scagnostics [7,30]. These approaches have the advantages of visual scalability but the potential disadvantages demonstrated by Anscombe's Quartet [1]. An overview of the metrics used in these approaches can be found in [3].

The alternative approach shows all data points. In this category, scatterplots and parallel coordinates have been shown most effective [12,22]. Since our approach follows this paradigm, we compare against these techniques.

2.2 Scatterplot

A Scatterplot (SCP) [5,20] is a simple plot of points used to investigate the relationships between 2 attributes [13]. The patterns of importance in this context are when the data points slope from lower left to upper right, suggesting positive correlation, and sloping from upper left to lower right suggests negative correlation. The direction of correlation (positive or negative) can be confusing to novice users. More importantly, the strength of correlation (high versus low) can at times be difficult to interpret.

For multi-attribute data, a Scatterplot Matrix (SPLOM) [13,18] shows the relationships of all pairs of attributes by organizing a grid of SCPs with each attribute occupying 1 row and 1 column. As the number of attributes increases, the number of plots grows quadratically making it difficult to present all of the data. This problem can be mitigated by approaches such as Corrgrams [10], which display a matrix of correlation glyphs. These glyphs scale well and give the user quick access to summary statistics, but they may hide important data features (e.g. Anscombe's Quartet). In other cases, navigation can be used to search larger spaces [8].

2.3 Parallel Coordinates Plot

Parallel Coordinates Plots (PCPs) [9,19] are another well-known visualization technique for exploring multi-attribute datasets, which display n parallel axes, 1 for each attribute. Data points map to vertices on each parallel axis and connect with line segments. For PCPs, in simple cases, the direction of correlation, though not necessarily intuitive, is easy to identify. Positive correlation appears as a series of parallel lines, while negative correlation appears as crossing lines.

In noisy cases, the ambiguity created by the crossing lines hides patterns but retains outlier visibility [33,34]. This makes correlation direction and strength difficult to interpret. Modifications to PCPs have been proposed by using color, opacity, smooth curves, frequency, density or animation [11,15,16] to partially address this. However, previous studies have shown that PCPs are slower and less accurate than SCPs for correlation tasks [12,22,23].

The advantage of a PCP is that it provides a continuous and comparative view across the axes, and the screen space needed for the visualization scales linearly with the number of attributes. At the same time, PCPs do not show all possible combinations of attribute pairs, requiring significant user interaction for exhaustive exploration. Using 3D parallel coordinates can enable exploration of the many-to-one relationship [21] with the traditional downsides of 3D—perspective effects and occlusion in large data. A PCP matrix [14] is another method that may help overcome this limitation.

3   Correlation Coordinates Plot

The task generality (i.e. the support for many tasks) plays as both an advantage and disadvantage for the SCP and PCP. Either method is capable of being used for correlation tasks, but they are not necessarily the most efficient methods available. This has led us to develop a new visual encoding focused specifically on correlation tasks, called Correlation Coordinate Plots (CCPs). The proposed method is centered on helping users quickly identify the existence, direction, and strength of pairwise correlations. The visual design is motivated by our desire to make the correlation task one of comparison using position along a common baseline, a highly effective visual channel [24].

For clarity in notation, we assume a dataset X contains n attributes and m data points, with X_i indicating a single data attribute of m values and X_{ij} indicating data point j of attribute i.

3.1   Coordinate System

We propose using a correlation coordinate system that differs from the Cartesian coordinate system, so as to highlight how well points adhere to the correlation. The coordinate system can be seen as a 1D parametrization of the data to an underlying model, in this case a line. The vertical position of a data point is the parameterization of the data. The position horizontally is more important, demonstrating the quality of the fit. Therefore, identifying correlation primarily relies on visibility of points to the left and right of the axis.

Transforming the data from a Cartesian domain into the correlation coordinate system is a two step process laid out in Fig. 3, with the top panel showing the positive relationships and the bottom panel demonstrating the negative relationships.

The first step is a scaling operation (Scl) that forces the data into a square region (see Fig. 3 panels 1 & 2). The process begins by normalizing the data to $[-1, 1]$,

$$Scl(X_i) = \frac{X_i - \arg\min_{X_i} X_{ij}}{\arg\max_{X_i} X_{ij} - \arg\min_{X_i} X_{ij}}. \tag{2}$$

The second step is the projection (P_{major} and P_{minor}) operation, which measures the location of the point relative to the positive correlation diagonal (lower left to upper right) or negative correlation diagonal (upper left to lower right). That measure is used to place the points into the CCP (see Fig. 3 panels 3 & 4). The now normalized location of a point i from attributes j and k determines the major (vertical) axis by:

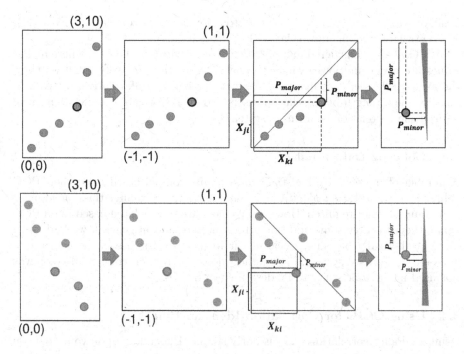

Fig. 3. Conversion to correlation coordinate system for positive (top) and negative (bottom) cases.

$$P_{major}(X_{ji}, X_{ki}) = X_{ki}. \tag{3}$$

The position on the minor axis is:

$$P_{minor}(X_{ji}, X_{ki}) = \begin{cases} \alpha \cdot (X_{ji} - X_{ki}) & \text{positive or no correlation} \\ \alpha \cdot (X_{ji} + X_{ki}) & \text{negative correlation} \end{cases} \tag{4}$$

The variable α is a scalar that effects the spread of data points when plotting. We selected a constant value based upon the width of the CCP.

The plot orientation was initially chosen to be vertical in order to pack many plots side by side on the display. Ultimately, the choice of a vertical plot is somewhat arbitrary and will be relaxed in forthcoming sections. Nevertheless, we present and evaluate our approach based upon the vertical orientation.

3.2 Coordinate Axis

We designed the coordinate axis to serves as a visual indicator of the existence and direction of correlation. For 2 attributes of a dataset, X_i and X_j, PCC is used to indicate positive correlation by $\rho(X_i, X_j) > \epsilon$, negative correlation by $\rho(X_i, X_j) < -\epsilon$, and uncorrelated by all other values. The major coordinate axis is laid out vertically and represented by a triangle whose base is at the top for positive correlation (Fig. 2(a)), the bottom for negative correlation (Fig. 2(c)), and a straight line for uncorrelated (Fig. 2(b)) data.

We have also considered mapping PCC to the width of the axis, where higher values are wider and lower values thinner. Due to the relatively small width of the axis, we decided this mapping was not particularly informative. Instead, to identify the strength of correlation, users should investigate the distribution of data in the correlation coordinate system.

3.3 Coloring Data Points

A number of figures have had their data points colored based upon their PCC value [{-1 : *blue*}, {0 : *black*}, {1 : *red*}]. Strictly speaking, this encoding is redundant and not required. However, if colors are interpolated based upon PCC value, they do carry some additional information, and in general, we find them more aesthetically pleasing. Because our focus is on the use of the coordinate axis and coordinate system, our method does not rely on color, and color was *not* used in the user study to be described in Sect. 8.

3.4 Using CCPs for Correlation Identification

Using CCPs for correlation tasks is fairly simple. Depending upon your goal, we suggest:

- First, use the axis to determine if the data is positive, negative, or uncorrelated.
- Next, use the shape of the data points to determine the basic relationship between the attributes (i.e. linear, nonlinear, etc.).
- Finally, the distance of the points from the axis can be used to estimate the strength of correlation, with small distances indicating high correlation, and other conditions such as outliers, noise, etc.

For example, in Fig. 2(c), by checking the axis, a negative correlation can be seen. By observing the closeness of the data points to the axis, a strong linear relationship with small amount of noise. On the other hand in Fig. 2(a), the axis indicates positive correlation. From the shape of the data, it is apparent that a nonlinear relationship exists with weak linear correlation properties.

4   Multi-attribute Visualization

Thus far, our approach can be used to investigate pairwise correlation. Our next goal was to develop an approach for investigating multi-attribute data. We began by looking at SPLOMs which have the advantage of showing all possible combinations of attributes at the cost of the number of plots needed growing at a rate of $O(n^2)$. This may leave little screen space for each individual plot. On the other hand, the number of plots in PCPs grow at a rate of $O(n)$ resulting in more available space for each.

4.1   Parallel CCP

In the PCP spirit, we first applied CCPs to multi-attribute data through a series of equally spaced vertical parallel CCP axes, as seen in Fig. 4. To explore additional combinations of attributes, users can drag an axis to configure the corresponding relationship.

Much like PCP, this approach does not provide immediate access to all attribute pairs, instead relying on user interaction to fully explore the data. As a compromise between the plot size benefits of PCPs and the comprehensiveness of SPLOMs, we developed a new correlation visualization layout, the Snowflake Visualization.

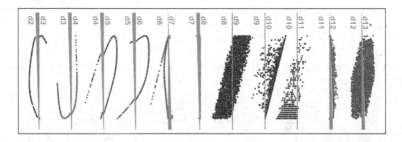

Fig. 4. Parallel CCP for 10 attributes data allow full exploration of the data, but, like PCP, it relies on heavy user interaction.

4.2   Snowflake Visualization

We focused on a radial based design due to their efficient use of space for multiple attribute visualizations [27]. As such, we have developed the Snowflake Visualization, which is constructed of a focus+context views.

Focus View. The focus view (Fig. 5(a)) enables investigating the correlation of 1 attribute to all other attributes. Given n attributes, there are $(n - 1)$ pairs laid out around the center of the circle with equal angular spacing. By default, the final attribute of data is the initial focus attribute. Attributes are sorted by ID but can be reordered with other sorting methods. The inner radius (the start of the CCP axes) is chosen such that none of the data points between CCPs will overlap. The outer radius (the end of the CCP axes) is adjustable as to give more or less space to the context views.

Context View. Given the attributes covered by the focus view, we designed the context view to give complete coverage of the remaining attribute pairs. These context views (Figs. 5(b) and (c)) are attached to the branches of the focus view. The objective is to prevent pairs of attributes from being repeated. This is done by organizing the pairings based on parity of n.

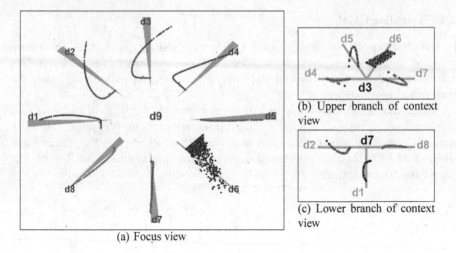

(a) Focus view

(b) Upper branch of context view

(c) Lower branch of context view

Fig. 5. A focus view (a) and multiple context views (b–c) for Snowflake Visualization.

Fig. 6. Branch attribute pairing matrix for n attributes, when n is odd, and the focus attribute is d_{2m}. Each row and column represent 1 data attribute. Pairings are found by selecting a column for the attribute and pairing with highlight attributes.

When n is odd ($m = (n-1)/2$), the organization, shown in Fig. 6, contains all pairwise correlations in data. Each pair (d_i, d_j), where $i = 0, 1, \ldots, 2m-1$ and $j = i+1, i+2, \ldots, 2m$, presents correlation between 2 attributes d_i and d_j. The red box in Fig. 6 contains all the attribute pairs that are presented in focus view, pairing the last attribute d_{2m} and all other attributes $(d_0, \ldots, d_{2m-1})$.

The context view has two groups—the upper branches and lower branches. In the upper branches, where $i = 0, \ldots, m-1$, the ith branch presents correlations between attribute d_i and m other attributes that are $(d_{i+1}, d_{i+2}, \ldots, d_{i+m+1})$. There are m pairwise correlations in each upper branch. We can see one upper branch in the Fig. 5(b) that has 4 pairwise correlations when the number of attributes n is 9 and m is 4.

In the lower branches, where $i = m, m+1, \ldots, 2m-1$, the ith branch presents correlation between attribute d_i and other attributes as shown in the

i^{th} column in Fig. 6. There are $(m - 1)$ attribute pairs in each lower branch. Figure 5(c) shows one lower branch that presents 3 pairwise correlations when number of attributes $n = 9$ and $m = 4$.

The organization is similar when n is even $(m = n/2)$. The focus view presents the correlations between last attribute d_{2m-1} and all other attributes in data $(d_0, d_1, \ldots, d_{2m-2})$. The context view has only a single branch type that has $m - 1$ attributes pairs in each upper or lower branch.

Detail View and Interaction. Typically a single large CCP detail view is also included with the Snowflake Visualization (a similar practice to SPLOMs). A few interactions are included with the Snowflake Visualization. These include:

- *Click-to-swap*: When the user clicks an attribute, it becomes the focus attribute. After swapping, outer attributes are reordered based upon a sorting criteria (i.e. by attribute ID).
- *Over-to-detail*: As the mouse moves over a plot, the detail view is updated to that pairing.

5 Multiway Attribute Correlations

Pairwise correlations are frequently important to understanding data. However, as the number of attributes increases, the desire to explore relationships of multiple attributes simultaneously increases as well. The Snowflake Visualization partially addressed the need by presenting many pairwise relationships simultaneously. Comparing 3 or more attributes requires looking at an exponentially increasing number of plots and mentally fusing the distributions. We can extend CCP design for presenting certain types of multi-attribute relationships.

To do this, we slightly modify visual metaphors of the CCP. First of all, we remove the positive/negative metaphor encoded via the axis. This is because multi-attribute relationships tend to not have a directional measure, only magnitude. Now, the parameterization model can be relaxed to any invertible function, $[s, t] = g(\bar{x})$. The vertical axis still represents a 1D parameterization of the data, s. The horizontal axis can now represent a secondary model parameterization, t. Finally, we represent information lost in this encoding via a series of partially transparent boxes, one per data point, that form a "haze" surrounding the data points. The size of the boxes found using the residual, $r = ||\bar{x} - g^{-1}(s,t)||$.

For our experiments we have used Principal Component Analysis (PCA) to parameterize the data. This could be replaced with any other model that fits our functional definition. Using PCA, we set $g(\bar{x})$ equal to the magnitude of the first two principal components of the data, and the size of the box is set to the residual. Figure 7 shows 2 examples. The SPLOM on the left (Fig. 7(a)) shows all of the attributes of the dataset. Two subsets have been selected in red and blue. The red subset are attributes that all appear pairwise linear. When we use the many-attribute CCP (Fig. 7(b)), we can see that all of the attributes are linear with respect to one another. On the other hand, the blue attributes appear nonlinear. When visualized with the many-attribute CCP (Fig. 7(c)), we can see a relatively simple nonlinear 2D pattern within the data.

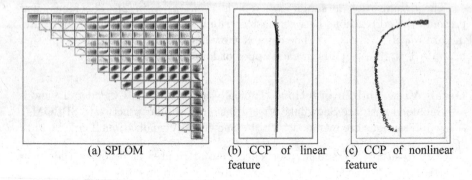

| (a) SPLOM | (b) CCP of linear feature | (c) CCP of nonlinear feature |

Fig. 7. CCP for multiple attributes using PCA. (b) The attributes in red are a linear feature. (c) The nonlinear feature in blue is 2D, with the residual visible in the red haze. (Color figure online)

6   Implementation

Algorithms 1, 2 and 3 contain pseudocode for the CCP and Snowflake Visualization. We have also included a sample visualization tool[1] that can be built in Processing.

Algorithm 1 is pseudocode to draw the CCP of two input data attributes with M items as we described in Sect. 3. Drawing the Snowflake Visualization is presented in 2 parts. The focus view, based on attribute j, can be draw using Algorithm 2 by drawing a series of CCPs plots around the center, with equal angular spacing. Algorithm 3 presents the method to draw context view of Snowflake Visualization, based on the parity of n.

Algorithm 1. Draw Correlation Coordinate Plot.

1: // Draw axis
2: **if** $PCC(X,Y) > \epsilon$ **then**
3:     $drawAxis$(upper-triangle-axis)
4: **else if** $PCC(X,Y) < \epsilon$ **then**
5:     $drawAxis$(lower-triangle-axis)
6: **else**
7:     $drawAxis$(straight-line-axis)
8: **end if**
9:
10: // Draw items
11: // Draw items

12: **for** $i = 1 : M$ **do**
13:     $[x_n, y_n] := normalize(X_i, Y_i)$
14:     $p_{major} := x_n$
15:     **if** $PCC(n-1, i) > \epsilon$ **then**
16:         $p_{minor} := \frac{y_n - x_n}{2.0f}$
17:     **else**
18:         $p_{minor} := \frac{y_n + x_n}{2.0f}$
19:     **end if**
20:     $drawPoint(p_{major}, p_{minor})$
21: **end for**

---

[1] CCPs: https://github.com/hoa84/CCPs_SnowflakeViz.

Algorithm 2. Draw Focus View of Snowflake.

1: // Draw attributes before focus j 6: // Draw attributes after focus j
2: **for** $i = 1 : j - 1$ **do**   7: **for** $i = j + 1 : N$ **do**
3:    $setPosition(cen, rad, (i-1) \cdot \frac{360^o}{n-1})$ 8: $setPosition(cen, rad, i \cdot \frac{360^o}{n-1})$
4:    $drawCCP(A_j, A_i)$ 9: $drawCCP(A_j, A_i)$
5: **end for**   10: **end for**

Algorithm 3. Draw Context View of Snowflake.

1: // Parity bit for even vs. odd n 17: $drawCCP(A_i, A_{i+j+1})$
2: $even = (n$ is even$) ? 1 : 0$ 18:   **end for**
3:     19: **end for**
4: // Draw attributes after focus j 20: **for** $i = range_0 + 1 : range_1$ **do**
5: $m = \lfloor n/2 \rfloor$ 21:   **for** j=0 to i-m+1-even **do**
6:     22: $setPosition(cen_i, ang_i + b_1 *$
7: // Loop ranges $(2m + j - i - 2))$
8: $range_0 := m - even$ 23: $drawCCP(A_i, A_j)$
9: $range_1 := 2m - 1 - even$ 24:   **end for**
10:     25:   **for** j=i+1 to 2m-2 **do**
11: // Angular separation for plots   26: $setPosition(cen_i, ang_i + b_1 *$
12: $b_0 = 180^o/(m - 1 - even)$ $(j - i - 1))$
13: $b_1 = 180^o/(m - 2)$ 27: $drawCCP(A_i, A_j)$
14: **for** $i = 0 : range_0$ **do**   28:   **end for**
15:    **for** $j = 0$ to $m - 2$ **do**   29: **end for**
16:     $setPosition(cen_i, ang_i + b_0 * j)$

---

7 Usage Examples

We applied three visualization methods, including the Snowflake Visualization, SPLOM, and PCP, to three publicly available datasets including Boston house price data[2], Pollen data[3], and Hurricane Isabel data[4].

7.1 Boston House Price

Boston housing data (see Fig. 8) is multivariate dataset containing 506 items across 14 attributes. The data contains several variables that try to explain variation in home values in the Boston area.

When comparing this dataset in a Snowflake Visualization and SPLOM, there are a number of features observable in both visualizations. For example, in both visualizations the Age/Rad pairing is fairly clearly a case for segmentation into two data groups. In the SPLOM, it will likely take longer.

A big advantage in Snowflake Visualization is that it makes way for exploiting additional visual channels. Take the Age/Ind pairing. In all visualization

---

[2] http://lib.stat.cmu.edu/datasets/boston.
[3] http://lib.stat.cmu.edu/datasets/pollen.data.
[4] http://vis.computer.org/vis2004contest/.

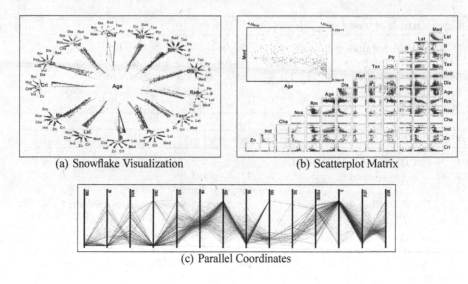

(a) Snowflake Visualization   (b) Scatterplot Matrix

(c) Parallel Coordinates

Fig. 8. Visualizations for Boston House data.

approaches, coloring scheme we have used makes it fairly easy to see that there is a strong positive correlation. However, without the coloring that might not be the case. If color had been used for some other purpose, classification for example, suddenly we lose the ability in SPLOMs to quickly determine correlation, while observing classification. Since CCPs do not rely on color to communicate correlation, we can encode other information in the color channel without significant loss of correlation information.

7.2 Pollen Data

The pollen data (Fig. 9) contains 3848 items each with 6 attributes. This dataset summarizes geometric features of pollen grains.

The nature of the data makes it difficult to use the PCP due to overdraw. Take the Ridge/Weight and Ridge/Density. Even though we can be fairly certainly that Ridge/Weight is more negatively correlated than Ridge/Density, any other detail is lost. We are unable to determine if it is due to outliers, nonlinearity, noise, etc. Techniques such as clustering, density, histogram PCPs can be used to further improve the representations. However, for correlation strength tasks, these approaches are not particularly beneficial.

For the Snowflake Visualization this data proves little trouble. When Ridge is selected as the focus parameter, Density and Weight can be compared in detail. The thinner spread of Ridge/Weight indicates a stronger linear relationship compared to Ridge/Density. In addition, the details available in the view confirm that any weakness in the correlation is due to noise.

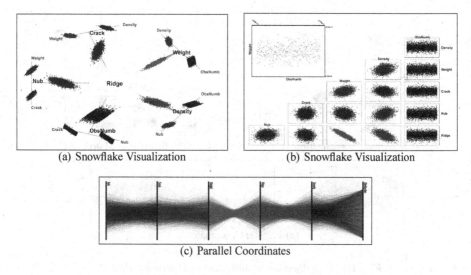

(a) Snowflake Visualization   (b) Snowflake Visualization

(c) Parallel Coordinates

Fig. 9. Visualizations for pollen data.

7.3 Hurricane Data

Hurricane Isabel (Fig. 10) data is provided as part of the IEEE Visualization 2004 contest. This dataset contains a variety of simulated variables related to Hurricane Isabel, a major Atlantic storm that occurred in September of 2003. Isabel data set consists of 48 timesteps, each containing measurements of 11 attributes with a spatial resolution of $500 \times 500 \times 100$. We also only show 7 of the more "interesting" attributes due to space considerations. Of the original data 25 million data items, we only use 10 million because approximately 15 million data items contain at least 1 invalid NaN field.

With 10 million data items in the Hurricane data, the overdraw in the PCP makes it hard to understand any relationships in the data. For example, the relationship Temp/Pres shows only the bowtie shape, losing the individual data patterns. In many ways, SCPs do a better job than PCPs. The Temp/Press relationship is visible with the SCP. However, clear interpretation is difficult, since as Temp increases, Press first decreases, then increases, and finally decreases.

Our approach presents these relationships more clearly. The direction and strength of relationship between Temp and Pres can be identified in Snowflake Visualization. The lower triangle shape of axis identifies the negative relationship. Additionally, the data points distribution, mostly being of similar distance to the axis with a few spread out, enables identifying that this relationship is not strongly negative and nonlinear.

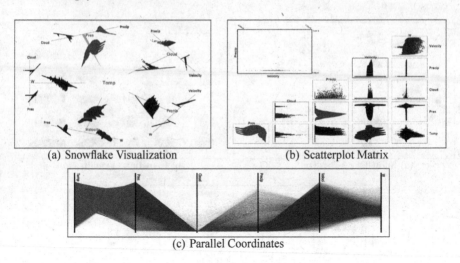

(a) Snowflake Visualization   (b) Scatterplot Matrix

(c) Parallel Coordinates

Fig. 10. Visualization techniques for Hurricane data.

8   User Study on Identifying Correlation

To further evaluate our visualization methods, we conducted a user study comparing CCP with SCP and PCP. In this study, we performed 3 experiments that ask subjects to perform correlation related tasks.

We invited 25 participants to take part in our study, 9 female and 16 male, all graduate students from a variety of science and engineering fields. Their ages range from 23 to 35 years old. We asked the subjects to self-report their level of familiarity with visualization—3 reported themselves as experts; 9 reported themselves as familiar; and 13 reported themselves as not familiar.

In each experiment, subjects started with a short set of slides and/or video to introduce the necessary background. Subjects were then given practice questions where, after answering, the correct answers were provided. They would then perform the experimental tasks. For each test, the subjects' answers and response times were recorded. Following the experiment, subjects completed a short survey. In total, the study lasted less than one hour, including training and testing. For all visualizations, gray color was used for axes and labels, black color was used to present data items.

The software for the user study was built using C++ and Qt, and run on a MacBook pro with a 2.5 GHz Intel Core i5, 4 GB RAM, and 512 MB Intel HD Graphics 4000. The study used a particle physics dataset containing 41 output attributes and 4000 data items per attribute. The data represents a parameter space search of 25 input attributes generated by a series of tools that simulate the theoretical physical properties of subatomic particles under the Supersymmetic extension of the Standard Model of particle physics.

The independent and dependent variables used in each experiment can be found in Table 1. We used a mixed experimental design using t-testing to

Table 1. Variables used to test hypotheses.

Independent variables	Potential values
Data [**H1** | **H2** | **H3** | **H4**]	2 random attributes from 41 attribute data
Data [**H5** | **H6** | **H7**]	10 or 21 attributes from 41 attribute data
Plot [**H1** | **H2** | **H3** | **H4**]	SCP/PCP/CCP
Plot [**H5** | **H6** | **H7**]	SPLOM/PCP/Snowflake Visualization
Question [**H1** | **H2**]	How are the 2 attributes correlated?
Question [**H3** | **H4**]	What is the type of correlation?
Question [**H5**]	How are the 2 attributes correlated?
Question [**H6**]	How many attributes are correlated to i?
Question [**H7**]	Which attributes are correlated to i?
Dependent variables	Potential values
Answer [**H1** | **H2** | **H5**]	High Positive Correlation
	Low Positive Correlation
	No Correlation
	Low Negative Correlation
	High Negative Correlation
Answer [**H3** | **H4**]	Nonlinear Correlation
	Linear Correlation
	No Correlation
Answer [**H6**]	Number of attribute
Answer [**H7**]	List of attribute
Response Time [all **H**]	Time recorded automatically

calculate t-value, p-value, mean difference, and 95% confidence interval to confirm our hypotheses. Only mean value and p-value are reported, but other data can be provided upon request.

8.1   Exp 1: Speed/Accuracy in Pairwise Correlation

When looking at SCP & PCP, 2 challenges persist. First, it can be confusing to determine positive versus negative correlations. Granted, for experts this is a trivial task, but for others, it can be confusing. In many ways the identification of correlation direction is easier with PCP than SCP—parallel lines positive and crossing lines negative. Second, there is some ambiguity when trying to identify the strength of correlation between 2 attributes. Ambiguity is a much larger problem for PCP. When the relationship is noisy or nonlinear, overlapping lines quickly obscure detail.

When comparing CCP with these other methods, CCP: (1) provides simple visual cues making identification of the direction of correlation fairly trivial; (2)

and reduces (not eliminates) the ambiguity by concentrating on correlation in the formulation of the coordinate system.

Given these factors, we developed 2 hypotheses as follows:

H1 | H2: *Using a Correlation Coordinates Plot will enable more accurate and faster identification in direction and strength of correlation between 2 attributes than a [H1: Scatterplot | H2: Parallel Coordinates Plot].*

Method. The experiment is summarized in Table 1 (**H1 & H2**). For a block of trials, we showed a participant a plot between 2 random attributes using either the SCP, PCP, or CCP method and asked a forced choice question. Subject accuracy and time were measured.

At the start of the experiment, participants were given an introduction to correlation, instructions on finding correlation in SCP, PCP, and CCP, and 6 training questions. Participants were then given 21 experimental questions (7 for each plot type, interleaved order).

Results and Discussion. The results of both the measured speed and accuracy of our experiments are shown in Fig. 11(a) and (b).

The results from Fig. 11(a) shows that when comparing accuracy, CCP showed improvement over SCP on average 91% compared to 69%, with statistical significance ($p = 0.001$). We also looked at subjects performance in just identifying the direction of correlation, where CCP had an accuracy of 99% compared to 79% for SCP, though not quite with statistical significance ($p = 0.06$). The response times (Fig. 11) showed similar results with CCP responses averaging 11.71 s compared to 23.4 s for SCP ($p = 0.001$). Given that in our experiments CCP outperformed SCP in both speed and accuracy, we consider **H1** confirmed.

A similar analysis shows that the accuracy CCP was 91% compared to 48% for PCP ($p = 0.001$). For identifying type only, CCP had an accuracy of 99% compared to 76% for PCP, though not with statistical significance ($p = 0.09$). The response times (Fig. 11) showed a similar result with CCP coming in on average 11.71 s compared to 24.5 s for PCP ($p < 0.001$). Given that CCP outperformed PCP in speed and accuracy, we consider hypothesis **H2** confirmed.

Although not explicitly selected as a hypothesis, we are also able to compare the performance of SCP and PCP. The results showed that SCP had a higher overall accuracy on average, 70%, compared to 58% for PCP ($p = 0.048$). However, when looking at type accuracy only, SCP had no statistical significance in average accuracy of 76% compared to 85% for PCP ($p = 0.25$). The results showed no statistical significance in average response times of SCP and PCP of 24.54 s and 25.78 s ($p = 0.774$), respectively. This result aligns with prior work [12,22,23].

The results of Exp. 1 confirmed the hypotheses **H1** and **H2**, indicating that using CCP subjects can identify correlation in less time and with higher accuracy compared to SCP and PCP. In our informal discussions with subjects after the experiment, they indicated that the shape of the axis and the distribution of

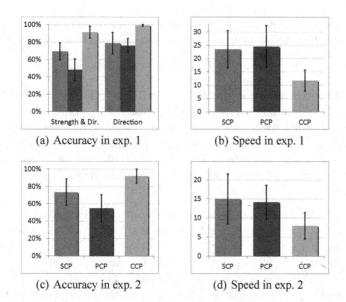

Fig. 11. Results of exp. 1 and exp. 2 show CCP (green, col. 3) outperforming SCP (blue, col. 1) and PCP (red, col. 2) in speed (sec) and accuracy (%). In all figures, error bars indicate standard deviation. (Color figure online)

points in CCP greatly assisted their comprehension of the correlation. Subjects complained that both the SCP and, in particular, the PCP were more difficult to distinguish positive and negative correlation in scenarios with low correlation. However, they found using CCP enabled them to easily recognize both the direction and strength.

8.2 Exp. 2: Differentiating Linear, Nonlinear, and Uncorrelated Relationships

Identifying nonlinear relationships between attributes can also be an important task. When comparing CCP with other methods, CCP provides simple visual cues making identification of correlation direction easier. Beyond that, CCP and SCP give similar visual cues (i.e. the tasks performed are basically the same) for the shape of the relationship, linear or nonlinear. This motivates our next hypothesis:

H3: *Using a Correlation Coordinates Plot and a Scatterplot will result in similar accuracy and speed for identification of linear, nonlinear, and uncorrelated relationships in 2 attributes.*

For PCP, identifying these relationships is far more challenging. The overdraw ambiguity that plagues linear correlations becomes significantly worse as even more lines overlap each other in nonlinear cases. This will slow and confuse users. This leads to our next hypothesis:

H4: *Using a Correlation Coordinates Plot will result in more accurate and faster identification of linear, nonlinear, uncorrelated relationships in 2 attributes than a Parallel Coordinates Plot.*

Method. The experiment is summarized in Table 1 (**H3** & **H4**). At the start of the experiment, participants were given instructions on linear and nonlinear correlation. Participants were then given 3 training questions followed by 9 experimental questions (3 for each plot type, interleaving order). For each question, participants saw a plot from 2 random attributes and were asked a forced choice question. Subject accuracy and time were measured.

Results and Discussion. The results of the measured speed and accuracy of our experiments are shown in Fig. 11(c) and (d), with all differences showing statistical significance ($p < 0.005$). The results of our experiment showed that CCP outperformed SCP. Our hypothesis **H3** however had predicted that the performance of CCP and SCP would be identical. This leads us to reject **H3**. In our discussions with subjects after the experiment, they indicated that the shape of axis and the distribution of points in SCP was more difficult to distinguish and that CCP assisted their comprehension of these specific types of correlation.

Due to CCP substantially outperforming PCP in both speed and accuracy, we consider hypothesis **H4** confirmed. As anticipated, participants complained that the overdraw problems made it difficult differentiate linear vs. nonlinear correlations in PCP.

8.3   Exp 3: Accuracy/Speed in Multi-attribute Datasets

The Snowflake Visualization was designed specifically for the task of quickly and accurately exploring pairwise correlations in multi-attribute data as compared with SPLOMs and PCP. As the number of attributes increases each SCP within a SPLOM becomes quite small and the number of plots becomes overwhelming. For PCP, as the number of attributes increases, the interaction required for many tasks puts increased pressure on the user to explore for features of interest. With these factors in mind, we developed 3 hypotheses:

H5: *Using a Snowflake Visualization will enable more accurate and faster identification of correlation between 2 attributes in multi-attribute data than a Scatterplot Matrix or Parallel Coordinates Plot.*

H6: *Using a Snowflake Visualization will enable more accurate and faster identification of how many attributes are correlated with a chosen attribute in multi-attribute data than a Scatterplot Matrix or Parallel Coordinates Plot.*

H7: *Using a Snowflake Visualization will enable more accurate and faster identification of which attributes are correlated with a chosen attribute in multi-attribute data than a Scatterplot Matrix or Parallel Coordinates Plot.*

Method. The experiment is outlined in Table 1 (**H5-H7**). Each participant was given an introduction and demo video for each visualization method and completed 12 sample questions using data unrelated to experimental trials. Then, each performed 21 experimental questions interleaving first between visualization types, then question types.

Results and Discussion

Identification of a Pairwise Correlation in Multi-attribute Data. The results of measured speed and accuracy in Fig. 12 (Type 1) show the Snowflake Visualization improved accuracy and speed over SPLOMs and PCPs with statistical significance (all $p < 0.05$). The average accuracy for Snowflake Visualization was 89% compared to 67% for SPLOM and 64% for PCP. The response times (Fig. 12) for Snowflake Visualization came in at an average of 19.9 s compared to 31.3 s for SPLOM and 26.1 s for PCP. Given that Snowflake Visualization outperformed SPLOM and PCP in speed and accuracy, we consider hypothesis **H5** confirmed.

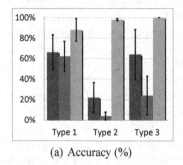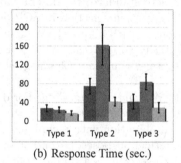

(a) Accuracy (%)　　　　　　(b) Response Time (sec.)

Fig. 12. Exp. 3 results show CCP (green, col. 3) outperformed SCP (blue, col. 1) and PCP (red, col. 2). (Color figure online)

Finding the Number of Correlated Attributes in Data. Again, the results of the experiments showed that the Snowflake Visualization improved accuracy and speed over SPLOMs and PCPs (see Fig. 12, Type 2) with statistical significance ($p < 0.05$). Therefore, we consider hypothesis **H6** confirmed.

Finding which Attributes are Correlated in Multi-attribute Data. The results of this final test also showed improved accuracy and speed over SPLOMs and PCPs (see Fig. 12, Type 3) with statistical significance ($p < 0.05$), leading us to also consider hypothesis **H7** confirmed.

　　The Snowflake Visualization's focus+context style greatly assisted subjects interactions and comprehension when working through multiple pairwise correlation questions. The participants complained that small SCPs made the SPLOM difficult to use, due to inability to see individual plots and difficulty tracking rows or columns of plots. Using PCP, participants complained that the number of dragging operations required to explore multiple correlations made it very difficult for them.

9   Discussion

User Study Task Selection. Selecting realistic tasks for a user study is a challenging problem when users are unfamiliar with the data and potentially visualization altogether. We have selected a number of simple tasks, which are building blocks for more complicated data analysis tasks that are commonly performed. The overall out performance of the CCP over SCP and PCP stands as evidence of its superiority, which should translate to more complex tasks.

Abstraction Selection. SCP and PCP have served a straw man role in our evaluation. There are any number of modifications that could be applied to either technique to better inform the user about correlation. However, since there is no single de facto standard, we did not want our evaluation to be clouded by questions of abstraction selection in SCP or PCP. Therefore, we stuck to the basic formulations of each approach. We hope this paper spurs the community to dig deeper into this subject and generate a more extensive evaluation of approaches, such as those of Harrison [12] and Kay [22].

Very High Attribute Count Data. For data with large numbers of attributes, we believe that approaches to extract the natural dimensionality of data, such as PCA, in combination with techniques such as CCP, will be critical in analysis. For all practical purposes, beyond 30 or 40 attributes, our approach is no longer viable. However, this is a similar limitation to SPLOMs and PCPs. We consider higher-dimensional cases to still be an open problem.

10   Conclusion

Correlation Coordinate Plots have been developed with the specific task of correlation identification. They have distinct advantages when compared to general task visualizations such as SCP and PCP. The advantages, as confirmed by our user study and real-world datasets, include:

- providing simple visual cues that make identification of the existence and direction of correlation fairly trivial;
- improving estimation of correlation strength by focusing the coordinate system on model fit; and
- improving identification of linear, nonlinear, and uncorrelated data by reducing ambiguity in the visualization.

In addition, the Snowflake Visualization showed significant performance improvements over SPLOMs and PCPs. The Snowflake Visualization is an efficient focus+context style layout representing a fair compromise between space efficient design, comprehensive visualization, and reduced user interaction for showing all pairwise correlations in multi-attribute data.

In conclusion, we believe that the CCP and Snowflake Visualization represent complementary approaches to existing techniques, replacing existing approaches only where correlation is the major feature of focus in data. We believe that more of these task specific approaches are on the horizon and will provide data analysts better, faster access to relevant information in their data.

Acknowledgements. We would like to thank our reviewers and colleagues who gave us valuable feedback on our approach. We would also like to thank our funding agents, NSF CIF21 DIBBs (ACI-1443046), NSF Core Program (IIS-1513616), Lawrence Livermore National Laboratory, and Pacific Northwest National Laboratory Analysis in Motion (AIM) Initiative.

References

1. Anscombe, F.J.: Graphs in statistical analysis. In: American Statistical Association, pp. 17–21 (1973)
2. Aris, A., Shneiderman, B.: Designing semantic substrates for visual network exploration. InfoVis **6**, 281–300 (2007)
3. Bertini, E., Tatu, A., Keim, D.: Quality metrics in high-dimensional data visualization: an overview and systematization. IEEE Trans. Vis. Comp. Graph. **17**(12), 2203–2212 (2011)
4. Bezerianos, A., Chevalier, F., Dragicevic, P., Elmqvist, N., Fekete, J.D.: Graphdice: a system for exploring multivariate social networks. Comput. Graph. Forum **29**(3), 863–872 (2010)
5. Buering, T., Gerken, J., Reiterer, H.: User interaction with scatterplots on small screens - a comparative evaluation of geometric-semantic zoom and fisheye distortion. IEEE Trans. Vis. Comp. Graph. **12**(5), 829–836 (2006)
6. Chen, Y.A., Almeida, J.S., Richards, A.J., Muller, P., Carroll, R.J., Rohrer, B.: A nonparametric approach to detect nonlinear correlation in gene expression. J. Comput. Stat. Graph. **19**(3), 552–568 (2010)
7. Dang, T.N., Wilkinson, L.: Transforming scagnostics to reveal hidden features. IEEE Trans. Vis. Comp. Graph. **20**(12), 1624–1632 (2014)
8. Elmqvist, N., Dragicevic, P., Fekete, J.D.: Rolling the dice: multidimensional visual exploration using scatterplot matrix navigation. IEEE Trans. Vis. Comp. Graph. **14**(6), 1148–1539 (2008)
9. Fanea, E., Carpendale, M.S.T., Isenberg, T.: An interactive 3D integration of parallel coordinates and star glyphs. In: InfoVis, pp. 149–156 (2005)
10. Friendly, M.: Corrgrams: exploratory displays for correlation matrices. Am. Stat. **56**(4), 316–324 (2002)
11. Geng, Z., Peng, Z., Laramee, R.S., Roberts, J.C., Walker, R.: Angular histograms: frequency-based visualizations for large, high dimensional data. IEEE Trans. Vis. Comput. Graph. **17**(12), 2572–2580 (2011)

12. Harrison, L., Yang, F., Franconeri, S., Chang, R.: Ranking visualizations of correlation using Weber's law. IEEE Trans. Vis. Comput. Graph. 20(12), 1943–1952 (2014)
13. Hartigan, J.A.: Printer graphics for clustering. JSCS 4(3), 187–213 (1975)
14. Heinrich, J., Stasko, J., Weiskopf, D.: The parallel coordinates matrix. In: EuroVis - Short Papers, pp. 37–41 (2012)
15. Heinrich, J., Weiskopf, D.: State of the art of parallel coordinates. In: Eurographics STAR, pp. 95–116 (2013)
16. Holten, D., van Wijk, J.J.: Evaluation of cluster identification performance for different PCP variants. In: EuroVis, vol, 29, no 3 (2010)
17. Hong, X., Wang, C.X., Thompson, J.S., Allen, B., Malik, W.Q., Ge, X.: On space-frequency correlation of UWB MIMO channels. IEEE Trans. Veh. Tech. 59(9), 4201–4213 (2010)
18. Huang, T.H., Huang, M.L., Zhang, K.: An interactive scatter plot metrics visualization for decision trend analysis. In: Conference on Machine Learning, Applications, pp. 258–264 (2012)
19. Inselberg, A.: The plane with parallel coordinates. Vis. Comput. 1(2), 69–91 (1985)
20. Jarrell, S.B.: Basic Statistics. W.C. Brow Comm, Reading (1994)
21. Johansson, J., Ljung, P., Jern, M., Cooper, M.: Revealing structure in visualizations of dense 2D and 3D parallel coordinates. Inf. Vis. 5(2), 125–136 (2006)
22. Kay, M., Heer, J.: Beyond Weber's law a second look at ranking visualizations of correlation. IEEE Trans. Vis. Comput. Graph. 22(1), 469–478 (2016)
23. Li, J., Martens, J.B., van Wijk, J.J.: Judging correlation from scatterplots and parallel coordinate plots. InfoVis 9, 13–30 (2010)
24. Mackinlay, J.: Automating the design of graphical presentations of relational information. ACM Trans. Graph. 5(2), 110–141 (1986)
25. Magnello, E., Vanloon, B.: Introducing Statistics: A Graphic Guide. Icon Books, London (2009)
26. Sharma, R.K., Wallace, J.W.: Correlation-based sensing for cognitive radio networks bounds and experimental assessment. IEEE Sens. J. 11(3), 657–666 (2011)
27. Tominski, C., Abello, J., Schumann, H.: Axes-based visualizations with radial layouts. In: ACM symposium on Applied computing, pp. 1242–1247. ACM (2004)
28. Wang, J., Zheng, N.: A novel fractal image compression scheme with block classification and sorting based on pearsons correlation coefficient. IEEE Trans. Image Process. 22(9), 3690–3702 (2013)
29. Wattenberg, M.: Visual exploration of multivariate graphs. In: SIGCHI, CHI 2006, pp. 811–819 (2006)
30. Wilkinson, L., Anand, A., Grossman, R.L.: Graph-theoretic scagnostics. InfoVis. 5, 21 (2005)
31. Xu, W., Chang, C., Hung, Y.S., Fung, P.C.W.: Asymptotic properties of order statistics correlation coefficient in the normal cases. IEEE Trans. Signal Process. 56(6), 2239–2248 (2008)
32. Yu, S., Zhou, W., Jia, W., Guo, S., Xiang, Y., Tang, F.: Discriminating DDoS attacks from flash crowds using flow correlation coefficient. IEEE Trans. PDS 23(6), 1073–1080 (2012)
33. Zhou, H., Cui, W., Qu, H., Wu, Y., Yuan, X., Zhuo, W.: Splatting lines in parallel coordinates. Comput. Graph. Forum 28(3), 759–766 (2009)
34. Zhou, H., Yuan, X., Qu, H., Cui, W., Chen, B.: Visual clustering in parallel coordinates. In: Computer Graphics Forum (2008)

Analysis and Comparison of Feature-Based Patterns in Urban Street Networks

Lin Shao[1]([✉]), Sebastian Mittelstädt[2], Ran Goldblatt[3], Itzhak Omer[3],
Peter Bak[4], and Tobias Schreck[1]

[1] Graz University of Technology, Graz, Austria
l.shao@cgv.tugraz.at
[2] University of Konstanz, Konstanz, Germany
[3] Department of Geography and Human Environment, Tel Aviv University,
Tel Aviv, Israel
[4] IBM Haifa Research Lab, Haifa, Israel

Abstract. Analysis of street networks is a challenging task, needed in urban planning applications such as urban design or transportation network analysis. Typically, different network features of interest are used for within- and between comparisons across street networks. We introduce StreetExplorer, a visual-interactive system for analysis and comparison of global and local patterns in urban street networks. The system uses appropriate similarity functions to search for patterns, taking into account topological and geometric features of a street network. We enhance the visual comparison of street network patterns by a suitable color-mapping and boosting scheme to visualize the similarity between street network portions and the distribution of network features. Together with experts from the urban morphology domain, we apply our approach to analyze and compare two urban street networks, identifying patterns of historic development and modern planning approaches, demonstrating the usefulness of StreetExplorer.

Keywords: Street network · Local patterns · Urban planning

1 Introduction

The analysis of network-oriented data is a recurring problem in many data analysis tasks, and to date knowledge discovery and visualization has provided many successful approaches to study network data. Many relevant phenomena can be described by network-oriented data structures, e.g., modeling social networks in social science, communication networks in data infrastructure, or gene regulation networks in bioinformatics applications.

One particular application of network-oriented data analysis arises in the investigation of street networks in urban areas as part of geographic data analysis. Street networks are an integral part of any urban structure, as they allow flows of traffic and pedestrians to connect and commute between different parts of the city. In conjunction with respective features within a city e.g., land usage

© Springer International Publishing AG 2017
J. Braz et al. (Eds.): VISIGRAPP 2016, CCIS 693, pp. 287–309, 2017.
DOI: 10.1007/978-3-319-64870-5_14

or traffic, street networks may determine important functional, social and perceptual properties of urban settlements, such as the efficiency of transportation and space utilization, social residential segregation and wayfinding. Recently, graph-theoretic methods have become popular to study topological properties of street networks (see Sect. 2.1). In practice, topological measures (also called features) like connectivity, integration or axiality can be computed for each street segment of a larger street network. Each of these features are typically given as real numbers, indicating e.g., how central a street segment is to the whole network. Experts are interested in investigating the properties of street networks, by inspecting the distribution of the different features across the street network, and correlating them with geometric and other properties of the street segments. However, the analysis and comparison of local patterns based on different street network's configurational attributes is a common problem to city planners and geographers. The similarity of these local patterns may depend on several criteria, e.g., extracted features, geometric properties or provenience, and the identification of outstanding pattern classes can help to enhance urban designs, city planning, transit-oriented development, or to study historic developments. For this purpose, it is useful to investigate meaningful patterns within and between cities, and apply the conclusions drawn from the analysis to future urban development.

We have worked with experts from urban morphology on the problem of analyzing topological and geometric features occurring in street networks. We identified the need for exploratory approaches to cope with the analysis problem, due to several factors, which make rule-based or purely automatic data analysis not fully effective. First, computational graph analysis provides a multitude of different topological features. Which of the feature is important for the current analysis is, however, not known a priori. Then, by consideration of geometric features, relevant patterns may occur at different scales with respect to the network, e.g., a smaller or larger local area may be of interest for comparison. Both calls for a highly interactive, exploratory approach to investigate street network features since it is not known a priori, which features are of interest, and at which scales.

We contribute an application for highly interactive visual exploration and comparison of local patterns across a network. We apply interactive search and appropriate visualization for local patterns of street features for a domain expert to investigate his data collection. At the heart of our approach is a flexible search function, by which the expert can quickly indicate for a specific network feature the regions of interest. We define a suitable similarity function to rank and compare street network properties, taking into account topological features, but also spatial properties of the network. We introduce a suitable color-mapping and boosting scheme, which allows visualizing local similarity to a user query in context of the overall feature distribution. It allows the user to quickly verify different hypothesis regarding to recurrent patterns, and arrive at meaningful findings on a given street network.

The remainder of this paper is structured as follows. In Sect. 2, we recall related work on analysis of urban street networks, pattern analysis in graph data, and on visualization of spatial data. In Sect. 3, we introduce the basic idea of our approach, based on two modes of query specification and result visualization. Then, in Sect. 4 we describe in detail our search methods and the similarity function behind the search. In Sect. 5, we substantiate our interaction choices by discussing them along the multi-level task abstraction of typology of Brehmer and Munzner and introduce a visual design to emphasize the matching street patterns. Further, in Sect. 6, we demonstrate the effectiveness of StreetExplorer by a use case application on a real street network analysis, conducted together with our co-author domain experts. We also provide a discussion of advantages and limitations of our approach. Finally, Sect. 7 concludes.

2   Related Work

We discuss related work in street network analysis, pattern extraction and visualization, and spatial visualization.

2.1   Analysis of Urban Street Networks

Street network patterns are a dominant component of a city's spatial properties. They have been shown to be significant for human spatial behavior, such as transportation mobility and accessibility [21] and vitality of urban life [39]. Urban street networks have been investigated with respect to their geometric and configurational attributes (e.g., number of intersections, number and size of blocks, connectivity, integration, fragmentation, etc.), their dynamics as well as their relations with other morphological components such as buildings, lots and the like. *Space syntax* is currently the dominant theoretical and methodological approach, which is based on configurational attributes of urban street networks [12,13]. This approach concentrates on the integration between urban streets (or places) and their relative accessibility and centrality in terms of intermediacy [18]. These configurational aspects have been found to be reliable indicators for purposes of comparison between street patterns [12,34] and for distinguishing street pattern development, e.g., self-organized versus planned-cities [15,29]. This body of research has produced classifications of patterns according to the spatial configuration attributes of street network. Moreover, prior research reveals that these configuration attributes are suitable to classify street patterns into different groups of historical urban planning approaches, namely pre-modern, modern and late modern patterns [14].

The space syntax attributes, which are based on axial maps, i.e., the smallest set of direct axial lines map of each investigated city [12,13], represent several aspects of accessibility and centrality at different scales that can be used for a classification of street networks. However, due to the multidimensional character of street networks' spatial configuration, the identification and classification of street patterns in urban areas is not a simple task. Space syntax studies have shown that street patterns can be identified by measuring axial lines or segments through a presentation of geographic distribution of spatial configuration attributes at different geographic scales. Such presentation can support the definition of spatial pattern patches in the street network at different scale in terms of their internal structure, contextual structure and relations between the two [42]. Such task requires interactive and complex actions by experts in the field of urban morphology.

2.2   Pattern Analysis in Graph Data

In graph and network analysis, similar to street network analysis, experts are interested in recognizing local patterns or subgraphs containing meaningful and relevant information. Since both research areas are highly connected they also share similar approaches to address related challenges. Tian et al. and Yang et al. use graph based approaches to identify significant patterns in street networks [32,41]. Street networks were represented as graphs, whereas streets are considered as edges and road junction as nodes. Heinzle et al. described the design and layout of basic street patterns and introduced an approach to detect patterns including strokes, grids and stars structures in street networks [11]. Luan and Yang in [20] use a density-based clustering method (DBSCAN) based on ε-neighborhood of each intersection (node) to identify complex road junction patterns. As another example, in [43], the authors investigated several existing approaches to automatically detect road intersection types (e.g., T-shaped, Y-shaped, X-shaped, Fork-shaped and so forth) in order to determine interchanges in street networks.

A large body of previous work addresses the visual analysis of graphs by search-based pattern exploration, of which we here can only give a small, illustrative sample. In [35] a visual analysis system was introduced, which automatically identifies common graph patterns (motifs) that are used as basis for navigation and exploration in graph data. In [40], a graph-substructure similarity search based on graph features was discussed. One key aspect of graph analysis is the representation of patterns in a suitable manner. Dunne introduced a motif simplification technique, in which common patterns are replaced by meaningful glyphs [8]. More generally, a survey of methods for visual analysis of graphs is given by [36].

2.3   Visualizations of Spatial and Movement Data

Many useful visualizations to date support interactive analysis of geospatial and movement data. An encompassing overview of visual analytics approaches and tools for movement data is covered in [1]. In our own previous work, in [3] we have applied visual cluster analysis to identify typical land usage patterns, and visually correlate these to properties of the underlying street network. In [26], we applied a similar Visual Analytics design to compare functional properties of cities with residential patterns. In [30], we presented an approach for visually correlating of the number of observed pedestrians with certain functional and street network properties of city spots. Common to these methods is that first, prominent patterns in the feature space were clustered. Then, cluster patterns can be effectively compared with other target variables of interest. Visual cluster analysis techniques are not only applicable on static data, but also temporal data may be of interest in many domains. To this end, in [2] we introduced a Visual Analytics approach to compare geospatial and temporal patterns with each other.

More specifically regarding transportation analysis, in [7,10] mobility patterns of taxi movements were investigated by extracting geographical information and using visualization techniques. Recent work in street network analysis [37] uses taxi GPS data to compute traffic flow rates and estimate traffic jams in the city. Finally, a visual analysis tool for support of urban space and place decision-making processes was developed in [28], relying on visualization techniques including time cubes, heat maps, and choropleth maps.

We build on these works for visualization of network and spatial data, contributing an approach for a explorative analysis of street network features based on adaptive selection-based search over geographic and topological properties of the street network, including appropriate data visualization.

3   Overview of Our Approach

The aim of StreetExplorer[1] is to support street network analysts during their investigation by supporting interactive search for similarities within the network, giving rise to potentially interesting local patterns. We rely on a wealth of currently existing computational methods for topological street network features, which we help to explore and understand by means of a search-based visual analysis system.

The exploration design of StreetExplorer is based on a two-step comparison approach that enables an investigation of street patterns on the *global* and *local* scale *within* and *between* urban networks. The design is depicted in Fig. 1. We adhere to Shneiderman's Information Visualization Mantra - Overview First, Zoom and Filter, Details on Demand [31] and propose to start the analysis session with a global overview comparison of features in a given network (shown in Fig. 2). By means of our small multiple view, analysts may recognize feature similarities, dependencies or correlations that help to find an interesting feature for further investigation.

We start the exploration by applying a global comparison of all available network features, to find a starting point for the exploration, and to identify interesting local patterns from a large number of pre-computed features. All available features are given by a list view to select from in the left hand side of Fig. 2. This view displays all features in a hierarchical structure based on the street-networks' configurational attributes and resolution levels. Thus, features containing a high number of resolutions will be aggregated in the list and provide a better overview than a plain list. From this list, several features may be chosen

---

[1] A demonstration of StreetExplorer can be found at: https://vimeo.com/149003539.

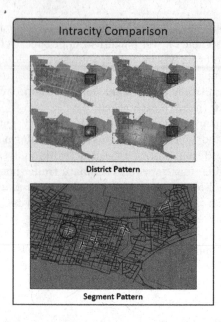

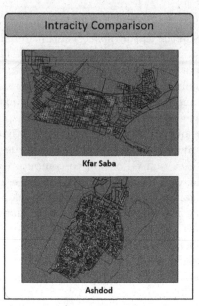

Fig. 1. Our proposed procedure to analyze street network patterns. The exploration starts with an interactive comparison of local patterns for a selected feature. This can be done by a district search or a user-defined search based on the segment level. Matching results will be highlighted on the map, and thus provide an overview of the availability, frequency and distribution of feature-based patterns. After interesting patterns were found, the analysis can be expanded to compare among several cities. (Color figure online)

and visualized in a small multiple view that represents the feature values by their original proportion of the city's street network. This allows for global comparison of the features. To support the global comparison, the features (or maps) can also be sorted according to the similarity of a selected one, by means of the below discussed similarity function (Sect. 4).

As a next step, the analyst could apply zooming and panning interactions (Zoom and Filter) to explore the features for certain regions of the city. To find locally interesting patterns, we support the query formulation by two selection-based methods, which are depicted on the left of Fig. 1 (district or segment pattern). The definition of query patterns can also directly be performed on the map by selecting certain street segments or a rectangular area of interest (district). Thereby, the analyst may define patterns of interest, which are composed of geometrical, topological and connectivity information of the individually chosen segments. The similarity search occurs interactively and the best matching results will be directly highlighted on the map.

Finally, after some interesting local patterns have been identified for a given street network, the analysis may continue by comparing the found pattern across different features and possibly, also across networks of different cities as depicted on the right of Fig. 1 (Details on Demand).

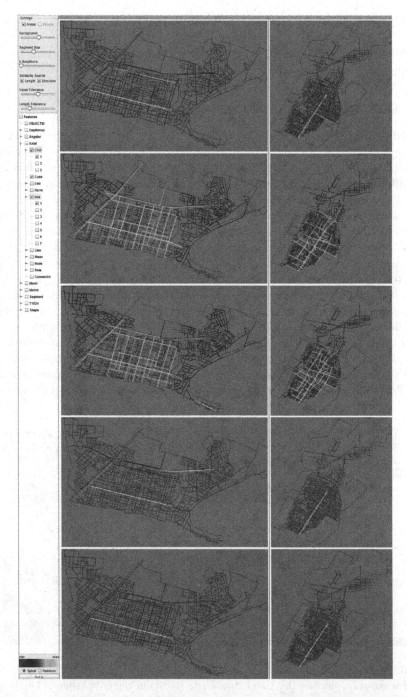

Fig. 2. Feature exploration in StreetExplorer to compare space syntax configurational attributes of two cities, namely Kfar Saba and Ashdod. The system creates a small multiple view of all selected features for comparison and further in-depth investigation for similarity in the street-network.

4   Exploration of Street Network Patterns

We describe the similarity functions of StreetExplorer as an important part of our supported exploration approach outlined in Fig. 1.

4.1   Global Street Network Comparison

According to our consultations with street network experts, a typical analysis session starts with the comparison of features in a given network. Therefore, analysts may select the desired features from the list view and visually compare the difference within the network configurations. To perform a global comparison - similarity search between different topological features and scales - we use a segment feature histogram, which relies on the street segment length and distribution of feature values. More precisely, we use an equal-width histogram of the street segment feature range and weight it by their segment lengths. Each bin represents a numerical range of the feature (e.g., Bin 1 contains the values from 0 to 9 - light green street segments) and reflect the total length of these street segments in the entire network. The histogram is normalized according to our color-mapping scheme (see Fig. 4 and Sect. 5.1). The bin size configuration is an ill-defined problem, which depends on the data set. This configuration can be adjusted by the analyst himself, but to achieve similar areas in accordance with the visual perception of the color-mapping, we suggest a minimum bin size that contains a stable color interval and covers all feature values that are visually in the same range. A demonstration of a global comparison is shown in

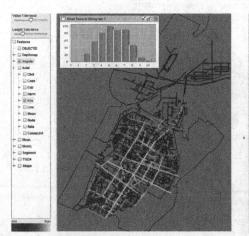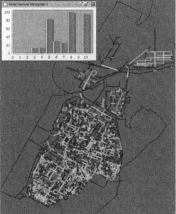

Fig. 3. Using Histogram Viewer to reveal the global distribution of the topological features *Axial Integration* (left) and *Angularity* (right) in the city Ashdod. The height of the histogram bins represents the aggregated segment length of the corresponding feature range. The streets of Ashdod have almost a normal (Gaussian) distribution of axial integrations and have a high number angular street segments. (Color figure online)

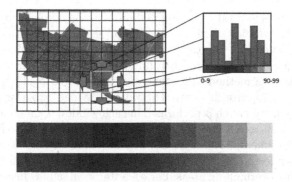

Fig. 4. Illustration of our histogram approach to compute district similarity. The histogram contains 10 bins that cover further 10 interpolated hues according to a spiral colormap. Street networks are partitioned into a regular grid and a sliding window approach iterates the search through the grid. (Color figure online)

Fig. 3. The *Histogram Viewer* can be used to identify the different distributions of topological features regardless of the geospatial position, whereas the small multiple views depict the difference on the segment level. Furthermore, we use this histogram descriptor to sort the small multiple view by computing the global similarity.

4.2   Comparison of District Patterns

StreetExplorer enables the search and comparison of feature similarities on the global and local scale within and between street networks. In case these properties cannot be seen at first sight, the interactive search methods of StreetExplorer can be utilized to explore the street network for locally interesting properties. We support the small multiple view by a concurrent search, meaning that a given query is run against all maps currently shown in the small multiple view, and matches are highlighted on all features maps at the same time. To this end, we provide two *local* search functions for street network patterns. The first search function is a *district comparison*, which identifies similar regions based on user-defined districts. Analysts may define a rectangular region by using a rubber band selection and detect similar districts based on geometric properties and feature values of the particular street segments. This is particularly suited if analysts want to approximately compare an area of interest e.g., east or west peripheral area of a city.

The technique we use to compute similarity between districts is adapted from image retrieval [19] and relies on the distribution of feature values as a basis for the similarity function. In our approach, we use again the equal-width histogram of all street segment features within the given selection, instead of the entire network as in the case of global comparison. We figured out that the good results were achieved by using a histogram with 10 bins (one bin for each color) that contain further 10 interpolated hues, as demonstrated in Fig. 4. Consequently, the histogram comprises in total a range of 100 units in the normalized feature interval between [0, 1]. We then store the length of all segments lying in the query district to the corresponding histogram bins and iterate the search over the entire map using a sliding window approach to find the most similar regions. The sliding window approach is a method that sequentially compares local regions for similarity search. In this approach, a street network is divided into a two-dimensional grid and the similarity search is performed in each window. Consequently, the histogram will take all intersecting segments into account and is weighted by the length of street segments. To assess the district similarity, we divide the street network into a uniform-sized grid and translate the rectangular district box over the grid including overlapping areas. For each iteration, a new histogram is computed and compared with the query district by the Euclidean distance. Due to the complex problem of scaling, the grid size can be interactively adjusted by the user, but at this point we would like to point out that a too small grid could cause many matching results at the same region (slightly shifted) and a too coarse grid could lead to missing associations. By default, the grid size will automatically be determined according to the query size, e.g., a smaller query district will also produce a smaller grid size. The grid size corresponds to a quarter of the original query size, and thus enables an overlapping degree of a half width and length. In the end, the regions with the most similar histograms will be emphasized on the map, as shown in Fig. 1 (District Pattern).

4.3   Comparison of Segment Patterns

To find more fine grained street patterns the second local search function of StreetExplorer can be used, which is shown in the lower left corner of Fig. 1 (segment pattern). It takes the connectivity as well as the length, direction and feature value of individual street segments into account. To distinguish the main segment orientations our domain experts defined 8 types of directions as shown in Fig. 5(a). Segments are considered as similar if they possess approximately equal direction, and are within a tolerance margin of length and feature value. This ensures that similar street patterns also consist of approximately the

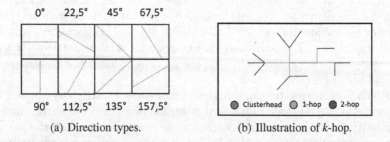

(a) Direction types.   (b) Illustration of k-hop.

Fig. 5. Illustration of our similarity properties for searching k-hop based patterns. (a) shows the eight different direction types; (b) illustrates the members of a street pattern based on 2-hops (blue and green segments). (Color figure online)

same segments irrespective of connection. Furthermore, we propose two different selection-based approaches to specify the neighborhood of a given segment and form a pattern.

The first selection is defined by a connected k-hop clustering, which uses a selected segment as clusterhead and groups several neighbors as members based on the hop distance. For instance, a 1-hop street pattern consists of one segment (clusterhead) and connected segments, in which the distance between the clusterhead and its members is 1 hop (junction). For better performance, we pre-calculated all segment connections in advance. An illustration of our k-hop selection is shown in Fig. 5(b). The orange marked segment denotes the selected segment, whereas the blue and green ones are the members that are reachable by a path of one and two hops respectively. By means of this selection approach a street pattern can easily be enlarged by increasing the number of k (shown in Fig. 8(a)–(c)–(e)). In this approach, the selection of the first segment is crucial since it builds the basis of the street patterns and initializes the similarity search based on its neighborhood. After a street segment has been defined as core segment (clusterhead), a single segment search over all target segments takes place, and computes an overall similarity score for its connected neighbors. Consequently, the analyst can quickly investigate similar local areas of a street network by changing the clusterhead segment. Since street patterns may have different neighborhood sizes, we compute a similarity score by comparing the average of feature value and the segment length of all segments that belong to one street pattern. Equation 1 shows the computation of the weighted average feature value where n is the number of neighboring segments of one street pattern. Accordingly, we determine the distance of the query to the target street patterns ($Feature_{avg}$) and eliminate those patterns, which exceed a certain threshold. Analysts may determine k for the neighborhood size as well as the tolerance margin for segment length and feature value to steer the analysis process.

$$Feature_{avg} = \frac{\sum_{i=1}^{n} SegmentLength_i \times FeatureValue_i}{\sum_{i=1}^{n} SegmentLength_i} \tag{1}$$

Alternatively, the analyst can also switch to a *free-form selection* in order to define an interesting pattern. By means of this selection approach, the analysts are able to specify even more accurate street patterns by selecting individual segments. In this way, it is possible to form complex street patterns that contain, e.g., stringy, circular or chain-like structures. The basic similarity search step is applied for each added segment and a *connectivity comparison* verifies whether the matched segments are connected correctly or not. Beginning from the first segment, the connectivity comparison stores the connection of each new added segment and observes the connectivity of query and target patterns in real time. Figure 6 demonstrates an example with four correct and one incorrect connected patterns referring to the query instance. After each new added segment, query and target patterns are compared for structural similarity of their neighboring segments. This means that every potentially similar segment is required to have an identical connected ancestors, otherwise it will be eliminated from the result set. For instance, all four correct target patterns in Fig. 6 have an equal connection to their ancestor segments (segment 1 is connected to segment 2 and segment 2 is again connected to segment 3), and thus are considered as similar. The problem with the negative example is that the last added segment (segment 3) is connected to both ancestor segments (segment 1 and segment 2), and are thus considered as dissimilar. To this end, our approach is invariant to rotation and allows slight variations of target patterns.

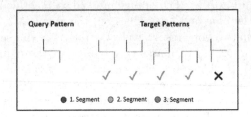

Fig. 6. Demonstration of our segment connectivity comparison for searching individual segment patterns. The color coding indicates the order of selected/matched segments. Based on the ancestor connection property, the search is invariant against pattern transformations. (Color figure online)

These local search functions can also be applied for intercity comparison. Basically, the search techniques and interaction possibilities for this application are the same as for intracity comparison. The only difference is that additional street networks are displayed in separate small multiple views and can include individual features (see Fig. 2). In this case, it might be beneficial to compare local patterns in different street networks (cities / features), which the analysts have considered as interesting. Hence, we designed a portable pattern search that can easily transfer street patterns (queries) of one city to another city. Accordingly, it enables additional comparison tasks and can reveal interesting findings, e.g., searching for certain patterns in all street networks; or comparison for spatially similar located patterns (northern part of the city or downtown); or searching for similar kinds of roads (main streets).

5   Interaction Design and Perceptually Supported Analysis

The workflow of StreetExplorer provides several search functions that were combined in an integrative exploratory system. In this section, we want to discuss the *why* and *how* of our interactions according to the multi-level typology of Brehmer and Munzner [5]. Further, we discuss novel colormapping techniques to enhance the readability of topological features in street networks and to support interactive highlighting.

System Interactions. The *Feature-Explorer* provides an overview of the topological features (shown on the left side of Fig. 2). It can be used to *browse* the different topological features and scales hierarchically, and interactively mark (*select*) features to specify the search space and further analysis. The selection occurs by clicking on the particular node of the *Feature-Explorer*. All selected features will be displayed in the small multiple view and reflect the candidates for inter-/intracity search. In order to provide a better overview (*explore*), displayed features can be reordered (*arrange*) based on similar properties. We realized the feature reordering function by using a Query-By-Example method to define a reference feature for sorting. For *selecting* (reference) a feature candidate the analysts can click on a map and *compare* it with all other selected features.

Interactions of the District Search. To explore (*discover*) similar urban districts based on topological features our district comparison method can be applied. In this respect, selection interactions are needed to express a region of interest. These regions can be *selected* by dragging the mouse over the map to frame a particular region. The best matching results will be highlighted (*select*) on the map to *compare* the spatial position of the similar districts. For local intracity comparison of feature-based patterns, we transfer the exact query region of the district search to all other selected features (maps) simultaneously. This ensures that the rubber band selection in all feature maps are equal. The additional Histogram Viewer (shown in Fig. 3) allows to *identify* the distribution of feature values. By clicking on certain histogram bins the particular segments will be emphasized (*select*) on the map.

Interactions of the Local Pattern Search. We designed several interaction possibilities in StreetExplorer to facilitate the usability and allow analysts to steer the analysis process. To generate (*produce*) different outcomes of street patterns, the parameter settings for the similarity search, such as feature value or segment length, can be modified (*manipulate*) by controllers to increase or decrease the similarity threshold. Other controllers are used to modify the k-hop level and brightness of the background. The parameter k can be varied from 0 (no hop) to 4 and *produces* various sizes of patterns for the search. Since, the background has a strong impact on the readability of pattern recognition (cf. Sect. 5.1), the controller can be used to ease the identification of patterns (*identify*). Furthermore, complete filtering steps, such as segment direction or length comparison, can be excluded by deactivating switchers to focus only on segment connections and feature values.

To formulate a query pattern for *search*, we implemented several techniques. The *selection* of individual street patterns takes place by a mouse over method to the corresponding segment. By hovering the mouse over street segments all similar segments and patterns respectively will appear. To modify an automatically generated neighborhood of the k-hop approach a deselection method was implemented. This might be interesting for analysts, who e.g., want to exclude dead ends or long main streets which have a major influence on the analysis. To generate (*encode*) an individual street pattern a *magnet selection* can be enabled by holding the "Ctrl" button and hovering over the desired segments.

(a) Without any boosting schemes.   (b) Highlighting with visual attention boosting.   (c) Highlighting with color contrast boosting.

Fig. 7. Demonstration of our visual boosting approach on the street network of Kfar Saba. To enhance visual perception of local patterns, we used a spiral colormap to show the feature values of street segments (a); reduced the saliency of unselected segments (b) and adapted the border color of located street patterns (c). (Color figure online)

Similar detected patterns will be highlighted by our Boosting scheme and facilitate the *presentation* of the result set. In order to *identify* and *compare* interesting patterns on the map, the analyst can *navigate* by using zooming and panning functions.

5.1 Visual Boosting

The visualization and selection of segments can be perceptually supported by novel colormapping and visual boosting approaches that we discuss in the following.

Colormapping. In [23], Mittelstädt et al. provide state-of-the-art guidelines and the tool ColorCAT to design colormaps for combined analysis tasks. In our application, the color encoded features are continuous and their interpretation requires the *identification* and *comparison* of metric quantities. We use Color-CAT to design our color encoding (see Fig. 4) for our application. The number of distinct colors is maximized for accurate *identification* of color values [17,24,38] and the perceptual linearity is preserved by linear increasing intensity, which is required for *comparing* color encoded data.

Boosting with Color Contrasts. If many streets are sharing dense areas of the display, it is hard for the user to perceive single streets. Studies showed that the visibility of low contrasts is reduced in high spatial frequency areas of the display [4]. The method of Mittelstädt et al. [24] compensates for harmful contrast effects in order to accurately visualize color encoded data. The method applies a perception model [9] in order to estimate the bias of contrast effects, which are amplified in the cones of our eye. With this approach it is possible to approximate a color c' that is in maximum color contrast to a target color c with $c'_x = D65_x - c_x$ (with x being the LMS channel in the CAT02 color space and $D65$ is the standardized reference light). Note, that this is only valid for saturated colors with one of the channels being close to zero. In order to visually boost the readability of streets in our visualization, we draw borders around the segments of streets. These borders are encolored with maximum color contrast to the segment color, which accords to *boosting with color* [25]. Further, we enable the user to control the segment and border sizes which enhances readability. On the one hand, these contrast effects can bias the user in reading color encoded features if the task is focused on a detailed analysis of specific (zoomed-in) data objects. On the other hand, this approach enhances the perception and recognition of streets in overviews, which enables us to read features even for dense areas on the screen. Therefore, we recommend the approach for tasks that require overviews of high frequency data. Further, this approach can be applied to highlight segments and street patterns (see Fig. 8).

Boosting with Visual Attention. The most common reason to highlight visual elements with color is to attract visual attention. Studies of Camgöz et al. show that humans are predominantly attracted by bright and saturated colors [6]. Since we use both to visualize metric quantities and want to accurately encode the elements in the selection, we reduce the visual saliency for unselected elements. We argue that these segments are still important for the context information of the selection but they need not to be visualized as prominently as the selected elements. Therefore, we reduce the intensity and saturation of these segments (we set both to 50% of the original color in the HSI color space [16]). To further decrease the visual saliency, we use the borders of unselected elements and decrease the contrast to the segment as well to the background by selecting the same color for the border but adjust the lightness and saturation of the border color to 50% of the original color. The low brightness and color contrasts reduce the visual saliency of unselected elements and steer our attention towards the selection.

Visual Boosting Effect. Figure 7 demonstrates the effect of our visual boosting approach. First, a spiral colormap is applied to visualize the continuous feature values of each street segment in the entire city (a). The colormap ranges from green/blue (low values), through purple/red to orange/yellow (high values). Consequently, analysts can easily detect streets with high feature values or areas including low feature values. Second, we reduced the visual saliency of inconsiderable street segments and thus, highlight local patterns resulting from similarity search (b). However, one remaining challenge is to perceive the patterns and their feature values, in particular in the case of rapid changes during the exploration. To tackle this issue, we finally support the highlighting by increasing the segment size and adapting the border color of detected local patterns (c). By default, the segment sizes are adjusted to the global street network view (without zoom level). We suggest to increase the segment size when focusing the analysis on a local area (with higher zoom level) to maintain the colors and feature values of the particular street patterns.

6   Case Study

In the previous section, we introduced the StreetExplorer system for search-based exploration of feature properties of urban street networks. We have worked closely with urban modeling experts from a university to use our system, and will describe the findings next. The expertise of the users reaches back two decades in research and consulting for urban planing and modeling, progressing the state-of-the-art in the domain significantly. Relevant application and research fields of the experts include defining effect of land-use types on ethnic residential integration, commercialization developments of urban neighborhoods, and predicting street segments' pedestrian friendliness and walkability. Understanding and analyzing the topology and configurational attributes of street networks are indispensable to these research and application domains.

6.1   Features Exploration

The network data, together with topological features from street-networks' configurational attributes has been provided to us in standardized shape files. The investigation was carried out by the experts and assisted by the developers. Measurement of space syntax configurational attributes is based on topological analysis of axial maps that treats individual axial lines (see Sect. 2.1) as nodes and axial line intersections as edges of a connectivity graph. The resulting graph provides the basis for several space syntax measures that describe the centrality of individual axial lines. In the current study, we have used the following measures: *Connectivity*, *Local Integration*, *Global Integration*, *Local Choice* and *Global Choice*. For any particular axial line, *Connectivity* denotes the number of directly linked axial lines. *Global Integration* indicates the closeness of an axial line to all other axial lines in the system by computing the shortest distance (or step depth) of the respective line from every axial line in the entire urban area. The *Local Integration* limits this computation to a certain neighborhood size, which is limited by a defined number of directional changes. Against to the above *to-movement* measures that represent the accessibility of a given axial line, the measures of *Global Choice* (which is equivalent to the graph-based centrality measures of betweenness) and *Local Choice* are through-movement measures. These measures indicate the number of times a location is encountered on a path from origin to destination for all pairs of axial lines in the entire urban area (global choice) or up to a defined topological distance (local choice).

In this study we chose to investigate two cities - Kfar Saba and Ashdod - which are representative of the street patterns of Israeli urban space [27]. We started our investigation by selecting the five space syntax measures, as described before, in order to reveal the global and local similarity between these cities. Results for the two cities are shown in a small multiple view (Fig. 2) for comparison. This representation clearly reflects the configurational differences between the cities, which has historical, developmental and topological reasons. Kfar Saba is a city with a nearly orthogonal street pattern that was established in the beginning of the 20th century and has developed mostly according to the pre-modern planning approach. In contrast, Ashdod is a relatively new city that was established in 1957 according to a comprehensive city plan based on modern planning approach. The city is characterized by a tree-like street layout associated with the idea of neighborhood units. These differences between these planning approaches and the associated street patterns are similar to other western cities (e.g. [21]).

Data on the cities' street networks were obtained as GIS layers for the year 2012 from MAPA company (http://www.gisrael.co.il). The Depthmap software [33] was used for constructing and analyzing the axial and segment maps and for computation of space syntax measures.

6.2   Interactive Search and Findings

We started our investigation with comparing similar local areas by using our district comparison tool. We realized very quickly that the distribution of similar local areas vary at most during the exploration of the feature *Local Integration* (second row of Fig. 2) and decided to use this feature for further investigation. Moreover, previous research indicates that this measure best distinguished cities from each other [27] and represent most effectively the differences in street patterns. Local integration measure describes integration only up to a defined number of changes of direction (topological distance), which is usually equal to three. The use of segments (the lines between junctions of axial lines) enables network analysis on a finer scale than using axial lines, it also extends with a consideration of angular distance (least angle distance) and metric distance that might be relevant to the purpose of investigation.

Our StreetExplorer implementation provided rapid response to a series of interactive queries we conducted, iteratively switching between selections of different segments in the overall street network, to find interesting repetitive structures in the map. We have been mainly focusing on areas that were established according to the modern planning approach. After a few trials, we selected one segment in the peripheral area of Kfar Saba (Fig. 8 - left hand side) and in Ashdod (Fig. 8 - right hand side). As a result, StreetExplorer highlighted all similar segments in the entire street-network for further investigation. Similar segments were found only on new areas that were developed in the second half of the 20th century: in the west and east peripheral areas of Kfar Saba, and almost in the entire area of Ashdod. We then increased the neighborhood level in a step-wise manner from no neighborhood (single segments are compared) to second and forth degrees of neighborhood (k-hop) included in the similarity function.

As shown in Fig. 8, the geographic distributions of similar patterns in both cities remained stable and increased consistently across the neighborhood level. Even though, when neighborhood size increases, some of the smaller patters that were close by, merge to larger ones, but some of them drop out, as their similarity falls below the defined threshold. This means, that by selecting the neighborhood size, the user in StreetExplorer can interactively determine the size of the pattern of interest and get immediate visual feedback on the search results.

In the latter query modality, the users selects only a single query segment, and the system increases the neighborhood size resolution along connected segments, thereby providing several query alternatives with different size. We note that the geometric similarity of the matches to the queries (perceived mainly by segment length and direction) is more prominent when the neighborhood size used in the

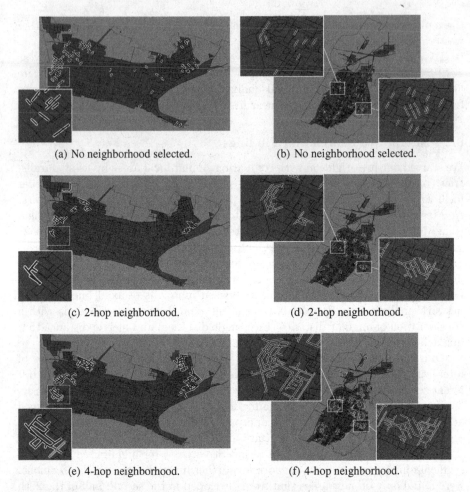

(a) No neighborhood selected.   (b) No neighborhood selected.

(c) 2-hop neighborhood.   (d) 2-hop neighborhood.

(e) 4-hop neighborhood.   (f) 4-hop neighborhood.

Fig. 8. Ashdod use case shows the distribution of similar segments in the city's periph-eral area. Geometric form, length and direction, are only marginally included in the similarity computation and only in the lowest neighborhood resolution. (Color figure online)

similarity computation is comparably low. From a domain analysis perspective this makes sense, as geometry and form are mainly considered more local config-urations of attributes, and rather not applicable to district and city levels.

Overall, this quite accurate pattern retrieval can be related to the considera-tion of topological spatial integration level at different network scales as well as to the geometric properties of segment length and direction. The revealed simi-lar patterns are characterized by the tree-like structure of the street network. In addition, distances between intersections on arterials are relatively larger than the distances in traditional patterns, and only streets at the same hierarchy or one above or below it, can intersect with the arterials. All these attributes do not exist in the traditional street patterns located in the older areas in the center of Kfar Saba.

Despite the similarity between street networks that are established at the same period in both cities regarding the location typical street patterns, they differ in the level of homogeneity within the street patterns. In Kfar Saba the neighborhoods are relatively homogeneous regarding their *Local Integration* value over all resolutions while in Ashdod the neighborhoods get quickly very heterogeneous when increasing the spatial resolution. This is visually salient in the number of distinct colors the segments have within a highlighted configuration. Moreover, while in Kfar Saba the similar neighborhoods remain in the periphery, independently of the resolution level, in Ashdod the neighborhoods spread to all districts of the city and quickly cover almost all the residential districts. The heterogeneity of the street patterns in Ashdod is a result of the hierarchical structure of the neighborhoods' street networks, which is reflected well in the spatial patterns of *Local Integration*, where arterial roads with the higher integration levels separate neighborhoods from each other. Against that, the hierarchy of importance in Kfar Saba is much weaker with no clear distinction between neighborhoods.

Thus, unlike previous space syntax works that were based on the spatial patterns of space syntax attribute values of individuals segment or axial lines (i.e. neighborhood at level 0), the investigation on pattern similarity in street networks here is conducted at different neighborhoods size and with simultaneous consideration of topological spatial integration and geometric aspects of street network (i.e. length and direction of segments). Due to these capabilities of this flexible search mechanism in StreetExplorer, we have identified street patterns that characterize urban planning approaches as well as sub-types street patterns, in this application, among modern street-based planning.

6.3   Expert Opinion

The unique contribution of StreetExplorer to the presentation and analysis of urban street network is twofold as stated by our co-author domain experts. At first, the simultaneous consideration of topological-visual (space syntax attributes) and geometric (i.e. length and direction of segments in Euclidean terms) aspects of spatial integration pattern in urban street network are innovative. Both aspects are essential for describing and identifying patterns in urban street networks [21] and should be considered together in street network analysis. For instance, in [22] topological features, such as continuity, connectivity and depth, of local street patterns were extracted and compared with a set of predefined patterns on a triangular projection space. StreetExplorer enables this simultaneous assessment and allows to represent user-defined patterns in their original spatial position. Secondly, for the analysis it is indispensable to represent similar patterns at different neighborhood sizes beyond the usual presentation at the level of individual segments. Consequently, the ability to identify street pattern pieces in urban street network as previously described enables the building of a typology of street patterns according to spatial cultures or urban planning approaches. This leads to exploration of street patterns for the emergence and design of urban phenomena such as walkable environment, retail activity and urban center, legible environments (wayfinding), residential segregation and crime. In addition to the current use of spatial integration values at the level of individual segments or axial lines in urban models (e.g. pedestrian or vehicle volume models) one could incorporate spatial integration values at the level of street pattern pieces that are defined by different neighborhood sizes. Finally, the experts stated, that they are not aware of any alternative tool or method that provides such interactive analytic functionality.

7   Summary and Future Work

We introduced StreetExplorer that helps analysts to explore a given space of network features in their context with a set of available topological street features provided by domain-specific software. The goal is to specifically support the exploration and visual comparison of street features on the global and local scale across and between features and regions in an effective way. We supported an interesting new application domain for visual analysis, and we provided three contributions in our work as follows. First, we defined a suitable similarity function to rank and compare street network properties, taking into account topological features, but also spatial properties of the network. Second, we defined suitable interaction functions, which allow the user to interactively select local areas of interest based on free-form selection and an adaptive neighborhood definition. Third, we defined a suitable color-mapping and boosting scheme, which allow the visualization of local similarity to a user query in context of the overall feature distribution. Additionally, we applied StreetExplorer together with domain experts, demonstrating the effectiveness and usefulness of the chosen designs by showing unexpected findings. While in the past the analysis has focused on predefined patterns like grid, star, or ring road patterns, now analysts are able to generate accurate patterns according to their interests. By means of StreetExplorer the domain experts were able to find the most informative feature and discovered interesting distributions of local patterns that can be traced back to historic development and modern planning of urban networks.

In the future, we plan to extend this work in different directions. We will enrich StreetExplorer on the metadata level by including additional features, such as landuse, barrier and general user comments, which can be extracted from openstreetmaps and provide an extensive search for local patterns in street networks. Moreover, StreetExplorer can be extended by different clustering methods for supporting comparison of interesting patterns. For instance, details-on-demand functions could be used to show the most significant or frequent patterns and propose further information for analysis. We want to investigate analytical methods to detect suitable parameter settings that reveal interesting patterns. This could be realized by using image-based techniques and comparison algorithms that consider several outcomes of parameter variations. We also would like to introduce functionalities for comparison of topological features and provide a semi-automatic feature selection approach based on statistical measurements. Furthermore, the visual representation and global arrangement of maps can be enhanced by specialized layouts that take the spatial properties of features into account.

References

1. Andrienko, G., Andrienko, N., Bak, P., Keim, D., Wrobel, S.: Visual Analytics of Movement. Springer Publishing Company, Heidelberg (2013)
2. Andrienko, G., Andrienko, N., Bremm, S., Schreck, T., von Landesberger, T., Bak, P., Keim, D.: Space-in-time and time-in-space self-organizing maps for exploring spatiotemporal patterns. Wiley-Blackwell Comput. Graph. Forum **29**(3), 913–922 (2010)
3. Bak, P., Omer, I., Schreck, T.: Visual analytics of urban environments using high-resolution geographic data. In: Painho, M., Santos, M., Pundt, H. (eds) Geospatial Thinking. LNGC, pp. 25–42. Springer, Heidelberg (2010). doi:10.1007/978-3-642-12326-9_2
4. Barten, P.G.: Contrast Sensitivity of the Human Eye and its Effects on Image Quality, vol. 72. SPIE Press, Bellingham (1999)
5. Brehmer, M., Munzner, T.: A multi-level typology of abstract visualization tasks. IEEE Trans. Vis. Comput. Graph. **19**(12), 2376–2385 (2013)
6. Camgöz, N., Yener, C., Güvenç, D.: Effects of hue, saturation, and brightness: part 2: attention. Color Res. Appl. **29**(1), 20–28 (2004)
7. Chu, D., Sheets, D., Zhao, Y., Wu, Y., Yang, J., Zheng, M., Chen, G.: Visualizing hidden themes of taxi movement with semantic transformation. In: 2014 IEEE Pacific Visualization Symposium (PacificVis), pp. 137–144, March 2014
8. Dunne, C., Shneiderman, B.: Motif simplification: improving network visualization readability with fan, connector, and clique glyphs. In: Proceedings of SIGCHI Conference on Human Factors in Computing Systems, pp. 3247–3256. ACM (2013)
9. Fairchild, M.D., Johnson, G.M.: iCAM framework for image appearance, differences, and quality. J. Electron. Imaging **13**(1), 126–138 (2004)
10. Ferreira, N., Poco, J., Vo, H.T., Freire, J., Silva, C.T.: Visual exploration of big spatio-temporal urban data: a study of New York city taxi trips. IEEE Trans. Vis. Comput. Graph. **19**(12), 2149–2158 (2013)
11. Heinzle, F., Anders, K., Sester, M.: Graph based approaches for recognition of patterns and implicit information in road networks. In: Proceedings of 22nd International Cartographic Conference, A Coruna, Spain (2005)
12. Hillier, B.: A theory of the city as object: or, how spatial laws mediate the social construction of urban space. Urban Des. Int. **7**(3), 153–179 (2002)
13. Hillier, B.: Space is the Machine: A Configurational Theory of Architecture. Cambridge University Press, Cambridge (2007)
14. Itzhak, O., Zafrir-Reuven, O.: The development of street patterns in Israeli cities. J. Urban Reg. Anal. **7**(2), 113–127 (2015)
15. Jiang, B.: A topological pattern of urban street networks: universality and peculiarity. Phys. A: Stat. Mech. Appl. **384**(2), 647–655 (2007)
16. Keim, D.: Designing pixel-oriented visualization techniques: theory and applications. IEEE Trans. Vis. Comput. Graph. **6**(1), 59–78 (2000)
17. Kindlmann, G., Reinhard, E., Creem, S.: Face-based luminance matching for perceptual colormap generation. In: Proceedings of the Conference on Visualization, pp. 299–306. IEEE Computer Society (2002)
18. Kropf, K.: Aspects of urban form. Urban Morphol. **13**(2), 105–120 (2009)
19. Liu, Y., Zhang, D., Lu, G., Ma, W.-Y.: A survey of content-based image retrieval with high-level semantics. Pattern Recogn. **40**(1), 262–282 (2007)
20. Luan, X., Yang, B.: Generating strokes of road networks based on pattern recognition. In: 13th Workshop of the ICA Commission on Generalisation and Multiple Representation, Zurich, Switzerland, vol. 12, p. 13 (2010)

21. Marshall, S.: Streets and Patterns. Spon Press, London and New York (2004)
22. Marshall, S.: Route structure analysis: a system of representation, calculation and graphical presentation. Working Paper (2008)
23. Mittelstädt, S., Jäckle, D., Stoffel, F., Keim, D.A.: ColorCAT: guided design of colormaps for combined analysis tasks. In: Proceedings of the Eurographics Conference on Visualization (EuroVis 2015: Short Papers), pp. 115–119 (2015)
24. Mittelstädt, S., Stoffel, A., Keim, D.A.: Methods for compensating contrast effects in information visualization. Comput. Graph. Forum **33**(3), 231–240 (2014)
25. Oelke, D., Janetzko, H., Simon, S., Neuhaus, K., Keim, D.A.: Visual boosting in pixel-based visualizations. In: Computer Graphics Forum, vol. 30, pp. 871–880. Wiley Online Library (2011)
26. Omer, I., Bak, P., Schreck, T.: Using space-time visual analytic methods for exploring the dynamics of ethnic groups' residential patterns. Taylor & Francis Int. J. Geograph. Inf. Sci. **24**(10), 1481–1496 (2010). Peer-reviewed article
27. Omer, I., Zafrir-Reuven, O.: Street patterns and spatial integration of Israeli cities. J. Space Syntax **1**(2), 295 (2010)
28. Pettit, C., Widjaja, I., Russo, P., Sinnott, R., Stimson, R., Tomko, M.: Visualisation support for exploring urban space and place. In: XXII ISPRS Congress, Technical Commission IV, vol. 25 (2012)
29. Porta, S., Crucitti, P., Latora, V.: The network analysis of urban streets: a primal approach. Environ. Plan. B: Plan. Des. **33**, 705–725 (2006)
30. Schreck, T., Omer, I., Bak, P., Lerman, Y.: A visual analytics approach for assessing pedestrian friendliness of urban environments. In: Vandenbroucke, D., Bucher, B., Crompvoets, J. (eds.) Proceedings of AGILE International Conference on Geographic Information Science. LNGC, pp. 353–368. Springer, Heidelberg (2013). doi:10.1007/978-3-319-00615-4_20
31. Shneiderman, B.: The eyes have it: a task by data type taxonomy for information visualizations. In: Proceedings IEEE Symposium on Visual Languages, pp. 336–343, September 1996
32. Tian, J., Ai, T., Jia, X.: Graph based recognition of grid pattern in street networks. In: Yeh, A., Shi, W., Leung Y., Zhou C. (eds.) Advances in Spatial Data Handling and GIS. LNGC, pp. 129–143. Springer, Heidelberg (2012). doi:10.1007/978-3-642-25926-5_10
33. Turner, A.: A program to perform visibility graph analysis. In: Proceedings of the 3rd Space Syntax Symposium, Atlanta, University of Michigan, pp. 1–31 (2001)
34. Vaughan, L., Jones, C.E., Griffiths, S., Haklay, M.M.: The spatial signature of suburban town centres. J. Space Syntax **1**(1), 77–91 (2010)
35. von Landesberger, T., Görner, M., Rehner, R., Schreck, T.: A system for interactive visual analysis of large graphs using motifs in graph editing and aggregation. In: VMV 2009, pp. 331–340 (2009)
36. von Landesberger, T., Kuijper, A., Schreck, T., Kohlhammer, J., van Wijk, J., Fekete, J.-D., Fellner, D.: Visual analysis of large graphs: state-of-the-art and future research challenges. Comput. Graph. Forum **30**(6), 1719–1749 (2011)
37. Wang, Z., Lu, M., Yuan, X., Zhang, J., Van De Wetering, H.: Visual traffic jam analysis based on trajectory data. IEEE Trans. Vis. Comput. Graph. **19**(12), 2159–2168 (2013)
38. Ware, C.: Color sequences for univariate maps: theory, experiments and principles. IEEE Comput. Graph. Appl. **8**(5), 41–49 (1988)
39. Wheeler, S.M.: The evolution of built landscapes in metropolitan regions. J. Plann. Educ. Res. **27**(4), 400–416 (2008)

40. Yan, X., Zhu, F., Yu, P.S., Han, J.: Feature-based similarity search in graph structures. ACM Trans. Database Syst. **31**(4), 1418–1453 (2006)
41. Yang, B., Luan, X., Zhang, Y.: A pattern-based approach for matching nodes in heterogeneous urban road networks. Trans. GIS **18**(5), 718–739 (2014)
42. Yang, T., Hillier, B.: The fuzzy boundary: the spatial definition of urban areas (2007)
43. Zhou, Q., Li, Z.: Experimental analysis of various types of road intersections for interchange detection. Trans. GIS **19**(1), 19–41 (2015)

Swarm-Based Edge Bundling Applied to Flow Mapping

Evgheni Polisciuc[(✉)] and Penousal Machado

CISUC, Department of Informatics Engineering,
University of Coimbra, Coimbra, Portugal
{evgheni,machado}@dei.uc.pt

Abstract. Applying flow maps in large datasets involves dealing with the reduction of visual cluttering. Nowadays, a technique known as edge bundling, which is geometric in nature, is often applied to reduce visual clutter and create meaningful traces that highlight the main streams of flow. This article presents an alternative approach of edge bundling for generating flow maps. Our approach uses a swarm-based system to reduce visual clutter, bundling edges in an organic fashion and improving clarity. The method takes into account the properties of data, edges and nodes, to bundle edges in a meaningful way while tracing lines that do not interfere visually with the nodes. Additionally, the Dorling cartograms technique is applied to reduce overlapping of graphical elements that render locations in geographic space. The method is demonstrated with application in the analysis of the US migration flow and transportation of products among warehouses and supermarkets of a major retail company in Portugal.

Keywords: Flow map · Edge bundling · Origin-destination map · Thematic map · Flow visualization · Geovisualization

1 Introduction

Flow mapping is a technique used to show the movement of objects from one location to another, such as the migration of people, the amount of goods being traded, the amounts of products being transported from warehouses to supermarkets, etc. Flow maps say little or nothing about the pathway, but include the information about what is flowing (moving, migrating, etc.), the direction of the flow, and how much is being transferred. In most cases, the data is represented using line width, color and spatial properties. Flow maps are advantageous in what regards to the visual clarity and ease of visual communication. This is achieved by merging edges that share similar destinations, or in some cases by tracing them through a similar path. However, this technique often fails when applied on large amounts of flow data. The visualization might become cluttered, making the map difficult to read, and difficult to distinguish the grouped individual streams.

© Springer International Publishing AG 2017
J. Braz et al. (Eds.): VISIGRAPP 2016, CCIS 693, pp. 310–326, 2017.
DOI: 10.1007/978-3-319-64870-5_15

In this work we describe a method for the generation of flow maps that is able to depict large amounts of transitions from one location to another (further expressed as *Origin-Destination* or simply *OD*). This method uses a customized swarming system to trace edges in an intuitive and organic fashion, reducing visual clutter. In particular, our system adapts the steering behaviors algorithm of autonomous agents presented by Reynolds [1], and the mechanism of indirect communication known as stigmergy. The agents of the systems communicate with each other through the environment, more precisely, the traces left by agents are used to stimulate the behavior of other agents. The method described in this article takes into account edge properties, such as directionality and the data attributes. Also, our method accounts for the presence of nodes and their properties. By routing the traces that avoid visual interference with nodes, the visualization provides clear distinction between traces and nodes. Additionally, our method employs graphic design decisions to promote clarity of visual communication in a high density environment. In order to improve clarity of representation of geographic locations, a technique, known as Dorling cartograms [2] was applied. With this technique, overlapped points were separated retaining some degree of spatial relationship. Finally, our approach supports mixed types of points – geo-referenced data points and those that have no fixed position in space. An example of the former are supermarkets with known location and of the later are warehouses with unknown geographic positions.

With that said, this work tackles the issue of depicting large amounts of products being transported from warehouses to hyper and supermarkets of a major retail company in Portugal. Our dataset is consisted of warehouse-to-supermarket transitions, over a time span of 6 months. The locations consist of approximately 60 warehouses, the majority of which located outside of Portugal and 1039 supermarkets in Portugal and 230 outside Portugal, from where only 680 are geo-located. The final graph used in this work, nonetheless comprises only the warehouses and supermarkets that are located on the continental part of Portugal.

In the following sections our approach is described in more detail. Section 3 presents the underlying idea and the method in detail. Section 4 describes an application of this technique on the available dataset. Finally, Sect. 5 presents a comparison of the results obtained by our approach and the existing one.

2   Background and Related Work

In the context of thematic mapping, an Origin-Destination representation commonly refers to the flow visualization (also known as flow maps), which is deeply rooted in the history of information visualization. Early examples, such as wine exports from France, produced by Minard [3, p. 25], represent the quantity as well as the direction of wine exports encoded by the thickness of the corresponding edges, which split from the parent edge when needed (Fig. 1). The work of Phan et al. [4] presents an automated approach to generate flow maps using a hierarchical clustering algorithm, given a series of nodes and a flow data. Generally, in the geographic context, a flow map depicts quantities of any type of objects that move from one location to another – e.g. migrations, transportation of goods, etc. The advantage of flow maps is that they reduce visual clutter by merging edges. However, when representing large amounts of data, this technique presents a series of problems, such as poor perception of the flow directionality, high degrees of visual clutter and overlapping of graphical elements that represent locations.

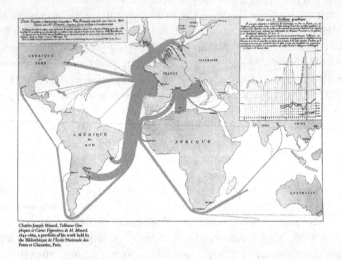

Charles Joseph Minard, *Tableaux Gra-
phiques et Cartes Figuratives de M. Minard,
1845–1869*, a portfolio of his work held by
the Bibliothèque de l'École Nationale des
Ponts et Chaussées, Paris.

Fig. 1. Export of French wine by Charles Minard, 1864.

Direct visualization of large amounts of Origin-Destination transitions can generate high degrees of visual clutter. In these cases a reduction strategy known as edge bundling can be applied. This is characterized not only by graph simplification, but also by the revelation of the principal streams of flow. Holten introduced edge bundling for compound graphs [5]. His work consisted in routing edges through a hierarchical layout using B-Splines. Nowadays, there are several variations of edge bundling, such as, force-directed edge bundling [6] or sophisticated kernel density estimation strategies [7]. More recently, approaches that account for edge properties have changed edge bundling into a meaningful, driven by data, edge tracing [8]. However, the way existing bundling techniques trace edges tends to be purely geometric in nature. Generally, edge bundling consists of drawing similar edges on the same path, i.e., edges that are related in geometry are routed along the same path. Ultimately, existing techniques tend to ignore nodes of the graph, or explicitly do not take them into account, which can result in a visual interfere of traces with nodes.

Another important characteristic of this work is the focus on nature-inspired approaches. The underlying idea is based on self-organizing systems, more precisely on the phenomenon of emergence in such systems. As the term indicates, self-organization is a property of some complex systems, in which the structure or organization appears without interference from external sources, typically resulting form a decentralized approach. Thus, self-organizing processes often result in the occurrence of emergent phenomena. More precisely, when the complex structure or behavior appears due to the simple low level interactions of a collection of individuals. There are several mechanisms of interaction among individuals, but the one that is used in this work is stigmergy, which consists of communication through the environment [9]. In the field of data visualization, there are techniques of graphical representation that are based on such systems. For instance *Geoboids* [10] employs a method to reveal patterns in spatial data through the use of a customized flocking system. In this system, each, so called geoboid explores the geographic space in accordance with the simple rules of interaction with other geoboids and the data found nearby. The visualization, which emerges from this simple process shows areas containing interesting information. Another series of works by Vande Moere exploit self-organization and emergence in information visualization. He introduces the idea of *infoticle*, which designates a particle that responds to data values and static forces in a particle system [11]. The visual output portrays the Internet file usage of a medium-sized company over time, conveying the patterns of file downloads. Another nature-inspired approach is *information flocking visualization* [12]. In this work, Vande Moere uses an artificial flocking system, originally proposed by [13], where the forces of attraction and repulsion are modified proportionally to the similarity between the data objects that each boid encodes. The emergent patterns analyzed at a higher-level, where each composition portrays short-term and long-term data tendencies in a time-varying dataset, convey meaningful changes over time.

3   Flow Map Flocking Model

In order to reflect the flowing nature of the data we resort to a customized flocking system. The underlying model to construct our flow map shares common characteristics with the work of Polisciuc et al. [14]. The idea is to get from point A to point B avoiding obstacles and, when possible, following existing trails that potentially lead to the destination. However, our new approach changes the way agents of the system interact with each other. Basically, the coordination is done through the environment, using stigmergy. Also, action selection rules were changed, in order to achieve meaningful behavior prioritizing obstacle avoidance and trail following.

The visualization itself can be seen as a directed graph composed by nodes and edges. The system consists of artificial agents (further referred to as *vehicles*), each one tracing an individual OD edge. A vehicle is characterized mainly by its position in space, direction and speed. During the simulation each vehicle leaves persistent traces, further referenced to as markers, which inherit the location and direction, at each simulation instant. Since the process is asynchronous, i.e., each trace is computed separately, the vehicle in the system interacts communicate with each other through the environment. While interacting with the markers, each vehicle follows simple rules: *attract* to the compatible markers; *avoid* obstacles, which are the nodes of the graph; arrive at the destination.

In order to determine the relationship between the vehicle and the marker we used a pairwise compatibility measures, including geometric properties and the weight of each one. With that said, this section describes our system in detail, including the strategy for steering behavior selection, compatibility metrics, and the modification of the environment. The output of the system is shown in Fig. 2.

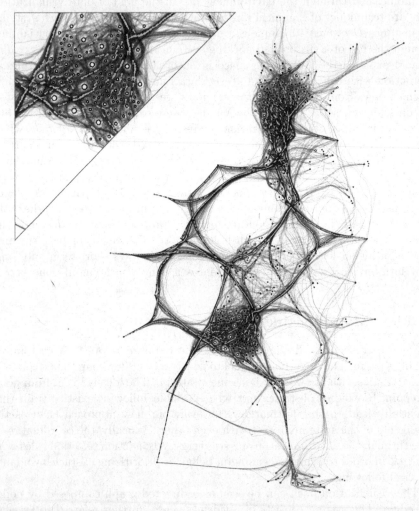

Fig. 2. Visual output from the system after 10 full cycles (bottom). Detail of the computed traces in black (top). Circles with black stroke represent nodes. White points with an arrow are markers.

3.1   Model

As previously mentioned, our system consists of a set of autonomous agents. Following Reynolds' ideas the motion behavior of an autonomous agent is defined by three layers: action selection, steering and locomotion [1]. Regarding the locomotion we use a simple vehicle model described in his work. The simple vehicle is characterized by a position in space, mass, velocity, maximum speed and maximum force. Using Euler integration the position is altered by adding the velocity, which is modified by applying steering forces that are truncated by the maximum force. Additionally, the velocity is truncated by the maximum speed. Regarding steering we used several behaviors, such as seek, arrive, obstacle avoidance, separation and cohesion. Finally, action selection uses the sensors of the vehicle to define the strategy, stetting the goal of the vehicle in response to the environment. For detailed information see the original Reynolds work [1].

In our systems each edge is traced by only one vehicle, starting at the origin node and finishing at the destination node, constantly leaving markers along its trail. Each trace is computed individually, and the computation of each one only starts when the previous one has finished, and so forth for all the traces. In each cycle the traces are updated according to the current state of the environment. More precisely, during the execution cycle each vehicle considers the markers left by other vehicles, and updates them if the specific conditions are met. This interaction determines the behavior of the vehicles. The process repeats until the visual result is acceptable and the user decides to stop it.

The behavior of our vehicles can be better understood by describing each step individually and their integration. There are four essential steps in our system: action selection, steering, locomotion and environment update.

Action Selection. The first step that defines the overall strategy of the vehicle's behavior is action selection. Using the attached sensors, vehicles perceive the environment and decide what action to take. In our model there are three types of sensors: one that detects the presence of obstacles, another that detects the presence of markers, and the last one that measures the similarity between an agent and a marker. The first sensor has the form of a cylinder, with a certain *radius* and *length*, and is aligned with the velocity vector. The length dictates how soon the obstacles will be detected and avoided. In order to create smooth traces the length should use reasonably high values, which depend on the scale of the image. The second sensor, which detects the presence of markers, is characterized by the field of sight, more precisely *angle* and *distance* of vision. The direction of the field of sight is also aligned with the velocity vector. If there are markers found in the field of sight they are considered for similarity comparison. Finally, the third sensor measures the similarity between the current vehicle and the markers in the field of vision.

The strategy of applying steering forces is based on the prioritization of behaviors and consists of simple rules. The very first step is to check if the vehicle is in its arrival distance. In the positive case the only behavior that is applied is arrive to the destination. Otherwise, we apply obstacle avoidance behavior. If the computed steering force is null, then we proceed to the second behavior, which is attraction. The attraction force consists of an average weighted vector pointing towards the markers in the field of view, using the compatibility measure described in the following paragraph. Finally, if the attraction force is null, we proceed to the calculation of the arrival behavior. In other words, the idea behind the vehicle's behavior is to avoid obstacles if there is the possibility of collision, otherwise it tries to follow existing potential trails that lead to the destination. If there are no trails in the field of vision, it just proceeds in the direction of the destination.

The compatibility measures are composed by the set of rules, which are based on the comparison of angles. In our approach we only consider the angle in the compatibility measure. While other solutions use a geometric approach, and, therefore, there is the need to use some edge compatibility measure to compute bundles, in our case bundles are the emergent output from the simple interaction among agents and their collective behavior. So, given a maximum angle, say θ, for each marker we apply the following rules: first, we calculate the angle between the vector pointing towards the marker and the vector pointing towards the destination node; second, we compute the angle between the vector pointing from the marker towards the vehicle's destination and the marker's direction. If both of these angles are bellow θ, then the attraction force is calculated and weighted by the marker's weight (Fig. 3). In order to get smooth approximation to the trail, the attraction vector points towards a location a few steps ahead the marker. In other words, the vehicle is attracted to the marker that points in the direction of vehicle's destination ensuring that it will reach its destination.

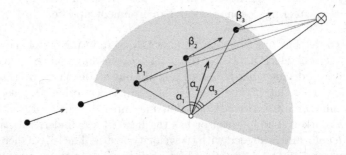

Fig. 3. Schematic representation of the vehicle and markers compatibility. Black circles represent markers. The white circle is a vehicle and the gray area is the field of sight. The white circle with a cross is the destination of vehicle. Arrows represent the direction of markers or velocity of the vehicle. Green lines are vectors pointing from the vehicle's location towards each marker. Red lines are vectors pointing from each marker's location towards the vehicle's destination. Assuming that the maximum allowed angle θ is 90 degrees, marker 1 is not considered for calculation, since the angle α is greater than θ. (Color figure online)

Steering and Locomotion. The steering behaviors used in our system are: arrive, avoid obstacles, and attract. These models and their implementation, as well as locomotion, strictly follow Reynolds ideas. However, there are some additional aspects that were introduced in order to achieve an intelligent edge bundling. In the beginning, the maximum speed limit is zero and increments with time.

This is done to achieve less chaotic behaviors near the origin location. The mass and the distance of vision are modified according to the edge's weight. So, the vehicle's that trace edges with higher weight are heavier and more "blind" than the ones that encode edges with low weights. In this case, the former traces become the central trails that suffer less modification. Finally, the attraction force is weighted according to the weight of the markers being considered in each step. Therefore, vehicles prioritize the trails with higher weights.

Environment Update. The environment of our system shares similar ideas of stigmergy, where vehicles communicate with each other through the environment. In contrast to other algorithms, such as ant colony, where there is one nest and various sources of food, in graphs there are various origins and destinations. In this case, in order to get to a specific destination there is the need to know an approximate indication about the direction of the trail. Therefore, we use the concept of marker, which is characterized not only by the location in space and its evaporation rate, but also by the direction. Each marker inherits some characteristics of the vehicle that deposited it – the location where the marker was left and the direction which corresponds to the velocity vector at that instance. With that said, the trails are updated as follows: with a certain rate, each vehicle leaves one marker; then we search for other markers that are already in the environment within a certain range; if no markers are found, the newly created marker is added to the environment; otherwise, for each found marker we check for compatibility by comparing their directions, and then update the location and the direction of each compatible marker. In this case there is no need to add the marker to the environment, since there is already one that is similar in location and direction. Both location and direction are updated with a certain rate. Otherwise, all the markers would quickly change their position, creating unwanted visual artifacts. The location is modified by pulling the marker towards the location of newly created markers, scaling down the attraction force. The direction modification is a simple summation of two directions, again scaling down the direction of the newly created marker.

An important aspect of environment modification is the rate of marker's fortification and the evaporation. At each frame (time unit) the weight of each marker in the environment is decreased by a certain degree. When the weight reaches zero, the marker is removed from the environment. Each time an existing marker is updated, the weight is increased by a certain degree. We used small values for fortification and lager ones for evaporation. In other words, the markers that are updated with frequency become more stable composing central trails of the environment. Also, there are upper limits of fortification that restrict continuous increase in the weight. Ultimately, the fortification and evaporation can be translated into the stabilization rate and how many bundling solutions are tested.

4 Application

In this section we describe a visualization application of our method for the flow map. We apply this technique to visualize the flow of products among warehouses and supermarkets, depicting movements of stocks of products of a major retail company in Portugal.

4.1 Data Description

Our dataset consists of warehouse-to-supermarket transitions of products over a time span of 6 months. Each transition has the following attributes: product id, quantity of products in transit (further referred to as *stock in transit* or *SIT*), quantity of products delivered (further referred to as *stock on hand* or *SOH*), warehouse id, supermarket id, and the date of transition. The locations consist of approximately 60 warehouses, some of which are located outside the Portugal and 1039 supermarkets in Portugal and 230 supermarkets outside the Portugal, from which only 680 are geo-located (see also [14–16]).

In order to get a graph representation of the data we proceeded with the calculation phase. First, the data were aggregated by days. Each day sums-up SIT quantities by aggregating pairs of warehouse-supermarket locations, which constitute the edges and the nodes of the graph. Finally, the total of SOH quantities per supermarket is calculated. Therefore, we get a weighted directed graph whose edges are directed from warehouse to supermarket nodes and weighted by SIT quantities. The nodes that represent supermarkets have an assigned SOH quantity, while the nodes that represent warehouse have none.

Finally, we filtered the nodes that only belong to the continental part of Portugal. We also excluded the supermarkets that have no geographic reference. For the sake of demonstration we focused only on the date of 1 of April, 2012. So, the final graph consists of 402 nodes and 2229 edges.

4.2 Flow of Products Visualization

This visualization depicts amounts of products that have been transported and the amounts of products that have been delivered during one day. It is important to describe the process to get a readable layout of a mixed graph. There are two challenges to visualize this particular graph: not all the locations have a geographical positions and the ones that do have may overlap. The first issue was solved by anchoring the position of the nodes that have a geographical location, and by using force-directed graph layout algorithm [17] to compute the location of other ones. This algorithm is efficient in what concerns the graph topology, since it considers clusters of nodes and not individual nodes. The second issue was solved by applying the Dorling cartogram technique over the pre-computed layout. The beauty of this algorithm is that it preserves the original relative positioning of geo-referenced elements, enabling the map of Portugal to be recognizable and making the locations distinguishable, increasing visual clarity (Fig. 4).

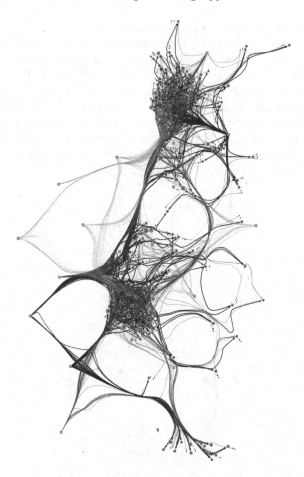

Fig. 4. Product transportation from warehouses and supermarket. (Color figure online)

In the visualization each trace represents the quantity of products in transit from warehouse to supermarket. These quantities are encoded by two means – *color* and *line thickness*. The two graphical elements are complimentary and make use of different mappings. The color uses a linear mapping. The values are mapped to a pale yellow-purple-dark blue gradient, being pale yellow and dark blue for lowest and highest values, respectively. The thickness variable, on the other hand, uses an exponential scale, emphasizing high values. Finally, we use an arrow, which is rendered at the end of a trace and scaled proportionally to the thickness of the line, in order to represent the direction of the flow (Fig. 4). The orientation of the arrow is determined in three steps. First, the length l and the with w of the arrow is computed. The w and l are equal to the 2x and 3x thickness, respectively. Second, the last vertex v of a trace that is located at the minimum distance from location of the node c plus its radius and the l is determined. Third, the arrow orientation is defined by the vector pointing from v to c.

In order to represent different types of nodes we use different graphical elements – *squares* and *circles* to represent warehouses and supermarkets, respectively. The nodes are colored according to the country they represent. Since our data contains several countries, the color could become ineffective. In this case, we calculate the total number of locations, aggregating by country, sort the countries in descendant order, and consider only the first two countries, which are Portugal and Spain, and treat the others as an unique instance. The colors used were green, red and blue for Portugal, Spain and others, respectively. Finally, the nodes that have geographic reference were rendered with the graphical shape of an X in the center of the node (Fig. 4).

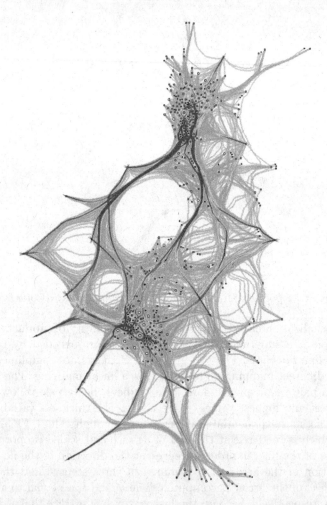

Fig. 5. A visualization of edge overlapping. The number of overlapped edges is represented with the color from pale green to intensive red. This image shows only the transitions from a selected warehouse. (Color figure online)

Density Visualization. Due to the high degree of overlapping traces, it is hard to identify main streams. For that reason, we proceed with a color-temperature visualization of the flow map. Using the same graph we apply different graphical elements. In this visualization the traces are colored according to the total number of overlapping elements. More precisely, we compute the number of overlapping segments that build-up the traces located within a defined range. Then at the render instance this value is mapped to a color scale from *intensive red* to a *pale green*, where red and green mean high and low degrees of overlaps. This representation is useful to get another perspective of such complex visualization (Fig. 5).

5   Comparison with Other Techniques

The visualizations shown in this section depict the US migration flow. The data covers migrations among cities of the United States during the period from 2014 to 2015. The graph consists of 377 nodes and 9780 edges. The weight of the edges correspond to the number of migrants normalized by the maximum value. It is important to mention that the cities not located in the continental part of the country were excluded. To encode the directionality of edges the color was used, more precisely the hue value [0, 360]. The absolute angle of the vector pointing from the origin node towards the destination was linearly mapped to the hue value.

With that said, this section proceeds with the comparison among straight line technique, force directed edge bundling (FDEB) method, and selected results of our technique. For the analysis of our approach we used pairs of low and high values for the parameters such as *maximum speed, maximum force, angle of vision, distance of vision* and *compatibility angle*. The maximum force can be translated into the agility of vehicles, which is easier to think of in the context of behavior. Higher maximum forces mean higher agility. Indicating actual values is meaningless, because they are relative to the scale of the graph. For reference, we used the values of 2 and 10 units for the graph with an approximate distance of 2500 pixels between the two most distant nodes. So, we generate all the permutations and identify combinations of values that can be divided in several groups. From the groups that achieved efficient results the representative images were selected and analyzed in detail.

The very first comparison reveals the efficiency of visual clutter reduction using the edge-bundling technique, comparing to the straight line representation (Fig. 6, image on the left). As can be observed, the force directed edge bundling (FDEB) method (Fig. 6, image in the middle) generates less visual clutter in comparison with our approach. However, the algorithms such as FDEB do not account for the presence of nodes, where traces can be drawn on top of nodes, and some of them neither account for the directionality of edges. In contrast, these properties are the emergent characteristic of our method. Since the vehicles in our system attempt to avoid obstacles, the nodes are clearly visible and do not visually interfere with the lines (see for instance Fig. 7). Also, as can be observed in the same figure, there are no traces with opposite directions that are routed through the same path. The only moment, where the traces can overlap, can be the split or the join of streams. Additionally, when using swarms, main streams of flow become visually distinct from each other, leaving enough space for the ones with less impact. Moreover, since our approach uses the edge weights in the calculation, the generated bundles result in a meaningful representation of the data being visualized (Figs. 7 and 8).

Fig. 6. Comparison between the techniques straight lines (left) and FDEB generated using 4 cycles, 50 iterations and stiffness 10 (right). The color represents the directionality of the edges.

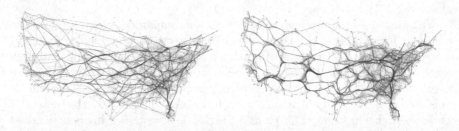

Fig. 7. Flow map generated by our method using: maximum speed equal to 10, maximum force equal to 12, mass equal to 1, vision angle equal to 135°, vision distance equal to 100 and compatibility angle equal to 45° (left); maximum speed equal to 10, maximum force equal to 12, mass equal to 8, vision angle equal to 135°, vision distance equal to 100 and compatibility angle equal to 90° (right). All values are in relative units, assuming that distance between two most distant nodes in the graph is approximately 2500 pixels.

Regarding the impact of parameters on the behavior of vehicles, and consequently over the final visual result, we established the following combinations. First of all it is important to mention that the analysis of results suggests that using low values for angle and distance of vision fails in generating an efficient visualization. In the later case the visualization becomes highly fragmented, and in many cases individual streams are hardly distinguishable, as can be observed in Fig. 8, image on the left, which was generated using low distance of sight.

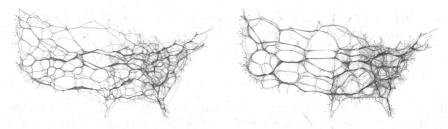

Fig. 8. Flow map generated by our method using: maximum speed equal to 2, maximum force equal to 12, mass equal to 1, vision angle equal to 135°, vision distance equal to 50 and compatibility angle equal to 90° (left); maximum speed equal to 2, maximum force equal to 2, mass equal to 1, vision angle equal to 135°, vision distance equal to 100 and compatibility angle equal to 90° (right).

Figure 7, image on the left, is the output from the system using *high* maximum speed and maximum force, and *low* mass and compatibility angle. As can be observed, meaningful bundles are clearly defined and distinguishable from each other. However, the resulting map is highly fragmented in comparison with examples that follow. Also, the representation presents constant changes between streams. This behavior is due to the low compatibility angle. The vehicles try to keep their velocity directions aligned with the destination. Finally, as can be noticed, due to high velocity and low agility the endings of traces are curly. This can be useful to distinguish origin and destination of flow.

The example shown in Fig. 7, image on the right, uses high values for all the parameters. In this example the main streams are clearly distinguishable from other traces. Due to high speed and low agility, vehicles tend to draw traces in a more organic manner. However, there is a loss in the detail of trace and overall visualization. Similar to the previous example the endings of traces present curly geometry.

The examples presented in Fig. 8 use low maximum speed. The distinction is in the agility: the example on the left uses low maximum force and the example on the right uses high maximum force. As can be observed, the traces in the representation on the left change their direction more abruptly than the ones of the image on the right. Additionally, it is clearly noticeable that curly endings disappear. This is because of low speed, therefore the vehicles have enough time to turn. Additionally, in the image on the left, we include the trails in order to illustrate the efficiency of the method to trace the streams of flow using swarms.

The method presented herein still has several limitations. The combination of high speed and mass with low agility can result in imperceptible visualization. As can be imagined with these parameters the vehicle advances fast and turns slowly, therefore making it harder to reach the destination. The vehicle spins around the destination node during a long time slowing down computational performance and creating visual artifacts. Another limitation, which can be advantageous in some cases, is that in highly dense graphs vehicles are unable to draw smooth traces. In particular, in the areas where nodes form clusters, the vehicles attempt to avoid the entire cluster, therefore creating visual artifacts.

6   Conclusions

As previously mentioned, applying flow maps to a large amount of data is challenging, since it involves dealing with high degrees of visual clutter. In this article we presented a method for generating flow maps that overcomes the cluttering issue in visualization. Our approach relies on a nature-inspired algorithm, which creates emergent visual patterns. This approach, results in edge bundling, reducing visual clutter and to promoting visual clarity in the representation. In addition, we explored different graphical languages applied on the generated graph to give diverse perspectives over the same dataset.

Our method consists of a set of vehicles that trace a path to represent each edge in the graph. Each single vehicle follows simple rules through the interaction with the neighboring trails. The compatibility between the edges and markers of the trails determines whether vehicles are attracted to markers or not. This makes the vehicles follow the trails that eventually take them to their destination. Furthermore, the vehicles not only contribute to the environment by leaving new markers, they also update existing ones. This is done for logical reasons and to reduce computational complexity. In addition, the vehicles that represent more products have higher impact on other members of the system. Finally, every vehicle attempts to avoid obstacles, i.e. nodes, creating clear visual distinction between traces and nodes.

We described two types of graphical representation. We presented the main visualization, which depicts transitions of products from warehouses to supermarkets. The total amounts of products being transported are represented with color and line thickness. The directionality of movement is indicated by an arrow at the end of each trace. The nodes use shape to indicate either if they represent a warehouse or a supermarket. Finally, the fixed nodes are marked with an "X" in the center of the node. Additionally, the sorting order of the edges reflects the emphasis on low or high values. Then, we presented a graphical approach to distinguish main streams of flow. This is achieved by coloring the edges by their degree of overlap. In this case, the red and green colors represent a high and low number of overlapped traces, giving a visual representation of the complexity of the graph.

Finally, we compared our method with the force-directed edge bundling technique and performed a study of the impact of systems parameters on the visual output. We distinguished several groups of parameter combinations that give similar results. Therefore, we established that using high values for maximum speed results in lower waviness of the streams, and in combination with high values for mass, the method produces smooth separation of traces from main streams. Also, using low values for maximum force results in a curly ending of traces, which can be used as an indicator to distinguish between the origin and destination of streams. On the other hand, low values of maximum speed result in a compact and definite bundling. However, the changes in the direction of streams become abrupt. Finally, low values for compatibility measure produce fragmented representation with a higher number of separate streams, but the traces become less "wavy" and the changes in the separation from main streams are less abrupt.

Acknowledgements. We would like to thank Catarina Maçãs, Hugo Amaro, Filipe Assunção, Antnio Cruz and Pedro Cruz for their contributions to this work. This project is partially funded by SONAE: Sonae Viz – Big Data Visualization for retail, and by Fundação para a Ciência e Tecnologia (FCT), Portugal, under the grant SFRH/BD/109745/2015.

References

1. Reynolds, C.W.: Steering behaviors for autonomous characters. In: Game Developers Conference, vol. 1999, pp. 763–782 (1999)
2. Dorling, D.: Area Cartograms: Their Use and Creation. Concepts and Techniques in Modern Geography, vol. 59. University of East Anglia: Environmental Publications (1996)
3. Tufte, E.R.: The Visual Display of Quantitative Information, vol. 2. Graphics Press, Cheshire (1983)
4. Phan, D., Xiao, L., Yeh, R.B., Hanrahan, P., Winograd, T.: Flow map layout. In: Stasko, J.T., Ward, M.O. (eds.) IEEE Symposium on Information Visualization (InfoVis 2005), 23–25 October 2005, MN, USA, p. 29. IEEE Computer Society, Minneapolis (2005)
5. Holten, D.: Hierarchical edge bundles: visualization of adjacency relations in hierarchical data. IEEE Trans. Vis. Comput. Graph. **12**, 741–748 (2006)
6. Holten, D., van Wijk, J.J.: Force-directed edge bundling for graph visualization. Comput. Graph. Forum **28**, 983–990 (2009)
7. Hurter, C., Ersoy, O., Telea, A.: Graph bundling by kernel density estimation. Comput. Graph. Forum **31**, 865–874 (2012)
8. Peysakhovich, V., Hurter, C., Telea, A.: Attribute-driven edge bundling for general graphs with applications in trail analysis. In: 2015 IEEE Pacific Visualization Symposium, PacificVis 2015, Hangzhou, China, 14–17 April 2015, pp. 39–46 (2015)
9. Di Marzo Serugendo, G., Gleizes, M.P., Karageorgos, A.: Self-organising systems. In: Di Marzo Serugendo, G., Gleizes, M.P., Karageorgos, A. (eds.) Self-Organising Software. Natural Computing Series, pp. 7–32. Springer, Heidelberg (2011). doi:10.1007/978-3-642-17348-6_2
10. Macgill, J., Openshaw, S.: The use of flocks to drive a geographic analysis machine. In: International Conference on GeoComputation (1998)
11. Vande Moere, A., Mieusset, K.H., Gross, M.: Visualizing abstract information using motion properties of data-driven infoticles. In: SPIE Proceedings Series, pp. 33–44 (2004)
12. Moere, A.V.: Time-varying data visualization using information flocking boids. In: Ward, M.O., Munzner, T. (eds.) 10th IEEE Symposium on Information Visualization (InfoVis 2004), 10–12 October 2004, TX, USA, pp. 97–104. IEEE Computer Society, Austin (2004)

13. Reynolds, C.W.: Flocks, herds and schools: a distributed behavioral model. In: Stone, M.C. (ed.) Proceedings of the 14th Annual Conference on Computer Graphics and Interactive Techniques, SIGGRAPH 1987, pp. 25–34. ACM (1987)

14. Polisciuc, E., Cruz, P., Amaro, H., Maçãs, C., Carvalho, T., Santos, F., Machado, P.: Arc and swarm-based representations of customer's flows among supermarkets. In: Proceedings of the 6th International Conference on Information Visualization Theory and Applications, pp. 300–306 (2015)

15. Maçãs, C., Cruz, P., Amaro, H., Polisciuc, E., Carvalho, T., Santos, F., Machado, P.: Time-series application on big data visualization of consumption in supermarkets. In: Proceedings of the 6th International Conference on Information Visualization Theory and Applications, IVAPP 2015, Berlin, Germany, 11–14 March 2015, pp. 239–246. SciTePress (2015)

16. Maçãs, C., Cruz, P., Martins, P., Machado, P.: Swarm systems in the visualization of consumption patterns. In: Proceedings of the Twenty-Fourth International Joint Conference on Artificial Intelligence, IJCAI 2015, Buenos Aires, Argentina, 25–31 July 2015, pp. 2466–2472. AAAI Press (2015)

17. Fruchterman, T.M., Reingold, E.M.: Graph drawing by force-directed placement. Softw. Pract. Exp. **21**, 1129–1164 (1991)

Computer Vision Theory and Applications

Relative Pose Estimation from Straight Lines Using Optical Flow-Based Line Matching and Parallel Line Clustering

Naja von Schmude[1,2(✉)], Pierre Lothe[1], Jonas Witt[3], and Bernd Jähne[2]

[1] Computer Vision Research Lab, Robert Bosch GmbH, Hildesheim, Germany
{najavon.schmude,pierre.lothe}@de.bosch.com
[2] Heidelberg Collaboratory for Image Processing,
Ruprecht-Karls-Universität Heidelberg, Heidelberg, Germany
{naja.von.schmude,bernd.jaehne}@iwr.uni-heidelberg.de
[3] Robert Bosch LLC, Palo Alto, CA, USA
jonas.witt@us.bosch.com

Abstract. This paper tackles the problem of relative pose estimation between two monocular camera images in textureless scenes. Due to a lack of point matches, point-based approaches such as the 5-point algorithm often fail when used in these scenarios. Therefore we investigate relative pose estimation from line observations. We propose a new algorithm in which the relative pose estimation from lines is extended by a 3D line direction estimation step. Using the estimated line directions, the robustness and computational efficiency of the relative pose calculation is greatly improved. Furthermore, we investigate line matching techniques as the quality of the matches influences directly the outcome of the relative pose estimation. We develop a novel line matching strategy for small baseline matching based on optical flow which outperforms current state-of-the-art descriptor-based line matchers. First, we describe in detail the proposed line matching approach. Second, we introduce our relative pose estimation based on 3D line directions. We evaluate the different algorithms on synthetic and real sequences and demonstrate that in the targeted scenarios we outperform the state-of-the-art in both accuracy and computation time.

Keywords: Relative pose estimation · Matching · Lines · Clustering · Monocular camera

1 Introduction

Relative pose estimation is the problem of calculating the relative motion between two or more images. It is a fundamental component for many computer vision algorithms such as visual odometry, simultaneous localization and mapping (SLAM) or structure from motion (SfM). In robotics, these computer vision algorithms are heavily used for visual navigation.

© Springer International Publishing AG 2017
J. Braz et al. (Eds.): VISIGRAPP 2016, CCIS 693, pp. 329–352, 2017.
DOI: 10.1007/978-3-319-64870-5_16

The classical approach to estimate the relative pose between two images combines point feature matches (e.g. SIFT [1]) and a robust (e.g. RANSAC [2]) version of the 5-point-algorithm [3]. This works well under the assumption that enough point matches are available, which is usually the case in structured and textured surroundings.

As our target application is visual odometry and SLAM in indoor environments (e.g. office buildings) where only little texture is present, point-based approaches do not work. But lots of lines are present in those scenes (cf. Fig. 1), hence, we investigate lines for relative pose estimation.

Fig. 1. A typical indoor scene with few texture in which a lot of lines are detected.

This paper recaps our previous work on relative pose estimation [4] and extends it by a novel line matching scheme which is based on optical flow calculation. We will present more details on the robust relative pose estimation framework and show in a thorough evaluation of the line matching and the relative pose estimation that we reach state-of-the-art results.

Our contributions are:

- a new line matching scheme in which optical flow estimation is used to find the correspondences
- a robust and fast framework combining all necessary steps for relative pose estimation using lines.

2   Related Work

The trifocal tensor is the standard method for relative pose estimation using lines. Its calculation requires at least 13 line correspondences across three views [5,6]. It is in general not possible to estimate the camera motion from two views as shown by Weng et al. [7] unless further knowledge of the observed lines is exploited. If for example different pairs of parallel or perpendicular lines are given, as it is always the case in the "Manhattan world"[1], the number of required views can be reduced to two.

---

[1] A scene complies with the "Manhattan world" assumption if it has three dominant line directions which are orthogonal and w.l.o.g. can be assumed to coincide with the x-, y- and z-axis of the world coordinate system.

As the number of views is reduced, the number of required line correspondences declines as well, because only five degrees of freedom have to be estimated (three for the rotation and two for the translation up to scale) compared to 26 for the trifocal tensor. Problems with small degrees of freedom are beneficial for robust methods like RANSAC as the number of required iterations can be lowered. As for the method proposed in this paper, only two views are required where the rotation is estimated from a minimal of two 3D line direction correspondences and the translation from at least two intersection points.

In their work, Elqursh and Elgammal [8] employ the "Manhattan world" assumption and require only two views as well. They try to find "triplets" of lines, where two of the lines are parallel and the third is perpendicular to the others. The relative pose can then be estimated from one triplet. The pose estimation process is split up into two steps: first, the vanishing point information of the triplet is used to calculate the relative rotation. Then, the relative translation is estimated using the already calculated rotation and intersection points between lines. The detection of valid triplets for rotation estimation is left over to a "brute force" approach, in which all possible triplet combinations are tested through RANSAC. As the number of possible triplets is in $O(n^3)$ for n lines, this computation is very expensive. Contrarily, our rotation estimation method is more efficient as we calculate it from 3D line directions and the number m of different 3D line directions per image is much smaller than the number of lines (in our cases $m < 10$ whereas $n > 100$). In addition, we do not need the restricting "Manhattan world" assumption which would require orthogonal dominant directions but a less stricter form where we allow arbitrary directions.

Similar approaches requiring two views were presented by Wang et al. [9] and Bazin et al. [10]. In both works, the pose estimation is split up into rotation and translation estimation as well, where the rotation calculation relies on parallel lines. Bazin et al. estimate the translation from SIFT feature point matches. Their approach is also optimized for omnidirectional cameras. Our method requires only lines and no additional point feature detection as we calculate the translation from intersection points.

There exists already several approaches to match lines across images: Prominent descriptor-based approaches are "Mean Standard Deviation Line Descriptor" (MSLD) [11], "Line-based Eight-directional Histogram Feature" (LEHF) [12] and "Line Band Descriptor" (LBD) [13,14]. These approaches follow the idea of describing the local neighborhood of a line-segment by analyzing its gradients and condensing their information into a descriptor vector. In the matching process, the descriptors are compared and the most similar descriptor decides the match. Often, techniques like thresholding the descriptor distance, "Left/Right Checking" (LRC) or "Nearest Neighbor Distance Ratio" (NNDR) are employed to robustify the matching. LRC ensures that the matching is symmetrical by only accepting matches where matching from "left" image to "right" image gives the same result as matching from "right" to "left". LRC handles therefore occlusions. NNDR is known from SIFT feature matching [1] and follows the idea that the descriptor distance for a correct match should be significantly smaller than the distance to the closest incorrect match.

As local neighborhoods of different lines are often not distinguishable, the resulting descriptors are similar and therefore not feasible for matching under severe viewpoint changes. Explicitly designed for the tracking of lines in sequences are the approaches from Deriche and Faugeras [15] and Chiba and Kanade [16]. Deriche and Faugeras propose a Kalman filter for predicting the line-segment's geometry in the next image, whereas Chiba and Kanade use optical flow for the prediction. Both approaches define a similarity function using only the geometry of the line to associate the predicted line with an observed one. As the changes between consecutive image frames are small, geometry is sufficient for the matching.

We present a novel matching technique which is also designed for the tracking of lines. Our algorithm is based on optical flow. We do not need to calculate descriptors as the optical flow vectors serve to associate line-segments in the images which saves valuable execution time. We can show that our approach shows better matching performance while requiering less computation time than state-of-the-art descriptor-based approaches in small baseline cases.

Computer vision systems requiring relative pose estimation are e.g. visual odometry, SLAM and SfM. Several line-based SfM methods exist like [6,7,17,18] which formulate the SfM problem as a nonlinear optimization. The initial configuration is calculated using the trifocal tensor or is derived from other input data. The approach from Schindler et al. [18] takes "Manhattan World" assumptions into account and includes a vanishing point clustering on pixel level. Our method could be used in these SfM algorithms as an alternative for initialization requiring only two views. An EKF-based SLAM method called "StructSLAM" [19] has recently been published which extends a standard visual SLAM method with "structure lines" which are lines aligned with the dominant directions. Witt and Weltin [20] presented a line-based visual odometry algorithm using a stereo camera setup. The relative pose is estimated using a nonlinear optimization of the 3D line reconstruction similar to ICP. Holzmann et al. [21] proposed a line-based visual odometry using a stereo camera, too. They present a direct approach and calculate the displacement between two camera poses by minimizing the photometric error of image patches around vertical lines.

The paper is structured as follows: in Sect. 3 we present our novel line matching algorithm based on optical flow. The line correspondences are then used in our relative pose estimation which is explained in Sect. 4. The paper is concluded with an extensive evaluation of the optical flow-based line matching (Sect. 5.1) and the relative pose estimation algorithm (Sect. 5.2) where we show that we outperform the current state-of-the-art.

3   Optical Flow-Based Line Matching

The proposed matching algorithm is explicitly designed for the matching of lines under small viewpoint changes such as consecutive frames in image sequences.

It follows the idea of Chiba and Kanade [16] in the point that optical flow calculation is exploited to generate the matches.

The algorithm consists of three main stages: first, the optical flow is calculated for points along the line-segments. Second, the flow vectors originating from the same line are checked for consistency. In the third step the flow vectors are used in a histogram-based approach to generate the matches. Every step is discussed in detail in the following sections.

3.1   Optical Flow Calculation

In the first step the optical flow is calculated using the method from Lucas and Kanade [22]. We name the point in the original image $\mathbf{p}$ and the point in the image after the motion $\mathbf{p}' = \mathbf{p} + \mathbf{v}$ where $\mathbf{v}$ is the optical flow vector.

In contrast to the matching procedure from Chiba and Kanade [16], we do not calculate the optical flow over a static grid on the whole image but consider only the image regions which are interesting for the matching of lines: the pixels belonging to the extracted line-segments. The question is now if all pixels belonging to line-segments should be used or if certain pixels are more appropriate for optical flow calculation then others?

Shi and Tomasi [23] tackled this question and analyzed which image regions are well suited for the Lucas-Kanade optical flow method. They found that the eigenvalues of the gradient matrix $\mathbf{G}$ of a pixel $\mathbf{p}$ are good indicators for the eligibility with

$$\mathbf{G} = \begin{pmatrix} g_x^2 & g_x g_y \\ g_x g_y & g_y^2 \end{pmatrix} \quad \text{and} \quad \mathbf{g} = \begin{pmatrix} g_x \\ g_y \end{pmatrix} \text{ the gradient at point } \mathbf{p}. \qquad (1)$$

If both eigenvalues are small, the pixel belongs to an uniform region which is uneligible for flow calculation. If one eigenvalue is small and the other large, the image region contains an edge. Edges are prone to the aperture problem so they are also unfit for flow estimation. Unfortunately, this is the most common case in our scenario as the lines are extracted from such image regions. It is best when both eigenvalues are large. Here, the image region is structured (e.g. contains a corner) and is therefore good for flow calculation. Shi and Tomasi propose to threshold the minimal eigenvalue to detect these suitable regions. We follow this idea and calculate the minimal eigenvalue for pixels belonging to lines. For each line, non-maximum suppression is applied on the minimal eigenvalues to keep only pixels which are local maximums. These pixels are then sorted by their eigenvalue and only the best 50% per line are kept. Figure 2(a) shows which "corner-like" points are selected for the optical flow calculation and Fig. 2(b) visualizes the result of the optical flow estimation on these points.

3.2   Consistency Check

The optical flow calculation is not error-free due to occlusion, noise etc. (cf. Fig. 2(b)). To mitigate the influence of these errors on the matching result, we introduce a filtering step where flow vectors which violate the "consistency" are discarded.

We define the consistency in two ways: first, the appearance of a point before and after the motion must stay the same - we call this the "appearance consistency". Second, we define the "consistent motion constraint" which states that points belonging to the same line must move in a consistent way.

We calculate the L1-norm between the image patch around the original point **p** in image I and the patch around the moved point **p**$'$ in image I' and use this value as measure for the "appearance consistency". We discard the 5% of flow vectors with the highest norm. For the second rule, we check if the points **p**$'$ originating from the same line l also form a line. We use a line fitting algorithm in RANSAC scheme to calculate the line which agrees best with the points **p**$'$. Then, all points are discarded which do not fit to this line. A point fits to this line if its point-line distance is less or equal to 1 px.

Figure 2(c) depicts which flow vectors are discarded by the "appearance consistency" and the "consistent motion constraint" and which are used for further processing.

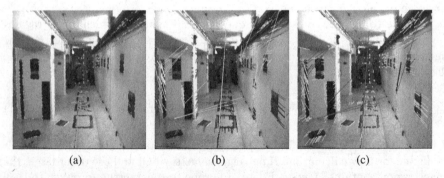

(a)   (b)   (c)

Fig. 2. (a) "Corner-like" points on the line segments for which the optical flow is estimated. (b) Resulting optical flow vectors. (c) Flow vectors in red are discarded due to the "appearance consistency", flow vectors in blue due to the violation of the "consistent motion constraint". Just the flow vectors drawn in green are considered for further processing. (Color figure online)

3.3   Histogram-Based Line Matching Using Optical Flow Vectors

After the optical flow result is filtered, the lines of image I and I' are finally associated. Due to noise and different line segmentation, the endpoints **p**$'$ of the flow vectors will not lie directly on the lines in image I'. To associate the endpoints **p**$'$ with lines l$'$ and then the lines l$'$ with lines l from image I, we designed an histogram-based approach where every optical flow endpoint **p**$'$ votes for its nearby lines l$'$. The votes are then accumulated over all points originating

Algorithm 1. Histogram-based line matching using optical flow vectors

Require:
    $\mathcal{L} \leftarrow$ lines in image I with $|\mathcal{L}| = n$
    $\mathcal{L}' \leftarrow$ lines in image I' with $|\mathcal{L}'| = m$
    $\mathcal{P} \leftarrow$ "corner-like" points for optical flow estimation on the lines $\mathcal{L}$
    $\mathcal{P}' \leftarrow$ endpoint of flow vectors, there is a 1:1 association between points in $\mathcal{P}$ and
    $\mathcal{P}'$.

    for all lines $l \in \mathcal{L}$ **do**
        $h(l') = 0$ set histogram entry for all lines $l' \in \mathcal{L}'$ to 0.
        $\mathcal{P}_l \leftarrow$ all points $\{\mathbf{p} | \mathbf{p} \in \mathcal{P} \wedge \mathbf{p}$ lies on line $l\}$
        for all points $\mathbf{p} \in \mathcal{P}_l$ **do**
            $\mathbf{p}' \leftarrow$ corresponding point to $\mathbf{p}$
            $\mathcal{L}'_{\mathbf{p}'} \leftarrow$ all lines $l' \in \mathcal{L}'$ which are in distance $\leq \delta$ to $\mathbf{p}'$
            for all lines $l' \in \mathcal{L}'_{\mathbf{p}'}$ **do**
                $h(l') = h(l') + \frac{1}{d}$ with d distance of $\mathbf{p}'$ to l'
            end for
        end for
        Save match (l, l') with $l' = \text{argmax}_\ell \, h(\ell)$
    end for

from the same line l. The line l' with the most votes is then the match of l. Algorithm 1 depicts this histogram-based matching process in detail.

The only adjustable parameter in this histogram-based matching process is δ which defines the search region for nearby lines around point $\mathbf{p}'$. We found experimentally that $\delta = 2\,\text{px}$ is a good value.

In Fig. 3 the resulting line matches are shown.

Fig. 3. Line matches generated with the optical-flow line matching algorithm. Corresponding lines share the same color in left and right image. (Color figure online)

4   Relative Pose Estimation from Lines

In the following, we recap our previously presented relative pose estimation [4]. First, we detail how to calculate the rotation from 3D line directions (Sect. 4.2).

Once the rotation between the two views is estimated, the translation is calculated using intersection points which is illustrated in Sect. 4.3. In Sect. 4.4 the robust relative pose estimation framework is presented which combines the rotation and translation estimation.

4.1 Notation

In relative pose estimation, we deal with entities in different coordinate systems: there are objects in the image (denoted by subscript i, like the line $\mathbf{l}_i$ or point $\mathbf{p}_i$) and in the camera frame (denoted by subscript c, like the 3D line direction vector $\mathbf{d}_c$).

A projective mapping is used to project a point $\mathbf{X}_c$ from the camera frame onto the image plane. $\mathbf{K}$ is the known 3×3 calibration matrix of the camera.

$$\mathbf{x}_i = \mathbf{K}\mathbf{X}_c \tag{2}$$

The transformation between two coordinate systems is defined by the matrix $\mathbf{T}_{to \leftarrow from} = [\mathbf{R}_{to \leftarrow from}, \mathbf{t}_{to \leftarrow from}]$ with $\mathbf{R}_{to \leftarrow from} \in \mathbb{SO}(3)$ the rotation and $\mathbf{t}_{to \leftarrow from} \in \mathbb{R}^3$ the translation. The transformation from a point $\mathbf{X}_{c_a}$ in camera a into the coordinate system of the camera b is done with

$$\mathbf{X}_{c_b} = \mathbf{R}_{c_b \leftarrow c_a}\mathbf{X}_{c_a} + \mathbf{t}_{c_b \leftarrow c_a}. \tag{3}$$

4.2 Rotation Estimation from 3D Line Directions

The transformation of a 3D line direction depends only on the rotation:

$$\mathbf{d}_{c_2} = \mathbf{R}_{c_2 \leftarrow c_1}\mathbf{d}_{c_1} \tag{4}$$

Given m corresponding (and possible noisy) directions, we want to find the rotation $\mathbf{R}_{c_2 \leftarrow c_1}$ between two cameras c_1 and c_2 which minimizes

$$\mathbf{R}_{c_2 \leftarrow c_1} = \underset{\mathbf{R}}{\operatorname{argmin}} \|\mathbf{D}_{c_2} - \mathbf{R}\mathbf{D}_{c_1}\| \tag{5}$$

where $\mathbf{D}_{c_1}$ and $\mathbf{D}_{c_2}$ are $3 \times m$ matrices which contain in each column the corresponding directions. This problem is an instance of the "Orthogonal Procrustes Problem" [24]. We employ the solution presented by Umeyama [25] which returns a valid rotation matrix as result by enforcing $\det(\mathbf{R}_{c_2 \leftarrow c_1}) = 1$:

$$\mathbf{R}_{c_2 \leftarrow c_1} = \mathbf{U}\mathbf{S}\mathbf{V}^\mathrm{T} \tag{6}$$

with

$$\mathbf{U}\mathbf{D}\mathbf{V}^\mathrm{T} = \operatorname{svd}\left(\mathbf{D}_{c_2}\mathbf{D}_{c_1}{}^\mathrm{T}\right) \tag{7}$$

$$\mathbf{S} = \begin{pmatrix} 1 & 0 & 0 \\ 0 & 1 & 0 \\ 0 & 0 & \operatorname{sgn}\left(\det(\mathbf{U})\det(\mathbf{V})\right) \end{pmatrix} \tag{8}$$

At least two non-collinear directions are required to calculate a solution.

The challenging task is now to extract 3D line directions from line observations in the two images and to bring these extracted directions in correspondence. We solve the direction estimation using a parallel line clustering approach which is explained now. Afterwards, we describe the matching of 3D line directions.

3D Line Direction Estimation by Parallel Line Clustering. The goal of this phase is to cluster lines of an image which are parallel in the world and to extract the shared 3D line direction for each cluster. This problem is closely related to the vanishing point detection, as the vanishing point $\mathbf{v}_i$ of parallel lines is the projection of the 3D line direction $\mathbf{d}_C$ [6]:

$$\mathbf{v}_i = \mathbf{K}\mathbf{d}_c \tag{9}$$

We suggest to work directly with 3D line directions $\mathbf{d}_c$ and not with the 2D vanishing points. Working in 3D space is beneficial, because it is independent from the actual camera (perspective, fisheye etc.) and allows an intuitive initialization of the clustering. In (9), we introduced how the vanishing point and the 3D line direction are related. Now, we have to transfer the line $\mathbf{l}_i$ into its corresponding 3D expression – its back-projection. The back-projection of an image line is the plane $\mathbf{\Pi}_c$ whose plane normal vector is given by (cf. [6]):

$$\mathbf{n}_c = \mathbf{K}^\mathsf{T}\mathbf{l}_i \tag{10}$$

Expectation-Maximization Clustering. Many of the vanishing point detection algorithms employ the Expectation-Maximization (EM) clustering method [26] to group image lines with the same vanishing point [27,28]. We got inspired by the work of Košecká and Zhang [28] and we adapt their algorithm so that it directly uses 3D line directions instead of vanishing points. This enables us to introduce a new and much simpler cluster initialization in which we directly set initial directions derived from the target environment.

The EM-algorithm iterates the expectation and the maximization step. In the expectation phase, the posterior probabilities $p(\mathbf{d}_c^{(k)}|\mathbf{n}_c^{(j)})$ are calculated. The posterior mirrors how likely a line $\mathbf{l}_i^{(j)}$ (with plane normal $\mathbf{n}_c^{(j)}$) belongs to a certain cluster k represented by direction $\mathbf{d}_c^{(k)}$. Bayes's theorem is applied to calculate the posterior:

$$p\left(\mathbf{d}_c^{(k)}|\mathbf{n}_c^{(j)}\right) = \frac{p\left(\mathbf{n}_c^{(j)}|\mathbf{d}_c^{(k)}\right) p\left(\mathbf{d}_c^{(k)}\right)}{p\left(\mathbf{n}_c^{(j)}\right)} \tag{11}$$

We define the likelihood as

$$p\left(\mathbf{n}_c^{(j)}|\mathbf{d}_c^{(k)}\right) = \frac{1}{\sqrt{2\pi\sigma_k^2}} \exp\left(\frac{-\left(\mathbf{n}_c^{(j)\,\mathsf{T}}\mathbf{d}_c^{(k)}\right)^2}{2\sigma_k^2}\right) \tag{12}$$

The likelihood reflects that a 3D line in the camera frame (and its direction $\mathbf{d}_c^{(j)}$) should lie in $\mathbf{\Pi}_c^{(j)}$ and is therefore perpendicular to the plane normal $\mathbf{n}_c^{(j)}$. If we substitute $\mathbf{n}_c^{(j)}$ and $\mathbf{d}_c^{(k)}$ with (9) and (10), we come to a likelihood term in the 2D image space which is the same as in the work of Košecká and Zhang [28].

In the maximization step, the probabilities from the expectation step stay fixed. The direction vectors are in this phase re-estimated by maximizing the objective function:

$$\underset{\mathbf{d}_c^{(k)}}{\operatorname{argmax}} \prod_j p\left(\mathbf{n}_c^{(j)}\right) = \underset{\mathbf{d}_c^{(k)}}{\operatorname{argmax}} \sum_j \log p\left(\mathbf{n}_c^{(j)}\right) \tag{13}$$

with

$$p\left(\mathbf{n}_c^{(j)}\right) = \sum_k p\left(\mathbf{d}_c^{(k)}\right) p\left(\mathbf{n}_c^{(j)}|\mathbf{d}_c^{(k)}\right) \tag{14}$$

As pointed out in [28], in the case of a Gaussian log-likelihood term, which is here the case, the objective function is equivalent to a weighted least squares problem for each $\mathbf{d}_c^{(k)}$:

$$\mathbf{d}_c^{(k)} = \underset{\mathbf{d}_c}{\operatorname{argmin}} \sum_j p\left(\mathbf{n}_c^{(j)}|\mathbf{d}_c\right) \left(\mathbf{n}_c^{(j)^{\mathrm{T}}}\mathbf{d}_c\right)^2 \tag{15}$$

After each EM-iteration, we delete clusters with less than two assignments to gain robustness.

Cluster Initialization. For initialization, we define a set of 3D directions which are derived from the targeted environment as follows: We apply our method in indoor scenes, hence we find the three dominant directions of the "Manhattan world". In addition, the camera is mounted pointing forward with no notable tilt or rotation against the scene, therefore we use the three main directions $(1\ 0\ 0)^{\mathrm{T}}$, $(0\ 1\ 0)^{\mathrm{T}}$, $(0\ 0\ 1)^{\mathrm{T}}$ for initialization. For robustness, we add all possible diagonals like $(1\ 1\ 0)^{\mathrm{T}}$, $(1\ -1\ 0)^{\mathrm{T}}, \ldots, (1\ 1\ 1)^{\mathrm{T}}$ (e.g. to capture the staircase in Fig. 4b)) and end up with overall 13 line directions. All line directions are normalized to unit length and have initially the same probability. The variance of each cluster is initially set to $\sigma_k^2 = \sin^2(1.5°)$ which reflects that the plane normal and the direction vector should be perpendicular up to a variation of $1.5°$.

Note that this derivation of initial directions is easily adoptable for other scenes or camera mountings. If the camera is for example mounted in a rotated way, we can simply rotate the directions accordingly. If such a derivation is not possible, we suggest to use the initialization technique proposed in [28] where the initial vanishing points are calculated directly from the lines in the image.

If an image sequence is processed, we propose to additionally use the directions estimated from the previous image in the initialization as "direction priors". In this case, we assign these priors a higher probability. We argue that the change between two consecutive images is rather small so the estimated directions from the previous image seem to be an valid initial assumption.

Results from the clustering step are visualized in Fig. 4.

(a)   (b)

Fig. 4. Results of the parallel line clustering. Lines with the same color belong to the same cluster and are parallel in 3D. Each cluster has a 3D direction vector $\mathbf{d}_c$ assigned which represents the line direction viewed from this image. (Color figure online)

3D Line Direction Matching. We need to establish correspondences between the 3D line directions (the clusters) of the two images in order to calculate the relative rotation. This is done using RANSAC.

The mathematical basis for the algorithm is that – as previously stated – the transformation of a direction $\mathbf{d}_{c_1}$ from the first camera to the direction $\mathbf{d}_{c_2}$ in the second camera depends only on the rotation $\mathbf{R}_{c_2 \leftarrow c_1}$ (cf. Eq. (4)). In the presence of noise, this equation does not hold, so we use the angular error ε between the directions:

$$\varepsilon = \arccos\left(\frac{\mathbf{d}_{c_2}{}^T \mathbf{R}_{c_2 \leftarrow c_1} \mathbf{d}_{c_1}}{\|\mathbf{d}_{c_1}\| \|\mathbf{d}_{c_2}\|}\right) \tag{16}$$

The RANSAC process tries to hypothesize a rotation which has a low angular error ε over a maximized subset of all possible correspondences. A rotation can be hypothesized from two randomly selected direction correspondences as described earlier. The idea behind this procedure is that only the correct set of correspondences yields a rotation matrix with small angular errors. Therefore, the correct rotation matrix only selects the correct correspondences into the consensus set.

The method has one drawback: it happens that different rotation hypotheses with different sets of correspondences result in the same angular error. This happens especially in an "Manhattan world" environment. This ambiguity is illustrated in Fig. 5.

As we have no other sensors in our system to resolve these ambiguities, we assume small displacements between the images and therefore restrict the allowed rotation to less than 45°. As we target visual odometry and SLAM systems, this is a valid assumption. Alternatively, if e.g. an IMU is present, its input could also be used.

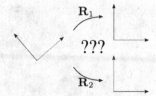

Fig. 5. Ambiguity in the direction matching. Which is the correct correspondence assignment for the directions from the left to the right? We could either rotate 45° clockwise in the plane ($\mathbf{R}_1$) or 45° counter-clockwise and then rotate around the green direction vector ($\mathbf{R}_2$). In this setting, the angular errors for $\mathbf{R}_1$ and $\mathbf{R}_2$ are the same. (Color figure online)

4.3 Translation Estimation from Intersection Points

The translation is estimated in the same way as proposed by Elqursh and Elgammal [8]. Intersection points of coplanar 3D-lines are invariant under projective transformation and therefore fulfill the epipolar constraint [6]:

$$\mathbf{p}_{i_2}{}^\mathrm{T}\mathbf{K}^{-\mathrm{T}}\left[\mathbf{t}_{c_2\leftarrow c_1}\right]_\times \mathbf{R}_{c_2\leftarrow c_1}\mathbf{K}^{-1}\mathbf{p}_{i_1}=0 \tag{17}$$

If two intersection point correspondences and the relative rotation are given, the epipolar constraint equation can be used to solve $\mathbf{t}_{c_2\leftarrow c_1}$ up to scale.

As we have no knowledge which intersection points from all $\frac{n(n-1)}{2}$ possibilities (with n the number of line correspondences) belong to coplanar lines, we follow the idea from [8] and use RANSAC to select the correct correspondences from all possible combinations while minimizing the Sampson distance defined in [6]. In contrast to [8], we can reduce the initial correspondence candidates as we take the clusters into account and only calculate intersection points between lines of different clusters.

4.4 Robust Relative Pose Estimation Framework

In this section, the explained algorithms are combined to an robust framework for relative pose estimation. The overall structure of the framework is depicted in Fig. 6.

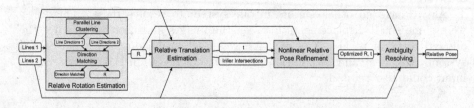

Fig. 6. The relative pose estimation framework.

The image line correspondences between the two images for which the relative pose is searched define the input of the framework. At first, the parallel line clustering is executed to estimate the 3D directions of the observed lines in both images. Each 3D line direction $\mathbf{d}_c$ is associated with a probability $p(\mathbf{d}_c)$ which reflects the size of the corresponding cluster. The more lines share the same direction, the higher the probability. We use this probability to guide the direction matching process: at the start, we select only 3D line directions from both images whose probability exceeds 0.1 and compute for them the direction matching with the RANSAC procedure as described. If the RANSAC process fails, the probability threshold is halved so more directions are selected and the RANSAC process is started again. This procedure in continued until a matching is found or all 3D line directions are selected and still no solution is obtained. In this case, the whole relative pose estimation process stops with an error. On success, the matched 3D directions and the corresponding rotation matrix is outputted. This rotation matrix is the rotation $\mathbf{R}_{c_2 \leftarrow c_1}$ between the two cameras.

Once the rotation is estimated, the translation is calculated from intersection points. Translation $\mathbf{t}_{c_2 \leftarrow c_1}$ and rotation $\mathbf{R}_{c_2 \leftarrow c_1}$ combined form the overall transformation $\mathbf{T}_{c_2 \leftarrow c_1}$ between the two cameras. As errors of the rotation directly influence the translation estimation, we use the inlier intersection points from the RANSAC step for a nonlinear optimization of the overall pose.

The optimized pose has still an ambiguity in the sign of the translation which needs to be resolved. We triangulate all line matches with respect to c_1. For every triangulated line $\mathbf{L}_{c_1}$ with endpoints $\mathbf{p}_L^{(i)}$, we measure the parallax ρ between its observations (this is the angle between the two back-projected planes $\mathbf{\Pi}_{c_1}$ and $\mathbf{\Pi}_{c_2}$ which intersect in the triangulated line) and check if the triangulation result is in front of the camera by calculating a visibility score ν:

$$\nu = \text{sgn}\left(\mathbf{r}_{c_1}^{(1)^\mathrm{T}} \mathbf{p}_L^{(1)}\right) + \text{sgn}\left(\mathbf{r}_{c_1}^{(2)^\mathrm{T}} \mathbf{p}_L^{(2)}\right) \tag{18}$$

with $\mathbf{r}_{c_1}^{(i)}$ are the back-projected rays of the endpoints of the line-segment seen in c_1. This value is weighted by the parallax angle and summed over all triangulated lines to form the ambiguity score A. To mitigate the influence of large parallax angles, we cut ρ if it becomes bigger than a certain threshold ρ_t:

$$\rho_c = \begin{cases} \rho & \text{if } \rho \leq \rho_t \\ \rho_t & \text{otherwise} \end{cases} \tag{19}$$

The overall ambiguity score is then

$$A = \sum_{\mathbf{L}_{c_1}} \rho_c \nu \tag{20}$$

If $A < 0$, the sign of the translation is changed, otherwise the translation stays as it is.

Once the ambiguity is resolved, the relative pose estimation is completed.

5   Experiments and Evaluation

In the following, we evaluate the presented optical flow-based line matching algorithm and the robust relative pose estimation approach. All experiments were conducted on a Intel® Xeon^{TM} CPU with 3.2 GHz and 32 GB RAM.

5.1   Evaluation of Line Matching

Now, we analyze the matching performance of the proposed matching algorithm and compare it to other state-of-the-art line matching approaches.

In detail, we use LEHF[2] from Hirose and Saito [12], MSLD[3] from Wang et al. [11] and LBD[4] from Zhang and Koch [13,14] for comparison. All three methods are combined with the LRC matching strategy and a global threshold on the descriptor distance. The threshold for LEHF and MSLD is set to 0.6 whereas LBD has a threshold of 52.

We use two common measures known from binary classification tasks to evaluate the matching performance: *precision* and *recall*:

$$\text{precision} = \frac{\text{correct matches}}{\text{all matches}} \tag{21}$$

$$\text{recall} = \frac{\text{correct matches}}{\text{all labeled matches}} \tag{22}$$

The execution time of the algorithms is also measured.

Figure 7 shows the matching precision and recall of the different matching algorithms on the matching test set which is shown in the Appendix A.

First of all, we observe that our proposed matching algorithm gives poor results on the "Facade01" and "Facade02" image pairs. These two image pairs clearly mark the limit of our approach as the images change a lot due to huge variation in viewpoint ("Facade01") or camera rotation (in "Facade02"). As consequence, the optical flow method does not succeed in calculating a correct flow which explains the low precision and recall values for our method. Besides that, the matching precision of our algorithm is comparable to the other approaches, in the case of the "Office03" image pair it is even 5% better then the next best (1.000 compared to 0.949 for MSLD). The recall increases in most cases between 4% for the "Warehouse01" image pair (from 0.925 for LEHF to 0.963) and 19% for the "Office03" image pair (from 0.804 for LEHF to 0.957). We conclude that our method is preferable to the other methods under small motion as we succeed in retrieving more correct matches with the same precision.

In Fig. 8 the average time over 50 executions of matching is plotted. As LEHF, MSLD, and LBD are descriptor-based, we additionally visualize the computation time to build the descriptors in both images.

---

[2] We thank the authors for providing us with their implementation.

[3] We use the implementation from https://github.com/bverhagen/SMSLD/tree/master/MSLD/MSLD/MSLD.

[4] We use the implementation of OpenCV 3.0.0.

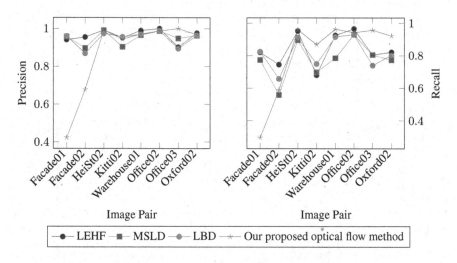

Fig. 7. Comparison of precision and recall of different matching approaches on the evaluation test set.

We observe that the matching process itself of MSLD and LEHF is very fast compared to our optical-flow based method. But the construction of the descriptors consumes most of the time. Especially the MSLD descriptor is expensive to compute which results in an accumulated matching and description time between 143 ms and 1064 ms. The LEHF descriptor is much more efficient and has accumulated runtimes between 9 ms and 59 ms which makes it comparable to the runtime of our optical flow-based matching with runtimes between 13 ms and 79 ms. The matching time of LBD is comparable to our algorithm but with the descriptor calculation the runtime increases and results in overall runtime between 26 ms to 140 ms.

From the evaluation of matching performance and runtime we conclude that our proposed optical flow-based matching technique outperforms the state-of-the-art methods in matching under small baseline. We achieve the same precision as the state-of-the-art while attaining a higher recall. The runtime is on the same level as the fastest state-of-the-art approach - LEHF - and clearly better than the other evaluated methods.

5.2 Evaluation of Relative Pose Estimation

Now, we evaluate our proposed robust algorithm for relative pose estimation. First, we introduce the datasets which we use for evaluation. Afterwards, we conduct our experiments and compare our algorithm to the state-of-the-art.

Datasets. For our experiments, we use synthetic and real image sequences. For the synthetic data, we created a typical indoor scene with a 3D wireframe model and generated images from it by projecting the line-segments from the model

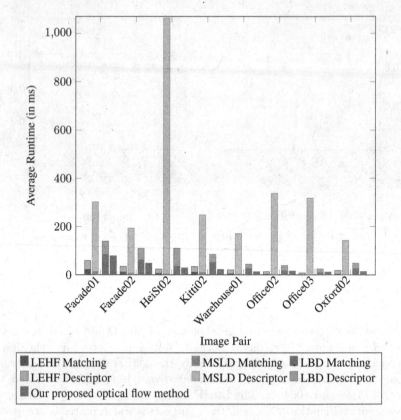

Fig. 8. Comparison of the execution time of different matching approaches on the evaluation test set. The plotted values are the mean over 50 executions. The time for descriptor calculation is additionally visualized for the descriptor-based approaches LEHF, MSLD and LBD.

into a virtual pinhole camera. The generated sequence consists of 1547 images. Example images are shown in Fig. 9.

To see how image noise affects the processing, we add Gaussian noise on the image lines. We do not add noise on the endpoints of the line-segments, because this effects segments of various length differently. We rather rotate the segments around their center point, where σ is the standard deviation of the rotation in degree.

For the experiments on real data, we use publicly available datasets with ground truth information of the camera poses. We select indoor sequences from robotics context as this is our targeted scenario but also evaluate on outdoor sequences from the automotive sector to test our algorithms in a variety of scenes. Robotic indoor scenes are represented by sequences from the "RGB-D SLAM benchmark" from TU Munich [29] and by the "Corridor" sequence from

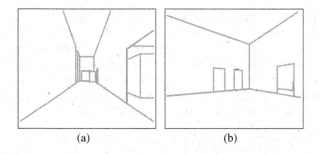

(a)   (b)

Fig. 9. A synthetic images showing a typical indoor scenario.

Oxford[5]. For outdoor scenes we use the "Stereo Ground Truth With Error Bars benchmark" from HCI [30] and the "KITTI Vision Benchmark Suite" from KIT [31]. The detailed overview of the used sequences is given in Table 1.

Table 1. List of sequences.

Sequence	Number of images	Image dimension	Trajectory length
Oxford Corridor	11	512 × 512	0.88 m
TUM fr2/pioneer_360	1185	640 × 480	17.98 m
TUM fr2/pioneer_slam	2861	640 × 480	44.19 m
TUM fr2/pioneer_slam2	2053	640 × 480	24.88 m
TUM fr2/pioneer_slam3	2524	640 × 480	27.63 m
HCI HeiSt Dataset 1	100	2560 × 1080	4.17 m
HCI HeiSt Dataset 2	600	2560 × 1080	21.68 m
KITTI Odometry Dataset 05	2761	1226 × 370	2205.58 m

As the datasets are now defined, we evaluate the accuracy of the resulting relative pose and compare it to the state-of-the-art algorithms using points or lines.

Comparison with State-of-the-Art. In this experiment, we compare the accuracy of our relative pose computation with the "Triplet" approach [8] – the state-of-the-art using lines – and with the "5-point" algorithm [3] as state-of-the-art representative for relative pose estimation using points.

The experimental setup is as follows: The line matching algorithm described in Sect. 3 is executed for each pair of consecutive images. For the 5-point algorithm, we detect and match SIFT features [1]. The detected image lines (or points) along with their correspondences define the input for the relative pose

---

[5] http://www.robots.ox.ac.uk/~vgg/data1.html.

estimation. The estimation algorithms are executed and the resulting relative poses $(\mathbf{R}_{est}, \mathbf{t}_{est})$ compared to the ground truth poses $(\mathbf{R}_{gt}, \mathbf{t}_{gt})$ of the test sequences. We evaluate the error in rotation and translation where the rotation error is the rotation angle of the axis-angle representation of $\mathbf{R}_{est}\mathbf{R}_{gt}^{\mathrm{T}}$. The translation error is expressed as angle between the ground truth translation and the estimated translation. We average the rotation and translation error angles α over each sequence using the following formula:

$$\alpha_{\varnothing} = \mathrm{atan2}\left(\frac{\sum \sin \alpha}{n}, \frac{\sum \cos \alpha}{n}\right) \qquad (23)$$

Additionally, we measure the execution time of each relative pose estimation algorithm.

We start the evaluation on the synthetic dataset using different noise levels σ. Since the synthetic dataset contains only lines, we just compare against the Triplet algorithm. The accuracy of the algorithms in term of rotation and translation error is shown in Fig. 10. The runtime behavior is presented in Fig. 11.

As expected, the accuracy of the relative pose estimation decreases while adding more and more noise to the date. Focusing on the rotation accuracy,

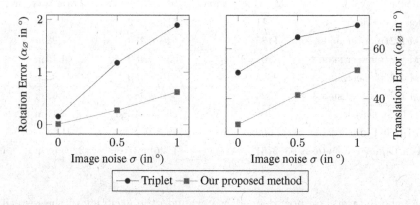

Fig. 10. Accuracy of the relative pose computation on the synthetic sequence.

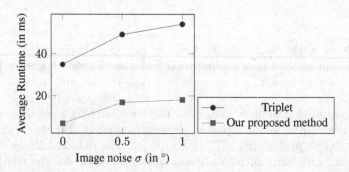

Fig. 11. Runtime analysis of the relative pose computation on the synthetic sequence.

our proposed algorithm seems to be more robust since the increase of the error is less compared to the Triplet approach. In the noise-free scenario, both approaches give similar results. Looking on the translation accuracy, the discrepancy between the results of both algorithms is much more severe (angle difference > 15°). On the first glance, this is surprising as both methods use intersection points to determine the translation. We give two explanations for this: first of all, the translation estimation is relying on the previously found rotation. As shown, the rotation estimated with our algorithms is more accurate so the translation estimation starts of better. Second, we only calculate intersection points for lines with different 3D direction which excludes vanishing points and favors real intersections. Even if vanishing points are theoretically inliers to the translation estimation, they are often far outside the image (e.g. the vanishing point of two parallel lines in the image) and affected by numerical errors. Also, slight measurement errors can lead to a huge error on the vanishing point's position so we decided to filter them out using the described procedure.

The execution time of the relative pose algorithm is depending on the number of input feature matches. The Triplet algorithm and our proposed method uses the same line matching input; for the synthetic dataset we have on average 31.10 matches per image pair. Despite the same input, our proposed algorithm needs only between 20% and 35% of the runtime. This is due to the fact that in the Triplet approach all possible $O(n^3)$ triplet combinations (n is the number of line matches) are generated and then tested in a RANSAC scheme to calculate the rotation which is very time consuming. Contrary to that, our estimation is based on the line directions calculated in the clustering step. In this sequence, we extract only around 3 different line directions per image, hence the rotation calculation is very fast. The number of generated intersection points is also lower as mentioned above which speeds up the translation estimation.

Now, we evaluate our algorithm on real data sequences and compare its accuracy against the Triplet approach as before and additionally against the 5-point approach. The accuracy on the real image datasets is visualized in Fig. 12, the runtime performance shown in Fig. 13.

This experiment confirms the previous results. Our proposed algorithm gives better results than the Triplet method on all sequences. On the four indoor sequences ("OxfordCorridor" and "TUMPioneer") our line based algorithm can compete with the 5-point algorithm. Both, rotation and translation accuracy are in a similar range, for the "TUMPioneerSLAM3" the translation error is even 8% better. But for the outdoor sequences ("HeiSt" and "KITTI"), the point-based method seems to be superior. Whereas for "HeiSt01" and "HeiSt02" our algorithm can correctly calculate the rotation, it cannot handle the "KITTI05" sequence and shows much more rotation error than the point-based approach. Also the translation error is much higher. One reason for the dominance of the 5-point algorithm on the outdoor sequences is that this setting is much more suited for the SIFT feature as a lot of texture is present in the images. This is not the case for the indoor sequences, especially the TUM dataset, where white walls and room structure are dominating.

348   N. von Schmude et al.

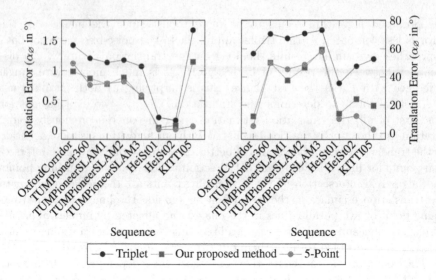

Fig. 12. Accuracy of the relative pose computation on the real sequence.

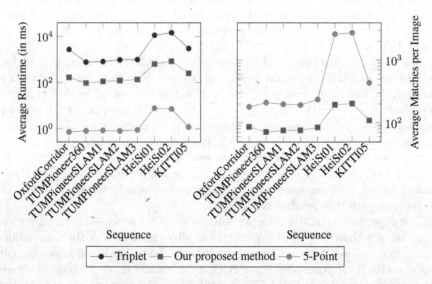

Fig. 13. Runtime analysis of the relative pose computation on the real image sequence.

Regarding the execution time, we clearly see that the more matches we have, the slower the algorithms. The triplet approach is in the order of one magnitude slower than our approach which is explained by their expensive triplet calculation as explained earlier. But our approach is still around two magnitudes slower that the 5-point algorithm. One reason is that the 5-point algorithm estimates rotation and translation in one step so only one RANSAC iteration is executed. The Triplet algorithm and our approach need two: one for the rotation estimation and one for the translation. Another important factor is the need of intersection points for translation estimation. The number of intersection points is in $O(n^2)$ for n matches. Even as we have less line matches then point matches (around 100 line matches compared to 200 point matches), the number of intersection points is much bigger.

Overall, this experiment shows that our presented algorithm is feasible for relative pose estimation. It is clearly superior to the Triplet method as average rotation and translation errors are smaller – and that with significantly lower computation time. In some scenarios it can compete with point-based approaches in terms of accuracy but the computation time is a limiting factor here.

6   Conclusion

In this paper, we presented a innovative relative pose estimation scheme using lines.

First, we introduced a novel line matching algorithm for small baseline displacement based on optical flow. We investigated line matching since relative pose estimation is directly influenced by the correspondences found. In a comparison to other state-of-the-art descriptor-based line matching approaches we could demonstrate that our proposal shows better recall by same precision with same or smaller computation time.

Second, we explained our proposed relative pose estimation framework. In our approach, we estimate the 3D line directions through a clustering step of parallel lines in the world and use this information throughout the whole processing pipeline. The direction information is used to calculate the relative rotation and to limit the intersection point calculation. Rotation and translation estimation are embedded into a robust relative pose estimation framework.

Finally, we compared our relative pose estimation to the state-of-the-art approach using lines from Elqursh and Elgammal [8] and against Nister's point-based algorithm [3]. We evaluated on publicly available indoor and outdoor datasets and showed that our method outperforms the line-based approach in terms of accuracy and runtime. Especially the runtime can be reduced from seconds to milliseconds. The point-based approach performs better as expected in highly textured outdoor scenes but our method can compete in textureless indoor scenarios.

In the future, we want to extend our approach to a complete SLAM system. To reach this goal, we need to relax the restriction to small motions in the direction matching. Also, line matching strategies supporting wide-baselines should be investigated.

A   Matching Test Set

The test set consists of 8 image pairs with small baseline displacement showing different indoor and outdoor scenes. For each image in the test set, we detect line-segments using the LSD algorithm[6] [32]. Then, we manually label corresponding

---

[6] We use the implementation in OpenCV 3.0.0 with default parameters.

lines in the image pairs and save them as ground truth matches. Note that a line-segment can correspond to multiple line-segments in the other image as the line segmentation may vary.

The complete test set is shown in Fig. 14. The image pairs "Facade01" and "Facade02" are taken from the matching evaluation from Zhang et al. [13,14], "HeiSt02" from the Heidelberger Stereo benchmark [30], "Kitti02" from the KITTI odometry benchmark [31] and "Oxford02" from the Oxford multiview dataset[7].

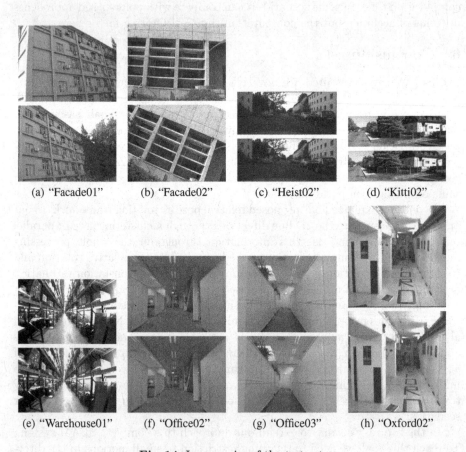

(a) "Facade01"   (b) "Facade02"   (c) "Heist02"   (d) "Kitti02"

(e) "Warehouse01"   (f) "Office02"   (g) "Office03"   (h) "Oxford02"

Fig. 14. Image pairs of the test set.

References

1. Lowe, D.G.: Distinctive image features from scale-invariant keypoints. Int. J. Comput. Vis. **60**, 91–110 (2004)

---

[7] http://www.robots.ox.ac.uk/~vgg/data1.html.

2. Fischler, M.A., Bolles, R.C.: Random sample consensus: a paradigm for model fitting with applications to image analysis and automated cartography. Commun. ACM **24**, 381–395 (1981)
3. Nister, D.: An efficient solution to the five-point relative pose problem. IEEE Trans. Pattern Anal. Mach. Intell. **26**, 756–770 (2004)
4. von Schmude, N., Lothe, P., Jähne, B.: Relative pose estimation from straight lines using parallel line clustering and its application to monocular visual odometry. In: International Conference on Computer Vision Theory and Applications, pp. 421–431 (2016)
5. Hartley, R.I.: Projective reconstruction from line correspondences. In: Proceedings of IEEE Conference on Computer Vision and Pattern Recognition, pp. 903–907 (1994)
6. Hartley, R.I., Zisserman, A.: Multiple View Geometry in Computer Vision, 2nd edn. Cambridge University Press, Cambridge (2004). ISBN 0521540518
7. Weng, J., Huang, T., Ahuja, N.: Motion and structure from line correspondences; closed-form solution, uniqueness, and optimization. IEEE Trans. Pattern Anal. Mach. Intell. **14**, 318–336 (1992)
8. Elqursh, A., Elgammal, A.: Line-based relative pose estimation. In: 2011 IEEE Conference on Computer Vision and Pattern Recognition (CVPR), pp. 3049–3056 (2011)
9. Wang, G., Wu, J., Ji, Z.: Single view based pose estimation from circle or parallel lines. Pattern Recogn. Lett. **29**, 977–985 (2008)
10. Bazin, J., Demonceaux, C., Vasseur, P., Kweon, I.: Motion estimation by decoupling rotation and translation in catadioptric vision. Comput. Vis. Image Underst. **114**, 254–273 (2010)
11. Wang, Z., Wu, F., Hu, Z.: MSLD: a robust descriptor for line matching. Pattern Recogn. **42**, 941–953 (2009)
12. Hirose, K., Saito, H.: Fast line description for line-based slam. In Bowden, R., Collomosse, J., Mikolajczyk, K. (eds.) 2012 British Machine Vision Conference, pp. 83.1–83.11 (2012)
13. Zhang, L., Koch, R.: Line matching using appearance similarities and geometric constraints. In: Pinz, A., Pock, T., Bischof, H., Leberl, F. (eds.) DAGM/OAGM 2012. LNCS, vol. 7476, pp. 236–245. Springer, Heidelberg (2012). doi:10.1007/978-3-642-32717-9_24
14. Zhang, L., Koch, R.: An efficient and robust line segment matching approach based on LBD descriptor and pairwise geometric consistency. J. Vis. Commun. Image Represent. **24**, 794–805 (2013)
15. Deriche, R., Faugeras, O.: Tracking line segments. In: Faugeras, O. (ed.) ECCV 1990. LNCS, vol. 427, pp. 259–268. Springer, Heidelberg (1990). doi:10.1007/BFb0014872
16. Chiba, N., Kanade, T.: A tracker for broken and closely-spaced lines. In. ISPRS International Society for Photogrammetry and Remote Sensing Conference, Hakodate, Japan, pp. 676–683 (1998)
17. Bartoli, A., Sturm, P.: Structure-from-motion using lines: representation, triangulation, and bundle adjustment. Comput. Vis. Image Underst. **100**, 416–441 (2005)
18. Schindler, G., Krishnamurthy, P., Dellaert, F.: Line-based structure from motion for urban environments. In: Third International Symposium on 3D Data Processing, Visualization, and Transmission (3DPVT 2006), pp. 846–853 (2006)
19. Zhou, H., Zou, D., Pei, L., Ying, R., Liu, P., Yu, W.: Structslam: visual slam with building structure lines. IEEE Trans. Veh. Technol. 1 (2015)

20. Witt, J., Weltin, U.: Robust stereo visual odometry using iterative closest multiple lines. In: 2013 IEEE/RSJ International Conference on Intelligent Robots and Systems (IROS 2013), pp. 4164–4171 (2013)
21. Holzmann, T., Fraundorfer, F., Bischof, H.: Direct stereo visual odometry based on lines. In: International Conference on Computer Vision Theory and Applications, pp. 474–485 (2016)
22. Lucas, B.D., Kanade, T.: An iterative image registration technique with an application to stereo vision. IJCAI **81**, 674–679 (1981)
23. Shi, J., Tomasi, C.: Good features to track. In: Proceedings of IEEE Conference on Computer Vision and Pattern Recognition, pp. 593–600 (1994)
24. Gower, J.C., Dijksterhuis, G.B.: Procrustes Problems, vol. 3. Oxford University Press, Oxford (2004)
25. Umeyama, S.: Least-squares estimation of transformation parameters between two point patterns. IEEE Trans. Pattern Anal. Mach. Intell. **13**, 376–380 (1991)
26. Dempster, A.P., Laird, N.M., Rubin, D.B.: Maximum likelihood from incomplete data via the EM algorithm. J. Roy. Stat. Soc. Ser. B (Methodol.) 1–38 (1977)
27. Antone, M.E., Teller, S.: Automatic recovery of relative camera rotations for urban scenes. In: IEEE Conference on Computer Vision and Pattern Recognition, CVPR 2000, pp. 282–289 (2000)
28. Košecká, J., Zhang, W.: Video compass. In: Heyden, A., Sparr, G., Nielsen, M., Johansen, P. (eds.) ECCV 2002. LNCS, vol. 2353, pp. 476–490. Springer, Heidelberg (2002). doi:10.1007/3-540-47979-1_32
29. Sturm, J., Engelhard, N., Endres, F., Burgard, W., Cremers, D.: A benchmark for the evaluation of RGB-D slam systems. In: 2012 IEEE/RSJ International Conference on Intelligent Robots and Systems (IROS 2012), pp. 573–580 (2012)
30. Kondermann, D., et al.: Stereo ground truth with error bars. In: Cremers, D., Reid, I., Saito, H., Yang, M.-H. (eds.) ACCV 2014. LNCS, vol. 9007, pp. 595–610. Springer, Cham (2015). doi:10.1007/978-3-319-16814-2_39
31. Geiger, A., Lenz, P., Urtasun, R.: Are we ready for autonomous driving? The Kitti vision benchmark suite. In: 2012 IEEE Conference on Computer Vision and Pattern Recognition (CVPR), pp. 3354–3361 (2012)
32. von Gioi, R., Jakubowicz, J., Morel, J.M., Randall, G.: LSD: a fast line segment detector with a false detection control. IEEE Trans. Pattern Anal. Mach. Intell. **32**, 722–732 (2010)

A Detailed Description of Direct Stereo Visual Odometry Based on Lines

Thomas Holzmann[✉], Friedrich Fraundorfer, and Horst Bischof

Institute for Computer Graphics and Vision,
Graz University of Technology, Graz, Austria
{holzmann,fraundorfer,bischof}@icg.tugraz.at

Abstract. In this paper, we propose a direct stereo visual odometry method which uses vertical lines to estimate consecutive camera poses. Therefore, it is well suited for poorly textured indoor environments where point-based methods may fail. We introduce a fast line segment detector and matcher detecting vertical lines, which occur frequently in man-made environments. We estimate the pose of the camera by directly minimizing the photometric error of the patches around the detected lines. In cases where not sufficient lines could be detected, point features are used as fallback solution. As our algorithm runs in real-time, it is well suited for robotics and augmented reality applications. In our experiments, we show that our algorithm outperforms state-of-the-art methods on poorly textured indoor scenes and delivers comparable results on well textured outdoor scenes.

1 Introduction

In the last years, Simultaneous Localization and Mapping (SLAM) and Visual Odometry (VO) became an increasingly popular field of research. In comparison to Structure from Motion (SfM), which is not constrained to run in realtime, SLAM and VO methods should be able to run in realtime even on lightweight mobile computing solutions.

Visual localization and mapping approaches are often used for small robotics devices (e.g. Micro Aerial Vehicles (MAVs)), where it is not possible to use heavier and more energy-consuming sensors like laser scanners. However, most of these devices are equipped with an Inertial Measurement Unit (IMU), which can be used in conjunction with vision methods to improve the results.

There exist widely used monocular SLAM/VO methods, which just use one camera as the only information source and therefore can be used on many different mobile devices. In contrast to monocular approaches, stereo VO methods using a stereo setup with a fixed baseline have some benefits, especially when used for Robotics: (1) Monocular approaches cannot correctly determine the metric scale of a scene, while stereo approaches do have a correct scale as the baseline of the camera setup is fixed. Additionally, monocular approaches also have a drift in scale. (2) Monocular approaches do have problems with pure

© Springer International Publishing AG 2017
J. Braz et al. (Eds.): VISIGRAPP 2016, CCIS 693, pp. 353–373, 2017.
DOI: 10.1007/978-3-319-64870-5_17

rotations as no new 3D scene information can be computed from images without translational movement. Contrary, stereo approaches do not have this problems, as a stereo rig already contains translationally moved cameras to triangulate scene information regardless of the motion of the cameras.

Based on the observation that man-made scenes usually contain strong line structures, like, e.g., doors or windows on white walls, we especially want to use these structures in our approach. Current feature-point based methods (e.g. [1]) have problems in these scenes, as the lack of texture leads to a low number of detected feature points. Direct pose estimation methods (e.g. [2]) directly use the image information to align consecutive frames and do not use explicitly detected feature point correspondences to estimate the camera pose. Therefore, direct methods perform better in poorly textured environments, as they use more image information and do not rely on detected feature points in the scene.

In this paper[1], we present a pose estimation method specifically targeted at poorly textured indoor scenes where other methods deliver poor results. We estimate the pose of a camera by minimizing the photometric error of patches around lines. We introduce a fast line detector which detects vertical lines aided by an inertial measurement unit (IMU). In case too few lines could be detected, we additionally detect point features and use patches around the detected points for direct pose estimation. In our experiments we show that our method outperforms state-of-the-art monocular and stereo approaches in poorly textured indoor scenes and delivers comparable results on well textured outdoor scenes.

To summarize, our main contributions are:

- A direct stereo VO method using lines running in real-time.
- A novel efficient line detection algorithm aided by an IMU, that detects vertical lines.
- A fast line matching technique using a lightweight line descriptor.

2   Related Work

A widely used approach to estimate the pose of a camera using the image data only is to use feature points in images. Feature points are detected and matched between subsequent images and used to estimate the pose of the camera. A popular monocular point-based SLAM-approach is PTAM [4]. It runs in real-time and is specifically designed to track a hand-held camera in a small augmented reality workspace. However, there also exist adoptions tailored at large-scale environments [5,6]. As it needs to detect and match feature points, the environment has to contain sufficient texture suited for the feature point detection algorithm. Additionally, as it is a monocular approach, it has difficulties in handling pure rotations and cannot estimate a correct scale of the scene. A feature-based approach, which uses a stereo camera as input and therefore does not have this problems is Libviso [1]. However, as it also uses explicitly detected feature points, the texture needs to be adequate for the detection algorithm. Additionally, as it

---

[1] This paper is a revised and extended version of [3].

is not a SLAM method like PTAM but just a VO method, it does not compute a global map of the environment but just computes the relative pose from frame to frame.

Instead of using points, also lines can be detected and matched in order to compute the pose of a camera. In [7], Elqursh and Elgammal propose to estimate the relative pose of two cameras by using three lines having a special primitive configuration. This has the advantage that no texture at all needs to be present, but just a special line configuration has to exist. However, as explicit line detection is relatively slow, this approach cannot be used on a low-end onboard computer.

Recently, direct pose estimation algorithms became popular. In comparison to feature point-based approaches, direct approaches do not explicitly detect and match features and compute the pose using the feature matches, but compute the new pose by minimizing the photometric error over the whole or over big parts of the image. For example, DTAM [8] tracks the pose of a monocular camera given a dense model of the scene, by minimizing the photometric error of the current image according to the whole model. As this is a computational expensive task, a high-performance GPU is necessary to compute the pose in real-time. Similarly, LSD-SLAM [2] computes the pose by minimizing the photometric error. However, to reduce computational cost, it just uses areas in the image where the image gradient is sufficiently high. Therefore, it runs in real-time even on hand-held devices like smartphones. Also an extension of LSD-SLAM to a stereo setup has been proposed recently [9]. In [10], an approach is proposed which uses both explicit feature point matching and direct alignment and is therefore called semi-direct visual odometry (SVO). It detects feature points at keyframes and computes the poses of images between keyframes by minimizing the photometric error of patches around the feature points. It runs very fast even on onboard computers and is specifically designed for a downwards looking camera of a micro aerial vehicle. In summary, direct methods perform direct image alignment to estimate the pose of a new frame with respect to previously computed 3D information by optimizing the pose parameters directly and by minimizing a photometric error. This is in contrast to feature-based methods, where features are detected and matched to compute the fundamental matrix, and optimization is just optionally used to refine the computed poses by using the reprojection error.

Our method is a direct method, but utilizes a stereo camera rig instead of a monocular camera. In contrast to all previous related work, we detect lines and estimate the pose by minimizing the photometric error of patches around the lines. Using our line detector, it is possible to detect lines very fast in constrained images. As in man-made environments most of the structure consists of lines, using patches around lines is usually sufficient to directly align consecutive images.

3   Direct Visual Odometry Based on Vertical Lines

In our approach, we explicitly use the structures contained in man-made environments to introduce a novel visual odometry approach.

Instead of directly aligning the whole image or explicitly detecting and matching feature points, we estimate the camera pose by aligning patches just around detected lines by minimizing the photometric error. In case no lines can be detected, feature points are detected and patches around the points are used for image alignment.

3.1   System Overview

Our approach is a keyframe-based stereo approach, which takes rectified stereo images and optionally synchronized IMU values as input. As preprocessing, we align the images according to the gravity direction using the IMU measurements. Without IMU measurements, our approach also works on images acquired with roughly gravity aligned cameras.

For every keyframe, we detect and match lines using gravity-aligned stereo images. If not enough lines can be matched, we additionally detect and match point features. We triangulate this information to get a mapping of the visible information in 3D. A new keyframe is selected when the photometric properties of the images changed too much or the movement w.r.t. the last keyframe is too big.

For every image, the pose gets estimated by direct image alignment: We project all triangulated information of the previous keyframe into the current image and estimate the pose by minimizing the photometric error of patches around the lines and points.

Additionally, for every consecutive keyframe, we project lines and points from the previous one into the current one to reuse already computed 3D information if the same lines or points are detected.

In Fig. 1, one can see a schematic overview of our processing pipeline.

3.2   Fast Line Detection and Matching

Areas around lines contain high gradient values which is an important property for direct image alignment. Therefore, we introduce a fast line detection and matching method, where the area around lines is used for direct image alignment afterwards. We compute the depth of the lines by computing the disparity of the lines using the stereo image pair.

Line Detection. Usually, line detection (like, e.g. [11]) is too slow for real-time applications [12]. To make the problem feasible, we focus on the detection of vertical lines, which can be done much faster. Man-made environments usually contain many vertical lines. Especially indoors, where the texture available very often mainly consists of lines, it is crucial for a visual odometry approach to

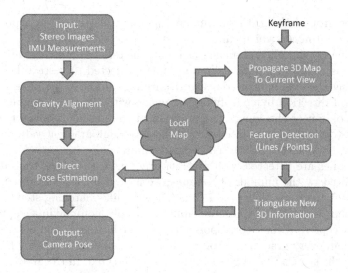

Fig. 1. Overview of the processing pipeline. Our algorithm directly estimates the pose of the camera by minimizing the photometric error of patches around vertical lines. For every keyframe, the 3D information from the previous keyframe gets projected into the current one, new lines and optionally points get detected and new 3D information is triangulated within the stereo setup.

use also this information. We introduce a detection algorithm aided by IMU information to make this challenging line detection step feasible: Having cameras with arbitrary poses and synchronized IMU information as input, the images can be aligned with the gravity direction using the simultaneously acquired IMU data (see Fig. 2). As the angular accuracy of IMUs is quite high (e.g., evaluation of the IMU of the Pixhawk autopilot [13]) and most of the mobile computing devices already contain built-in IMUs, this preprocessing step can be done with many available mobile computing devices.

Even if the alignment is not completely perfect, it does not influence the pose estimation step, as the lines are not directly used for pose estimation but instead

Fig. 2. Image alignment according to the gravity direction. *Left:* an unaligned image from a forwards looking camera *Right:* image aligned using the IMU. As one can see, having an image aligned according to the gravity direction, straight, vertical lines in the 3D scene are mapped to straight, vertical lines in the image.

the patches around them. Errors in the alignment just result in a poorer line detection performance which can lead to shorter detected lines.

Assuming gravity aligned images as input, the task of detecting vertical lines gets easier as real-world vertical lines are now projected to vertical lines in the image. Using this assumption, our algorithm first computes the image gradient in horizontal direction by convolving the image with the Sobel operator. Next, for every column of the image, a histogram of the gradient values is created. When a histogram bin associated with sufficiently high gradient values contains enough elements, a line is detected in this column.

As the lines are detected with pixel accuracy and, therefore, the number of lines is limited by the number of columns, we have a small and limited number of lines which increases the speed of the following line matching step.

After the detection of lines, we apply a simple non-maxima suppression: Lines get accepted, if their corresponding gradient histogram bin contains more elements than every gradient histogram bin corresponding to nearby lines.

Finally, the start and end point of the line is detected. In the gradient image, we search along the associated column for the first pixel belonging to the same bin as the corresponding line bin. This defines the start point of the line. Next, we search for a pixel belonging to a different bin. This defines the end point of the line. In order to be robust against noise and short discontinuities of the line, pixels with a similar histogram bin and discontinuities of a few pixels are also assigned to the line. Figure 3 shows a result of the line detection.

Fig. 3. Vertical lines detected with our line detection algorithm. One can see that the lines are not detected over the whole length all the time. This is due to slight gravity misalignment of the images or due to changing light conditions.

Line Matching. We iterate over all lines in the left image and try to find corresponding lines in the right image. Between a predefined minimum and maximum disparity, we store all lines as potential matches which were detected having a similar gradient histogram bin as the left line. Then, we compute the Sum of

Absolute Differences (SAD) of a patch around the line in the left image and the potential line matches in the right image.

For this, we need a equally-sized patch around the line in the left and right image. However, as the epipolar constraints do not hold anymore for the gravity aligned images (the images have been rotated after rectification), it is not possible to use just the maximum and minimum y-positions of the lines, as it would be possible in the original, not gravity-aligned, rectified left and right image. Instead, we first have to project the line start and end points back to the unaligned images, select the maximum and minimum line start and end point and project the points back to compute a correct equally-sized patch for both lines.

Having computed the SAD for all potential matches, the line with the lowest SAD value in the right image is accepted as correct match for the line in the left image, if the SAD value is lower than a threshold.

We assume that the detected vertical lines are projections of vertical lines in the 3D scene to the image. Therefore, to compute the depth of the line, we just have to compute one disparity value for the whole line. However, as the epipolar constraints do not hold for the aligned image, we again project the lines into the unaligned images to compute the line disparity.

3.3 Supporting Points

When lines exist in the scene, they deliver a good texture for direct image alignment. However, also in man-made environments consisting mainly of lines, there are areas where no or just a few lines exist. Therefore we need a fall-back solution for the direct image alignment for images where not many or no lines can be detected. We selected to use patches around feature points in this case.

Feature Point Detection. If no sufficient area of the image is covered by detected lines, we detect point features in areas of the image which are not covered by line features. To determine the coverage of features over the image, the image is partitioned into a 5×5 grid of equally sized cells. Every cell which contains a line or parts of a line is marked as covered, all other cells are marked as uncovered. If less than 20 of the 25 cells are marked as covered, point detection is applied. However, as in cells marked as covered already information from lines exist, we just extract feature points in the uncovered cells.

For every uncovered cell, we detect FAST corners [14] using an adaptive threshold. If the number of detected corners falls under 10, we decrease the corner threshold, while if the number of detected corners exceeds 20, we increase the threshold. Then, we store the determined thresholds for each cell for the next image to be processed.

Feature Point Matching. Having corresponding feature points in two images and a calibrated stereo setup, one can triangulate a 3D point. Therefore, we need to find points in the right image corresponding to points in the left image.

To find point matches, we proceed as follows: For every detected feature point in the left image, we search for detected feature points in the right image lying on the corresponding epipolar line. Therefore, the points have to be projected back into the original unaligned image, as the epipolar constraint does not hold anymore for the aligned images.

We search for points between a minimum and maximum disparity value to ignore too far and too near points. Having found some potential matches, we compare the point from the left image with the potential matching points in the right image by computing the SAD for a patch around the point. The corresponding point with the lowest SAD is detected as match, if the SAD value is under a fixed threshold.

3.4   Pose Estimation by Direct Image Alignment

Given 3D lines and points from the previous keyframe, our pose estimation algorithm estimates the pose by aligning the patches around the lines and points in the keyframe with the patches around the lines and points projected into the current frame (see Fig. 4). Alignment is done by non-linear least squares optimization on the intensity image.

Fig. 4. Direct image alignment using patches around lines. *Left:* keyframe, where lines are detected and patches around lines (red boxes) are extracted. *Right:* image to align. Red boxes are shifted according to the proposed new pose until a minimal photometric error is reached. (Color figure online)

Definitions and Notation. In this section, we state some definitions and notation conventions which are used in the following algorithm description.

In terms of notation, we denote matrices by bold capital letters (e.g., $\boldsymbol{T}$) and vectors by bold lower-case letters (e.g., $\boldsymbol{\xi}$).

A rigid body transformation $\boldsymbol{T} \in SE(3)$ is defined as

$$\boldsymbol{T} = \begin{pmatrix} \boldsymbol{R} \ \boldsymbol{t} \\ \boldsymbol{0} \ 1 \end{pmatrix} \textit{ with } \boldsymbol{R} \in SO(3) \textit{ and } \boldsymbol{t} \in \mathbb{R}^3. \tag{1}$$

Having a transformation $T_{0,i}$ which transforms the origin 0 into camera i and maps points from the coordinate frame of i to the origin 0, the same transformation can be assembled having a transformation $T_{0,i-1}$ from a previous camera $i-1$ and a transformation $T_{i-1,i}$ describing the transformation from the camera $i-1$ to camera i:

$$T_{0,i} = T_{0,i-1} \cdot T_{i-1,i}. \tag{2}$$

For the optimization, we need a minimal description of transformations. Therefore, we use the Lie algebra $\mathfrak{se}(3)$, which is the tangent space of the matrix group $SE(3)$ at the identity, and its elements are denoted as twist coordinates

$$\xi = \begin{pmatrix} \omega \\ v \end{pmatrix} \in \mathbb{R}^6, \tag{3}$$

where ω defines the rotation and v defines the translation. As stated in [15], twist coordinates can be mapped onto $SE(3)$ by using the exponential map:

$$T(\xi) = exp(\xi). \tag{4}$$

Non-linear Least Squares Optimization on Image Intensities. To estimate the 6 DoF pose of a new image frame given the image patches around lines and points from the previous keyframe, we minimize the photometric error for the twist ξ by applying non-linear least squares optimization. We keep the 3D lines and points fixed in the optimization step, as we assume to have already accurately triangulated 3D information from using the calibrated stereo setup. The quadratic photometric error is defined as follows:

$$E(\xi) = \sum_{i=1}^{n} (I_{kf}(p_i) - I(\pi(p_i, d_{kf}(p_i), \xi)))^2 = \sum_{i=1}^{n} r_i^2(\xi), \tag{5}$$

where $I_{kf}(p_i)$ is the intensity value of position p_i in the keyframe to align to, $\pi(p_i, d_{kf}(p_i), \xi)$ denotes the warping function which warps point p_i of the keyframe into the current image and I is the current intensity image. This function computes all n residual values $r_i(\xi)$ for every pixel i.

In order to decrease the influence of outliers, i.e. patches with very high residual values, we use a Cauchy Loss function to down-weight high residuals.

For performance reasons, we do not warp all pixels of the patches in image I_{kf} to image I but just the central top and bottom points of the patches. For residual computation we then compare patches around the warped points in I with the patches from I_{kf}. This simplification is valid for small patches and small motion between frames and also similarly used by others (e.g. [10]).

For all of the image patches, we seek to find a solution for ξ which minimizes the total squared photometric error

$$\xi = \arg\min_{\xi} \frac{1}{2} E(\xi). \tag{6}$$

The minimum value of $\boldsymbol{\xi}$ occurs when the gradient is zero:

$$\frac{\partial E}{\partial \boldsymbol{\xi}_j} = 2 \sum_{i=1}^{n} r_i \frac{\partial r_i}{\partial \boldsymbol{\xi}_j} = 0, (j = 1, \ldots, 6) \tag{7}$$

As Eq. 6 is non-linear, we iteratively solve the problem. In every iteration, we linearize around the current state in order to determine a correction $\Delta\boldsymbol{\xi}$ to the vector $\boldsymbol{\xi}$:

$$E(\boldsymbol{\xi} + \Delta\boldsymbol{\xi}) \approx E(\boldsymbol{\xi}) + \boldsymbol{J}(\boldsymbol{\xi})\Delta\boldsymbol{\xi}, \tag{8}$$

where $\boldsymbol{J}(\boldsymbol{\xi})$ is the Jacobian matrix of $E(\boldsymbol{\xi})$. It has a size of n x 6 with n as the number of residual values and is given by:

$$\boldsymbol{J}_{i,j}(\boldsymbol{\xi}) = \frac{\partial E_i(\boldsymbol{\xi})}{\partial \boldsymbol{\xi}_j}. \tag{9}$$

Using this linearization, the problem can be stated as a linear least squares problem:

$$\Delta\boldsymbol{\xi} = \arg\min_{\Delta\boldsymbol{\xi}} \frac{1}{2}(\boldsymbol{J}(\boldsymbol{\xi})\Delta\boldsymbol{\xi} + E(\boldsymbol{\xi})). \tag{10}$$

In our approach, we use the Levenberg-Marquardt algorithm [16,17] to solve the linear least squares problem stated in Eq. 10.

By rearrangement and plugging Eqs. 8 and 9 into Eq. 7, we obtain the following normal equations:

$$\boldsymbol{J}(\boldsymbol{\xi})^T \boldsymbol{J}(\boldsymbol{\xi})\Delta\boldsymbol{\xi} = -\boldsymbol{J}(\boldsymbol{\xi})^T E(\boldsymbol{\xi}). \tag{11}$$

We solve this equation system by using QR decomposition.

Having computed the twist $\boldsymbol{\xi}$, we can use Eqs. 2 and 4 to compute an updated camera pose

$$\boldsymbol{T}_{0,i} = \boldsymbol{T}_{0,kf} \cdot \boldsymbol{T}(\boldsymbol{\xi}), \tag{12}$$

where $\boldsymbol{T}_{0,i}$ is the current pose, $\boldsymbol{T}_{0,kf}$ is the pose of the last keyframe and $\boldsymbol{T}(\boldsymbol{\xi})$ is the movement from the last keyframe to the current frame.

If the IMU measurements used for the gravity alignment of the images were exact all the time, the problem would be reducible to 4DoF, as then roll and pitch would be constantly 0. However, as the IMU measurements contain errors, also roll and pitch can vary. Therefore, we don't fix roll and pitch but just allow small rotations.

Even though patches around lines do not contain a lot horizontal structure, the horizontal structure contained in the patches is enough to align the images correctly also in vertical direction.

3.5   Keyframe Selection and Map Extension

After a certain amount of motion, a new keyframe has to be selected and consecutively new lines and possibly points have to be matched and triangulated.

We use a simple measure for keyframe selection, which proved to work well in our experiments: After the pose estimation step, we compute the average photometric error per pixel. If this error exceeds a certain threshold, the current frame is selected as new keyframe. Additionally, if the translation and rotation w.r.t. the last keyframe exceeds a threshold, the current frame is selected as new keyframe.

For every keyframe, new lines and points are detected. However, before triangulating new lines or points, the already triangulated lines and points of the previous keyframe get projected into the current keyframe. If they are near a detected line/point in the current keyframe, the already computed depth value gets assigned to the lines/points in the current keyframe. All remaining detected lines get matched and triangulates as described in Sect. 3.2 and additional points are used if necessary as described in Sect. 3.3.

4   Experiments and Results

In this section, we first describe our implementation used for the experiments in more detail and then evaluate our method on our own challenging indoor dataset acquired with a flying MAV, on the publicly available Rawseeds dataset [18,19] and on a subset of the KITTI dataset [20]. We show that our approach outperforms state-of-the-art methods on the first two challenging indoor datasets and delivers comparable results on the last dataset.

4.1   Implementation Details

We implemented our algorithm in C++ using OpenCV as image processing library and Ceres Solver [21] for the optimization tasks.

In our evaluation, we used the images in full size, as downsampled images with a small stereo setup baseline significantly decreased the quality of the triangulated 3D information.

For line detection, we set the gradient histogram bin size to 20 which has shown to deliver good detection results. In the non-maximum suppression step, we take into account lines with a distance not farther than 1.5% of the image width. For the matching, a line is matched with another one if their gradient histogram bins do not differ by more than 2. Then, a match is accepted if the photometric error of the patch around the line has a smaller average photometric error per pixel than 15.

In the optimization step, we initialize the pose estimation of a new frame with the pose of the last frame. In order to avoid random jumps and rotations when too few or erroneous measurements are used in the optimization, we set upper and lower bounds for maximum rotational and translational movements w.r.t. the last keyframe: The maximum translational movement is set to 1 m and the maximum rotational movement is set 35°. These thresholds multiplied by 0.9 are also used for keyframe selection. In patch alignment, bilinear image interpolation is used to determine the interpolated image intensity and gradient values. For lines, we use a patch width of 21 pixels and height according to the line length. For points we use a patch size of 21 × 21 pixels.

4.2 Evaluation Setup

We evaluated our algorithm on a desktop PC which is an Intel Core i7-4820K CPU having 4 cores with 3.7 GHz, up to 8 threads and 16 GB of RAM. However, our algorithm just used 4 threads for computation.

We compared our approach against Libviso [1], which is a sparse feature-point based stereo visual odometry approach. We used it with subpixel refinement, set bucket height and width to 100px, maximum features per bucket to 10 and match radius to 50, as these settings have shown to deliver good results on our MAV dataset and the Rawseeds dataset. For one test sequence, we also compared with LSD-Slam [2], which is a direct monocular SLAM approach. However, we didn't use LSD-SLAM in the other experiments, as the sequences are not suited for a monocular approach, which cannot handle pure rotations and has problems with pure forward motion. We disabled loop closure and global map optimization in LSD-SLAM by setting the *doSLAM* option to *false*, as these trajectory improvement techniques are also not used in our approach.

4.3 Own Test Sequence

To evaluate the performance of our algorithm in challenging indoor environments where just few texture is available, we captured an image sequence by flying with an MAV in a room containing few texture elements. As capturing system, we used an Asctec Pelican equipped with two forward and slightly downwards look-ing Matrix Vision BlueFox-MLC202b cameras with a baseline of approximately 13.5 cm (see Fig. 5). In this setup, each camera has a horizontal field of view of 81.2° and acquires images with 20 frames per second (FPS) at a resolution of 640 × 480 pixels. In order to capture synchronized images and IMU measure-ments, we externally trigger the cameras with the Asctec Autopilot. Simultane-ously to the trigger signal, the Asctec Autopilot captures a timestamped IMU

Fig. 5. The Asctec Pelican used to acquire our own test dataset. It is equipped with an IMU (included in the autopilot), a stereo camera and an Odroid XU3 Lite processing board.

measurement. Both, the images and the IMU measurements are then stored on the Odroid XU3 Lite computation board onboard the MAV. We calibrated both the stereo camera setup and the rotation from IMU to camera with the Kalibr toolbox [22].

The sequence is captured while flying with the MAV in front of a wall in a broad hallway. The trajectory is similar to a rectangle, while both cameras are looking in the direction of the wall all the time. The start and end points are nearly identical, therefore the absolute trajectory error can be observed. This scenario is similar to an augmented reality application, where, for example, an MAV equipped with a laser projector should project something onto the wall.

As can be seen in the input images in Fig. 6, the scene mostly consists of white, untextured walls. Only few structure elements are visible (doors, recycle bin).

Fig. 6. Images of the sequence acquired with the flying MAV. As can be seen, the scene mostly consists of white walls with little texture.

We compare this sequence with Libviso [1] and LSD-SLAM [2]. As LSD-SLAM is a monocular approach and therefore does not provide a metric scale, we manually aligned the trajectory scale with the stereo approaches. However, generally this trajectory should be well suited for a monocular approach, as there are no pure rotations and no forward movement in the initialization phase.

The computed trajectories are plotted in Fig. 7. In comparison to Libviso (blue), our approach (red) does not have a big drift. The start/end point difference of our approach is approximately 2.5 m while Libviso has a difference of approximately 7 m. Libviso has already a big translational drift when the MAV is standing still on the floor at the beginning of the trajectory. While flying, Libviso accumulates a big rotational drift, which leads to big changes in the

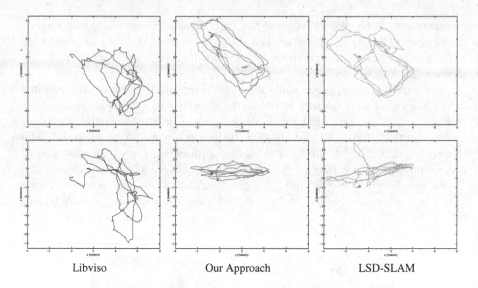

Libviso   Our Approach   LSD-SLAM

Fig. 7. Estimated trajectories of Libviso (*left, blue*), our approach (*middle, red*) and LSD-SLAM (*right, green*) of the test sequence captured with an MAV. *Top row:* x- and y-axes (top view). *Bottom row:* x- and z-axes (side view). The start point is at (0,0,0) and the trajectory end point, which should be near the start point is indicated with a circle. Even though our approach (red) has a slight drift (start point is not identical with end point), Libviso (blue) has a much higher rotational drift. LSD-SLAM works well for some parts of the sequence. However, occasionally it does some random jumps and moves far away from the desired trajectory. (Color figure online)

z-dimension. Contrary, LSD-SLAM estimates parts of the trajectory quite well. However, some random pose jumps happen in the estimation, which yields to a trajectory which is partly far away from the correct solution. In total, our approach performs best compared to the others, as it does not have a big drift and no big outliers.

Lines vs. Points. We additionally use point features in cases where insufficiently many lines were detected. Using lines-only would decrease the pose estimation results in these sections of the trajectory. However, using points-only would lead to poorer estimation results in other parts of the trajectory. In this section, we will discuss failure cases of our algorithm when using both lines-only and points-only.

In Fig. 8, one can see two subsequences of our own test sequence, where using points-only leads to a significantly higher difference to the combined version (top left) and where using lines-only leads to a significantly higher difference to the combined version (bottom left). For both subsequences, the combined feature approach (using lines and points), which is a subset of the estimated trajectory of our approach in Fig. 7, can be defined as best trajectory estimate by visually comparing with the input images.

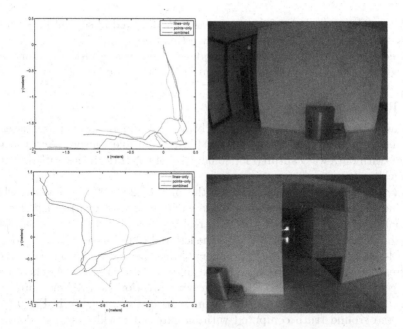

Fig. 8. Comparison of trajectories of subsequences using lines-only, points-only and a combination of both. Both trajectories start at $(0,0)$. *Top left:* subsequence where the difference between points-only (red) and the combined version (blue) is significantly higher than the difference between lines-only (green) and the combined version. Before starting the movement to the right, points-only drifts towards the wrong direction and at approx. $x = -1$ (corresponding to the image at *top right*) points-only does a jump. *Bottom left:* subsequence where lines-only (green) has a higher difference with respect to the combined version than points-only (red). When starting the movement backwards (which corresponds to the image at *bottom right*), lines-only jumps towards a wrong direction. (Color figure online)

In the subsequence at the top left, points-only (red) has a high drift to the right shortly before moving to the right in front of the wall. Then, at approx. $x = -1$, points-only makes a jump. These two errors lead to a big trajectory length difference in this subset compared to lines-only (green) and combined (blue). At the top right, one can see an input image at approx. $x = -1$, where points-only makes a jump. As nearly only the white wall and some vertical structure elements are visible, too few point features can be detected in order to compute a correct trajectory estimate. Such failure cases of using points-only can be compensated frequently by using lines in poorly textured man-made environments.

In the subsequence illustrated in the bottom left, lines-only (green) makes a jump directly when starting to move backwards (away from the wall). An example input image from this position can be seen at the bottom right. The error is generated by not sufficiently enough detected and matched lines, which is also caused by an incorrect image gravity alignment leading to shorter line detections. Contrary, point features can still be detected in several parts of the image, which leads to a better trajectory estimation result.

Generally, the combined version produces the most robust trajectory estimates due to the usage of multiple visual information sources.

4.4   Rawseeds

The Rawseeds datasets [18, 19] are publicly available indoor and outdoor datasets captured with a ground robot equipped with multiple sensors. Additionally, ground truth data was captured for some parts of the trajectories with an external tracking system.

In our experiments, we chose to use the Bicocca_2009-02-27a dataset, which is an indoor dataset captured in a static environment with mixed natural and artificial lightning. Note that this is a really large-scale dataset, as the total trajectory length is 967.18 m. We used the left and right images of the trinocular forward-looking camera for our experiments, which have a resolution of 640×480 pixels each, a baseline of 18 cm and were captured with 15 FPS. As the images were acquired roughly aligned with the gravity direction, no IMU measurements are needed for our algorithm. Example input images can be seen in Fig. 9.

As the ground truth computed with an external tracking system does not cover the whole trajectory, we use the (publicly available) extended ground truth in our experiments. This ground truth also includes trajectory parts estimated with data of an onboard laser scanner.

In the experiments for this dataset, we compare our approach with Libviso. In Table 1 one can see the relative pose errors as proposed in [23]. Our approach gives better results for all computed relative pose errors. However, due to the difficulty of the sequence, which also contains images with no texture at all (see Fig. 9), some parts of the trajectory also have a high error when using our algorithm. Additionally, the frame rate should be higher for our approach. As this dataset is just captured with 15 FPS, sometimes the scene just moves too fast in front of the camera and the camera pose cannot be tracked exactly. However, in the parts of the sequences where there exist some vertical line structure (e.g. doors, windows or shelves), our approach works better and therefore the errors are lower.

Table 1. Relative translational and rotational error of the Rawseeds sequence as defined in [23] per seconds of movement of our approach and Libviso. In all metrics, our approach performs better.

	Our approach	Libviso
Transl. RMSE (m/s)	**0.1498**	0.1721
Transl. Median (m/s)	**0.0599**	0.0825
Rot. RMSE (deg/s)	**3.4610**	5.3659
Rot. Median (deg/s)	**0.0093**	0.0094

Fig. 9. Images of the Rawseeds Bicocca_2009-02-27a sequence. *Top left:* a big part of the images look similar to this, where the robot moves through a narrow corridor. *Top right:* at some corridor intersections, where nearly pure rotations are performed, just the white wall is captured with the cameras, which is extremely difficult for a VO algorithm. *Bottom row:* also parts with better input data for VO exist, e.g., wider hallways and a library.

4.5   KITTI

The KITTI dataset [20] is currently a popular dataset for visual odometry evaluation. It is acquired with a stereo camera mounted on a car which drives on public streets, mostly through cities. We compare our algorithm against Libviso [1] on a subset of the dataset (sequence 07).

The images have a resolution of 1241×376 pixels and a framerate of 10 FPS (example input images can be seen in Fig. 10). As the images are roughly aligned with the gravity direction, there is no need to incorporate IMU measurements. Due to the low capture framerate and the relatively fast movement of the car, the image content can change quite rapidly. This is a problem for direct VO methods which deliver good results when having small inter-frame motions. Additionally, it contains enough texture for feature-based methods to work properly. However, we want to show that our approach works also well on datasets for which it is not specifically designed for.

To overcome the low framerate, we had to change the patch size of the patches around lines and points used in the optimization from 21 to 31. Additionally, we had to increase the maximal translational movement for the optimization step. Using this settings we get results comparable to state-of-the-art methods with the drawback of higher runtimes compared to the other datasets used in our

Fig. 10. Images of the KITTI visual odometry dataset (sequence 07). In this sequence, there are some lines detectable in some parts (e.g. in top image). In these parts, the detected lines are beneficial for the VO algorithm. However, in some images there exist no lines (e.g. image bottom) and our algorithm also depends on the detected feature points.

evaluation. Additionally, as more reliable points can be detected in this scene, we decreased the line matching threshold so that more points are used.

Also for Libviso, we changed the parameter settings: We used it with default settings and subpixel refinement.

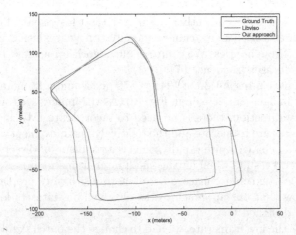

Fig. 11. Evolving trajectory of the KITTI sequence. One can observe that our approach (red) performs comparable to Libviso (blue). (Color figure online)

In Fig. 11 the computed trajectories of Libviso and our approach are plotted. As one can see, the absolute trajectory error is slightly worse in our approach. Also the rotational and translational errors as defined in the KITTI vision benchmark suite are worse than Libviso (see Table 2). However, our algorithm still works acceptable on a dataset like this one for which it is not designed for.

Table 2. Relative translational and rotational error of the KITTI sequence 07 computed with the KITTI vision benchmark suite [20] as mean over all possible subsequences of length $(100,\ldots,800)$ m.

	Our approach	Libviso
Transl. error (%)	8.0558	**2.3676**
Rot. error (deg/m)	0.000744	**0.000321**

4.6   Runtime

The mean computation time per frame for the first two datasets is 49.3 ms, which is a framerate of 20.3 FPS. With this computation speed, both the Rawseeds dataset and our own MAV dataset can be processed in real-time. Due to changed parameters for the KITTI dataset, the computation speed for the KITTI dataset is much slower and needs approximately 3 times the processing time of the other datasets.

Most of the processing time is needed for the pose estimation step, which could be optimized by a multi-scale approach, using SIMD instructions (SSE, NEON) or writing an own optimizer to overcome the overhead of Ceres.

5   Conclusion

We have presented a direct stereo visual odometry method, which detects and matches lines at every keyframe using our novel fast line detection algorithm and estimates the pose of consecutive frames by direct pose estimation methods using patches around vertical lines. These patches have proven to be a good selection for indoor environments, which are poorly textured, but where a lot of vertical structure exists. In our experiments, we have shown that our algorithm delivers better results than state-of-the-art methods in challenging indoor environments and comparable results in well textured outdoor environments. As our implementation runs in real-time, it is suitable for various robotics and augmented reality applications. Future work will include subpixel refinement of the line detection step, global map optimization to minimize drift errors and to evolve it to a complete SLAM system, a multi-scale approach for the pose estimation step to deliver better accuracy for large inter-frame movements and improve computation speed, and further speed improvements (e.g., using SSE and NEON instructions) to make it real-time capable also on small robotic platforms.

Acknowledgements. This project has been supported by the Austrian Science Fund (FWF) in the project V-MAV (I-1537).

References

1. Geiger, A., Ziegler, J., Stiller, C.: StereoScan: dense 3D reconstruction in real-time. In: IEEE Intelligent Vehicles Symposion (2011)
2. Engel, J., Schöps, T., Cremers, D.: LSD-SLAM: large-scale direct monocular slam. In: Proceedings of European Conference on Computer Vision (2014)
3. Holzmann, T., Fraundorfer, F., Bischof, H.: Direct stereo visual odometry based on lines. In: 11th International Conference on Computer Vision Theory and Application (VISAPP) (2016)
4. Klein, G., Murray, D.: Parallel tracking and mapping for small AR workspaces. In: Proceedings of International Symposium on Mixed and Augmented Reality (2007)
5. Mei, C., Sibley, G., Cummins, M., Newman, P., Reid, I.: RSLAM: a system for large-scale mapping in constant-time using stereo. Int. J. Comput. Vis. **94**, 198–214 (2010)
6. Weiss, S., Achtelik, M.W., Lynen, S., Achtelik, M.C., Kneip, L., Chli, M., Siegwart, R.: Monocular vision for long-term micro aerial vehicle state estimation: a compendium. J. Field Robot. **30**, 803–831 (2013)
7. Elqursh, A., Elgammal, A.M.: Line-based relative pose estimation. In: Proceedings of IEEE Conference Computer Vision and Pattern Recognition, pp. 3049–3056. IEEE Computer Society (2011)
8. Newcombe, R.A., Lovegrove, S.J., Davison, A.J.: DTAM: dense tracking and mapping in real-time. In: Proceedings of International Conference on Computer Vision, pp. 2320–2327 (2011)
9. Engel, J., Stückler, J., Cremers, D.: Large-scale direct SLAM with stereo cameras. In: International Conference on Intelligent Robots and Systems (2015)
10. Forster, C., Pizzoli, M., Scaramuzza, D.: SVO: fast semi-direct monocular visual odometry. In: International Conference on Robotics and Automation (2014)
11. Grompone, R., Jakubowicz, J., Morel, J.M., Randall, G.: LSD: a fast line segment detector with a false detection control. IEEE Trans. Pattern Anal. Mach. Intell. **32**, 722–732 (2010)
12. Hofer, M., Maurer, M., Bischof, H.: Improving sparse 3D models for man-made environments using line-based 3D reconstruction. In: International Conference on 3D Vision (2014)
13. Cortinovis, A.: PIXHAWK - attitude and position estimation from vision and IMU measurements for quadrotor control. Technical report, Computer Vision and Geometry Lab, Swiss Federal Institute of Technology (ETH) Zurich (2010)
14. Rosten, E., Drummond, T.: Fusing points and lines for high performance tracking. In: Proceedings of International Conference on Computer Vision (2005)
15. Ma, Y., Soatto, S., Kosecka, J., Sastry, S.S.: An Invitation to 3-D Vision: From Images to Geometric Models. Springer, Heidelberg (2003)
16. Levenberg, K.: A method for the solution of certain non-linear problems in least squares. Q. Appl. Math. **2**, 164–168 (1944)
17. Marquardt, D.: An algorithm for least-squares estimation of nonlinear parameters. SIAM J. Appl. Math. **11**(2), 431–441 (1963)

18. Bonarini, A., Burgard, W., Fontana, G., Matteucci, M., Sorrenti, D.G., Tardos, J.D.: RAWSEEDS: robotics advancement through web-publishing of sensorial and elaborated extensive data sets. In: International Conference on Intelligent Robots and Systems (2006)
19. Ceriani, S., Fontana, G., Giusti, A., Marzorati, D., Matteucci, M., Migliore, D., Rizzi, D., Sorrenti, D.G., Taddei, P.: Rawseeds ground truth collection systems for indoor self-localization and mapping. Auton. Robots **27**, 353–371 (2009)
20. Geiger, A., Lenz, P., Urtasun, R.: Are we ready for autonomous driving? The KITTI vision benchmark suite. In: Proceedings of IEEE Conference Computer Vision and Pattern Recognition (2012)
21. Agarwal, S., Mierle, K., et al.: Ceres solver. http://ceres-solver.org
22. Furgale, P., Rehder, J., Siegwart, R.: Unified temporal and spatial calibration for multi-sensor systems. In: International Conference on Intelligent Robots and Systems (2013)
23. Sturm, J., Engelhard, N., Endres, F., Burgard, W., Cremers, D.: A benchmark for the evaluation of RGB-D SLAM systems. In: International Conference on Intelligent Robots and Systems, pp. 573–580 (2012)

Consumer-Level Virtual Reality Motion Capture

Catarina Runa Miranda$^{(\boxtimes)}$ and Verónica Costa Orvalho

Instituto de Telecomunicações, Universidade do Porto, Porto, Portugal
catarina.runa@gmail.com, veronica.orvalho@gmail.com

Abstract. Virtual Reality (VR) is creating a new paradigm in humans' communication. Today, we can enter in a virtual environment and interact with each other through 3D characters. However, VR headsets occlude user's face, limiting the Motion Capture (MoCap) of facial expressions and, thus, limiting the introduction of this non-verbal component. The unique solution available is not suitable for consumer-level applications, relying on complex hardware and calibrations. In this work, we deliver consumer-level methods for facial MoCap under VR environments. We developed an occlusions-support method compatible with generic facial MoCap systems. Then, we extract facial features and deploy Random Forests algorithms that accurately estimate emotions and upper face movements occluded by the headset. Our VR MoCap methods are validated and a facial animation use case is provided. With our novel methods, both calibration and hardware is reduced, making possible face-to-face communication in VR environments.

Keywords: Facial Motion Capture · Emotion and expressions recognition · Virtual Reality · Facial animation

1 Introduction

In the last two decades, we lived a revolution of global digital interactions and communication between humans [11]. We erased geographic barriers and started communicating with each other through phones, computers and, more recently, inside virtual environments using Virtual Reality (VR) headsets. Oculus VR company was the responsible by bringing this hardware to consumer-level making this way of interaction more appealing to common users (Oculus VR 2014). However, VR communications remain a challenge. Human communication strongly rely on a synergistic combination of verbal (e.g. speech) and non-verbal (e.g. facial expressions and gestures) signals between interlocutors [11]. Past communication technologies, like phones and computers, adopted the image stream (e.g. webcams) coupled with speech to transmit both signals creating more realistic and complete experiences [13]. In VR scenarios, we cannot use image stream since we are interacting with the virtual world embodied in 3D characters [2,28]. As result, the demand for on-the-fly algorithms for 3D characters animation and interaction is even higher. Ahead of unlocking both communication channels

© Springer International Publishing AG 2017
J. Braz et al. (Eds.): VISIGRAPP 2016, CCIS 693, pp. 374–394, 2017.
DOI: 10.1007/978-3-319-64870-5_18

(i.e. verbal and non-verbal), the believable animation of 3D characters using user's movements enhance the three components of the sense of embodiment in VR environments: self-location, agency and body ownership [2,12]. Even with technological advances in Computer Vision (CV) and Computer Graphics (CG), the reproduction of human's facial expressions as facial animation of 3D characters is still hard to achieve [24]. To automatise facial animation, facial Motion Capture (MoCap) has been widely used to trigger animation [5,6,15,29,30]. However, these approaches are not suitable for consumer-level VR applications, requiring or expensive setups [29], manual complex calibrations [6,15,30] or do not support the persistent partial occlusion of the face produced by VR headsets [5].

To overcome the tracking problem created by persistent partial occlusions, Li *et al.* [14] proposed a hardware based solution using a RGB-D camera for capture and strain gauges (i.e. flexible metal foil sensors) attached to VR headset to measure the upper face movements that are occluded. But again, this approach is not suitable for general user. It requires a complex calibration composed by hardware calibration to user and a blendshapes calibration to trigger animation. At the moment, this is the unique on-the-fly facial animation with MoCap solution compatible to VR environments.

Contributions: This work delivers and validates consumer-level real-time methods for: (i) facial MoCap method for persistent partial occlusions created by VR headsets and (ii) facial expressions prediction algorithms of occluded face region using movements tracked in non-occluded region. Compared to literature, we reduce user-dependent calibration and hardware requirements, requiring only a common RGB camera for capture. Our methods make current facial MoCap approaches compatible to VR environments and enable the extraction of key facial movements of bottom and upper face regions. The movements tracked and emotions detected can be combined to: trigger on-the-fly facial animation, enabling non-verbal communication in VR scenarios; as input for emotion-based applications, like emotional gaming (e.g. Left 4 Dead 2 by Valve). As an extension, we added an use case regarding the facial animation of 3D characters as proof of concept of this application.

2   Background

In this section, we aim to study the literature regarding two different topics: (i) facial MoCap solutions for persistent partial occlusions created by VR Head Mounted Displays (HMD) and (ii) partial occlusions impact in facial expressiveness. The first topic presents state of the art facial MoCap solutions to overcome the persistent occlusions' issue. Then, in (ii), we explore how these occlusions restrict face-to face communication and their impact in face expressiveness. By the end, we search for a connection between occluded and non-occluded facial parts used as guide for methodology definition.

2.1   Persistent Partial Occlusions: A Today's Problem

In literature, we are able to find several promising solutions for real-time automatic facial MoCap [5,6,15,29,30]. However, the arise of VR commercial approaches of consumer-level HMD's (Oculus VR 2014), raised a new issue: the real-time automatic tracking of faces partially occluded by hardware (i.e. persistent partial occlusions of face) [28]. Current MoCap approaches adopt model-based trackers, which produce cumulative errors in presence of persistent partial occlusions [5]. Therefore, due to the absence of VR devices in mass-market, this issue was almost ignored for years. This resulted in a lack of technological solutions for face-to-face communication for VR environments. Only in 2015, Li et al. [14] highlighted this problem and proposed a hardware based tracking solution. This solution uses an RGB-D camera combined with eight ultra-thin strain gauges placed on the foam liner for surface strain measurements to track upper face movements, occluded by the HMD. The first limitation of this approach is the long initial calibration required to fit the measures to each individual's faces using a training sequence of FACS [8]. Also, in subsequent wearings by the same person, a smaller calibration is needed to re-adapt the hardware measures. This training step allows the detection of user's upper and bottom face expressions and activate a blendshape's rig containing the full range of FACS shapes [8]. Besides the manipulation complexity, the solution also presents drifts and decrease of accuracy due to variations in pressure distribution from HMD placement and head orientation. As consequence, HMD straps positioning influence eyebrows' movement detection [14]. Li et al. solution is currently the only one available to overcome the persistent partial occlusions issue, making this an open research topic in CV algorithms for facial MoCap.

2.2   Partial Occlusions and Expressiveness

Everyday, humans' communication use facial expressions and emotions to transmit and enhance information not provided by speech [13]. Even through technology, we always search for a way to use the non-verbal communication channel. As example, using video stream of our faces; virtual representations, like *emotion smiles*, cartoons or 3D characters with pre-defined facial expressions, etc. Understanding facial expressions and improve their representation in 3D characters is one of the key challenges of CG and plays an important role in digital economy [11]. This role is even more relevant now, with recent advances in VR communications at consumer level [2]. *But how can we use the common solutions of facial animations, like MoCap, if user's face is occluded? Are we able to represent faces using information only from bottom of the face?* To answer these questions, we make a literature overview regarding several face regions impact in non-verbal communication. The goal is to understand how a partial occlusion of the face affects communication. We also researched for a relationship between occluded and non-occluded facial parts through emotion-based and biomechanics studies. This information was used to build one of this work hypothesis.

In a study about face perception [10], we concluded that humans have independent shape representations of upper and bottom parts of the face. Similar conclusions are found in emotion perception's literature, where mouth and eyes play different roles [3,7,13]. In [3,7] it is shown that according to the emotion detected participants used information from eyes, or mouth or both. More precisely, in happy expressions participants used information from the mouth; for sad and angry, from eyes; and to fear and neutral, both mouth and eyes are used. For additional information about non-verbal communication, we forward the reader to [13]. Taking these statements into account, if we occlude certain region of the face, face-to-face communication is affected and we may not be able to decode expressions properly. Subsequently, the tracking of only certain facial regions, like mouth, is not enough for emotion recognition, for proper communication and to generate believable facial animation of 3D characters.

From the biomechanical point of view, we know that facial muscles work synergistically to create expressions. The muscles interweave with one another, being difficult to decode their boundaries, since their terminal ends are interlaced with other muscles. A detailed research about facial anatomy and biomechanics can be accessed at Chap. 3 of the book *Computer Facial Animation* [23]. Several studies in CG applied the biomechanical approach to create coding systems. These coding systems parameterize human face enabling a faster generation of facial expressions in 3D characters [8,18,21]. Although, they do not provide a clear solution for facial expressions estimation constrained to certain regions of the face. Furthermore, the definition and prediction of facial expressions is even harder when the diversity of facial expressions is considered. Scott McCloud [20] explains the infinite possibilities of facial expressions combinations (i.e. the way mixing any two of universal emotions can generate a third expression, which, in many cases, is also distinct and recognizable enough to earn its own name) [20].

Then, analyzing literature, we are able to attain that occlusions generated by VR devices affect communication and using only the information of non-occluded regions is not enough to animate a 3D character. However, biomechanics and facial animation coding systems show a connection between the different facial regions and how diverse and complex is the world of possible expressions. Using these statements, we describe a novel methodology to overcome occlusions problem of facial MoCap and then, to assess facial expressions using non-occluded face information.

3   Methodology

The literature overview of previous section allowed us to formulate the following hypothesis:

> *to create a method estimate facial expressions of upper face and emotions using only bottom face's movement.*

Therefore, we deliver VR consumer-level methods that:

– overcomes the persistent partial occlusions issue in MoCap, making possible the bottom face's movements tracking;
– recognizes universal emotions, plus neutral [9,11] using bottom face's movements;
– estimates upper face's movements (i.e. eyebrows movements) using information tracked from bottom part of the face.

Figure 1 shows the connection between our VR methods. We start by presenting a method to make generic MoCap systems compatible to persistent partial occlusions produced by VR headsets. Then, applying this algorithm, we are able to track properly the bottom face's features and use them to develop methods that predict the following facial expressions: (i) universal emotions, plus neutral [9,11] and (ii) eyebrows movements. Combining aforementioned methods, we make possible the MoCap of upper and bottom face movements and estimation of facial emotions under persistent partial occlusions created by VR headsets.

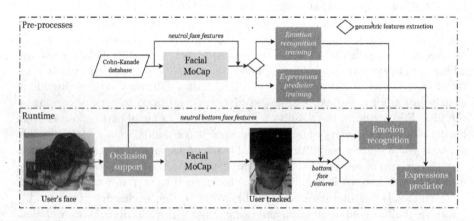

Fig. 1. VR methods' framework.

As setup, we suggest the usage of a Head Mounted Camera (HMC) combined with the VR HMD (see Fig. 2). At first, we justify the adoption of HMC as capture hardware: When the user is inside the VR environment he is not aware of the space around him. The VR devices precisely substitute the user's sensory input and transform the meaning of their motor outputs with reference to an

Fig. 2. VR setup definition.

exactly knowable alternate reality [28]. Hence, the user moves and reacts to impulses from VR environment. If we want to capture his face, we have to attach a capture device (i.e. camera) to his body and the device should follow user's movements (see HMC on Fig. 2). It is not possible to use a static camera, because the user is not going to be able to place himself in a position proper for capture. A similar setup was also proposed by Li *et al.* [14], but we removed the strain sensors.

In the next subsections, we provide a complete description of the VR methods.

3.1 VR Persistent Partial Occlusions: A Novel Method

To deploy our occlusion support method for facial MoCap, we used the following statement: we know the kind of occlusion created by HMD, so we know which part of the face is occluded. We also know that MoCap algorithms fail in these situations because they use a face model. When the face is occluded this model starts not to fit since there is not a full face being captured. As a solution, we use the knowledge that the region occluded is the upper part of the face to "re-create" the whole face.

Our novel method overlays the upper part of the face captured on a neutral pose during calibration. Firstly, we assume that the higher visible point of the face is the nose and define it as cut point (i.e. this point can be changed to fit the occlusion created by certain HMD). Then, we detect the cut point with the MoCap and we cut the upper part of the calibration image (i.e. frame streamed) from the nose up, and use it to overlay to all the next camera/video frames. Hence, now the occluded part of the face is replaced with a static neutral face. The MoCap system is now able to detect the features in the combined half static/half expressive face (see Fig. 3). We ensure a proper re-creation of a face

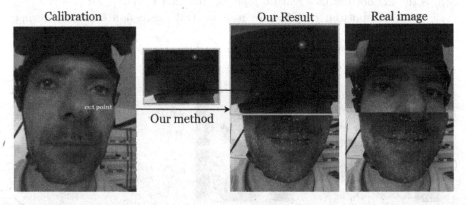

Fig. 3. VR method: persistent partial occlusions. From left to right: calibration image without VR HMD; our method uses cut point (red circle) to cut image an overlay at subsequent images: at left, what facial MoCap method see is a full face and, at right, the real image. (Color figure online)

since we use a HMC that removes the user's head movements, i.e. user's face is in the same position during calibration and next streamed images.

3.2   VR Assessing Facial Expressions

During the development of VR facial expressions method, we applied face features and machine learning know-how from our past real-time emotion recognition research [16]. In this novel method, we set the following goals: real-time emotion recognition of universal emotions [9] and upper face expressions prediction under VR scenarios. We aim to track facial expressions ahead of only emotions, in order to get a wide change of facial expressions and better cover and representation of the diversity of faces [19]. In opposition to the emotion classification method [16], where we needed to reduce the number of features tracked, in VR scenarios we have to maximize the information tracked in the bottom part of the face. Therefore, the feature extraction method should be able to retrieve enough information to allow an accurate prediction of facial expressions by the machine learning algorithm.

As a solution, we propose to use all the features tracked of bottom face region (see Fig. 4 blue rectangle) and apply a geometrical features extraction algorithm. This algorithm is defined as the Euclidean distance between neutral face features (stored during calibration step of previous persistent partial occlusions method) and current frame (i.e. instant in time) features. Summarizing, to each feature tracked p in certain instant i, we calculate the distance $D(p_i, p_c)$:

$$D(p_i, p_c) = \sqrt{\frac{((p_i(x) - p_c(x))^2 + (p_i(y) - p_c(y))^2}{\|p_i - p_c\|}},$$

where:

p_i is the 2D bottom face feature p at the instant i in time;
p_c is the 2D bottom face feature p of neutral expression captured during calibration;
$\|p_i - p_c\|$ is the norm between p_i and p_c in Cartesian space.

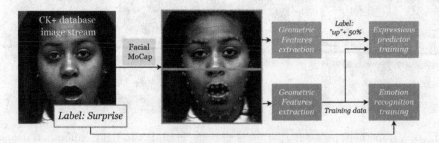

Fig. 4. VR methods: expressions predictor training (purple) and emotion predictor training (blue) with CK+ database. (Color figure online)

Since the occlusion produced varies according to VR headset used, we also created machine learning models to assess facial expressions using the bottom face features information including and excluding nose features. The bottom face features without nose feature can be used by the different kinds of HMD, since the nose region is the one affected by the device size.

To create the machine learning models to predict the emotions and upper face expressions, we used the Cohn-Kanade (CK+) database [17]. CK+ database contains posed and spontaneous sequences from 210 participants (i.e. cross-cultural adults of both genres). Each sequence starts with a neutral expression and proceeds to a peak expression. This sequences are FACS coded and emotion labeled. The transition between neutral and a peak expression allowed us to detect spontaneous expressions and not only pure full expressions.

To implement the algorithms, we adopted a GPU version of Random Forest [4] of OpenCV [1] to generate respective machine learning models for real-time prediction. As facial MoCap testing approach, we deployed the Saragih *et al.* [27] system. (see Fig. 4 tracking landmarks in green).

VR Emotion Recognition: Novel Method. As preprocessing stage, we create the Random Forests model that is used to predict emotions in real-time [16]. To build the model for emotion classification, to each database's sequence we applied the facial MoCap method and extracted bottom face features. Using the first frame of the sequence as neutral expression, to subsequent frames in the sequence, we calculate the distance $D(p_i, p_c)$, between bottom face features of current frame and neutral expression's frame. Thus, to train the machine learning model for emotion recognition we used aforementioned geometrical extraction algorithm: distance $D(p_i, p_c)$ of bottom face's features of each frame. As response value, to each distance calculated, we used respective CK+ emotion label (see Fig. 4 blue processes).

As observed in the Fig. 2, in runtime, we apply once our occlusions support method and store neutral face features. This step is only execute one time per user. After, in runtime, the adapted facial MoCap system delivers bottom face's movements and distance $D(p_i, p_c)$ is calculated to each feature p. The group of distances are used as input in the Random Forests classifier that predicts the user's emotion represented by that distances and respective accuracy's percentage.

VR Facial Expressions Predictor: Novel Method. To build the upper face expressions model, we also applied the distance of neutral and expression bottom face features as geometric extraction algorithm. However, we have to define the movements that we wanted to predict in order to create specific tags to the training process. For simplicity, we set as upper face expressions the prediction of eyebrows movements, i.e. the detection if eyebrows are going up or down, and the "how much" they are moving compared to a neutral position. This last parameter is measured as a percentage of movement up/down compared to neutral expression. Similarly to assumption made in [10], we assume symmetry of the eyebrows movements. To define the tags, we calculated the Euclidean distance $D(p_i, p_c)$ between neutral position of eyebrows and the expression positions in the other frames of the sequence. If the average of the eyebrows features indicated that they are going up, we tagged "up"; the opposite if the eyebrows went down we tag "down" (i.e. we used image coordinate system, so this distance was negative when eyebrows go up and vice-versa). Simultaneously to each frame of the sequence tagged we saved the percentage of movement compared to neutral position (up or down). As result, to each frame of the sequence of each participant in CK+ database we tagged: eyebrows "up" or "down", plus percentage of movement. In Fig. 4 with purple processes, the reader can observe an example of method's framework.

At preprocessing stage, we trained two Random Forests models with the same input data: the distances $D(p_i, p_c)$ between neutral and current bottom face features; but using one of the following response values:

- "up" and percentage of movement, if eyebrows are rising
- "down" and percentage of movement, if eyebrows are descending,

to each frame of each sequence of CK+ database.

Since we are using a GPU approach of the classifier, with high computational performance, to maximize the prediction accuracy of eyebrows movements, we trained two models: one to predict the rise movement and, other, to predict the opposite. In runtime, we apply the defined geometrical features extraction to the bottom face's features tracked by the adapted MoCap. The extracted features are used as input in both Random Forests classifiers, to retrieve one of the predictions:

1. **eyebrows "rising"** and percentage of movement;
2. **eyebrows "descending"** and percentage of movement.

Since we are using two different classifiers, there is a probability of confusion of both models return simultaneously an "up" and "down" movement. As a solution, our method compares the accuracies of prediction from the two classifiers' predictions, and the result delivered is the one with higher accuracy.

4   Results and Validation

In this section, we show the results and statistical validation of the methods proposed. Statistical analysis was performed using R software [25].

4.1   VR Persistent Partial Occlusions

To test our occlusions method, we applied it to Saragih et al. [27] and Cao et al. [5] MoCap systems (see Figs. 5 and 6, respectively). At the Fig. 7, we test a generic partial occlusion created by a piece of paper.

As observed in the Figs. 5, 6 and 7, our occlusion-support method adapts to MoCap systems making them compatible to persistent partial occlusions. The

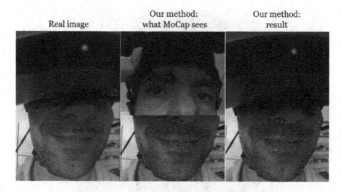

Fig. 5. VR method results: persistent partial occlusions method applied to Saragih *et al.* [27] MoCap. The real image (left), our method result and what MoCap processes (middle) and final result from our method (right).

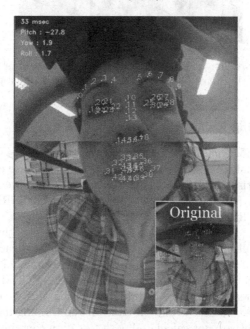

Fig. 6. VR method results: persistent partial occlusions method applied to Cao *et al.* [5] MoCap.

"paper" test case represented a generic occlusion created by a random VR device. As conclusion, our method is not only adaptable to MoCap, but it could be also used to generic partial occlusions created by different VR HMD's.

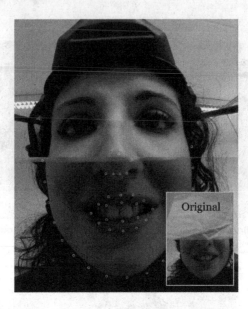

Fig. 7. VR method results: persistent partial occlusions method applied to Cao *et al.* MoCap algorithm [5] to overcome a general occlusion created by a piece of paper.

4.2   VR Assessing Facial Expressions

We divided the validation of our prediction methods in three steps: (i) statistical validation, (ii) visual validation and (iii) facial animation use case.

To validate statistically our machine learning classifiers we adopted a k-Fold Cross Validation (k-Fold CRM) with k = 10 [26]. The k-Fold CRM, after iterating the process of dividing the input data in k slices for k times, trains a classifier with k-1 slices. The remaining slices are used as test sets on their respective k-1 trained classifier, allowing us to calculate the accuracy of each one of the k-1 classifiers. The final accuracy value is given by the average of the k calculated accuracies. Though, to each method we analyze k-Fold CRM accuracy to the methods under different scenarios. We highlight that this validation procedure ensures that the test dataset is not the same of the training dataset. Therefore, prediction accuracies are not calculated with test data contained in the training dataset.

Furthermore, we provide a statistical analysis of sensitivity versus specificity and positive versus negative predictive value (i.e. pred. in Tables) [22]. The sensitivity measures the performance of the classifier in correctly predicting the actual class of an item, while specificity measures the same performance but in not predicting the class of an item that is of a different class. Summarizing, sensitivity and specificity measure the true positive and true negative performance, respectively. We added the positive and negative predictive value analysis because these values reflect the probability that a true positive/true negative is correct given knowledge about the prevalence of each class in the data analyzed.

By the end of this section, we validated visually our VR methods regarding: occlusions, emotion and facial expressions prediction. The visual validation data was acquired in our laboratory and is not part of the training dataset (learning made CK+ database). The visual data was not acquired with HMC, but we asked to the participants to avoid extreme head movements. As result, we were able to test our VR method of occlusion-support and the facial expressions methods simultaneously. As an application example of our methodology, we added a facial animation use case to show how it can be easily included in VR MoCap facial animation pipelines.

VR Emotion Recognition. Using the k-Fold CRM, we executed a method's validation to two emotion recognition scenarios: (i) six universal emotions of Ekman and Friesen [9], plus neutral; (ii) four universal emotions of Jack and Jack [11], plus neutral. The six universal emotions [9] are the commonly used and accepted by literature studies. However, recent advances in psychology of the emotions show that these emotions are not reproducible throughout different cultures. The non-universality of Ekman's emotions is explored by the survey [11]. This complete study defends that only a subset of the six "universal" emotions is universally recognized, i.e. Joy/Happy, Surprise, Anger and Sad/Sadness. This subset excludes fear and disgust, since these emotions present low recognition cross-culturally being biologically adaptive movements from the emotions surprise and anger, respectively [11].

Therefore, the Table 1 shows the k-Fold CRM accuracies to the two scenarios.

Table 1. k-Fold CRM accuracy comparison to scenario (i) and to the scenario (ii). Results in percentage (%).

Emotions	k-fold accuracy (%)	95% confidence interval
Six [8]	64.80	[61.72;67.79]
Four [11]	69.07	[65.59;72.40]

In the Table 1, we observe an increase of the accuracy detection when recognizing four emotions, compared to six emotions classification. This result is not surprising, since we are reducing the number of emotions predicted. In addition, we detect that the bottom features of the face allow a weak recognition of face emotions, resulting in accuracies lower than 70%.

More in detail, we report in the Tables 2 and 3, a statistical analysis of each emotion recognition obtained with Random Forests classifier to scenario (i) and (ii), respectively.

Both statistical analysis resulted in a p-value lower than $2.2 \times e^{-}16$ to a significance level of 5%, which validates our method's hypothesis: classifying the six/four universal emotions using bottom of face features tracking. Specifically, to scenario (i) at the Table 2, we observe an overall low sensitivity to emotions

Table 2. Statistical analysis of scenario (i) - results in percentage (%).

	Anger	Disgust	Fear	Joy	Sadness	Surprise	Neutral
Sensitivity	53.15	39.44	26.09	81.29	12.70	59.40	90.80
Specificity	86.55	97.70	95.84	95.17	99.13	96.35	85.39
Positive pred.	40.21	57.14	39.34	75.90	50.00	71.82	75.51
Negative pred.	91.56	95.40	92.62	96.45	94.31	93.81	94.92

Table 3. Statistical analysis of scenario (ii) - results in percentage (%).

	Anger	Joy	Sadness	Surprise	Neutral
Sensitivity	75.50	77.85	13.80	68.75	80.09
Specificity	76.16	95.14	99.07	98.39	91.34
Positive pred.	45.06	81.46	66.67	88.51	80.44
Negative pred.	92.31	94.00	89.52	94.59	91.16

classified (with exceptions to Joy/Happy and Neutral). The opposite is observed to specificity. This indicates that the method does not have high accuracy to detect a certain class, however, does not predict incorrectly. The predictive values weighted using information about the class prevalence in population, show an overall increase of accuracy for true positive and maintain to negative. Therefore, as example to Surprise, despite our classifier only being able to positively identify surprise in 59.40% of the time there is a 71.82% chance that, when it does, such classification is correct. Looking to Table 3, compared to previous results of scenario (i) at Table 2, we observe an increase of sensitivity, while maintaining an high accuracy of specificity. In general, the same is observed in positive and negative predictive values. This is expected, since decreasing the number of classes of emotions will decrease the degree of confusion that lead to a better split between classes, resulting in a better emotion recognition method. These results confirm the statement of Background section, i.e. bottom face features provide incomplete information about face expression of emotions. Though, our method presents better performance when four universal emotions [11] are classified.

VR Facial Expressions Predictor. To analyze and validate the VR facial expressions predictor, we executed the k-Fold cross-validation to the classifier eyebrows "rising" and to classifier eyebrows "descending". Taking into account the variance of nose tracking with the type of HMD used, we propose to study the influence of tracking these features (subset $S1$) and not tracking the nose features (subset $S2$) in the prediction of eyebrows' movements. Average K-Fold CRM accuracies and respective confidence intervals can be accessed in the Table 4.

In the Table 4, we observe a small decrease of accuracy when the nose features tracking is removed. Although, the confidence intervals show that this decrease is only significant in eyebrows "up" detection. Our method allows an

Table 4. k-Fold CRM accuracy comparison facial expressions assessed (Eyebrows-Up or Down) with subset $S1$ and $S2$. Results in percentage (%).

Eyebrows movements	k-fold accuracy(%)	95% Confidence interval
Up $S1$	91.47	[89.76;92.98]
Up $S2$	87.02	[84.97;88.89]
Down $S1$	70.63	[67.99;73.18]
Down $S2$	69.13	[66.40;71.76]

high performance of eyebrows "up" estimation (at least, 85%) compared to eyebrows "down" estimation (at least, 66%). The different results arise from the fact that we are using an emotion database for training, where there is more data describing the "rising" movement than the opposite (i.e. only anger and sadness emotions usually present this facial expression behavior [8]).

Similarly to emotion recognition method, we present the statistical analysis of sensitivity/specificity and positive/negative predictive values to both eyebrows movements using the subsets $S1$ and $S2$ (Table 5).

Table 5. Eyebrow up prediction - statistical analysis to subsets $S1$. Results in percentage (%).

Eyebrows up	$S1$	$S2$
Sensitivity	97.34	96.27
Specificity	71.79	59.18
Positive pred.	92.04	87.65
Negative pred.	92.31	84.06

Both p-values of further analysis are lower than the significance level (i.e. p-value equal to $2.2 \times e^{-}16 < 0.05$). Therefore, both methods are suitable for eyebrows movement estimation using bottom face's movements. Table 4 shows that the method is able to classify the eyebrows "up" movement accurately, with exception for specificity using the subset $S2$. So, the removal of nose features tracking leads, essentially, to a decrease in accuracy of the classifier in not giving incorrect predictions. However, when we take in to account the prevalence of the class in population, the overall accuracy of prediction to both positive and negative values increase, presenting values above 84.04%.

Table 6 contains the statistical analysis to the prediction of eyebrows "descending" movement with ($S1$) and without ($S2$) nose features tracking.

Observing the Table 6, we observe that our method predicts correctly the "descending" movements of the eyebrows, at least, 73.18% of the time and does not predict incorrectly this movements in at least, 63.97% of the time. The lower values are obtained to the subset $S2$, however, the differences between

Table 6. Eyebrow down prediction - statistical analysis to subsets *S1*. Results in percentage (%).

Eyebrows down	S1	S2
Sensitivity	77.13	73.18
Specificity	62.73	63.97
Positive pred.	71.57	72.09
Negative pred.	69.28	65.23

subsets performance are not significant. Similar behavior is beheld taking into account the prevalence of the class in the population. The positive/negative predictive values are not significantly different between sensitivity/specificity. As expected by previous k-Fold CRM results, prediction of the "descending" movement presents lower performance compared to prediction of the opposite movement. Again, this result occurred due to the low prevalence of the "down" class in population. This statement is confirmed by the lower influence shown in positive and negative predictive values when compared to sensitivity and specificity, respectively.

Summarizing, our methods of facial expressions prediction are suitable for the estimation of eyebrows movements using features from the bottom of the face, specially in estimation of the "rising" movement. This conclusion corroborates the hypothesis of this work: our results traduce a connection between bottom and upper face behaviors.

VR Assessing Facial Expressions: Visual Validation. Applying the methods to videos where the participants expressed emotions [9], we are able to check visually the performance of the methods: occlusions support, emotion recognition and expressions prediction. We chose a non-VR scenario in order to verify if the upper face movements and emotions predicted (using only bottom face's movements) match the original facial expressions. Results can be observed in the Figs. 8, 9, 10 and 11

Looking throughout the Figures, we verify that our occlusion method is able to "re-create" the face even not using a HMC. Regarding emotion recognition using only the facial features (green' dots), in the Fig. 8, 9 and 10, we show three examples of correct classification. Figure 11 presents an example of a wrong emotion recognition. The classifier returned Anger when the user's emotion label of the video was Sad. This confusion is predicted since the bottom features inherent to Anger and Sad emotions are identical [9].

Regarding the facial expressions prediction method, in the Figs. 8 and 11 we observed that the algorithm correctly estimates eyebrows "down", which is confirmed by the original images. The same is detected in the Fig. 9 for eyebrows "up" predictor. Moreover, in the Fig. 10, comparing eyebrows of image analyzed and original image, we observe no movement, which traduced in a correct no estimation of movement from both predictors.

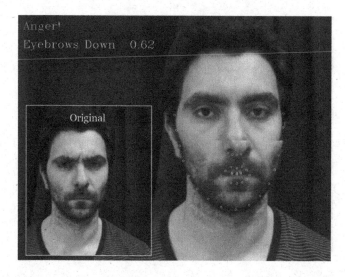

Fig. 8. VR assessing facial expressions: emotion recognition result (blue) and expression predictor result (green). Check that our emotion and prediction match original image eyebrows movements (green box). (Color figure online)

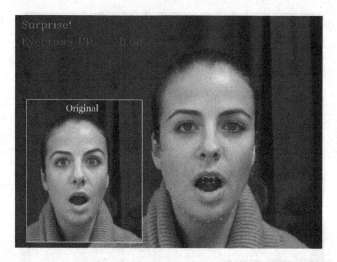

Fig. 9. VR assessing facial expressions: emotion recognition result (blue) and expression predictor result (red). Check that our emotion and prediction match original image eyebrows movements (green box). (Color figure online)

VR Assessing Facial Expressions: Facial Animation Use Case. This facial animation use case was added as proof of an easy integration of our methodology to trigger facial animation. The 3D character's rig was bone-based and, for simplicity, the association between the bones and the rig and the output from our VR MoCap method was done by directly associating each landmark

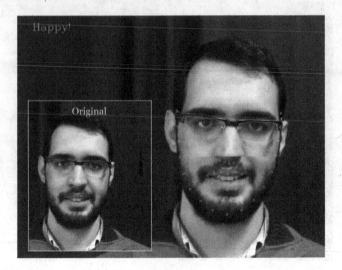

Fig. 10. VR assessing facial expressions: correct emotion recognition result (blue) and no expression predictor result, since there is not movement. Check original image in green box. (Color figure online)

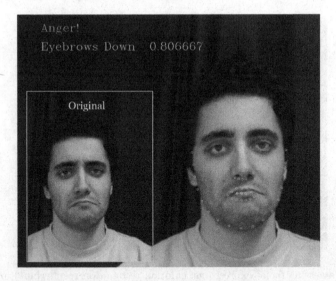

Fig. 11. VR assessing facial expressions: incorrect emotion recognition result (blue) and expression predictor result. Check original image to see that expression predictor is correct (green box). (Color figure online)

tracked to a certain bone in the rig. This approach was adopted to reduce the influence of a mapping algorithm that may mask the impact of our MoCap approach in animation. In detail, we associated directly the movement tracked in the bottom region of the face and predicted to the upper part of the face. To be able to evaluate the tracking of the upper part of the face, we adopted the same approach of Visual Validation subsection using non-occluded faces to be able to visualize the eyebrows. Results are shown in Figs. 12 and 13.

Figure 12 mirrors an example of correct prediction of eyebrows "down" and bottom region tracking and their correct reproduction in the 3D character. On opposite, the Fig. 13 shows an example of an eyebrows "up" behavior. Hence, to this use case our VR MoCap methods allowed the suitable creation of 3D character's facial animation in real-time.

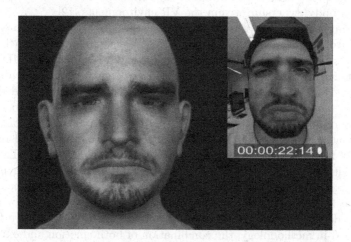

Fig. 12. Facial animation of 3D character (left) from face tracked using our VR MoCap method (right).

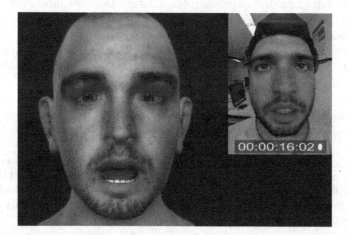

Fig. 13. Facial animation of 3D character (left) from face tracked using our VR MoCap method (right).

5 Conclusions

This work delivers VR consumer-level methods to achieve the three goals: make MoCap systems compatible to persistent partial occlusions, real-time recognition of universal emotions and real-time prediction of upper face movements using bottom face features tracking. Combining the three methods deployed, we are able to track in real-time facial expressions from non-occluded and occluded facial regions. The development of these methods lead to improvement in the three components of sense of embodiment, i.e. enhances the sense of self-location, agency and body ownership within the VR environments [12].

Analyzing the results, we conclude that the three goals proposed where achieved. We deliver a method to make MoCap systems able to track bottom face features under generic partial occlusions created by different HMD's. Note, we do not deliver a method that is able to overcome generic and unpredicted facial occlusions, since we require the knowledge of the area occluded. Then, using these facial features, we were able to define methodologies to real-time recognition of four universal emotions [11] with an accuracy of 69.07% and prediction of facial movements in the occluded regions, i.e. eyebrows "rising" with accuracy of 91.47% and "descending" with an accuracy of 70.63%. The results obtained with the facial expressions prediction method confirmed our method's hypothesis. Therefore, besides bottom features of the face being not enough to describe the six emotions of Ekman and Friesen [9], our predictor of facial expression decode a connection between bottom face and upper face features. As explained in methodology, the combination of both emotion and expressions tracked/predicted make us able to access a wide range of facial expressions enabling us to represent the diversity of faces [19]. This conclusion opens new lines of research to predict more complex movements of the face, even when we are not able to track them using CV algorithms. Furthermore, and as explored in our additional use case, our methods outputs enable the real-time animation of 3D characters. Ahead of 3D characters animation, our methods are suitable for emotion-based applications, like affective virtual environments, advertising or emotional gaming.

As further research, we aim to explore a more complex use case with more users and 3D character's rigs to statistically validate our methodology applied to 3D character's animation scenarios.

Acknowledgements. This work is supported by Instituto de Telecomunicações (Project Incentivo ref: Projeto Incentivo/EEI/LA0008/2014 and project UID ref: UID/EEA/5008/2013) and University of Porto. The authors would like to thanks Elena Kokkinara from Trinity College Dublin and Pedro Mendes from University of Porto for their support given in the beginning of the project.

References

1. OpenCV (2014)
2. Biocca, F.: The Cyborg's dilemma: progressive embodiment in virtual environments. J. Comput.-Mediated Commun. **3**(2) (1997)
3. Bombari, D., Schmid, P.C., Schmid Mast, M., Birri, S., Mast, F.W., Lobmaier, J.S.: Emotion recognition: the role of featural and configural face information. Q. J. Exp. Psychol. **66**(12), 2426–2442 (2013)
4. Breiman, L.: Random forests. Mach. Learn. **45**(1), 5–32 (2001)
5. Cao, C., Hou, Q., Zhou, K.: Displaced dynamic expression regression for real-time facial tracking and animation. ACM Trans. Graph. (TOG) **33**(4), 43 (2014)
6. Cao, C., Weng, Y., Lin, S., Zhou, K.: 3d shape regression for real-time facial animation. ACM Trans. Graph. **32**(4), 41 (2013)
7. Eisenbarth, H., Alpers, G.W.: Happy mouth and sad eyes: scanning emotional facial expressions. Emotion **11**(4), 860 (2011)
8. Ekman, P., Friesen, W.: Facial Action Coding System: A Technique for the Measurement of Facial Movement. Consulting Psychologists Press, Palo Alto (1978)
9. Ekman, P., Friesen, W.V.: Unmasking the Face: A Guide to Recognizing Emotions from Facial Cues. Malor Books, Cambridge (1975)
10. Fuentes, C.T., Runa, C., Blanco, X.A., Orvalho, V., Haggard, P.: Does my face fit?: a face image task reveals structure and distortions of facial feature representation. PLoS ONE **8**(10), e76805 (2013)
11. Jack, R.E., Jack, R.E.: Culture and facial expressions of emotion Culture and facial expressions of emotion. Vis. Cogn. 1–39 (2013)
12. Kilteni, K., Groten, R., Slater, M.: The sense of embodiment in virtual reality. Presence: Teleoperators Virtual Environ. **21**(4), 373–387 (2002)
13. Lang, C., Wachsmuth, S., Hanheide, M., Wersing, H.: Facial communicative signals. Int. J. Social Robot. **4**(3), 249–262 (2012)
14. Li, H., Trutoiu, L., Olszewski, K., Wei, L., Trutna, T., Hsieh, P.-L., Nicholls, A., Ma, C.: Facial performance sensing head-mounted display. ACM Trans. Graph. (Proceedings SIGGRAPH 2015) **34**(4), 47 (2015)
15. Li, H., Yu, J., Ye, Y., Bregler, C.: Realtime facial animation with on-the-fly correctives. ACM Trans. Graph. **32**(4) (2013)
16. Loconsole, C., Miranda, C.R., Augusto, G., Frisoli, G., Orvalho, V.C.: Real-time emotion recognition: a novel method for geometrical facial features extraction. In: 9th International Joint Conference on Computer Vision, Imaging and Computer Graphics Theory and Applications (VISAPP 2014), vol. 1, pp. 378–385 (2014)
17. Lucey, P., Cohn, J.F., Kanade, T., Saragih, J., Ambadar, Z., Matthews, I.: The extended cohn-kanade dataset (CK+): a complete dataset for action unit and emotion-specified expression. In: 2010 IEEE Computer Society Conference on Computer Vision and Pattern Recognition Workshops (CVPRW), pp. 94–101. IEEE (2010)
18. Magnenat-Thalmann, N., Primeau, E., Thalmann, D.: Abstract muscle action procedures for human face animation. Vis. Comput. **3**(5), 290–297 (1988)
19. McCloud, S.: Understanding Comics: The Invisible Art. Kitchen Sink Press, Northampton (1993)
20. McCloud, S.: Making Comics: Storytelling Secrets of Comics, Manga and Graphic Novels. William Morrow, William Morrow Paperbacks, New York (2006)
21. Pandzic, I.S., Forchheimer, R.: MPEG-4 Facial Animation: The Standard Implementation and Applications. Wiley, Hoboken (2003)

22. Parikh, R., Mathai, A., Parikh, S., Sekhar, G.C., Thomas, R.: Understanding and using sensitivity, specificity and predictive values. Indian J. Ophthalmol. **56**(1), 45 (2008)
23. Parke, F.I., Waters, K.: Computer Facial Animation, vol. 289. AK Peters Wellesley, Natick (1996)
24. Pighin, F., Lewis, J.: Performance-driven facial animation. In: ACM SIGGRAPH (2006)
25. R Core Team: R: A Language and Environment for Statistical Computing. R Foundation for Statistical Computing, Vienna (2013). ISBN 3-900051-07-0
26. Rodriguez, J., Perez, A., Lozano, J.: Sensitivity analysis of k-fold cross validation in prediction error estimation. IEEE Trans. Pattern Anal. Mach. Intell. **32**(3), 569–575 (2010)
27. Saragih, J.M., Lucey, S., Cohn, J.F.: Deformable model fitting by regularized landmark mean-shift. Int. J. Comput. Vis. **91**(2), 200–215 (2011)
28. Slater, M.: Grand challenges in virtual environments. Front. Robot. AI **1**, 3 (2014)
29. von der Pahlen, J., Jimenez, J., Danvoye, E., Debevec, P., Fyffe, G., Alexander, O.: Digital ira and beyond: creating real-time photoreal digital actors. In: ACM SIGGRAPH 2014 Courses, p. 1. ACM (2014)
30. Weise, T., Bouaziz, S., Li, H., Pauly, M.: Realtime performance-based facial animation. ACM Trans. Graph. (TOG) **30**(4), 77 (2011)

Ground-Truth Tracking Data Generation Using Rotating Real-World Objects

Zoltán Pusztai[1,2] and Levente Hajder[1,2(✉)]

[1] Distributed Events Analysis Research Laboratory, MTA SZTAKI,
Kende utca 13-17, Budapest 1111, Hungary
{pusztai.zoltan,hajder.levente}@sztaki.mta.hu
[2] Eötvös Loránd University, Budapest, Hungary
http://web.eee.sztaki.hu/

Abstract. Quantitative comparison of feature matchers/trackers is essential in 3D computer vision as the accuracy of spatial algorithms mainly depends on the quality of feature matching. This paper shows how a structured-light applying turntable-based evaluation system can be developed. The key problem here is the highly accurate calibration of scanner components. The ground truth (GT) tracking data generation is carried out for seven testing objects. It is shown how the OpenCV3 feature matchers can be compared on our GT data, and the obtained quantitative results are also discussed in detail.

1 Introduction

Developing a realistic 3D approach for feature tracker evaluation is very challenging since realistic moving 3D objects can simultaneously rotate and translate, moreover, occlusion can also appear in the images. It is not easy to implement a system that can generate ground truth (GT) data for real-world 3D objects. The aim of this paper is to present a novel structured-light reconstruction system that can produce highly accurate feature points of rotating spatial objects.

The Middlebury database[1] is considered as the state-of-the-art GT feature point generator. The database itself consists of several datasets that have been continuously developed since 2002. In the first period, they generated corresponding feature points of real-world objects on stereo images [37]. The first Middlebury dataset can be used for the comparison of feature matchers. Later on, this stereo database was extended with novel datasets using structured-light [38] or conditional random fields [31]. Even subpixel accuracy can be achieved in this way as it is discussed in [36].

However, our goal is to generate tracking data via multiple frames, the stereo setup is a very strict limitation for us. The description of the optical flow datasets of Middlebury database was published in [5]. It was developed in order to make

---

[1] http://vision.middlebury.edu/.

© Springer International Publishing AG 2017
J. Braz et al. (Eds.): VISIGRAPP 2016, CCIS 693, pp. 395–417, 2017.
DOI: 10.1007/978-3-319-64870-5_19

the optical flow methods comparable. The latest version contains four kinds of video sequences:

1. *Fluorescent Images*: Nonrigid motion is taken by a color and a UV-camera. Dense ground truth flow is obtained using hidden fluorescent texture painted on the scene. The scenes are slowly moved, at each point capturing separate test images in visible light, and GT images with trackable texture in UV light.
2. *Synthesized Database*: Realistic images are generated by an image syntheses method. The tracked data can be computed by this system as all parameters of the cameras and the 3D scene are known.
3. *Imagery for Frame Interpolation*. GT data is computed by interpolating the frames. Therefore the data is computed by a prediction from the measured frames.
4. *Stereo Images of Rigid Scenes*. Structured light scanning is applied first to obtain stereo reconstruction. (Scharstein and Szeliski 2003). The optical flow is computed from ground truth stereo data.

The main limitation of the Middlebury optical flow database is that the objects move approximately linearly, there is not any rotating objects in the datasets. This is a very strict limitation as tracking is a challenging task mainly when the same texture is seen from different viewpoint.

It is interesting that the Middlebury multi-view database [39] contains ground truth 3D reconstruction of two objects, however, the ground truth tracking data were not generated for these sequences. Another limitation of the dataset is that only two low-textured objects are used.

It is obvious that tracking data can also be generated by a depth camera [40] such as Microsoft Kinect, but its accuracy is very limited. There are other interesting GT generators for planar objects such as the work proposed in [13], however, we would like to obtain the tracked feature points of real spatial objects.

Due to these limitations, we decided to build a special hardware in order to generate ground truth data. Our approach is based on a turntable, a camera, and a projector. They are not too costly, nevertheless, the whole setup is extremely accurate as it is shown here.

Accurate Calibration of Turntable-Based 3D Scanners. The application of structured-light scanner is a relatively cheap and accurate solution to build a real 3D scanner as it is discussed in the latest work of [26]. Another possibility for a very accurate 3D reconstruction is laser scanning [8], however, the accurate calibration of the turntable is not possible using a laser stripe since it can only reconstruct a 2D curve at a moment. For turntable calibration, the reconstruction of 3D objects is a requirement since the axis of the rotation can be computed by registrating the point clouds of the same rotating object.

Moreover, the calibration of the camera and projector intrinsic and extrinsic parameters is also crucial. While the camera calibration can be accurately carried out by the well-known calibration method of [45], the projector calibration is a more challenging task. The projector itself can be considered as an inverse camera: while the camera projects the 3D world to the 2D image, the projector projects the planar image onto the 3D world. For this reason, the corresponding points of the 3D world and the projector image cannot be matched. Therefore, firstly the pixel-pixel correspondences have to be detected between the camera and the projector. The application of structured light was developed in order to efficiently realize this correspondence detection [38].

Many projector calibration methods exist in the field. The first popular class of existing solutions [21,35,44] is to (i) use a calibrated camera to determine the world coordinate, (ii) then a pattern is projected onto the calibration plane, the corners are detected and locations are estimated in 3D, (iii) the 3D → 2D correspondences are given by running the [45] calibration. The drawback of this kind of approaches is that its accuracy is relatively low since the projected 3D corner locations are estimated, and these estimated data are used for the final calculation.

Another possible solution is to ask the user to move the projector at different positions [3,16]. It is not possible for our approach as the projector is fixed. Moreover, the accuracy of these kind of approaches is also low.

There are algorithms where both projected and printed pattern are used [4,24]. The main idea here is that if the projected pattern is iteratively adjusted until it fully overlaps the printed pattern, then the projector parameters can be estimated. Color patterns can also be applied for this purpose [32]. However, we found that this quite complicated method is not required to calibrate the camera-projector system.

Our calibration methods for both the camera and projector use a simply chessboard plane. The proposed algorithms are very similar to those of [26]. As it is shown later, we calibrate the camera first by the method of [45]. Then the point correspondences between camera and projector pixels are determined by robustly estimating the local homography close to the chessboard corners. The intrinsic projector parameters can be computed by [45] as well. The extrinsic parameters (relative translation and orientation between the camera and the projector) is given by a stereo calibration problem. For this purpose, there are several solutions as it is discussed in [15] in detail. However, we found that the accuracy of stereo calibration is limited, therefore we propose a more sophisticated estimation here.

Contribution of This Study. The main novelty of this paper is that we show here how very accurate GT feature data can be generated for rotating object if a camera-projector system with a turntable is applied. To the best of our knowledge, our approach is the first system that can yield such accurate GT tracking data. The usage of a turntable for 3D reconstruction itself is not a novel idea, but its application for GT data generation is. The calibration algorithms within the system have a minor and a major improvements:

- The camera-projector correspondence estimation is based on a robust (RANSAC-based) homography estimation.
- The turntable calibration is totally new: while usual turntable calibrators [17] compute the axis by performing a usual chessboard-based calibration

method [45] for the rotating chessboard plane, and the axis of the rotation is computed from the extrinsic camera parameters, we propose a novel optimization algorithm that minimizes the reprojection for the corners of the rotating chessboard. We found that the accuracy of this novel algorithm is significantly higher. During the turnable calibration, the extrinsic parameters of the camera and projector are also obtained.

Another significant contribution of our study is that the feature matchers implemented in OpenCV3 are quantitatively compared on our GT data.

Structure of the Paper. First we deal with the calibration of the components: the camera, projector and turntable calibration is described one by one in Sects. 2.1, 2.2, and 2.3, respectively. Section 3 shows how accurate GT data can be generated by the developed equipment. Feature matchers of OpenCV3 are compared on this novel GT dataset. Finally, Sect. 4 concludes the work and discusses the limitations.

2   Proposed Equipment and Algorithms

Our 3D scanner consists of three main parts. It is visualized in Fig. 1. The left plot is the schematic setup, while the right one shows a snapshot of the camera when the object is illuminated by the structured light. The main components of the equipment are the camera, the projector, and the turntable. Each of the above needs to be calibrated correctly to reach high accuracy in 3D scanning. The camera and the projector are fixed to their arms, but the turntable can move[2]: it is able to rotate the object to be reconstructed.

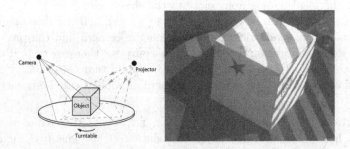

Fig. 1. Left: setup of structured light scanning. Right: camera snapshot of our realized scanner.

The bottleneck of the proposed approach is the calibration of the components. In this section, it is overviewed how the camera, the projector, and the axis of the rotating table can be accurately calibrated.

---

[2] These arms are can also move, but their calibration is not considered here, it is a possible future work.

2.1   Camera Calibration

For representing the camera, we choose the pinhole model with radial distortion. Assuming that the coordinate system is aligned to the camera, the projection of the point $X \in \mathbb{R}^3$ onto the camera plane is $\tilde{u} = [\tilde{u}_x, \tilde{u}_y, 1]^T \in \mathbb{R}^3$, which can be described by the following equation:

$$\tilde{u} \sim K_c X, \quad K_C = \begin{bmatrix} f_x & \gamma & p_x \\ 0 & f_y & p_y \\ 0 & 0 & 1 \end{bmatrix},$$

where K_C stands for the camera matrix, f_x and f_y are the focal length, (p_x, p_y) is the principal point, and γ is the shear, while the operator $\sim$ denotes equality up to a scale.

Furthermore, the standard radial distortion model [15] is also used to obtain the distorted 2D coordinates in vector $u \in \mathbb{R}^2$. In our case, we use only radial distortion which can be described by two parameters: (k_1, k_2). The applied distortion model here is as follows:

$$u = \begin{bmatrix} \tilde{u}_x[1 + k_1 r^2 + k_2 r^4] \\ \tilde{u}_y[1 + k_1 r^2 + k_2 r^4] \end{bmatrix}, \quad r^2 = \tilde{u}_x^2 + \tilde{u}_y^2.$$

The camera matrix and the distortion parameters are together called the intrinsic parameters of the camera.

A black and white chessboard is held in the field of view of the camera at arbitrary positions. Images were taken and the chessboard corners were found on the images, and they refined to reach sub-pixel precision. Then we can compute the intrinsic parameters of the camera by the method of [45].

2.2   Projector Calibration

Since the projector can be viewed as an inverse camera, it can be represented by the same model applied for the camera before. However, finding the right projector pixels, which through the chessboard corners are seen from the viewpoint of the projector is not so obvious. To overcome this problem, a structured light sequence is projected onto the scene. It precisely encodes the pixel coordinates in the projector image. For each scene point, the projected codes has to be decoded. From now, the chessboard has to be placed in a position that can be viewed from both the camera and projector.

The structured light we used for the calibration is based on the binary Gray code since it is the most accurate coding for structured light scanning as it is discussed in [38]. In addition, we project inverse images after every single one, meaning that every pixel on the images is reversed. But before structured light utilized, full black and white images are projected for easier object recognition, and for easier decoding of the structured light.

Since the resolution of our projector is 1024×768, the number of the projected images are 42 for each chessboard orientation. The projected sequence consists of 2 pure black and white images, 10 images for encoding the horizontal, and 10 for encoding the vertical coordinates of each projector pixel. Additionally, the inverse images have to be inserted into the sequence as well. These images are taken from one viewpoint, and they are called as the image set.

After all the images are taken, one can begin the decoding of the structured light. First of all, we calculate the direct and indirect intensity of the light, pixel by pixel for each image set. The full method is described in [28]. Then the minimum and maximum intensities are determined per pixel and the direct and indirect values are given by the equations as follows:

$$L_D = \frac{L_{max} - L_{min}}{1 - B}, \quad L_I = \frac{2(L_{min} - B * L_{max})}{1 - B^2},$$

where B is the amount of light emitted by a turned-off projector pixel. We needed to separate these two components from each other, because we are only interested in the direct intensities illuminated by the projector.

Then we need to classify the pixels on each image pair, consisting the image given by the structured light and its inverse. There are 3 clusters to classify into: (1) The pixel is illuminated on the first image. (2) The pixel is not illuminated on the first image. (3) Cannot be determined.

The classification rules are as follows:

– $L_D < M \implies$ the pixel is in the third class,
– $L_d > L_I \wedge P_1 > P_2 \implies$ the pixel is illuminated,
– $L_d > L_I \wedge P_1 < P_2 \implies$ the pixel is not illuminated,
– $P_1 < L_D \wedge P_2 > L_I \implies$ the pixel is not illuminated,
– $P_1 > L_I \wedge P_2 < L_I \implies$ the pixel is illuminated,
– Otherwise it cannot be determined.

The pixel intensity in the first and inverse images are denoted by P_1, and P_2, while M is a user-defined threshold: $M = 5$ is set in our approach.

For further reading about the classification, we recommend to read the study of [43].

Since the chessboard consists of alternating black and white squares, decoding near the chessboard corners can resolve errors. To avoid these errors, we calculate local homographies around the corners. We use 11 pixel-wide kernel window and every successfully decoded projector pixel is in consideration. For the homography estimation, a RANSAC-based [10] DLT homography estimator is applied in contrast to the work of [26] where robustification is not dealt with. We found that the accuracy is increased when RANSAC-scheme is applied. After the homography is computed among the camera and projector pixels, we use this homography to transform the camera pixels to the projector image. In this way we get the exact projector pixels we needed and we can use the method of [45] to calibrate the projector. Remark that the extrinsic projector calibration is refined later, but the intrinsic parameters are not.

2.3   Turntable Calibration

The aim of the turntable calibration is to compute the axis of the turntable. It is represented by a point and a direction. Therefore, the degree of freedom of a general axis estimation is four (2 DoFs: position of a plane; other 2 DoFs: direction).

Fortunately, the current problem is constrained. We know that the axis is perpendicular to the plane of the turntable. Thus, the direction is given, only the position should be calculated within the turntable plane.

The turntable is calibrated if we know the centerline around which the table is turning. Two methods was used to calculate this 3D line. First we place the chessboard on the turntable, and start rotating it. Images are taken between the rotations, and the extrinsic parameters can be computed for each image since the camera is already calibrated. This motion is equivalent with the motion of a fixed chessboard and a moving camera. The circle that the camera follows has the same centerline as the turntable. Thus fitting a circle to the camera points gives the centerline of the turntable [17].

However, we found that this method is not accurate enough. Therefore, we developed a novel algorithm that is overviewed in the rest of this section.

Problem Statement of Turntable Axis Calibration. Given a chessboard with known size, for which the corners can be easily detected by widely used pattern recognition algorithms, the goal is to estimate the axis of the turntable. This is part of the calibration of a complex structured-light 3D reconstruction system that consists of one camera, one projector, and one turntable. The latter one is driven by a stepping motor, the angle of the rotation can be very accurately set. The camera and projector intrinsic parameters are also known, in other words, they are calibrated.

The input data for axis calibration comes from detected chessboard corners. The chessboard is placed on the turntable. Then it is rotated and images are taken with different rotational axis. The corners are detected on all of these images. The chessboard is placed in a higher position on the turntable, but the new plane orientation is also parallel to the turntable. Then the chessboard is rotated, and the corners are detected again. The chessboard can be placed in arbitrary altitudes. We only use two different values, but the proposed calibration method can work with arbitrary number of positions.

If we consider the case when the planes of the chessboard and the turntable are parallel, the distance between them is h, then the chessboard corners can be written as

$$
\begin{bmatrix} X \\ Y \end{bmatrix} = \begin{bmatrix} \cos\alpha & -\sin\alpha \\ \sin\alpha & \cos\alpha \end{bmatrix} \begin{bmatrix} x - o_x \\ y - o_y \end{bmatrix} + \begin{bmatrix} o_x \\ o_y \end{bmatrix} = \\
\begin{bmatrix} x\cos\alpha - y\sin\alpha + o_x\left(1 - \cos\alpha\right) + o_y\sin\alpha \\ x\sin\alpha + y\cos\alpha - o_x\sin\alpha + o_y\left(1 - \cos\alpha\right) \end{bmatrix}, \tag{1}
$$

where α denotes the current angle of the rotation. Note that altitude h does not influence the relationship. Also remark that capital X and Y denote spatial coordinates, while their lowercase letters (x and y) are 2D coordinates in image space. Remark that the third spatial coordinate is not mentioned above as the world coordinate system is selected as the chessboard plane is $Z = 0$.

Proposed Algorithm. The proposed axis calibration consists of two main steps:

1. Determination of the axis center $[o_x, o_y]^T$ on chessboard plane, and
2. computation of the camera and projector extrinsic parameters.

Axis Center $[o_x, o_y]^T$ Estimation on Chessboard Plane. The goal of the axis center estimation is to calculate the location $[o_x, o_y]^T$. We propose an alternation-type method with two substeps:

(1) Homography-step. The plane-plane homography is estimated for each image. The 2D locations of the corners in the images are known. The 2D coordinates can be determined in the chessboard plane by Eq. 1. If the homogeneous coordinates are used, the relationship becomes

$$\begin{bmatrix} u \\ v \\ 1 \end{bmatrix} \sim H \begin{bmatrix} x\cos\alpha - y\sin\alpha + o_x\,(1 - \cos\alpha) + o_y\sin\alpha \\ x\sin\alpha + y\cos\alpha - o_x\sin\alpha + o_y\,(1 - \cos\alpha) \\ 1 \end{bmatrix}. \tag{2}$$

We apply the standard normalized direct linear transformation (normalized DLT) with a numerical refinement step [15] in order to estimate the homography. It minimizes the linearized version of Eq. 2:

$$E\,(\alpha, x, y, o_x, o_y) = E_1\,(\alpha, x, y, o_x, o_y) + E_2\,(\alpha, x, y, o_x, o_y),$$

where

$$\begin{aligned} E_1\,(\alpha, x, y, o_x, o_y) = {} & uh_{31}\,(x\cos\alpha - y\sin\alpha + o_x\,(1 - \cos\alpha) + o_y\sin\alpha) + \\ & uh_{32}\,(x\sin\alpha + y\cos\alpha - o_x\sin\alpha + o_y\,(1 - \cos\alpha)) + uh_{33} - \\ & h_{11}\,(x\cos\alpha - y\sin\alpha + o_x\,(1 - \cos\alpha) + o_y\sin\alpha) - \\ & h_{12}\,(x\sin\alpha + y\cos\alpha - o_x\sin\alpha + o_y\,(1 - \cos\alpha)) - h_{13}, \end{aligned}$$

and

$$\begin{aligned} E_2\,(\alpha, x, y, o_x, o_y) = {} & vh_{31}\,(x\cos\alpha - y\sin\alpha + o_x\,(1 - \cos\alpha) + o_y\sin\alpha) + \\ & vh_{32}\,(x\sin\alpha + y\cos\alpha - o_x\sin\alpha + o_y\,(1 - \cos\alpha)) + vh_{33} - \\ & h_{21}\,(x\cos\alpha - y\sin\alpha + o_x\,(1 - \cos\alpha) + o_y\sin\alpha) - \\ & h_{22}\,(x\sin\alpha + y\cos\alpha - o_x\sin\alpha + o_y\,(1 - \cos\alpha)) - h_{23}. \end{aligned}$$

This is a linear problem. The center and the scale of the applied coordinate system for the images can be arbitrary chosen. As it is discussed in [15], the mass center and quasi-uniform scale is the most accurate choice. The error function $E(\alpha, x, y, o_x, o_y)$ can be written for every chessboard corner point for every rotational angle. Therefore, the minimization problem is formulated as

$$\arg\min_{H} \sum_{i=1}^{G_x} \sum_{j=1}^{G_y} \sum_{k=1}^{N} E(\alpha_k, x_{i,\alpha}, y_{j,\alpha}, o_{x,\alpha}, o_{y,\alpha}).$$

where $a_k \in [0, 2\pi]$, $x_i \in [0, G_x]$, and $y_i \in [0, G_y]$, and G_x, G_y are the dimensions of chessboard corners, respectively. (Possible values for (x_i, y_j) are $(1,1),(1,2),(2,1),...$etc. This problem remains an over-constrained homogeneous linear one that can be optimally solved.

(2) Axis-step. Its goal is to estimate the axis location $[o_x, o_y]^T$. The above two equations are linear with respect to the center coordinates. Therefore, the equations form a homogeneous linear system of equations $A[o_x, o_y]^T = b$, where $A = \begin{bmatrix} a_{11} & a_{12} \\ a_{21} & a_{22} \end{bmatrix}$ and

$$a_{11} = h_{11} - h_{11}\cos\alpha - h_{12}\sin\alpha - u(h_{31} - h_{31}\cos\alpha - h_{32}\sin\alpha),$$
$$a_{12} = h_{11}\sin\alpha + h_{12} - h_{12}\cos\alpha - u(h_{31}\sin\alpha + h_{32} - h_{32}\cos\alpha),$$
$$a_{21} = h_{21} - h_{21}\cos\alpha - h_{22}\sin\alpha - v(h_{31} - h_{31}\cos\alpha - h_{32}\sin\alpha),$$
$$a_{22} = h_{21}\sin\alpha + h_{22} - h_{22}\cos\alpha - v(h_{31}\sin\alpha + h_{32} - h_{32}\cos\alpha).$$

Furthermore, $b = \begin{bmatrix} b_{11} - b_{12}, & b_{21} - b_{22} \end{bmatrix}^T$, where

$$b_{11} = h_{13} + h_{11}(x\cos\alpha - y\sin\alpha) + h_{12}(y\cos\alpha + x\sin\alpha),$$
$$b_{12} = u(h_{33} + h_{31}(x\cos\alpha - y\sin\alpha) + h_{32}(y\cos\alpha + x\sin\alpha),$$
$$b_{21} = h_{23} + h_{21}(x\cos\alpha - y\sin\alpha) + h_{22}(y\cos\alpha + x\sin\alpha),$$
$$b_{22} = v(h_{33} + h_{31}(x\cos\alpha - y\sin\alpha) + h_{32}(y\cos\alpha + x\sin\alpha)).$$

The above equations can be written for all corners of the chessboard for all rotated positions. Therefore, both the homography- and the axis-steps are extremely over-constrained, thus the parameters can be very accurately estimated. It is interesting that the homography and the axis location estimations are homogeneous, and inhomogeneous linear problems, respectively. They can also solved for the over-determined case using the Moore-Penrose pseudo-inverse of the coefficient matrix as it is well-known [7].

The two substeps have to be run one after the other. Both steps minimize the same algebraic error, therefore the method converges to the closest (local) minimum. Unfortunately, global optimum cannot be theoretically guaranteed. But we found that the algorithm converges to the correct solution. The speed of the convergence is relatively fast, 20–30 iterations are required to reach the minimum.

Parameter Initialization. The proposed alternation method requires initial values for o_x and o_y. It has been found that the algorithm is not too sensitive to the locations of the initial values. The center of the chessboard is an appropriate solution for o_x and o_y. Moreover, we have tried more sophisticated methods. If the camera centers are estimated by a Perspective n Point (PnP) algorithm such as [18], then the camera centers for the rotating sequence form a circle [17] as it is mentioned in the first part of this section. The center of this circle is also a good initial value. However, we found that the correct solution is reached as well if the initial center is an arbitrary point within the chessboard region.

Axis Center Estimation in the Global System. The first algorithm estimates the center of the axis in the coordinate system of the chessboard. But the chessboard are placed in different positions with different altitudes. The purpose of the algorithm discussed in this section is to place the rotated chessboard in the global coordinate system and to determine the extrinsic parameters (location and orientation) of the projector. The global system is fixed to the camera, therefore, the camera extrinsic parameters have not to be estimated.

In our calibration setup, only two chessboard sequences are taken. The extrinsic position can be easily determined. If the 3D coordinates of the plane are known, and the 2D locations are detected, then the estimation of the projective parameters is called the PnP problem. Mathematically, the PnP optimization can be written as

$$\arg\min_{R,t} \sum_{i=1}^{G_x} \sum_{j=1}^{G_y} \sum_{k=1}^{N} Rep\left(R, t, \begin{bmatrix} u_{i,\alpha} \\ v_{j,\alpha} \end{bmatrix}, \begin{bmatrix} x'_{i,\alpha} \\ h \end{bmatrix} \right)$$

where the definition of the function Rep is as follows:

$$Rep\left(R, t, \begin{bmatrix} u_i \\ v_j \end{bmatrix}, \begin{bmatrix} x'_i \\ y'_j \\ h \end{bmatrix} \right) = \left\| DeHom\left(R \begin{bmatrix} x'_i \\ y'_j \\ h \end{bmatrix} + t \right) - \begin{bmatrix} u_i \\ v_j \end{bmatrix} \right\|_2^2 .$$

The applied comma (') means that the origin of the coordinate system for chessboard corners are placed at $[o_x, o_y]^T$. Function $DeHom$ gives the dehomogeneous 2D vector of a spatial vector as $DeHom([X, Y, Z]^T) = [X/Z, Y/Z]^T$.

There are PnP methods that can cope with planar points. We used the EPnP [18] algorithm for our approach. At this point, the relative transformation between the chessboard planes and the camera can be calculated. They are denoted by $[R^1, t^1]$, and $[R^2, t^2]$. The altitude of the chessboard can be measured. Without loss of generalization, altitude of the first plane can be set to zero: $h_1 = 0$. (The simplest way is to set the first chessboard to the turntable. Then the altitude of the second chessboard can be easily measured with respect to the turntable.)

The final task is to calculate the relative angle $\Delta\alpha$ between the two rotating chessboard planes. The estimation of one parameter is relatively simple. We solve it by exhaustive search. The best value is given by the rotation for which the reprojection error of the PnP problem is minimal:

$$\arg\min_{R,t} \sum_{i=1}^{G_x} \sum_{j=1}^{G_y} \sum_{k=1}^{N} \left(Rep^1 + Rep^2\right)$$

where

$$Rep^1 = Rep\left(R^1, t^1, \begin{bmatrix} u_{i,\alpha_k} \\ v_{j,\alpha_k} \end{bmatrix}, \begin{bmatrix} x'_{i,\alpha_k} \\ y'_{j,\alpha_k} \\ 0 \end{bmatrix}\right), Rep^2 = Rep\left(R^2, t^2, \begin{bmatrix} u_{i,k} \\ v_{i,k} \end{bmatrix}, \begin{bmatrix} x'_{i,\alpha+\alpha_k} \\ y'_{j,\alpha+\alpha_k} \\ h \end{bmatrix}\right).$$

The upper index denotes the number of the chessboard. The relationship between the left and right terms are that the spatial points have to rotated with the same angle, but a fix angular offset $\Delta\alpha$ has to be added to each rotation for the second chessboard plane with respect to the first one. The impact of $\Delta\alpha$ for the second rotation matrix is written as follows:

$$R^2 = \begin{bmatrix} \cos\Delta\alpha & -\sin\Delta\alpha & 0 \\ \sin\Delta\alpha & \cos\Delta\alpha & 0 \\ 0 & 0 & 1 \end{bmatrix} R^1 \tag{3}$$

The minimization problem is also a PnP one, therefore it can be efficiently solved by the algorithm of [18]. The estimation of $\Delta\alpha$ is obtained by an exhaustive search.

Finally, the extrinsic parameters of the projector are computed by running the PnP algorithm again for the corners detected in the projector images. The obtained projector parameters have to be transformed by the inverse of the camera extrinsic parameters since our global coordinate system is fixed to the camera.

2.4   Object Reconstruction

The object reconstruction looks very similar to the projector calibration. In this case, an object is placed on the turntable instead of the chessboard. Structured light is projected onto it, images are taken, then the object is rotated. This procedure is repeated until the object returns to the starting position. Then we decode the projector pixels from the projected structured light in each image set. After it is done, we use the Hartley-Strum triangulation technique [14] for corresponding camera-projector pixels due to its accuracy to determine the object points from one viewpoint. We calculate these for each viewpoint, and then we can combine the point sets together, which results a 3D points set of the full object.

3   Comparison of Feature Matchers Implemented in OpenCV3

The main advantage of our method is that the whole GT data generation is totally automatic. Therefore, arbitrary number of objects can be reconstructed. We show here seven typical objects that have well trackable feature points. They are as follows:

Dinosaur. A typical computer vision study deals with the reconstruction of a dinosaurs as it is shown in several scientific papers, e.g [11]. It has a simple diffuse surface that is easy to reconstruct in 3D, hence the feature matching is possible. For this reason, a dino is inserted to our testing dataset. **Flacon.** The plastic holder is another smooth and diffuse surface. A well-textured label is fixed on the surface. **Plush Dog.** The tracking of the feature point of a soft toy is a challenging task as it does not have a flat surface. A plush dog is included into the testing database that is a real challenge for feature trackers. **Poster.** The next sequence of our dataset is a rotating poster taken from a motorcycle magazine. It is a relatively easy object for feature matchers since it is a well-textured plane. The pure efficiency of the trackers can be checked in this example due to two reasons: (i) there is no occlusion, and (ii) the GT feature tracking is equivalent to the determination of plane-plane homographies. **Cube.** This object is manually textured, its material is cardboard. The texturing is quite sparse, only a little part of the object area can be easily tracked. **Bag.** A canvas bag with beautiful and well-textured decoration is a good test example. It seems to be as easy as the Poster, however, its shape is more varied. **Books.** Different books are placed on the turntable. The book covers are well-textured.

During the test, the objects were rotated by the turntable, the difference of the degree of two subsequent positions was set to 3°. Our GT tracking data generator has two modes. (i) The first version regularly generates the feature points in the first image. The feature points are located across a regular grid in the valid region of the camera image. (ii) The points in the first image is determined by the applied feature generator.

Then the generated feature points were reconstructed in the first image using the structured light. These 3D reconstructed point coordinates were rotated around the turntable axis with the known rotating axis, and projected to the next image. This procedure was repeated for all the images of the test sequence. The 2D feature coordinates after projection give the final GT for quantitative feature tracker comparison.

Input images and the corresponding (reconstructed) 3D point clouds of all testing sequences except 'Poster' are visualized in Fig. 2. The 3D models are represented by colored point clouds, however, the color itself does not influence the reconstruction. It is only painted due to its spectacularity. Sequence 'Poster' is missing as it is a planar paper and its 3D model is not interesting.

The computed ground truth data for the sequence 'Poster' are pictured in Fig. 3. The first row shows the tracked points when they are selected across a grid. The second row of Fig. 3 consist of images with the tracked GT SIFT feature points (yellow dots).

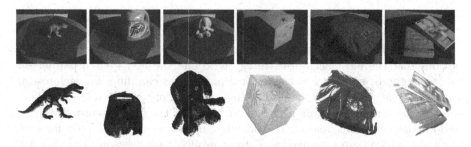

Fig. 2. Top row: camera snapshots for the input sequences. Bottom: reconstructed 3D point clouds. Test sequences from left to right: 'Dinosaur', 'Flacon', 'Plush dog', 'Cube', 'Bag', and 'Books'. (Color figure online)

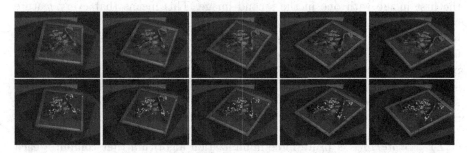

Fig. 3. The visualized ground truth tracking data drawn on images of the 'Poster' sequence. Top row: features generated by a grid within valid image region. Bottom row: features generated by SIFT method. The images are best viewed in color.

The obtained ground truth data were visually checked by us and we have not found any inaccuracy on it. We think that the accuracy is below pixel, in other word, subpixel accuracy was reached. This is extremely low as the camera resolution is 2592×1936 (5 Mpixel). **Compared Methods.** Firstly, the possibilities is overviewed that OpenCV can give about feature tracking. The currently supported feature detectors in OpenCV are as follows: AGAST [23], AKAZE [30], BRISK [19], FAST [33], GFTT [42] (Good Features To Track – also known as Shi-Tomasi corners), KAZE [2], MSER [25], ORB [34].

However, if one compiles the contrib(nonfree) repository with the OpenCV, the following detectors can also be used: SIFT [22], STAR [1], and SURF [6].

We use our scanner to take 20 images of a rotating object. After each image taken, a structured light sequence is projected in order to make the reconstruction available for every position. (Reconstructing the points only in the first image is not enough.)

Then we start searching for features in these images using all feature detectors. After the detection is completed, it is required to extract descriptors. Descriptors are needed for matching the feature points in different frames. The following descriptors are used (each can be found in OpenCV): AKAZE [30], BRISK [19], KAZE [2], ORB [34]. If one compiles the contrib repository, he/she can also get SIFT [22], SURF [6], BRIEF [9], FREAK [29], LATCH [20], DAISY [41] descriptors[3].

Another important issue is the parameterization of the feature trackers. It is obvious that the most accurate strategy is to find the best system parameters for the methods, nevertheless the optimal parameters can differ for each testing video. On the other hand, we think that the authors of the tested methods can set the parameters more accurately than us as they are interested in good performance. For this reason, the default parameter setting is used for each method, and we plan to make the dataset available for everyone and then the authors themselves can parameterize their methods.

After the detection and the extraction are done, the matching is started. Every image pair is taken into consideration, and match each feature point in the first image with one in the second image. This means that every feature point in the first image will have a pair in the second one. However, there can be some feature locations in the second image, which has more corresponding feature points in the first one, but it is also possible that there is no matching point.

The matching itself is done by calculating the minimum distances between the descriptor vectors. This distance is defined by the used feature tracking method. The following matchers are available in OpenCV:

- L_2 – BruteForce: a brute force minimization algorithm that computes each possible matches. The error is the L_2 norm of the difference between feature descriptors.
- L_1 – BruteForce: It is the same as L_2 – BruteForce, but L_1 norm is used instead of L_2 one.
- Hamming – BruteForce: For binary feature descriptor (BRISK, BRIEF, FREAK, LETCH,ORB,AKAZE), the Hamming distance is used.
- Hamming2 – BruteForce: It is a variant of the Hamming distance. The difference between Hamming and Hamming2 is that the former considers every bit as element of the vector, while Hamming2 use integer number, each bit pair forms a number from interval $0 \ldots 3$[4].
- Flann-Based: FLANN (Fast Library for Approximate Nearest Neighbors) is a set of algorithms optimized for fast nearest neighbor search in large datasets and for high dimensional features [27].

It is needed to point out that one can pair each feature detector with each feature descriptor but each feature matchers is not applicable for every descriptor. An exception is thrown by OpenCV if the selected algorithms cannot work together. But we try to evaluate every possible combination.

---

[3] The BRIEF descriptor is not invariant to rotation, however, we hold it in the set of testing algorithms as it surprisingly served good results.

[4] OpenCV's documentation is not very informative about Hamming2 distance. They suggest the usage of that for ORB features. However, it can be applied for other possible descriptors, therefore all possible combinations are tried in our tests.

The comparison of the feature tracker predictions with the ground truth data is as follows: The feature points are reconstructed first in 3D using the images and the structured light. Then, because it is known that the turntable was rotated by $3°$ per images, the projections of the points are calculated for all the remaining images. These projections were compared to the matched point locations of the feature trackers and the L_2 norm is used to calculate the distances.

Evaluation Methodology. The easiest and usual way for comparing the tracked feature points is to compute the summa and/or average and/or median of the 2D tracking errors in each image. This error is defined as the Euclidean distance of the tracked and GT locations. This methodology is visualized in the left plot of Fig. 4.

However, this comparison is not good enough because if a method fails to match correctly the feature points in an image pair, then the feature point moves to an incorrect location in the next image. Therefore, the tracker follows the incorrect location in the remaining frames and the new matching positions in those images will also be incorrect.

To avoid this effect, a new GT point is generated at the location of the matched point even if it is an incorrect matching. The GT location of that point can be determined in the remaining frames since that point can be reconstructed in 3D as well using the structured light scanning, and the novel positions of the new GT point can be determined using the calibration data of the test sequence.

Then the novel matching results are compared to all the previously determined GT points. The obtained error values are visualized in the right plot of Fig. 4.

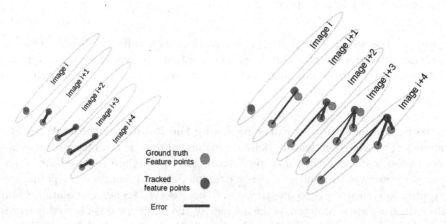

Fig. 4. Error measurement based on simple Euclidean distances (left). Applied, more sophisticated error measurement (right).

The error of a feature point for the i-th frame is the weighted average of all the errors calculated for that feature. For example, there is only one error value for the second frame as the matching error can only be compared to the GT location of the feature detected in the first image. For the third frame, there are two GT locations since GT error generated on both the first (original position) and second (position from first matching) image. For the i-th image, $i - 1$ error values are obtained. the error is calculated as the weighted average of those. It can be formalized as $Error_{p_i} = .\sum_{n=1}^{i-1} \frac{||p_i - p'_{i,n}||_2}{i-n}$, where $Error_{p_i}$ is the error for the i-th frame, p_i the location of the tested (detected) feature, while $p'_{i,n}$ is the GT location of the feature points reconstructed from the n-th frame. The weights of the distances is $1/(i - n)$ that means that older GT points has less weights. Remark that the Euclidean (L_2) norm is chosen in order to measure the pixel distances.

If a feature point is only detected in one image and was not being followed in the next one (or was filtered out in the fundamental-matrix-based filtering step), then that point is discarded. After the pixel errors are evaluated for each point in all possible images, the minimum, maximum, summa, average, and median error values of every feature points are calculated per image. The number of tracked feature points in the processed image is also counted. Furthermore, the average length of the feature tracks is calculated which shows in how many images an average feature point is tracked through.

3.1   Results

The purpose of this section is to show the main issues occurred during the testing of the feature matchers. Unfortunately, we cannot show to the Reader all the charts due to the lack of space.

General Remark. The charts in this section show different combinations of detectors, descriptors, and matchers. The method 'xxx:yyy:zzz' denotes in the charts that the current method uses the detector 'xxx', descriptor 'yyy', and matcher algorithm 'zzz'.

Feature Generation and Filtering Using the Fundamental Matrix. The number of the detected feature points is examined first. It is an important property of the matcher algorithms since many good points are required for a typical computer vision application. For example, at least hundreds of points are required to compute 3D reconstruction of the observed scene. The matched and filtered values are calculated as the average of the numbers of generated features for all the frames as features can be independently generated in each image of the test sequences. Table 1 shows the number of the generated features and that of the filtered ones. The filtering method is based on the standard normalized RANSAC-based robust eight-point fundamental matrix estimation method implemented in OpenCV. There are a few interesting behaviors within the data:

Table 1. Mean number of generated feature points (#Fea.) and that of inliers (#Inl.). Maximal values denoted by bold font.

Detector	Plush dog		Poster		Flacon		Dinosaur		Books		Cube		Bag	
	#Fea	#Inl	#Fea	#Inl	Fea	#Inl	#Fea	#Inl	#Fea	#Inl	#Fea	#Inl	#Fea	#Inl.
BRISK	21.7	16.9	233.6	188.8	219.7	161.0	21.6	14.8	144.2	110.2	9.75	5.8	8.3	5.2
FAST	19.7	9.4	224.8	139.2	387.0	275.4	51.1	27.0	490.3	305.2	28.8	17.9	61.0	34.6
GFTT	1000	38.2	956.7	618.8	1000	593.4	1000	92.0	1000	703.7	1000	65.5	903.3	232.0
KAZE	68.6	40.8	573.5	469.1	484.1	387.9	58.6	33.9	302.5	256.2	17	12.5	24.5	20.0
MSER	**5321**	10.6	**4864**	40.3	**3664**	31.7	**5144**	17.9	**5092**	85.6	**6062**	24.3	**4528**	41.4
ORB	42.3	34.1	259.5	230.8	337.7	287.4	67.1	45.9	253.7	206.2	28.5	23.2	52.7	41.3
SIFT	67.7	42.8	413.4	343.1	348.2	260.9	52.8	35.0	311.7	52.2	32.9	23.9	27.1	14.5
SURF	514.1	**326.0**	1877	**1578**	953.0	**726.8**	277.0	**132.6**	2048.7	**1463**	180.6	**105.4**	396.7	**261.5**
AGAST	22.5	11.8	275.8	200.3	410.2	303.5	55.0	29.9	591.3	390.5	33.1	22.7	65.8	38.2
AKAZE	144.0	101.7	815.0	761.4	655.0	553.1	89.1	59.2	282.1	260.9	18.4	15.3	23.0	20.2

- The best images for feature tracking are obtained when the poster is rotated. The runner up is the sequence 'Book'. The feature generators give significantly the most points in these cases when the scenes consist of large well textured planes. It is a more challenging task to find good feature points for the rotating non-planar objects such as the dog or dinosaur. It is because the area of these objects in the images are smaller than that a book or a poster. Another interesting behavior that only a few outliers are retrieved for the sequence 'Cube' due to the lack of large well-textured areas.
- It is clearly seen that number of SURF feature points are the highest in all test cases after outlier removal. This fact suggests that they will be the more accurate features.
- The MSER method gives the most number of feature points, however, more than 90% of those are filtered. Unfortunately, the OpenCV3 library does not contain   sophisticate matchers for MSER such as [12], therefore its accuracy is relatively low[5].
- Remark that the GFTT algorithm usually gives 1000 points as the maximum number was set to thousand for this method. It is a parameter of OpenCV that may be changed, but we did not modify this value.

Matching Accuracy. Two comparisons were carried out for the feature tracker methods. In the first test, every possible combination of the feature detectors and descriptors is examined, while the detectors are only combined with their own descriptor in the second test.

It is important to note that not only the errors of feature trackers should be compared, we must also pay attention to the number of features in the images and the feature track lengths[6]. A method with less detected features usually

---

[5] Many researchers have informed us that the OpenCV MSER implementation is not perfect.

[6] Feature track length is defined as the number of images on which the feature appears.

obtains better results (lower error rate) than other methods with higher number of features. The mostly used chart is the AVG-MED, where the average and the median of the errors are shown.

Testing by All Possible Algorithms. As it is seen in the left plot of Fig. 5 (sequence 'Plush Dog'), the SURF method dominates the chart. With the usage of SURF, DAISY, BRIEF, and BRISK descriptors more than 300 feature points remained and the median values of the errors are below 2.5 pixels, while the average is around 5 pixels. Moreover, the points are tracked through 4 images in average which yields pretty impressive statistics for the SURF detector.

The next test object was the 'Poster'. The results are visualized in the right plot of Fig. 5. It is interesting to note that if the trackers are sorted by the number of the outliers and plot the top 10 methods, only the AKAZE detector remains where more than 90% of the feature points was considered as inlier. Besides the high number of points, average pixel error is between 3 and 5 pixels depending on the descriptor and matcher type. In the test where the 'Flacon' object was used, we got similar results as in the case of 'Poster'. Both of the objects is rich in features, but the 'Flacon' is a spatial object. However, if we look at Fig. 6 where the methods with the lowest 10 median value were plotted,

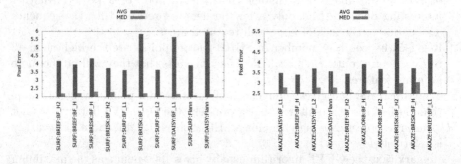

Fig. 5. Average and median errors of top 10 methods for sequences 'Plush Dog' (left) and 'Poster' (right).

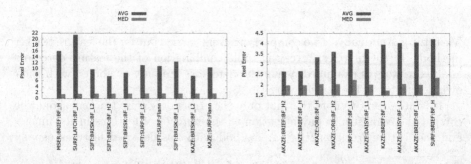

Fig. 6. Top 10 method with the lowest median for sequence 'Flacon'. Charts are sorted by median (left) and average (right) values.

one can see that KAZE and SIFT had more feature points and can track these over more pictures than MSER or SURF after the fundamental filtering. Even though they had the lowest median values, the average errors of these methods were rather high.

However, if one takes a look at the methods with the lowest average error, then he/she can observe that AKAZE, KAZE and SURF present in the top 10. These methods can track more points then the previous ones and the median errors are just around 2.0 pixels.

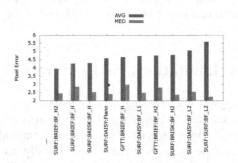

Fig. 7. Top 10 methods (with lowest average error) on sequence 'Dinosaur'.

For the sequence 'Dinosaur' (Fig. 7), the test object is very dark which makes feature detection hard. The number of available points is slightly more than 100. In this case, the overall winner of the methods is the SURF with both the lowest average and median errors. However, GFTT also present in the last chart too.

In the upper comparisons only the detectors were mentioned against each other. As one can see in the charts, most of the methods used either DAISY, BRIEF, BRISK or SURF descriptors. From the perspective of matchers, it does not really matter which type of the matcher is used for the same detector/descriptor type. However, if the descriptor gives a binary vector, then obviously the hamming distance outperforms the L2 or L1 ones. But there are just slightly differences between the L1–L2 and H1–H2 distances.

Testing of Algorithms with Same Detector and Descriptor. In this comparison, only the detectors that have an own descriptor are tested. Always the best matchers re selected for which the error are minimal for the observed detector/descriptor. As it can be seen in the log-scale charts in Fig. 8, the median error is almost the same for the AKAZE, KAZE, ORB and SURF trackers, but SURF is considered with the lowest average value. The tests 'Flacon' and 'Poster' result the lower pixel errors. On the other hand, the rotation of the 'Bag' was the hardest to track, it resulted much higher errors for all trackers comparing to the other tests. We think that this effect occurred because the scene contains huge amount of non-textured areas.

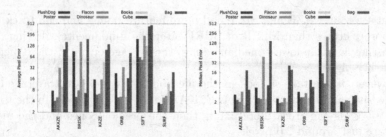

Fig. 8. Overall average (top) and median (bottom) error values for all trackers and test sequences. The detectors and descriptors were the same. Charts are best viewed in color.

4   Conclusions, Limitations, and Future Work

We have proposed a novel GT tracking data generator here that can automatically produce very accurate tracking data of rotating real-world spatial objects. The main novelty of our approach is that it consists of a turntable, and we showed how this turntable can be accurately calibrated. Finally, the validation of our equipment was shown on the quantitative comparison of OpenCV 3 matchers. It was justified that the proposed structured-light 3D scanner can produce accurate tracking data as well as realistic 3D point clouds. The GT tracking data are public, they are available at our web page[7].

The main goal of the approach proposed here is to be able to generate ground truth tracking data of real-world rotating objects. Therefore, the turntable-based equipment is unable to simulate moving cameras. However, other databases (e.g. the famous Middlebury one) can do that, thus our approach should be unified with existing datasets. Nevertheless, our equipment contains two moving arms for both the camera and projector, therefore novel viewpoints can be added to the system. It is possible if the arms are very accurately calibrated. This is a possible feature work of our GT generation project.

Acknowledgement. This work was partially supported by the Hungarian National Research, Development and Innovation Office under the grant VKSZ_14-1-2015-0072.

References

1. Agrawal, M., Konolige, K., Blas, M.R.: CenSurE: center surround extremas for realtime feature detection and matching. In: Forsyth, D., Torr, P., Zisserman, A. (eds.) ECCV 2008. LNCS, vol. 5305, pp. 102–115. Springer, Heidelberg (2008). doi:10.1007/978-3-540-88693-8_8
2. Alcantarilla, P.F., Bartoli, A., Davison, A.J.: KAZE features. In: Fitzgibbon, A., Lazebnik, S., Perona, P., Sato, Y., Schmid, C. (eds.) ECCV 2012. LNCS, vol. 7577, pp. 214–227. Springer, Heidelberg (2012). doi:10.1007/978-3-642-33783-3_16

---

[7] http://web.eee.sztaki.hu.

3. Anwar, H., Din, I., Park, K.: Projector calibration for 3D scanning using virtual target images. Int. J. Precis. Eng. Manuf. **13**(1), 125–131 (2012)
4. Audet, S., Okutomi, M.: A user-friendly method to geometrically calibrate projector-camera systems. In: Computer Vision and Pattern Recognition Workshops, pp. 47–54 (2009)
5. Baker, S., Scharstein, D., Lewis, J., Roth, S., Black, M., Szeliski, R.: A database and evaluation methodology for optical flow. Int. J. Comput. Vis. **92**(1), 1–31 (2011)
6. Bay, H., Ess, A., Tuytelaars, T., Van Gool, L.: Speeded-up robust features (surf). Comput. Vis. Image Underst. **110**(3), 346–359 (2008)
7. Björck, Å.: Numerical Methods for Least Squares Problems. SIAM, Philadelphia (1996)
8. Bradley, C., Vickers, G., Tlusty, J.: Automated rapid prototyping utilizing laser scanning and free-form machining. CIRP Ann. - Manuf. Technol. **41**(1), 437–440 (1991)
9. Calonder, M., Lepetit, V., Strecha, C., Fua, P.: BRIEF: binary robust independent elementary features. In: Daniilidis, K., Maragos, P., Paragios, N. (eds.) ECCV 2010. LNCS, vol. 6314, pp. 778–792. Springer, Heidelberg (2010). doi:10.1007/978-3-642-15561-1_56
10. Fischler, M., Bolles, R.: Random sampling consensus: a paradigm for model fitting with application to image analysis and automated cartography. Commun. Assoc. Comp. Mach. **24**, 358–367 (1981)
11. Fitzgibbon, A.W., Cross, G., Zisserman, A.: Automatic 3D model construction for turn-table sequences. In: Koch, R., Gool, L. (eds.) SMILE 1998. LNCS, vol. 1506, pp. 155–170. Springer, Heidelberg (1998). doi:10.1007/3-540-49437-5_11
12. Forssén, P.-E., Lowe, D.G.: Shape descriptors for maximally stable extremal regions. In: ICCV. IEEE (2007)
13. Gauglitz, S., Höllerer, T., Turk, M.: Evaluation of interest point detectors and feature descriptors for visual tracking. Int. J. Comput. Vis. **94**(3), 335–360 (2011)
14. Hartley, R.I., Sturm, P.: Triangulation. Comput. Vis. Image Underst.: CVIU **68**(2), 146–157 (1997)
15. Hartley, R.I., Zisserman, A.: Multiple View Geometry in Computer Vision. Cambridge University Press, Cambridge (2003)
16. Draréni, J., Roy, P.S.S.: Geometric video projector auto-calibration. In: Proceedings of the IEEE International Workshop on Projector-Camera Systems, pp. 39–46 (2009)
17. Kazo, C., Hajder, L.: High-quality structured-light scanning of 3D objects using turntable. In: IEEE 3rd International Conference on Cognitive Infocommunications (CogInfoCom), pp. 553–557 (2012)
18. Lepetit, V., Moreno-Noguer, F., Fua, P.: Epnp: an accurate o(n) solution to the pnp problem. Int. J. Comput. Vis. **81**(2), 155–166 (2009)
19. Leutenegger, S., Chli, M., Siegwart, R.Y.: BRISK: binary robust invariant scalable keypoints. In: Proceedings of the 2011 International Conference on Computer Vision, ICCV 2011, pp. 2548–2555 (2011)
20. Levi, G., Hassner, T.: LATCH: learned arrangements of three patch codes. CoRR (2015)
21. Liao, J., Cai, L.: A calibration method for uncoupling projector and camera of a structured light system. In: IEEE/ASME International Conference on Advanced Intelligent Mechatronics, pp. 770–774 (2008)

22. Lowe, D.G.: Object recognition from local scale-invariant features. In: Proceedings of the International Conference on Computer Vision, ICCV 1999, pp. 1150–1157 (1999)

23. Mair, E., Hager, G.D., Burschka, D., Suppa, M., Hirzinger, G.: Adaptive and generic corner detection based on the accelerated segment test. In Proceedings of the 11th European Conference on Computer Vision: Part II, pp. 183–196 (2010)

24. Martynov, I., Kamarainen, J.-K., Lensu, L.: Projector calibration by "Inverse Camera Calibration". In: Heyden, A., Kahl, F. (eds.) SCIA 2011. LNCS, vol. 6688, pp. 536–544. Springer, Heidelberg (2011). doi:10.1007/978-3-642-21227-7_50

25. Matas, J., Chum, O., Urban, M., Pajdla, T.: Robust wide baseline stereo from maximally stable extremal regions. In: Proceedings of BMVC, pp. 36.1–36.10 (2002)

26. Moreno, D., Taubin, G.: Simple, accurate, and robust projector-camera calibration. In: 2012 Second International Conference on 3D Imaging, Modeling, Processing, Visualization & Transmission, Zurich, Switzerland, 13–15 October 2012, pp. 464–471 (2012)

27. Muja, M., Lowe, D.G.: Fast approximate nearest neighbors with automatic algorithm configuration. In: VISAPP International Conference on Computer Vision Theory and Applications, pp. 331–340 (2009)

28. Nayar, S.K., Krishnan, G., Grossberg, M.D., Raskar, R.: Fast separation of direct and global components of a scene using high frequency illumination. ACM Trans. Graph. 25(3), 935–944 (2006)

29. Ortiz, R.: FREAK: fast retina keypoint. In: Proceedings of the 2012 IEEE Conference on Computer Vision and Pattern Recognition (CVPR), pp. 510–517 (2012)

30. Pablo Alcantarilla (Georgia Institute of Technology), Jesus Nuevo (TrueVision Solutions AU), A.B. Fast explicit diffusion for accelerated features in nonlinear scale spaces. In Proceedings of the British Machine Vision Conference. BMVA Press (2013)

31. Pal, C.J., Weinman, J.J., Tran, L.C., Scharstein, D.: On learning conditional random fields for stereo - exploring model structures and approximate inference. Int. J. Comput. Vis. 99(3), 319–337 (2012)

32. Park, S.-Y., Park, G.G.: Active calibration of camera-projector systems based on planar homography. In: ICPR, pp. 320–323 (2010)

33. Rosten, E., Drummond, T.: Fusing points and lines for high performance tracking. In: Internation Conference on Computer Vision, pp. 1508–1515 (2005)

34. Rublee, E., Rabaud, V., Konolige, K., Bradski, G.: ORB: an efficient alternative to sift or surf. In: International Conference on Computer Vision (2011)

35. Sadlo, F., Weyrich, T., Peikert, R., Gross, M.H.: A practical structured light acquisition system for point-based geometry and texture. In: 2005 Proceedings of Symposium on Point Based Graphics, Stony Brook, NY, USA, pp. 89–98 (2005)

36. Scharstein, D., Hirschmüller, H., Kitajima, Y., Krathwohl, G., Nesic, N., Wang, X., Westling, P.: High-resolution stereo datasets with subpixel-accurate ground truth. In Proceedings of Pattern Recognition - 36th German Conference, GCPR 2014, Münster, Germany, 2–5 September 2014, pp. 31–42 (2014)

37. Scharstein, D., Szeliski, R.: A taxonomy and evaluation of dense two-frame stereo correspondence algorithms. Int. J. Comput. Vis. 47, 7–42 (2002)

38. Scharstein, D., Szeliski, R.: High-accuracy stereo depth maps using structured light. In: CVPR, vol. 1, pp. 195–202 (2003)

39. Seitz, S.M., Curless, B., Diebel, J., Scharstein, D., Szeliski, R.: A comparison and evaluation of multi-view stereo reconstruction algorithms. 2006 IEEE Computer Society Conference on Computer Vision and Pattern Recognition (CVPR 2006), 17–22 June 2006, pp. 519–528, NY, USA, New York (2006)

40. Sturm, J., Engelhard, N., Endres, F., Burgard, W., Cremers, D.: A benchmark for the evaluation of RGB-D SLAM systems. In: Proceedings of the International Conference on Intelligent Robot Systems (IROS) (2012)
41. Tola, E., Lepetit, V., Fua, P.: DAISY: an efficient dense descriptor applied to wide baseline stereo. IEEE Trans. Pattern Anal. Mach. Intell. **32**(5), 815–830 (2010)
42. Tomasi, C., Shi, J.: Good features to track. In: IEEE Conference Computer Vision and Pattern Recognition, pp. 593–600 (1994)
43. Xu, Y., Aliaga, D.G.: Robust pixel classification for 3d modeling with structured light. In: Proceedings of the Graphics Interface 2007 Conference, 28–30 May 2007, pp. 233–240. Montreal, Canada (2007)
44. Yamauchi, K., Saito, H., Sato, Y.: Calibration of a structured light system by observing planar object from unknown viewpoints. In: 19th International Conference on Pattern Recognition, pp. 1–4 (2008)
45. Zhang, Z.: A flexible new technique for camera calibration. IEEE Trans. Pattern Anal. Mach. Intell. **22**(11), 1330–1334 (2000)

The Sliced Pineapple Grid Feature for Predicting Grasping Affordances

Mikkel Tang Thomsen[✉], Dirk Kraft, and Norbert Krüger

Maersk Mc-Kinney Moller Institute, University of Southern Denmark,
Campusvej 55, 5230 Odense M, Denmark
mtt@mmmi.sdu.dk

Abstract. The problem of grasping unknown objects utilising vision is addressed in this work by introducing a novel feature, the Sliced Pineapple Grid Feature (SPGF). The SPGF encode semi-local surfaces and allows for distinguishing structures such as "walls", "edges" and "rims". These structures are shown to be important when learning successful grasping affordance predictions. The SPGF feature is used in combination with two different grasp affordance learning methods and achieve grasp success-rates of up to 87% for a combined varied object set. For specific object classes within the object set, success-rates of up to 96% is achieved. The results also show how two different grasp types can complement each other and allow grasping of objects that are not graspable by one of the types.

Keywords: Computer vision · Robotics · Grasp affordance learning

1 Introduction

An important problem that is being addressed in computer vision and robotics is the ability for agents to interact in previously unseen environments. This is a challenging problem as the sheer amount of potential actions and objects is infeasible to model. A way to overcome this infeasibility is to introduce and learn generic structures in terms of visual features and action representations to be reused over multiple actions and over different objects. It is well known that such reuseabliliy is occurring in the human brain, where the cognitive vision system have a generic feature representation with features of different sizes and level of abstractions that can be used as it fits, see [1], also for a general overview of the human visual system from a computer-vision/machine learning perspective.

In this paper, we learn grasping affordances based on a novel semi-local shape-based descriptor named the Sliced Pineapple Grid Feature (SPGF). The descriptor is derived by k-means clustering [2] on radially organized surface patches with a defined centre surface patch (see Fig. 1b). The descriptor is able to represent both sides of a surface as well as non-existence of shape information. Both aspects are important when we want to code grasps as these are strong cues for potential grasp points.

© Springer International Publishing AG 2017
J. Braz et al. (Eds.): VISIGRAPP 2016, CCIS 693, pp. 418–438, 2017.
DOI: 10.1007/978-3-319-64870-5_20

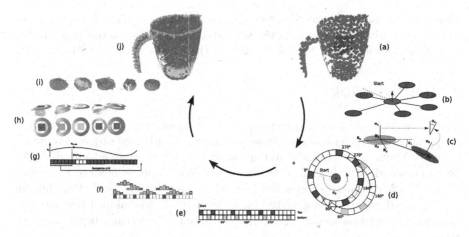

Fig. 1. Overview of the feature creation and learning process for the SPGF feature. (a) A scene represented by τ features with a selected one (red) and its neighbours (blue) within a radius, r. (b) Selected feature and its neighbours. (c) Spatial relations between the central feature (red) and its neighbours. (d) Organization of the neighbours in two circular grid structures introduced in the plane of the central feature, one grid describing features with the normal in the same direction as the central one and one with normals pointing in the opposite direction of the central one. (e) Unfolding of the circular grids. (f) Performing a weighted moving average filter to fill up empty cells in the grids. (g) Alignment of the grids to the direction of highest curvature. (h) k-means clustering in the grid feature space resulting in a finite set of learned SPGF features. (i) Features with associated grasp affordances. (j) Inferred features on object. (Color figure online)

By means of our descriptor and unsupervised learning, we are able to learn a discrete set of relevant semi-local descriptors that covers semantically distinct surface categories—which we call shape particles—such as "wall", "rim", "surface" (see Fig. 1h). In a second step, we associate grasp affordances to the shape particles by two different methods. Method 1, a probabilistic voting scheme introduced in [3] which results in shape-grasp particles. These shape-grasp particles allow us to probabilistically code the success likelihood of grasps in relation to the shape particles (see Fig. 1i). Method 2, where grasp signatures of SPGF features are extracted with respect to grasps and used to learn a logistic regression model.

We evaluate our system on an object set covering three categories in a simulation environment. We show that we are able to reliable predict grasps with a success-rate of up to 96% for individual object classes and 87% for the full object set, when utilising two complimentary grasp types namely a narrow- and wide two finger pinch grasp.

The paper is organised as follows: In Sect. 2, we relate our work to state-of-the-art in terms of feature- and model-based grasping of novel objects. Next in Sect. 3, we introduce the SPGF shape descriptor and relate it to two grasp affordance methods. We present the acquired results in Sect. 5 both quantitatively and qualitatively based on the simulated experimental setup presented in Sect. 4. In the conclusion, Sect. 6, we discuss the results as well as future work.

2   Related Work

Within vision based robotic grasping of unknown objects, two approaches are prevalent. First, model based approaches where the unknown object is approximated by simple shapes like bounding boxes (e.g. [4]) or more advanced shapes like super-quadratics (e.g. [5]). Other model based approaches are for example in [6] where object shapes are learned as prototypical parts that human demonstrated grasps are associated to. In work by [7] a combined contact- and hand-model based on visual appearance is learned for selecting successful grasp poses.

The other main branch in vision based grasping are feature based approaches, where visual features are either used as cues or in a combined way used as input for making grasp predictions. [8] showed how features from 2D images could be used to find reliable grasp points for a dishwasher emptying scenario. In work by [9], it was shown how simple surface and edge features could be used for predicting grasps with a reasonable probability of success. In [3], visually triggered action affordances were learned by associating related pairs of small surface patches with successful grasping actions. Another feature-based approach is proposed in work by [10]. Here the visual feature representations were learned unsupervised using deep learning techniques as a preliminary step towards grasp learning. This work is of particular interest as it showed superior performance compared to a previous paper with fundamentally the same grasp learning approach but where the visual features were hand selected [11]. Other approaches that utilise deep learning techniques for unsupervised feature learning and later for grasp selection are work by [12] where AlexNet [13] has been adopted to use RGB-D data as input. In work by [14], Superpixel Hierarchical Matching Pursuit has been proposed and used to learn geometric visual features on RGB-D data on which tool affordance learning has been applied. For an extensive review of the work performed in the robotic grasping domain see [15].

In our work, we propose a novel semi-local shape descriptor, SPGF, aimed at grasp affordance learning. The SPGF feature allows for encoding of semantically rich local surface structures, including gaps and walls that can be found in multiview or SLAM [16] acquired scenes. When utilised for grasp affordance learning and prediction on previously unseen objects, the learned features demonstrate high performance. As the feature types are learned in an unsupervised way using k-means clustering, they are not strictly bound to the grasping actions and can therefore be utilised for different actions although this utility is only weakly exploited in this paper.

3   Method

The aim of the proposed SPGF feature is to provide a solid foundation for reliable grasp affordance learning and prediction. To achieve this, a number of desirable properties have been identified, that should be captured by the feature:

1. Encoding of local shape geometry in SE(3).
2. Encoding of double sided structures.
3. Encoding of gaps.
4. Rotation invariance.

An overview of the process is shown in Fig. 1, where the steps from object (Fig. 1a) to clustered reference features (Fig. 1h), denoted prototypes, to grasp association (Fig. 1i) can be followed. In the following subsections, first the feature creation process will be explained (Fig. 1a–g and Sect. 3.1), next the learning process of extracting a small finite set of descriptive feature prototypes will be introduced (Fig. 1h and Sect. 3.2). In Sect. 3.3, the feature inference process, that allows for using the prototypes on novel situations will be addressed (Fig. 1j) and finally in Sect. 3.4, the learned features are linked to grasping poses and grasp affordance learning (Fig. 1i).

As a starting point and input to the feature learning system, a set of scenes, represented by small surface patch descriptors (concretely texlets3D [17]) are used, see Fig. 1a for an example. As a general notation the base features are described by a position $\mathbf{X}$ and a surface normal vector $\mathbf{n}$:

$$\tau = \{\mathbf{X}, \mathbf{n}\} = \{x, y, z, n_x, n_y, n_z\}, \quad |\mathbf{n}| = 1 \tag{1}$$

3.1   Feature Creation

We start out with a scene representation consisting of the above mentioned surfaces features (τ) and for each surface feature we follow the steps sketched below:

1. Find all the neighbours, within an Euclidean radius r, see Fig. 1a, b. This leads to a context-dependent number of neighbours (J).
2. For each of the J neighbours compute pairwise spatial relations between the centre feature (red) and the neighbour (blue). This will result in J pairwise relations, see Fig. 1c.
3. Split the neighbours into two sets based on the relation of their normals with respect to the centre feature; surface patches oriented in the same direction make up one set of relations (r_t), the others the second (r_b). Order the neighbours into two circular discretized grid structures based on the rotation around the normal of the centre feature, see Fig. 1d and e.
4. Fill out empty grid locations by applying a weighted moving averaging filter over the grid structures to combat sampling artefacts, see Fig. 1f.

5. To achieve rotation invariance, the start point is moved simultaneously for the two grids to the grid-cell of highest curvature. In addition, the 6D pose of the centre feature is corrected to align one of the in plane axis with that direction, see Fig. 1g.
6. Finally, the top and bottom layer grids are concatenated into a feature vector, f, consisting of the aligned 6D pose of the centre surface feature and all the sorted relational values (r'_t, r'_b).

In the following subsections specific details are given for the spatial relationships (step 2, Sect. 3.1) and the grid organizing (steps 3–6, Sect. 3.1) procedures.

Spatial Relationship. The relational descriptor used in this work is based on a set of pairwise relations between surface patch features of the type described in Eq. 1. The pairwise relations resembles the ones proposed in [18,19]. The three different angular relations $(\alpha_1, \alpha_2, \alpha_3)$ are visualised in Fig. 2 and described by Eqs. 2–4.

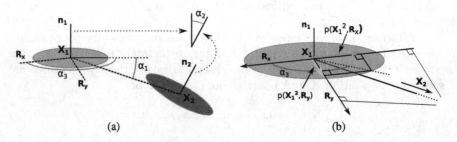

(a)   (b)

Fig. 2. (a) The pairwise spatial relationships between the centre feature (feature one) and feature two utilised in this work. α_1 depict the angle between the plane of the centre feature (red) and the vector connecting the positions of the features $(\mathbf{X}_1^2)$, α_2 depicts the angle between the surface normals of the two features, α_3 depicts the rotation angle around the normal of the centre feature $(\mathbf{n_1})$ from a reference direction, $\mathbf{R_x}$, on the plane to the projection of the connecting vector onto the plane. (b) A detailed view on how the angle α_3 is derived from the projection of the vector $\mathbf{X}_1^2$ onto the reference directions $\mathbf{R_x}$ and $\mathbf{R_y}$ (green and yellow arrows), from which α_3 can be derived. (Color figure online)

Before defining these angles a reference coordinate system $(\mathbf{R_x}, \mathbf{R_y}, \mathbf{n_1})$ for the centre patch needs to be defined. We choose the patch's normal $(\mathbf{n_1})$ as z-axis. $\mathbf{R_x}$ is chosen to be an arbitrarily direction in the plane of the first surface patch. The final axis $(\mathbf{R_y})$ can then be computed using the cross product of $\mathbf{n_1}$ and $\mathbf{R_x}$. Furthermore the vector connecting the two features $(\mathbf{X}_1^2 = \mathbf{X_1} - \mathbf{X_2})$ and the projection of this vector on an axis $(p(\mathbf{X}_1^2, \mathbf{R_x}) = \frac{(\mathbf{R_x} \cdot \mathbf{X}_1^2)}{||\mathbf{X}_1^2||})$ are needed to then define the angles between two features (τ_1, τ_2) as follows:

$$\alpha_1 = \frac{\pi}{2} - acos(\frac{\mathbf{X}_1^2}{||\mathbf{X}_1^2||} \cdot \mathbf{n_1}) \tag{2}$$

$$\alpha_2 = acos(\mathbf{n_1} \cdot \mathbf{n_2}) \tag{3}$$

$$\alpha_3 = atan2(p(\mathbf{X}_1^2, \mathbf{R_x}), p(\mathbf{X}_1^2, \mathbf{R_y})) \tag{4}$$

Together with the organisation in a grid structure, presented next, these measures are used to form the relational descriptor.

Organizing the Neighbourhood in a Sliced Pineapple Grid Structure. The next step is to organize the information about the neighbouring features into two circular grid structures. The two grid structures represent respectively a top layer (r_t) and a bottom layer (r_b). The top layer grid describes the neighbourhood of features with the normal in the same direction ($\alpha_2 < 90°$) as the centre feature and the bottom layer describes features with the normal in the opposite direction of the centre feature ($\alpha_2 > 90°$).

Both layers are then discretized into a circular grid, see Fig. 3(a), with a resolution of N_g. All neighbours are projected into the plane of the centre feature (see Fig. 3(c)) and their (α_1, α_2) are then placed into the rotational bin that corresponds to the specific α_3 value (see Fig. 3(d)). If multiple neighbours fall within the same cell, the average value of the relations are used. These steps are performed for both, the top layer and the bottom layer, resulting in two circular grids representing the neighbourhood of a feature. The grids are then unfolded to two flat grids.

Weighted Moving Average. Depending on the number of neighbours within the radius, r, and the discretization of the grid, the grid structures will consist of a substantial amount of cells where no data is found. These undefined cells are considered to be of two types.

1. They are a result of the general low density of the underlying feature representation within a small radius.
2. They are real gaps depicting a direction where no visual data exist.

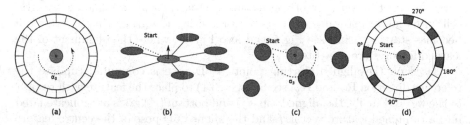

(a)   (b)   (c)   (d)

Fig. 3. Introduction of the grid organization for a single layer ($\alpha_2 > 90°$ or $\alpha_2 < 90°$). (a) The proposed grid structure discretized into N_g cells. (b) An example with six neighbours with an arbitrary start direction. (c) Top view of the projection of the six neighbour positions in the plane of the red feature. (d) The six neighbours organised in the proposed discretized grid structure based on their angle around the normal of the red feature, α_3. (Color figure online)

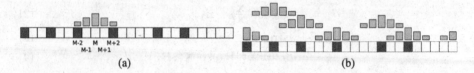

Fig. 4. Weighted moving average of the grid structure. (a) Illustrates how the cell at M is equally contributed to by the cell at M+2 and M−2, practically resulting in the average of the two. (b) A practical illustration of how the cells with datapoints (coloured ones) affect the surrounding cells with the weight depicted by the size of the bars. (Color figure online)

The second type is of specific interest, as gaps can be a strong visual cue (e.g., for affording pinch grasps or indicating open structures), whereas the first type is to be avoided. To address this artefact of sampling, a weighted moving average (WMA) is performed on the feature vectors to fill gaps. In addition, the WMA improves the robustness of the feature representation.

In Fig. 4, the principle is illustrated by showing the contribution that the existing datapoints give to neighbouring cells in the grid, hereby filling out small gaps in the representation. The WMA for the cells is performed for the two relations (α_1, α_2) independently. It should be noted, that if no value exists in a cell, it will not contribute to the average. The length of the filter n, determines the amount of smoothing.

After the moving average operation, gaps can still exist depending on the length of the averaging filter. These gaps are considered to be the "real" gaps. To encode the real gaps in a meaningful way that can be handled by the vector based k-means clustering, they are described with saturated values of the parameters. For α_1 this means −90° and for α_2 this means 90° and 180° for the top and bottom grid respectively.

Alignment of Grid. As a final step, the feature grid is aligned to make the representation rotation invariant. The selected alignment is at the place of highest curvature on the top part of the feature vector. This equals finding the grid cell with the largest value of α_2 and reorganising the grid such that this cell becomes starting point, see Fig. 5 and also Fig. 11, where the alignment of the learned features are presented.

Based on the aligned starting point, a 6D pose is computed (update the reference direction $\mathbf{R_x}$ and $\mathbf{R_y}$, see Subsect. 3.1) to place the feature with respect to the world. Finally, the aligned top (r_t') and bottom (r_b') grids are concatenated into a combined feature vector f and the aligned 6D pose of the centre feature is added.

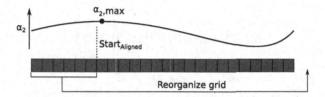

Fig. 5. Alignment of the feature grid starting point based on the maximum value of α_2, $\alpha_{2,max}$, of the top layer grid. The variation of α_2 over the grid is visualised above the grid. Both top layer and bottom layer are aligned to this new starting point and the grids are reorganized accordingly.

3.2 Feature Learning

Once this representation is designed, we are interested in finding reference features, called prototypes, that are able to describe the features present in the training object scenes in the best way. For that the following two steps are taken.

1. Perform the feature creation steps, see Sect. 3.1, for the full object set, resulting in a large set of feature vectors, $f_i : i = 1, ..., L$
2. Perform a k-means Euclidean clustering [2] in the relational space of the f's (only the r'_t, r'_b parts, not the position or orientation of the centre). K defines the number of prototypes (P_{id}) that the learned dictionary, ($\mathbb{P}$) Eq. 6, should consist of. The prototypes are described by a point in the relational-vector space, see Eq. 5 and Fig. 1h.

$$P_{id} = \{\alpha^0_{1,id}, \alpha^0_{2,id}, \alpha^1_{1,id}, \alpha^1_{2,id}, \alpha^{2N_g-1}_{1,id}, \alpha^{2N_g-1}_{2,id}\} \in \mathbb{R}^{2 \cdot 2N_g} \tag{5}$$

$$\mathbb{P} = \{P_1, P_2, ..., P_K\} \tag{6}$$

3.3 Feature Inference

Provided a set of feature prototypes has been learned, the inference process of utilising them to describe novel objects is introduced with the following steps:

1. Perform the feature creation process, see Sect. 3.1, for an object where features are to be inferred. This results in a set of feature vectors $\mathbb{F}$.
2. For every $f \in \mathbb{F}$, find the closest prototypes P_{id} in the set of learned prototypes $\mathbb{P}$ using the Euclidean distance on the relation part (only r'_t, r'_b) of f, see Eq. 7. Given this id, a new feature T is created that consists of the feature pose (position $\mathbf{X}$ and quaternion $\mathbf{q}$, given by f) and the computed id. See also Fig. 1j for an example object with inferred features.

$$D(f, P_{id}) = \sqrt{\sum_{i=0}^{2N_g-1} (\alpha^i_{1,f} - \alpha^i_{1,id})^2 + (\alpha^i_{2,f} - \alpha^i_{2,id})^2} \tag{7}$$

$$id = \operatorname*{argmin}_{id}(D(f, P_{id}) : P_{id} \in \mathbb{P}) \tag{8}$$

$$T = \{\mathbf{X}, \mathbf{q}, id\} \tag{9}$$

3.4   Grasping Affordance Learning

In order to utilize the learned SPGF features for grasp affordance learning (Fig. 1i), grasping poses are related to the visual features. Two methods are presented for relating grasps and visual features, a shape-grasp particle approach and a grasp signature based approach utilising additional spatial structure.

The basis for learning the grasp affordances is simulated grasping representations consisting of grasps G described by a pose with respect to the world coordinate system (position vector $\mathbf{X}$ and orientation vector $\mathbf{q}$ represented by a unit quaternion) and an outcome s whether the grasp was successful or not, see Eq. 10.

$$G = \{\mathbf{X}, \mathbf{q}, s\} \tag{10}$$

The second input for the grasp affordance learning is objects represented by inferred SPGF features T. The SPGF feature's poses are defined with respect to the same world coordinate system as the grasps.

Method 1 (Shape Grasp Particles). In order to utilize the learned SPGF features for grasp affordance learning (Fig. 1i), the pose of the visual feature is linked to the pose of a grasp by a 6D pose transformation, see Fig. 6. We use the method introduced in [3] and give here a brief overview of the method.

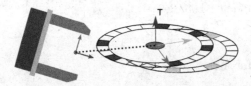

Fig. 6. Linking of a grasping action and a visual feature, T, described by the 6D pose transformation, a grasp outcome (success/failure) and an id (relating to the corresponding P_{id}).

The shape-grasp space is occupied by a large set of shape-grasp particles describing how individual actions relate to individual features. To condense the information for reliable action predictions, a neighbourhood analysis is performed to estimate the success probability of a given point in the space as well as the amount of similar points (called support). Next, the learned shape-grasp particle space is condensed to only consist of points that have a significant support. This knowledge is then used to vote for grasps in novel situations with the probability associated during learning. This results in a set of grasps, that can be ranked based on the predicted success probability and the amount of votes.

Method 2 (Grasp Signature). The second method takes a different more pragmatic approach towards grasping affordance learning. The approach relies on histogram signatures that map grasps and visual features to outcomes. This information is then used to train a logistic regression model. The basis for creating grasp signatures are objects represented by grasps G and features T. Given these two representations, a grasp signature is computed for each of the grasps associated to an object. The proposed grasp signature is explained below.

Grasp Signature. The grasp signature approach is based on an organized spatial structure consisting of 19 cubes aligned to the pose of the manipulator as shown in Fig. 7. The centre of the cube structure is positioned in the centre of the gripper grasping point, which is defined to be between the two finger of the gripper. The cube structure is defined by a distance d_c (default $= 0.033$ m) that depict the length of the sides of the individual small cubes.

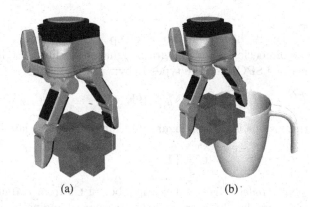

(a)   (b)

Fig. 7. Aligned structure of 19 cubes to the pose of a two finger gripper.

Given a grasp G, with respect to an object represented by SPFG features T, the grasp signature can be computed by the following steps, see also Fig. 8 for an overview:

1. Find all the SPGF features within a positional Euclidean radius r_G of the grasp G. The radius is set such that the all points on the combined cube structure is inside:
$$r_G = \sqrt{2(1.5d_c)^2 + (0.5d_c)^2} \tag{11}$$

2. Recompute the pose of the found SPGF features such that they are defined with respect to the coordinate system of the grasp rather than the world coordinate system.
3. Collect the SPGF features that fall within each of the 19 cubes (Fig. 8b).
4. For each of the 19 cubes, compute a frequency histogram of the number of different types of SPGF features (Fig. 8c) and normalize, such that the sum of the histogram bins equals 1.0.

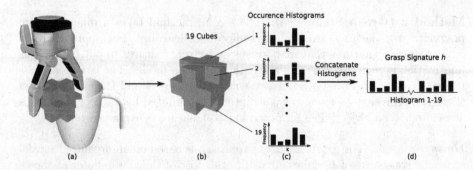

Fig. 8. Extraction of grasp signatures by means of a cube organization and histograms. (a) A grasp pose with respect to an object and the associated cube structure. (b) The 19 cubes cube structure. (c) From each of the 19 cubes, the frequency of found SPGF features is counted and a histogram is made. (d) The histogram values from all 19 cubes are concatenated into a grasp signature vector **h**.

5. Concatenate the 19 histograms to a grasp signature vector denoted **h** (Fig. 8d). The dimension N_h of the grasp signature is directly related to the number of learned SPGF feature types K by:

$$N_h = 19K \tag{12}$$

The result of these steps is a grasp signature that can be associated to a grasp and form:

$$\hat{G} = \{\mathbf{X}, \mathbf{q}, s, \mathbf{h}\} \tag{13}$$

Logistic Regression Model. Given a large number of $\hat{G}$s from training objects, a logistic regression model [20] is trained that learns a mapping from the grasp signatures **h** to the outcomes s. Given the trained model, the probability for a new grasp to be successful can be predicted using the grasp signature. As opposed to the resulting predictions of method 1, the logistic regression model both classifies whether a grasp signature suggest a successful or a failed grasps and also gives an associated probability.

4   Experiments

The experiments performed in this work are based on a simulated set-up created in RobWorkSim [21], see Fig. 9. The set-up consists of simulated RGB-D sensors and the object of interest in the centre, from which a visual object representation is acquired, see Figs. 9a and 1a for an extracted scene. It should be noted that modelled sensor noise is added and that the four views still lead to incomplete models that the four views introduce. In addition to the scene rendering, the simulation environment is also utilised for performing grasping simulations by means of the simulation framework RobWorkSim. The simulator is based on a dynamics engine that simulates object dynamics in terms of the contact forces

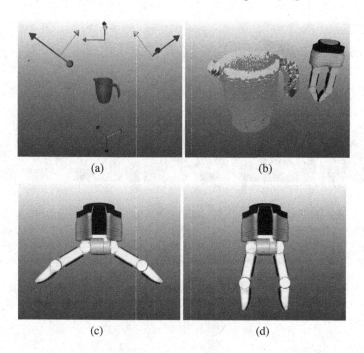

Fig. 9. Experimental setup. (a) Visualisation of the simulated sensor set-up. Four RGB-D sensors, represented by the four frames, surround the object. (b) Outcome of grasp simulations performed in RobWorkSim [21] with the NPG (Narrow Pinch Grasp), see (d). Red and pink equals failed grasps and green depicts successful grasps. (c) The Wide Pinch Grasp (WPG). (Color figure online)

that emerge during a grasping execution. Grasping is performed with a simulated version of the Schunk SDH-2 hand with two different grasp preshapes, a wide pinch grasp (WPG) and a narrow pinch grasp (NPG), see Fig. 9(c) and (d). The grasping affordance learning methods relies on a set of candidate grasps that can used for training or prediction. As a means to provide a meaningful set of candidate grasps, a weak visual bias based on the small surface patches, τ. the grasp representation shown in Fig. 9(b) is based on such a weak bias. The experiments are performed on an object set consisting of 50 objects, see Fig. 10 for a subset. The object are classified into three classes, containers (20 objects), boxes (14 objects) and curved (16 objects) objects.

Evaluation Score. A 5-fold cross-validation is performed to evaluate the learned prototypes for prediction of grasps on novel objects. The learning and evaluation of features and grasps are performed on the full object set. However, in order to compare the achieved results with work of [3] on the same dataset the object class-wise grasp prediction performance is also presented. To measure the performance of the methods and in particular the ability to suggest some suitable grasps in an application context, the N most likely grasps to be successful are

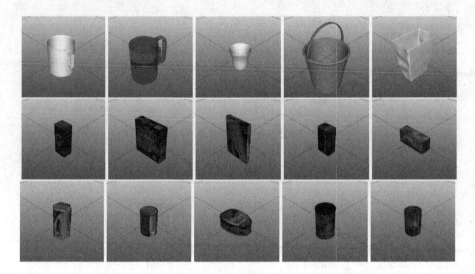

Fig. 10. A subset of the 50 objects used in the experimental work. The 3D models have been taken from the KIT database [22] and from archive3d [23]. See [3] for the full object set.

taken out. The motivation for taking out N grasps rather than all is that only one grasp can effectively be used to grasp an object. By introducing N grasps there is multiple choices if some external factors constraints the use of some of the grasps.

Method 1 Score. To measure the performance of shape-grasp particle method, we use the success-rate of the highest ranked grasps from the votes of shape-grasp particles. The N highest ranked grasps (in a combined selection of support and probability, here the support depict that a grasp have achieved a significant amount of votes) are compared to the actual outcome of the grasps and the success-rate is computed as follows:

$$\text{success-rate} = \frac{N_{success}}{N} \tag{14}$$

Method 2 Score. In order to measure the performance of the signature based grasp affordance learning, a slightly different success-measure is used. The N highest ranked grasps (according to the predicted probability of the logistic regression model) is compared to the actual outcome of the grasps with the condition that the grasps are predicted to be successful. This means that if no grasps are predicted to be successful for an object, this will not be taken into account for computing the average success-rate. To illustrate that not all objects are used for the success-rate computation, an additional object ratio score is plotted. This score depict how many of the objects that have contributed to the success-rate, which translates into how many objects have been given grasp suggestions.

Feature Learning Parameters. For the feature learning part five parameters need to be set. A radius, r, of 0.03 m is used. The discretization of the circular grid, N_g, is set to 36, equalling slices of 10°. The WMA filter is set to be of length 6 resulting in an average over 60° in each direction. Finally, the amount of clusters, (K learned features), are varied between 1 and 25 in steps as a part of the experimental results. In the presented results N = 10 is used.

5   Results

First, a qualitative assessment is presented of a set of learned features and learned grasping affordances for method 1 (Sect. 5.1). Secondly an evaluation is presented of the grasp prediction performance for different amounts of learned features (different K's) for the two methods (Sect. 5.2).

5.1   Qualitative Assessment

In Fig. 11, the learned features from one of the folds are visualised for K = 10. In Fig. 12, a subset of the novel objects in the same fold is shown with inferred features. The second figure can be useful to understand what the features describe. Multiple meaningful structures can be identified when assessing the features qualitatively, see Fig. 11. There are wall features with different curvature (1 and 2). Walls at an edge (7). Wall that have a gap which is identified as a rim (8). A surface feature with slight curvature (3) and surface edge structures (4, 5 and 6) with a varying degree of curvature.

Fig. 11. Visualisation of a learned set of visual features with K = 10. The features are denoted 1 to 10 from left to right. The bottom and top rows show the same features from different angles. In addition to the actual inclination of the outer ring feature, the colour also denote the angle difference to the normal of the centre feature, green/cyan depict strong curvature whereas red depict none or little curvature. The orientation of the features are described by the inlaid frames (red, green and blue sticks). The colour in the centre of the features is used for encoding the inferred features on the novel objects, see Fig. 12. (Color figure online)

Fig. 12. Example objects with inferred learned features. See the features in Fig. 11 and the corresponding colour coding to interpret the objects. (Color figure online)

Fig. 13. Visualisation of the visual features with the learned grasp affordances for the NPG grasp. The features are denoted 1 to 5 in the three leftmost columns from top to bottom and 6–10 in the three rightmost columns from top to bottom. Red depict low probability (0.0) of success. The colour changes towards green that depict a success probability of 1.0. First column show the features, the second column show the features with associated grasp from a perspective view and the third column show the features from a top view. (Color figure online)

Grasping Affordances Learned with Method 1. In Figs. 13 and 14, the learned visual features are shown with the associated learned grasp affordances for method 1, in terms of coloured stick figures that depict the orientation of the grasps with respect to the feature. When assessing the features from the NPG (Fig. 13), only feature eight seems to be a reliable predictor for successful grasps. This feature describes the rim of a wall which explains the good performance as it intuitively is a good place to try a narrow pinch grasp. Some of the other features show some performance potential, for instance feature one, that describes a wall structure. A wall is intuitively graspable by a NPG grasp, but because it is found in a uniform area, one DOF in the pose is ill defined resulting in a low but rather uniform success probability around this DOF.

For the case of the WPG grasp, visualised in Fig. 14, the results are quite different. For the mostly flat structures (1 and 3), areas of high probability are not found. However for the highly curved features that somehow relate to edges (2, 4, 6 and 10) there are structures that suggest very high likelihood of grasp success. For features (2, 4 and 10) high probability areas are found in a somehow box structure aligned to the feature and below the features (the third column shows the bottom view). For feature six, the high probability areas are also found

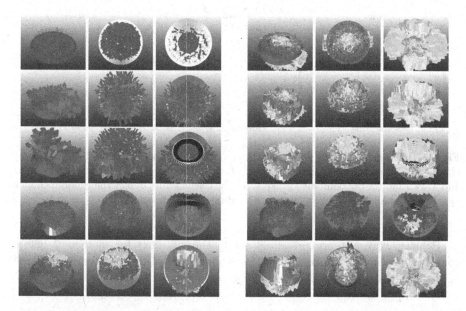

Fig. 14. Visualisation of the visual features with the learned grasp affordances for the WPG grasp. The features are denoted 1 to 5 in the three leftmost columns from top to bottom and 6–10 in the three rightmost columns from top to bottom. Red depict low probability (0.0) of success. The colour changes towards green that depict a success probability of 1.0. First column shows the features from a perspective view. The second column, from a top view and the third column, from a bottom view. See also the first column of Fig. 13 for a view of the features without grasps. (Color figure online)

towards the normal of the centre feature. When this is matched with the inferred features on the fourth object in Fig. 12, this seems reasonable.

Finally, the qualitative result show that a diverse set of features are needed for predicting even slightly different actions with high rates of success. Exemplified by the fact that only a single prototype is good for the NPG grasp whereas others are suitable for the WPG grasp. This is an indication that our unsupervised approach for prototype learning based on the occurrence in the object set is a reasonable way to go.

5.2   Grasp Prediction Evaluation

In Figs. 15 and 16, the results are presented for the two grasp affordance learning methods. In each of the figures, the results are shown for the individual object classes (red, green and blue) for the two different grasp types (NPG, WPG) (see Fig. 9) as well as for the grasp performance when the highest ranked of the two grasp types are chosen (CG). These measures are also presented for the full object set (orange). The success-rates are computed for five different values of K (5, 10, 15, 20, 25).

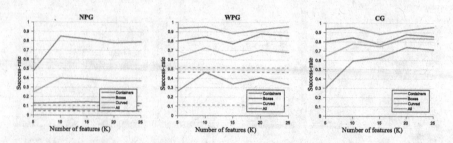

Fig. 15. The results for the grasp predictions, using method 1, for the different object classes as well as for all the objects, presented for the two different grasp types NPG and WPG and for the combined grasp CG. The dashed lines depict the random chance of selecting a successful grasp for the individual object classes.

Below the results are presented for method 1, next for method 2 before comparing the performance of the two methods. Finally, the results are compared to the results acquired in [3] which are summarized in Table 1.

Method 1. The results for using grasping affordance learning method 1, is shown in Fig. 15. For the NPG type, the performance for the container objects are high for K >= 10, whereas the performance for the curved and the box objects are low. For the WPG type, the result is opposite as the performance for the curved and box objects are very good and the performance for the container objects are mediocre. For the CG, the curved and box objects exhibit similar good performance as for the WPG. The result for the container objects with the CG is in between the result of the results of the NPG and the WPG. The performance for all the objects on the CG are in general better than the best of the NPG and the WPG.

Method 2. The results for using grasping affordance learning method 2, are shown in Fig. 15. For the NPG type, the performance for the container objects are high for K >= 10, whereas the performance for the curved and box objects are more mixed. An important point to notice is that successful NPG's are only predicted for a limited set of the curved and box objects, see the object-ratio scores. For the WPG, the performance for the curved and box objects is stable around 85% whereas the performance of the container objects are mediocre. For the case of CG, the performance is consistently high for all three object classes approximately around 85% which resembles the best performance for the container objects with NPG and the best performance for the curved and box objects with WPG. Finally it should be noted that the object-ratio's for the CG is equal to one which means that every object has a grasp that is predicted to be successful. From an application point of view this means that we can suggest a grasp for every object.

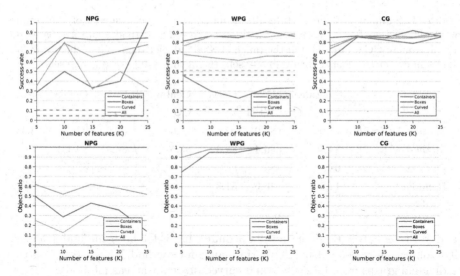

Fig. 16. The results for the grasp predictions, using method 2, for the different object classes as well as for all the objects, presented for the two different grasp types NPG and WPG and for the combined grasp CG. The top row depict the achieved success-rate for the objects where successful grasps was predicted and the dashed lines depict the random chance of selecting a successful grasp for the individual object classes. The bottom row shows the ratio of objects in which successful grasps where found.

Comparison of Method 1 and Method 2. When comparing the results of Figs. 15 and 16 a couple of things is noted. For the narrow pinch grasp the performance with the container object class is similar, whereas the performance for the box and curved objects are better for method 2. This difference is due to that method 2 also gives a classification whether a grasp is predicted successful and if not they are disregarded from the success-rate computation. For the WPG, the performance of method 1 is better for the curved objects whereas the box object have similar performance. For the CG, there is a significant difference between the results. For method 2, the results are very consistent around a success-rate of 85% for the individual object classes and for all objects. This compares to the larger spread in performance seen with method 1, where the container objects perform significantly worse, the box objects show similar performance as with method 1 and the curved objects perform better. The CG performance for all objects is slightly better for method 2 as compared to method 1.

Comparison with the Results in [3]. When comparing the results of method 1 and 2 with the achieved scores in [3], see Table 1, it becomes visible that improved performance is achieved. Despite the fact that the conditions for the results acquired in this work is significantly more difficult due to that visual and action training is performed on the full object set as compared to the class-wise approach in [3]. The performance on the individual object classes for the NPG

Table 1. Summary of the best results acquired in [3] and a subset of the results acquired with method 1 and method 2. The chosen K for method 1 and 2 is based on the maximum CG score found in the range of Ks. [*] depict the results achieved in [3]. Random depict the chance for randomly selecting a successful grasp in the sets.

	Containers	Boxes	Curved	Containers	Boxes	Curved	All
	NPG [%]	WPG [%]		CG [%]			
Method 1 (K = 20)	78.0	87.9	92.5	74.0	87.9	92.5	83.4
Method 2 (K = 25)	84.4	86.4	88.8	85.5	86.4	89.4	87.0
[*]	68	84	84	-	-	-	-
Random	11.5	46.6	51.3	-	-	-	-

and WPG is better than the results acquired in [3] for both method 1 and 2. The object class dependent CG performance for method 1 is better than the individual grasps from [3] for box and curved objects whereas method 2 outperforms the best score for the container objects. Finally, the highest achieved score of 83.4% (Method 1) and 87% (Method 2) probability for selecting a successful grasp, CG, for the full object set is comparable or better than the highest achieved score (84%) for the individually grasp types in [3]. Finally, the ability of method 2 to predict failures as well as successes allows it to make more useful predictions as compared to both method 1 and [3].

6   Conclusion

In this paper, we have proposed, the Sliced Pineapple Grid Feature (SPGF), a novel semi-local shape-based descriptor, with the aim of utilising it for grasp affordance learning. The descriptor has a number of key properties such as its ability to encode double sided structures, encoding of gaps as well as being rotational invariant. As the extraction of a specific discrete set of shape descriptors is based on an unsupervised approach, the amount of features can be tuned for different applications or object sets.

The visual features were utilised for two grasp affordance learning and prediction methods. By utilising the methods to predict grasps on novel objects, our systems where able to predict grasp success with a rate of up to 96% for individual object classes (Method 1) and up to 87% (Method 2) when an object set of three classes was used. By utilising the learned features, the performance is better or comparable for both methods when compared to the best results achieved in [3] performed on the same dataset despite that the learning conditions in this work are significantly more difficult.

Regardless of the respectable performance of the grasp affordance system of method 1, the results indicated that he potential of the system was not yet fully realized, primarily illustrated by the fact that the CG performance for the container objects is well below the highest achieved score for the individual grasp types WPG and NPG. However, by introducing the signature based grasping affordance learning (Method 2) this issue was solved and the combined best performance almost resembled the best performance acquired for the individual grasp types.

From a qualitative perspective, the learned features exhibit similarity to structures that can be identified as building blocks of objects such as "walls", "rims", "edges", "surfaces" and others. Furthermore, the qualitative results of the learned grasping affordances for method 1 indicate that different features are suitable for different affordances, in our work demonstrated by the two grasp types.

Acknowledgement. The research leading to these results has received funding from the European Community's Seventh Framework Programme FP7/2007-2013 (Specific Programme Cooperation, Theme 3, Information and Communication Technologies) under grant agreement no. 270273, Xperience.

References

1. Krüger, N., Janssen, P., Kalkan, S., Lappe, M., Leonardis, A., Piater, J., Rodríguez-Sánchez, A.J., Wiskott, L.: Deep hierarchies in the primate visual cortex: what can we learn for computer vision? IEEE PAMI **35**, 1847–1871 (2013)
2. Lloyd, S.: Least squares quantization in PCM. IEEE Trans. Inf. Theor. **28**, 129–137 (2006)
3. Thomsen, M., Kraft, D., Krüger, N.: Identifying relevant feature-action associations for grasping unmodelled objects. Paladyn J. Behav. Robot. **6**(1), 85–110 (2015)
4. Curtis, N., Xiao, J., Member, S.: Efficient and effective grasping of novel objects through learning and adapting a knowledge base. In: IEEE International Conference on Robotics and Automation (ICRA), pp. 2252–2257 (2008)
5. Huebner, K., Ruthotto, S., Kragic, D.: Minimum volume bounding box decomposition for shape approximation in robot grasping. In: IEEE International Conference on Robotics and Automation, ICRA 2008, pp. 1628–1633. IEEE (2008)
6. Detry, R., Ek, C.H., Madry, M., Kragic, D.: Learning a dictionary of prototypical grasp-predicting parts from grasping experience. In: IEEE International Conference on Robotics and Automation (2013)
7. Kopicki, M., Detry, R., Schmidt, F., Borst, C., Stolkin, R., Wyatt, J.L.: Learning dexterous grasps that generalise to novel objects by combining hand and contact models. In: IEEE International Conference on Robotics and Automation (2014, to appear)
8. Saxena, A., Driemeyer, J., Ng, A.Y.: Robotic grasping of novel objects using vision. Int. J. Robot. Res. **27**, 157–173 (2008)
9. Kootstra, G., Popović, M., Jørgensen, J.A., Kuklinski, K., Miatliuk, K., Kragic, D., Krüger, N.: Enabling grasping of unknown objects through a synergistic use of edge and surface information. Int. J. Robot. Res. **31**(10), 1190–1213 (2012)
10. Lenz, I., Lee, H., Saxena, A.: Deep learning for detecting robotic grasps. CoRR (2013)
11. Jiang, Y., Moseson, S., Saxena, A.: Efficient grasping from RGBD images: learning using a new rectangle representation. In: ICRA 2011, pp. 3304–3311 (2011)
12. Redmon, J., Angelova, A.: Real-time grasp detection using convolutional neural networks. CoRR abs/1412.3128 (2014)

13. Krizhevsky, A., Sutskever, I., Hinton, G.E.: Imagenet classification with deep convolutional neural networks. In: Advances in Neural Information Processing Systems, pp. 1097–1105 (2012)
14. Myers, A., Teo, C.L., Fermüller, C., Aloimonos, Y.: Affordance detection of tool parts from geometric features. In: ICRA (2015)
15. Bohg, J., Morales, A., Asfour, T., Kragic, D.: Data-driven grasp synthesis a survey. IEEE Trans. Robot. **30**, 289–309 (2014)
16. Durrant-Whyte, H., Bailey, T.: Simultaneous localization and mapping: part i. IEEE Robot. Autom. Mag. **13**, 99–110 (2006)
17. Kraft, D., Mustafa, W., Popovic, M., Jessen, J.B., Buch, A.G., Savarimuthu, T.R., Pugeault, N., Krüger, N.: Using surfaces and surface relations in an early cognitive vision system. In: Computer Vision and Image Understanding (2014)
18. Wahl, E., Hillenbrand, U., Hirzinger, G.: Surflet-pair-relation histograms: a statistical 3D-shape representation for rapid classification. In: Proceedings of Fourth International Conference on 3-D Digital Imaging and Modeling, 3DIM 2003, pp. 474–481. IEEE (2003)
19. Mustafa, W., Pugeault, N., Krüger, N.: Multi-view object recognition using viewpoint invariant shape relations and appearance information. In: IEEE International Conference on Robotics and Automation (ICRA) (2013)
20. Freedman, D.A.: Statistical Models: Theory and Practice. Cambridge University Press, Cambridge (2009)
21. Jørgensen, J.A., Ellekilde, L.P., Petersen, H.G.: RobWorkSim - an open simulator for sensor based grasping. In: Robotics (ISR), 2010 41st International Symposium on and 2010 6th German Conference on Robotics (ROBOTIK), pp. 1–8 (2010)
22. Kasper, A., Xue, Z., Dillmann, R.: The KIT object models database: an object model database for object recognition, localization and manipulation in service robotics. Int. J. Robot. Res. **31**(8), 927–934 (2012)
23. Archive3D: Archive3D free online cad model database. http://www.archive3d.net

Extending Guided Image Filtering for High-Dimensional Signals

Shu Fujita[1,2(✉)] and Norishige Fukushima[2]

[1] Nagoya University, Nagoya, Japan
s.fujita@fujii.nuee.nagoya-u.ac.jp
[2] Nagoya Institute of Technology, Nagoya, Japan
fukushima@nitech.ac.jp

Abstract. This paper presents an extended method of guided image filtering (GF) for high-dimensional signals and proposes various applications for it. The important properties of GF include edge-preserving filtering, local linearity in a filtering kernel region, and the ability of constant time filtering in any kernel radius. GF can suffer from noise caused by violations of the local linearity when the kernel radius is large. Moreover, unexpected noise and complex textures can further degrade the local linearity. We propose high-dimensional guided image filtering (HGF) and a novel framework named combining guidance filtering (CGF). Experimental results show that HGF and CGF can work robustly and efficiently for various applications in image processing.

1 Introduction

Recently, edge-preserving filtering has been attracting increasing attention and has become as fundamental tool in image processing. The filtering techniques such as bilateral filtering [33] and guided image filtering (GF) [17] are used for various applications including image denoising [4], high dynamic range imaging [8], detail enhancement [3,10], flash/no-flash photography [9,27], super resolution [23], depth map denosing [14,24], guided feathering [17,22] and haze removal [19].

One representative formulation of edge-preserving filtering is weighted averaging, i.e., finite impulse response filtering, based on space and color weights that are computed from distances among neighborhood pixels. When the distance and weighting function are Euclidean and Gaussian, respectively, the formulation becomes a bilateral filter [33], which is a representative edge-preserving filter. The bilateral filter has useful properties, but is known as time-consuming; thus, several acceleration methods have been proposed [5,13,26,28,29,31,35,36]. Another formulation uses geodesic distance, representative examples being domain transform filtering [15], and recursive bilateral filtering [34,37]. These are formulated as infinite impulse response filtering, represented by a combination of horizontal and vertical one-dimensional filtering. These methods can, therefore, efficiently smooth images.

© Springer International Publishing AG 2017
J. Braz et al. (Eds.): VISIGRAPP 2016, CCIS 693, pp. 439–453, 2017.
DOI: 10.1007/978-3-319-64870-5_21

A guided image filter [17,18], which is an efficient edge-preserving filter, is based on an assumption different from those of previously introduced filtering methods. The guided image filter assumes a local linear model in each local kernel. This property is convenient and essential for several applications in computational photography [8,17,19,23,27]. Furthermore, the guided image filter can efficiently compute in constant time, with the result that the computational cost is independent of the size of the filtering kernel. This fact is also useful for fast visual correspondence problems [20]. However, the local linear model can be violated by unexpected noise, such as Gaussian noise, and different types of textures. Such situations often arise when the kernel is large. Then, the resulting image can contain noise. Figure 1 demonstrates feathering [17], where the result of GF (Fig. 1(c)) contains noise around the border of an object.

For noise-robust implementation, several studies have used patch-wise processing, such as non-local means filtering [4] and, discrete cosine transform (DCT) denoising [12,38]. Patch-wise processing gathers intensity or color information in each local patch to channels or dimensions of a pixel. In particular, non-local means filtering obtains filtering weights from the gathered information between the target and reference pixels. Since patch-wise processing utilizes richer information, it can work more robustly for noisy information compared to pixel-wise processing. Extension to high-dimensional representations

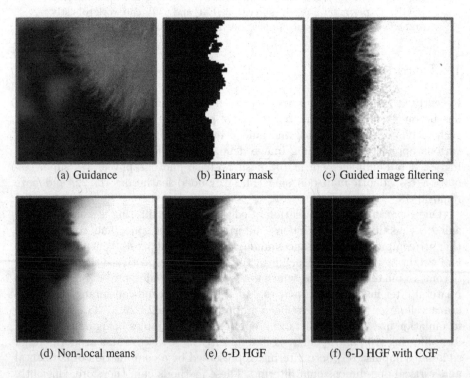

| (a) Guidance | (b) Binary mask | (c) Guided image filtering |
| (d) Non-local means | (e) 6-D HGF | (f) 6-D HGF with CGF |

Fig. 1. Guided feathering results. (c) contains noise around object boundaries, while our results (e) and (f) can suppress such noise.

such as high-dimensional Gaussian filtering has also been discussed [1,2,13,16]. However, these previous filters for high-dimensional signals cannot support GF. Figure 1(d) shows the result of non-local means filtering extended to joint filtering for feathering. The result has been over-smoothed because of the loose of the local linearity.

Therefore, we propose a high-dimensional extension of GF to obtain robustness. We call the extension high-dimensional guided image filtering (HGF). We first extend GF to ensure the filter can handle high-dimensional information. In this regard, assuming d as the number of dimensions of the guidance image, the computational complexity of HGF becomes $O(d^{2.807\cdots})$ as noted in [16]. We then introduce a dimensionality reduction technique for HGF to reduce the computational cost. Furthermore, we introduce a novel framework for HGF, called *combining guidance filtering (CGF)*, which builds a new guidance image by combining the HGF output with the guidance image, and then re-executes HGF using the combined guidance image. This framework provides more robust performance to HGF and utilizes the HGF characteristics that can use high-dimensional information. Figures 1(e) and (f) show our results. HGF suppresses the noise, and HGF with CGF further reduces the noise.

This paper is an extended version of our conference paper [11]. The main extension part is the section on CGF and the associated experimental results.

2   Related Works

We discuss several acceleration methods for high-dimensional filtering in this section.

Paris and Durand [26] introduced the bilateral grid [6], which is a high-dimensional space defined by adding the intensity domain to the spatial domain. We can obtain edge-preserving results using linear filtering on the bilateral grid. The bilateral grid is, however, computationally inefficient because the high-dimensional space is huge. As a result, the bilateral grid requires down-sampling of the space for efficient filtering. However, the computational resources and memory footprints are expensive, especially when the dimension of guidance information is high. Gaussian kd-trees [2] and permutohedral lattice [1] focus on representing the high-dimensional space with point samples to overcome these problems. These methods have succeeded in alleviating computational complexity when the filtering dimension is high. However, since they still require a significant amount of calculation and memory, they are not sufficient for real-time applications.

Adaptive manifolds [16] provides a slightly different approach. The three methods described above focus on how to represent and expand each dimension. In contrast, the adaptive manifolds samples the high-dimensional space at scattered manifolds adapted to the input signal. The method thereby avoids having pixels are enclosed into cells to perform barycentric interpolation. This property enables us to efficiently compute a high-dimensional space and reduces the memory requirement. This is why the use of the adaptive manifolds is more efficient than that of other high-dimensional filtering methods [1,2,26]. Its accuracy is, however, lower than theirs. The adaptive manifolds causes quantization artifacts depending on the parameters.

3 High-Dimensional Guided Image Filtering

We introduce our high-dimensional extension techniques for GF [17,18] in this section. We first extend GF to high-dimensional information. Next, a dimensionality reduction technique is introduced to increasing computing efficiency. We finally present CGF, which is a new framework for HGF, to further suppress noise caused by violation of the local linearity.

3.1 Definition

The guided image filter assumes a local linear model between an input guidance image I and an output image q. The assumption of the local linear model is also invariant for our HGF. Let J denote an n-dimensional guidance image. We assume that J is generated from the guidance image I using a function f:

$$J = f(I). \tag{1}$$

The function f constructs a high-dimensional image from the low-dimensional image signal I; for example, the function might use a square neighborhood centered at a focusing pixel, DCT or principle components analysis (PCA) of the guidance image I.

HGF uses this high-dimensional image J as the guidance image; thus, the output q is derived from a linear transform of J in a square window ω_k centered at a pixel k. When we let p be an input image, the linear transform is represented as follows:

$$q_i = a_k^T J_i + b_k. \quad \forall i \in \omega_k. \tag{2}$$

Here, i is a pixel position, and a_k and b_k are linear coefficients. In this regard, J_i and a_k represent $n \times 1$ vectors. Moreover, the linear coefficients can be derived using the solution used in [17,18]. Let $|\omega|$ denote the number of pixels in ω_k, and U be an $n \times n$ identity matrix. The linear coefficients are computed by

$$a_k = (\Sigma_k + \epsilon U)^{-1} \left(\frac{1}{|\omega|} \sum_{i \in \omega_k} J_i p_i - \mu_k \bar{p}_k \right) \tag{3}$$

$$b_k = \bar{p}_k - a_k^T \mu_k, \tag{4}$$

where μ_k and Σ_k are the $n \times 1$ mean vector and $n \times n$ covariance matrix of J in ω_k, ϵ is a regularization parameter, and $\bar{p}_k (= \frac{1}{|\omega|} \sum_{i \in \omega_k} p_i)$ represents the mean of p in ω_k.

Finally, we compute the filtering output by applying the local linear model to all local windows in the entire image. Note that the q_i values in all local windows including a pixel i are not necessarily the same. Therefore, the filter output is computed by averaging all possible values of q_i as follows:

$$q_i = \frac{1}{|\omega|} \sum_{k:i\in\omega_k} (a_k J_i + b_k) \tag{5}$$

$$= \bar{a}_i^T J_i + \bar{b}_i, \tag{6}$$

where $\bar{a}_i = \frac{1}{|\omega|}\sum_{k\in\omega_i} a_k$ and $\bar{b}_i = \frac{1}{|\omega|}\sum_{k\in\omega_i} b_k$.

The computation time of HGF does not depend on the kernel radius that provides the inherent ability of GF. HGF consists of many instances of box filtering and per-pixel small matrix operations. The box filtering can compute in $O(1)$ time [7], but the number of instances of box filtering linearly depends on the dimensions of the guidance image. In addition, the order of the matrix operations exponentially depends on the number of dimensions.

3.2 Dimensionality Reduction

For efficient computing, we use PCA for dimensionality reduction. The dimensionality reduction technique was proposed in [32] for non-local means filtering or high-dimensional Gaussian filtering. The approach aims for finite impulse response filtering with Euclidean distance. We extend the dimensionality technique for HGF.

For HGF, the guidance image J is converted into new guidance information, which is projected onto the lower-dimensional subspace determined by PCA. Let Ω be a set of all pixel positions in J. To conduct PCA, we must first compute the $n \times n$ covariance matrix Σ_Ω for the set of all guidance image pixels J_i. The covariance matrix Σ_Ω is computed as follows:

$$\Sigma_\Omega = \frac{1}{|\Omega|} \sum_{i\in\Omega}(J_i - \bar{J})(J_i - \bar{J})^T, \tag{7}$$

where $|\Omega|$ and $\bar{J}$ are the number of all pixels and the mean of J in the whole image, respectively. After that, pixel values in the guidance image J are projected onto d-dimensional PCA subspace by the inner product of the guidance image pixel J_i and the eigenvectors e_j ($1 \le j \le d, 1 \le d \le n$, where d is a constant value) of the covariance matrix Σ_Ω. Let J^d be a d-dimensional guidance image, then the projection is performed as:

$$J_{ij}^d = J_i \cdot e_j, \quad 1 \le j \le d, \tag{8}$$

where J_{ij}^d is the pixel value in the jth dimension of J_i^d, and $J_i \cdot e_j$ represents the inner product of the two vectors. We show an example of the PCA result of each eigenvector e in Fig. 2.

In this manner, we can obtain the d-dimensional guidance image J^d, which is used to replace J in Eqs. (2), (3), (5), and (6). Moreover, each dimension in J^d can be weighted by the eigenvalues λ, where is a $d \times 1$ vector, of the covariance

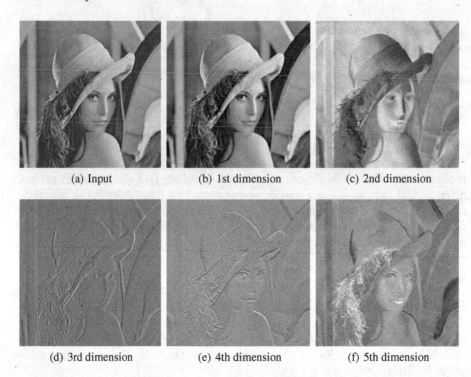

| (a) Input | (b) 1st dimension | (c) 2nd dimension |
| (d) 3rd dimension | (e) 4th dimension | (f) 5th dimension |

Fig. 2. PCA result. We construct the color original high-dimensional guidance image from 3×3 square neighborhood in each pixel of the input image. We reduce the dimension $27 = (3 \times 3 \times 3)$ to 5. (Color figure online)

matrix Σ_Ω. Note that the eigenvalue elements from the $(d+1)$th to nth are discarded because HGF uses only d dimensions. Hence, the identity matrix U in Eq. (3) can be weighted by the eigenvalues $\boldsymbol{\lambda}$. Then, we take the element-wise inverse of eigenvalues $\boldsymbol{\lambda}$:

$$\boldsymbol{E}_d = diag(\boldsymbol{\lambda}^{inv}) \tag{9}$$

$$= \begin{bmatrix} \frac{1}{\lambda_1} & & \\ & \ddots & \\ & & \frac{1}{\lambda_d} \end{bmatrix}, \tag{10}$$

where $\boldsymbol{E}_d$ represents a $d \times d$ diagonal matrix, $\boldsymbol{\lambda}^{inv}$ represents the element-wise inverse eigenvalues, and λ_x is the xth eigenvalue. Note that we take the logarithms of the eigenvalues $\boldsymbol{\lambda}$ depending on the applications and normalize the eigenvalue based on the first eigenvalue λ_1. We take the element-wise inverse of $\boldsymbol{\lambda}$ to use the smallest ϵ for the dimension having the large eigenvalue as compared to the small eigenvalue, because the elements of $\boldsymbol{\lambda}$ satisfy $\lambda_1 \geq \lambda_2 \geq \cdots \geq \lambda_d$, and the eigenvector whose eigenvalue is large is more important. As a result, we can preserve the characters of the image in the principal dimension.

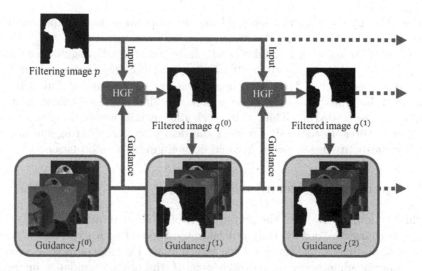

Fig. 3. Overview of CGF using our HGF. This figure shows the case of $d = 3$, i.e., the initial guidance images are three.

Therefore, we can obtain the final coefficient a_k instead of using Eq. (3) in the high-dimensional case as follows:

$$a_k = (\Sigma_k^d + \epsilon E_d)^{-1}(\frac{1}{|\omega|} \sum_{i \in \omega_k} J_i^d p_i - \mu_k^d \bar{p}_k), \qquad (11)$$

where and μ_k^d and Σ_k^d are the $d \times 1$ mean vector and the $d \times d$ covariance matrix of J^d in ω_k, respectively.

3.3 Combining Guidance Filtering

Our extension of HGF can utilize high-dimensional signals. In other words, HGF can use multiple single-channel images as the guidance information by using the function f as merging multiple image channels. Utilizing this property and extending HGF, we present the novel framework CGF.

An overview of CGF is shown in Fig. 3. CGF involves three main steps: (1) computing a filtered result using initial guidance information $J^{(0)}$, (2) generating new guidance information $J^{(t)}$ by combining the filtered result $q^{(t)}$ with the initial guidance information $J^{(0)}$, and (3) re-executing HGF using the combined guidance image $J^{(t)}$. Here, steps (2) and (3) are repeated, and t is the number of iterations. According to our preliminary experiments, two to three iterations are sufficient to obtain adequate results. Note that the filtering target image is not changed from the initial input image to avoid an over-smoothing problem. This framework works well in recovering edges from additional guidance information as guided feathering [17]. This is because the additional guidance information is not discarded and is added to new guidance information. Moreover, the filtered guidance image added to the new guidance information plays an important role to suppress noise.

Our CGF framework is similar to the framework of rolling guidance image filtering proposed by Zhang *et al.* [39]. Rolling guidance image filtering is an iterative processing with a fixed input image and updated guidance. This filtering framework is limited to direct filtering, i.e., the filter does not utilize joint filtering, such as feathering. Thus, their work aims at image smoothing to remove detailed textures. In contrast, our work can deal with joint filtering and mainly aims primarily at edge recovery from additional guidance information.

4   Experimental Results

In this section, we evaluate the performance of HGF in terms of efficiency and verify its characteristics by using several applications. In our experiments, each pixel of high-dimensional images J has multiple pixel values comprising a fixed-size square neighborhood around each pixel in the original guidance image I. The dimensionality is reduced using the PCA approach discussed in Sect. 3.2.

We first reveal the processing time of HGF. Our proposed and competing methods are implemented in C++ with Visual Studio 2010 on Windows 7 64 bit. The code is parallelized using OpenMP. The CPU for the experiments is a 3.50 GHz Intel Core i7-3770K. The input images, with resolution of one-megapixel, i.e., 1024×1024, are grayscale or color images.

Figure 4 shows the result of the processing time. The processing time of HGF increases exponentially as the guidance image dimensionality increases. This cost increasing result makes the dimensionality reduction essential for HGF. In addition, the computational cost of PCA is small as compared with the increase in the filtering time caused by increase in dimensionality. Therefore, although the computational cost becomes high from increasing the dimensionality, the problem is not significant. Tasdizen [32] remarked that the performance of dimensionality reduction peaks at approximately six. This fact is also shown in our experiments.

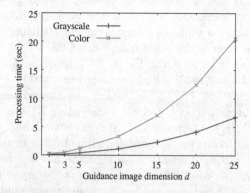

Fig. 4. Processing time of high-dimensional guided image filtering with respect to guidance image dimensions.

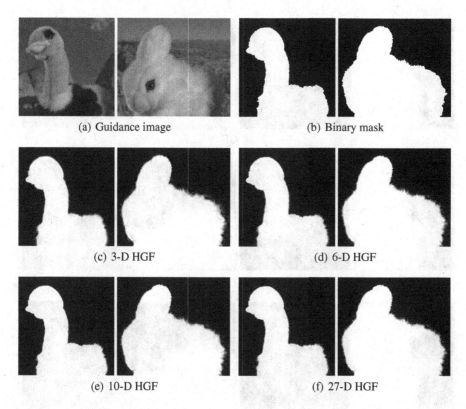

(a) Guidance image

(b) Binary mask

(c) 3-D HGF

(d) 6-D HGF

(e) 10-D HGF

(f) 27-D HGF

Fig. 5. Dimension sensitivity. The color patch size for high-dimensional image is 3×3, i.e., the complete dimension is 27. The parameters are $r = 15$, $\epsilon = 10^{-6}$. (Color figure online)

Figure 5 shows the result of the dimension sensitivity of HGF. We obtain the binary input mask using GrabCut [30]. We can improve the edge-preserving effect of HGF by increasing the dimension. However, the amount of the improvement is slight in the case of more than ten dimensions. Thus, we do not need to increase the dimensionality.

Next, we compare the characteristics of GF and HGF. As mentioned in Sect. 1, GF can transfer detailed regions such as feathers, but may simultaneously cause noise near the object boundary (see Fig. 1(c)). In contrast, HGF can suppress noise while the detailed regions are transferred, as shown in Fig. 1(e). This noise suppression ability can be further improved by using CGF, as shown in Fig. 1(f). We use two iterations for CGF, i.e., we set $t = 2$. Therefore, we can apply CGF if we desire better results.

We also show the detailed results of guided feathering and alpha matting in Fig. 6. All guidance images and initial masks used in this experiment are the same as those in Fig. 5. The result of GF causes noise and color mixtures near the object boundary. HGF can alleviate these problems and suppress noise and

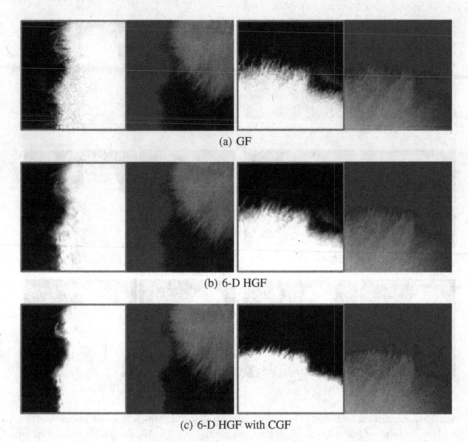

(a) GF

(b) 6-D HGF

(c) 6-D HGF with CGF

Fig. 6. Guided feathering and matting results using different methods. The parameters are the same as Fig. 5.

color mixtures. However, some noise and blurs remain near the object boundary. These problems can be improved by using CGF. The use of HGF with CGF further suppresses noise and results in clear boundaries compared to the other methods, as shown in Fig. 6(c).

Figure 7 shows the image abstraction results. The result requires three iterations of filtering. As shown in Figs. 7(b) and (d), since the local linear model is often violated in filtering with large kernels, the pixel values are scattered. In contrast, HGF can smooth the image without such problems (see Figs. 7(c) and (e)).

HGF also shows excellent performance for haze removal [19]. The haze removal results and the transition maps are shown in Fig. 8. In the case of GF, the transition map preserves major textures, while there are over-smoothed regions near the detailed regions or object boundaries, e.g., between trees or branches. The over-smoothing effect affects haze removal in such regions. In our case, the transition map of HGF preserves such detailed textures; thus, HGF can remove

(a) Input image   (b) GF

(c) 6-D HGF   (d) Detail of (b)   (e) Detail of (c)

Fig. 7. Image abstraction. The local patch size for the high-dimensional image is 3×3. The parameters for GF and HGF are $r = 25$, $\epsilon = 0.04^2$.

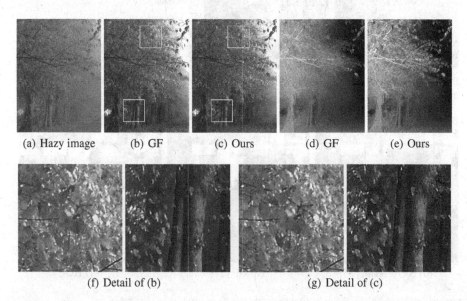

(a) Hazy image   (b) GF   (c) Ours   (d) GF   (e) Ours

(f) Detail of (b)   (g) Detail of (c)

Fig. 8. Haze removal. The bottom row images represent transition maps of (b) and (c). The local patch size for high-dimensional image is 5×5. The parameters for GF and HGF are $r = 20$, $\epsilon = 10^{-4}$.

the haze more effectively than GF in the detailed regions. These results show that HGF is effective for preserving detailed areas or textures.

An other application for HGF is image classification with a hyperspectral image. A hyperspectral image contains considerable wavelength information, which is useful for distinguishing different objects. Although we can obtain good results using support vector machine classifiers [25], Kang et al. improved the accuracy of image classification by applying GF [21]. They created a guidance image using PCA from a hyperspectral image; however, most of the information was unused because GF cannot utilize high-dimensional data. Our extension has an advantage in such cases. Since HGF can utilize high-dimensional data, we can further improve classification accuracy by adding the remaining information.

Figure 9 and Table 1 show the result of classifying the Indian Pines dataset, which was acquired using an Airborne Visible/Infrared Imaging Spectrometer (AVIRIS) sensor. We objectively evaluate the classification accuracy by using three metrics: the overall accuracy (OA), the average accuracy (AA), and the kappa coefficient, which are widely used for evaluating classification. OA is the ratio of correctly classified pixels, AA is the average ratio of correctly classified pixels in each class, and the kappa coefficient denotes the ratio of correctly

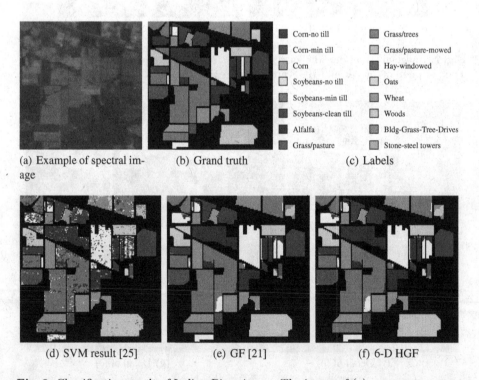

(a) Example of spectral image

(b) Grand truth

(c) Labels

Corn-no till   Grass/trees
Corn-min till   Grass/pasture-mowed
Corn   Hay-windowed
Soybeans-no till   Oats
Soybeans-min till   Wheat
Soybeans-clean till   Woods
Alfalfa   Bldg-Grass-Tree-Drives
Grass/pasture   Stone-steel towers

(d) SVM result [25]   (e) GF [21]   (f) 6-D HGF

Fig. 9. Classification result of Indian Pines image. The image of (a) represents a spectral image that the wavelength is $0.7\,\mu m$. The parameters for GF and HGF are $r = 4$, $\epsilon = 0.15^2$.

classified pixels corrected by the number of pure agreements. We can confirm that HGF achieves a better result than GF. In particular, the detailed regions are improved using our method. The accuracy is further improved objectively, as shown in Table 1.

Table 1. Classification accuracy [%] of the classification results shown in Fig. 9.

Method	OA	AA	Kappa
SVM	81.0	79.1	78.3
GF	92.7	93.9	91.6
HGF	**92.8**	**94.1**	**91.8**

5 Conclusion

We proposed high-dimensional guided image filtering (HGF) by extending guided image filtering [17,18]. The extension allows the guided image filter to utilize high-dimensional signals, e.g., local square patches and hyperspectral images, and suppress unexpected noise that is a limitation of guided image filtering. Our high-dimensional extension has a limitation that the computational cost increases as the number of dimensions increases. To alleviate this limitation, we simultaneously introduce a dimensionality reduction technique for efficient computing. Furthermore, we presented a novel framework named as combining guidance filtering (CGF) in this study, to further exploit the HGF characteristics that can utilize high-dimensional information. As a result, HGF with CGF obtains robustness and can further suppress noise caused by violation of the local linear model. Experimental results showed that HGF can work robustly in noisy regions and transfer detailed regions. In addition, HGF can compute efficiently by using the dimensionality reduction technique.

Acknowledgement. This work was supported by JSPS KAKENHI Grant Number 15K16023.

References

1. Adams, A., Baek, J., Davis, M.A.: Fast high-dimensional filtering using the permutohedral lattice. Comput. Graph. Forum **29**(2), 753–762 (2010)
2. Adams, A., Gelfand, N., Dolson, J., Levoy, M.: Gaussian KD-trees for fast high-dimensional filtering. ACM Trans. Graph. **28**(3), 21 (2009)
3. Bae, S., Paris, S., Durand, F.: Two-scale tone management for photographic look. ACM Trans. Graph. **25**(3), 637–645 (2006)
4. Buades, A., Coll, B., Morel, J.M.: A non-local algorithm for image denoising. In: Proceedings of IEEE Conference on Computer Vision and Pattern Recognition (CVPR) (2005)

5. Chaudhury, K.: Acceleration of the shiftable O(1) algorithm for bilateral filtering and nonlocal means. IEEE Trans. Image Process. **22**(4), 1291–1300 (2013)
6. Chen, J., Paris, S., Durand, F.: Real-time edge-aware image processing with the bilateral grid. ACM Trans. Graph. **26**(3), 103 (2007)
7. Crow, F.C.: Summed-area tables for texture mapping. In: Proceedings of ACM SIGGRAPH, pp. 207–212 (1984)
8. Durand, F., Dorsey, J.: Fast bilateral filtering for the display of high-dynamic-range images. ACM Trans. Graph. **21**(3), 257–266 (2002)
9. Eisemann, E., Durand, F.: Flash photography enhancement via intrinsic relighting. ACM Trans. Graph. **23**(3), 673–678 (2004)
10. Fattal, R., Agrawala, M., Rusinkiewicz, S.: Multiscale shape and detail enhancement from multi-light image collections. ACM Trans. Graph. **26**(3), 51 (2007)
11. Fujita, S., Fukushima, N.: High-dimensional guided image filtering. In: Proceedings of International Conference on Computer Vision Theory and Applications (VISAPP) (2016)
12. Fujita, S., Fukushima, N., Kimura, M., Ishibashi, Y.: Randomized redundant DCT: efficient denoising by using random subsampling of DCT patches. In: Proceedings of ACM SIGGRAPH Asia Technical Briefs (2015)
13. Fukushima, N., Fujita, S., Ishibashi, Y.: Switching dual kernels for separable edge-preserving filtering. In: Proceedings of IEEE International Conference on Acoustics, Speech and Signal Processing (ICASSP) (2015)
14. Fukushima, N., Inoue, T., Ishibashi, Y.: Removing depth map coding distortion by using post filter set. In: Proceedings of IEEE International Conference on Multimedia and Expo (ICME) (2013)
15. Gastal, E.S.L., Oliveira, M.M.: Domain transform for edge-aware image and video processing. ACM Trans. Graph. **30**(4), 69 (2011)
16. Gastal, E.S.L., Oliveira, M.M.: Adaptive manifolds for real-time high-dimensional filtering. ACM Trans. Graph. **31**(4), 33 (2012)
17. He, K., Shun, J., Tang, X.: Guided image filtering. In: Proceedings of European Conference on Computer Vision (ECCV) (2010)
18. He, K., Shun, J., Tang, X.: Guided image filtering. IEEE Trans. Pattern Anal. Mach. Intell. **35**(6), 1397–1409 (2013)
19. He, K., Sun, J., Tang, X.: Single image haze removal using dark channel prior. In: Proceedings of IEEE Conference on Computer Vision and Pattern Recognition (CVPR) (2009)
20. Hosni, A., Rhemann, C., Bleyer, M., Rother, C., Gelautz, M.: Fast cost-volume filtering for visual correspondence and beyond. IEEE Trans. Pattern Anal. Mach. Intell. **35**(2), 504–511 (2013)
21. Kang, X., Li, S., Benediktsson, J.: Spectral-spatial hyperspectral image classification with edge-preserving filtering. IEEE Trans. Geosci. Remote Sens. **52**(5), 2666–2677 (2014)
22. Kodera, N., Fukushima, N., Ishibashi, Y.: Filter based alpha matting for depth image based rendering. In: IEEE Visual Communications and Image Processing (VCIP) (2013)
23. Kopf, J., Cohen, M., Lischinski, D., Uyttendaele, M.: Joint bilateral upsampling. ACM Trans. Graph. **26**(3), 96 (2007)
24. Matsuo, T., Fukushima, N., Ishibashi, Y.: Weighted joint bilateral filter with slope depth compensation filter for depth map refinement. In: International Conference on Computer Vision Theory and Applications (VISAPP) (2013)

25. Melgani, F., Bruzzone, L.: Classification of hyperspectral remote sensing images with support vector machines. IEEE Trans. Geosci. Remote Sens. **42**(8), 1778–1790 (2004)
26. Paris, S., Durand, F.: A fast approximation of the bilateral filter using a signal processing approach. Int. J. Comput. Vis. **81**(1), 24–52 (2009)
27. Petschnigg, G., Agrawala, M., Hoppe, H., Szeliski, R., Cohen, M., Toyama, K.: Digital photography with flash and no-flash image pairs. ACM Trans. Graph. **23**(3), 664–672 (2004)
28. Phạm, T.Q., Vliet, L.J.V.: Separable bilateral filtering for fast video preprocessing. In: Proceedings of IEEE International Conference on Multimedia and Expo (ICME) (2005)
29. Porikli, F.: Constant time O(1) bilateral filtering. In: Proceedings of IEEE Conference on Computer Vision and Pattern Recognition (CVPR) (2008)
30. Rother, C., Kolmogorov, V., Blake, A.: GrabCut: interactive foreground extraction using iterated graph cuts. ACM Trans. Graph. **23**(3), 309–314 (2004)
31. Sugimoto, K., Kamata, S.I.: Compressive bilateral filtering. IEEE Trans. Image Process. **24**(11), 3357–3369 (2015)
32. Tasdizen, T.: Principal components for non-local means image denoising. In: Proceedings of IEEE International Conference on Image Processing (ICIP) (2008)
33. Tomasi, C., Manduchi, R.: Bilateral filtering for gray and color images. In: Proceedings of IEEE International Conference on Computer Vision (ICCV) (1998)
34. Yang, Q.: Recursive bilateral filtering. In: Fitzgibbon, A., Lazebnik, S., Perona, P., Sato, Y., Schmid, C. (eds.) ECCV 2012. LNCS, vol. 7572, pp. 399–413. Springer, Heidelberg (2012). doi:10.1007/978-3-642-33718-5_29
35. Yang, Q., Ahuja, N., Tan, K.H.: Constant time median and bilateral filtering. Int. J. Comput. Vis. **112**(3), 307–318 (2014)
36. Yang, Q., Tan, K.H., Ahuja, N.: Real-time O(1) bilateral filtering. In: Proceedings of IEEE Conference on Computer Vision and Pattern Recognition (CVPR) (2009)
37. Yang, Q.: Recursive approximation of the bilateral filter. IEEE Trans. Image Process. **24**(6), 1919–1927 (2015)
38. Yu, G., Sapiro, G.: DCT image denoising: a simple and effective image denoising algorithm. Image Process. On Line **1**, 292–296 (2011)
39. Zhang, Q., Shen, X., Xu, L., Jia, J.: Rolling guidance filter. In: Fleet, D., Pajdla, T., Schiele, B., Tuytelaars, T. (eds.) ECCV 2014. LNCS, vol. 8691, pp. 815–830. Springer, Cham (2014). doi:10.1007/978-3-319-10578-9_53

Exemplar-Based Image Inpainting Using an Affine Invariant Similarity Measure

Vadim Fedorov[1(✉)], Pablo Arias[2], Gabriele Facciolo[3], and Coloma Ballester[1]

[1] University Pompeu Fabra, Barcelona, Spain
{vadim.fedorov,coloma.ballester}@upf.edu
[2] CMLA, ENS Cachan, Cachan, France
pablo.arias@cmla.ens-cachan.fr
[3] LIGM, UMR 8049, École des Ponts, UPE, Champs-sur-Marne, France
facciolo@cmla.ens-cachan.fr

Abstract. Patch-based approaches are used in state-of-the-art methods for image inpainting. This paper presents a new method for exemplar-based image inpainting using transformed patches. The transformation is determined for each patch in a fully automatic way from a surrounding texture content. We build upon a recent affine invariant patch similarity measure that performs an appropriate patch comparison by automatically adapting the size and shape of the patches. As a consequence, it intrinsically extends the set of available source patches to copy information from. We incorporate this measure into a variational formulation for inpainting and present a numerical algorithm for optimizing it. We show that our method can be applied to complete a perspectively distorted texture as well as to automatically inpaint one view of a scene using other view of the same scene as a source. We present experimental results both for gray and color images, and a comparison with some exemplar-based image inpainting methods.

1 Introduction

Image inpainting, also known as image completion, disocclusion or object removal, refers to the recovery of occluded, missing or corrupted parts of an image in a given region so that the reconstructed image looks natural. It has become a key tool for digital photography and movie post-production where it is used, for example, to eliminate unwanted objects that may be unavoidable during filming.

Automatic image inpainting is a challenging task that has received significant attention in recent years from the image processing, computer vision, and graphics communities. Remarkable progress has been achieved with the advent of exemplar-based methods, which exploit the self-similarity of natural images by assuming that the missing information can be found elsewhere outside the inpainting domain. Roughly speaking, these methods work by copying patches taken from the known part of the image and pasting them smartly in the inpainting domain. These methods can obtain impressive results but many of them rely

© Springer International Publishing AG 2017
J. Braz et al. (Eds.): VISIGRAPP 2016, CCIS 693, pp. 454–474, 2017.
DOI: 10.1007/978-3-319-64870-5_22

on the assumption that the required information can be copied as it is, without any transformations. Therefore, applicability of such methods is limited to the scenes in which objects are in a fronto-parallel position with respect to the camera.

In the image formation process, textured objects may appear distorted by some complex transformation (see Fig. 1). This is a pervasive phenomenon in our daily life. In fact, any person can mentally fill-in occluded parts of an image, even if the missing information is available to them under a different perspective. Our brain is able to appropriately transform the available information to match the perspective of the occluded region. For instance, in Fig. 1 one can easily infer what is hidden behind the red rectangle in the graffiti scene on the left, or use the non-trivially distorted context in the right image to fill-in the hole.

Fig. 1. Self-similarity under different distortions. On the left: two views of the same scene related by a projective transformation. On the right: self-similar texture underwent a severe fish-eye lens distortion.

In this work we address this issue by transforming known patches before pasting them in the inpainting domain. The transformation is determined for each patch in a fully automatic way. Moreover, instead of searching for an appropriate transformation in a high dimensional space, our approach allows us to determine a single transformation from the surrounding texture content. As opposed to some previous works which only consider rotations and scalings, we can handle full affinities, which in principle extends the applicability of the method to any transformation that can be locally approximated by an affinity, such as perspective distortion.

We follow the approach recently proposed in [1], where affine covariant structure tensor fields computed a priori in each image are used to define an *affine invariant similarity measure* between patches. We incorporate this measure into a variational inpainting formulation. The affine covariant structure tensors determine elliptical patches at each location of the image domain. Due to the affine covariance property of the structure tensors, these patches transform appropriately when computed on an affinely transformed version of the image. Figure 2 illustrates the patches defined by the affine covariant structure tensors of [1], computed for a set of corresponding points in two images related

Fig. 2. Affine covariant neighborhoods (patches) computed at corresponding points in two images taken from different viewpoints. Despite the change in appearance, patches capture the same visual information.

by a homography. Note that even though the transformation is not an affinity, the patches still match, since a homography can be locally approximated by an affinity.

The paper is organized as follows. In Sect. 2 we review the related work. Then in Sect. 3 we summarize the results of [1] which motivates the definition of the similarity measure that we use. Section 4 is devoted to the proposed inpainting method. Section 5 covers some implementation details. In Sect. 6 we present experiments, asserting the validity of our theoretical approach, together with a comparison with well-known exemplar based methods. Finally, Sect. 7 concludes the paper.

2 Related Work

Most inpainting methods found in the literature can be classified into two groups: geometry- and texture-oriented, depending on how they characterize the redundancy of the image.

The geometry-oriented methods formulate the inpainting problem as a boundary value problem and the images are modeled as functions with some degree of smoothness expressed, for instance, in terms of the curvature of the level lines [2–5], with propagation PDE's [6], or as the total variation of the image [7]. These methods perform well in propagating smooth level lines or gradients, but fail in the presence of texture or big inpainting domains.

Texture-oriented (also called exemplar-based) methods were initiated by the work of Efros and Leung [8] on texture synthesis. In that work the idea of self-similarity is exploited for direct and non-parametric sampling of the desired texture. The self-similarity prior is one of the most influential ideas underlying the recent progress in image processing and has been effectively used for different image processing and computer vision tasks, such as denoising and other inverse problems [9–13]. It has also found its application to inpainting: the value of each target pixel x in the inpainting domain can be sampled from the known part of the image or even from a vast database of images [14].

The exemplar-based approach to inpainting has been intensively studied [15–20]. However, many such methods are based on the assumption that the information necessary to complete the image is available elsewhere and can be copied *without any modification but a translation.*

Some works consider a broader family of transformations. Drori et al. [21] used heuristic criteria to vary the scale of patches. Mansfield et al. [22] and Barnes et al. [23] extended the space of available patches by testing possible rotations and scales of a source patch. The search in the space of available patches is usually performed by a collaborative random search. However, this implies that for each query patch, the position of the matching patch as well as the parameters of the transformation (scale, rotation angle, tilt, etc.) must be determined. The high dimensionality of the parameter space makes the search problem very computationally expensive and the excessive variability of candidates may lead to unstable results. In order to restrict the search space, the authors of [5] propose to combine an exemplar-based approach that includes all rotated patches, with a geometric guide computed by minimizing Euler's elastica of contrasted level lines in the inpainted region.

Several authors [24, 25] have addressed this issue using some user interaction to guide the search process. For example, the user provides information about the symmetries in the image, or specifies 3D planes which are then used for rectification and the rectified planes in turn are used to look for correspondences. Recently, Huang et al. [26] proposed a method for automatic guidance that searches for appropriately transformed source patches. It starts by detecting planes and estimating their projection parameters, which are then used to transform the patches. This allows one to handle perspective transformations, in situations when representative planes can be detected.

Most of those works use a similarity measure, either explicitly or implicitly, to compute a matching cost between patches. We propose to use an affine invariant similarity measure which automatically distorts the patches being compared [1]. Our method considers a rich patch space that includes all affine-transformed patches, furthermore, for each pair of patches the transformations are uniquely determined using the image content. This effectively limits the search space, making the method more stable. Since the patch distortions depend on the texture content of the image, our technique is related in that sense to a shape-from-texture approach [27–29].

In this paper we extend the variational framework described in [17, 18, 20] proposing a new energy and an optimization algorithm for affine invariant exemplar-based inpainting.

Let us finally note that [30] proposed a self-similarity measure for image inpainting, comparing dense SIFT descriptors on square patches of a fixed size. However, the method is not fully affine invariant, for example, neither the dense SIFT descriptors nor the square patches are scale invariant. Several authors have addressed the affine distortion and affine invariance problem in other contexts such as image comparison [31], object recognition [32], and stereo [33].

3 An Affine Invariant Similarity Measure

Non-local self-similarity is an accepted prior for natural images. To formalize it, a patch similarity or comparison measure is needed. Let us consider the general problem of comparing patches on two images $u : \Omega_u \to \mathbb{R}$ and $v : \Omega_v \to \mathbb{R}$, $\Omega_u, \Omega_v \subseteq \mathbb{R}^2$ (for simplicity, we can assume the image domains to be $\mathbb{R}^2$). A widely used comparison measure between two patches centered respectively at x and y is the weighted squared Euclidean distance

$$\mathcal{D}(t, x, y) = \int_{\mathbb{R}^2} g_t(h)(u(x + h) - v(y + h))^2 \, dh, \tag{1}$$

where g_t is a given window that we assume to be Gaussian of variance t. The Gaussian g_t represents a weighted characteristic function of both patches being compared and determines the size of the patches or, in other words, the scale.

In many occasions, similar patches exist in the image but have undergone a transformation, for example due to a different position with respect to the camera. The Euclidean distance is not appropriate for detecting these similarities. Consider for example a simple case in which v is a rotated version of image u. If the rotation is known, we should use the Euclidean distánce between patches in u and rotated patches in v, namely

$$\mathcal{D}^R(t, x, y) = \int_{\mathbb{R}^2} g_t(h)(u(x + h) - v(y + Rh))^2 \, dh. \tag{2}$$

In a more realistic scenario, one does not know the appropriate transformation that matches both patches being compared and even whether it exists. Some previous works addressed this issue by searching among all possible transformations [22,23] which involves probing of all the parameters (scale, rotation angle, etc.). The high dimensionality of the parameter space makes the problem very difficult. In this paper we use an affine invariant similarity measure, introduced in [1], that automatically deduces this transformation from the local texture context.

The similarity measure defined in [1] is based on affine covariant structure tensor fields *a priori* computed in each image. It was derived as an approximation to a more general framework introduced in [34], where similarity measures between images on Riemannian manifolds are studied.

In the remainder of this section we present an alternative, self-contained overview of this similarity measure. We first briefly discuss the concept of affine covariant structure tensors. Then we describe an algorithm to compute them. Finally, we show how they are used to define the affine invariant similarity measure and establish the relation between our derivation and the theory of [1,34].

3.1 Affine Covariant Structure Tensors

Given a real-valued image u, we consider an image-dependent structure tensor field T_u as a function that associates a structure tensor (a symmetric, positive semi-definite 2×2 matrix) to each point x in the image domain. As before, for simplicity we assume the image domain to be $\mathbb{R}^2$. The structure tensor field is said to be *affine covariant* if, for any affinity A,

$$T_{u_A}(x) = A^T T_u(Ax)A, \tag{3}$$

where $u_A(x) := u(Ax)$ denotes the affinely transformed version of u. Given a structure tensor $T_u(x)$ we can associate with it an elliptical region of "radius" r centered at x

$$B_u(x,r) = \{y : \langle T_u(x)(y - x), (y - x) \rangle \le r^2\}. \tag{4}$$

When the structure tensor is affine covariant, we have that

$$AB_{u_A}(x,r) = B_u(Ax,r).$$

This implies that the structure tensors can be used to define regions that transform appropriately via an affinity (Fig. 2).

As shown in [1], given two affine covariant structure tensors we can extract the affine transformation between the corresponding elliptical patches up to some rotation. Indeed, for any affine transformation A, there exists an orthogonal matrix R such that

$$A = T_u(Ax)^{-\frac{1}{2}} R T_{u_A}(x)^{\frac{1}{2}}. \tag{5}$$

This last equation provides an intuitive geometric relationship between the structure tensors, the associated elliptical regions and the affinity. Consider a point x and the corresponding affine covariant elliptic neighborhood $B_{u_A}(x)$. Mapping $B_{u_A}(x)$ by the affinity yields $B_u(Ax)$. The application of A can be decomposed in three steps. First, applying $T_{u_A}(x)^{\frac{1}{2}}$, we transform $B_{u_A}(x)$ into a circle or radius r. We refer to the resulting patch as a *normalized patch*. Then, a rotation is applied to the normalized patch. Finally, $T_u(Ax)^{-\frac{1}{2}}$ maps the rotated normalized patch to the elliptical neighborhood $B_u(Ax)$.

To fully determine the affinity A, one needs to find the rotation R. Any rotation would yield an affinity that maps the elliptical neighborhood associated with T_{u_A} at x to the one associated with T_u at Ax. For a wrong value of the rotation, the image content inside both neighborhoods will not match. Therefore, the right value for the rotation can be computed by aligning the image content of both patches. For this aim, we decompose the rotation as

$$R = R_u(Ax)R_{u_A}^{-1}(x), \tag{6}$$

where $R_u(Ax)$ and $R_{u_A}(x)$ are estimated from the image content in the patches. In practice, we calculate them by aligning the dominant orientation of the normalized patches to the horizontal axis. To compute the dominant orientation we use histograms of gradient orientations as in the SIFT descriptors [35].

3.2   Computation of Affine Covariant Structure Tensors

The following iterative algorithm introduced in [1] allows us to compute a dense field of affine covariant structure tensors and the associated neighborhoods on an image u:

$$T_u^{(k)}(x) = \frac{\int_{B_u^{(k-1)}(x,r)} Du(y) \otimes Du(y)\, dy}{\text{Area}(B_u^{(k-1)}(x,r))}, \tag{7}$$

where u is the given image and $B_u^{(k)}$ is the elliptical patch related to $T_u^{(k)}$, defined by

$$B_u^{(k)}(x,r) = \{y : \langle T_u^{(k)}(x)(y-x), (y-x)\rangle \le r^2\} \tag{8}$$

for $k \ge 1$, and

$$B_u^{(0)}(x,r) = \{y : |Du(x)(y-x)| \le r\} \tag{9}$$

for $k = 0$.

Throughout this paper we follow the notation of [1] and denote by $T_u(x)$ the affine covariant structure tensor $T_u^{(k)}(x)$ for a fixed value of k ($k = 30$ in all the experiments) and a given value of r ($r > 0$ is a free parameter which is in range $[250, 350]$ in our experiments). We denote by $B_u(x)$ the affine covariant neighborhood $B_u^{(k)}(x,r)$.

Notice that the structure tensor (7) is guaranteed to be affine covariant at any iteration of the scheme, therefore, the purpose of it is not to enforce affine covariance property, but rather to diminish dependency on the very first iteration.

3.3   An Affine Invariant Patch Similarity

Previously in this section we were considering two images u and u_A, related by a global affinity. For the patch comparison problem we can generalize our reasoning and consider two arbitrary images u and v. Let x and y be two given points in images u and v respectively. The structure tensors $T_u(x)$ and $T_v(y)$ define elliptical patches $B_{T_u}(x)$ and $B_{T_v}(y)$ around these points. In order to compare the patches, Eqs. (5) and (6) suggest the following mapping between the elliptical patches:

$$P(x,y) = T_v(y)^{-\frac{1}{2}} R_v(y) R_u^{-1}(x) T_u(x)^{\frac{1}{2}}. \tag{10}$$

We can interpret $P(x,y)$ as an affinity, mapping the elliptical patch associated with $T_u(x)$ into the one associated with $T_v(y)$. If u in the vicinity of x is an affinely transformed version of v in the vicinity of y, then $P(x,y)$ recovers the true affinity. An affine invariant patch similarity measure could be built by computing the distance between the elliptical patch at y and the elliptical patch at x transformed by $P(x,y)$. In practice, it is more convenient to transform both elliptical patches to the circle of radius r (Fig. 3) and compare the aligned normalized patches:

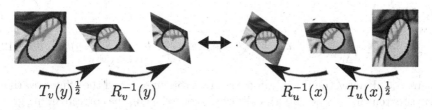

$$T_v(y)^{\frac{1}{2}} \qquad R_v^{-1}(y) \qquad\qquad R_u^{-1}(x) \qquad T_u(x)^{\frac{1}{2}}$$

Fig. 3. An affine invariant patch comparison which is achieved by normalizing the patches to discs and aligning them with suitable rotations.

$$\mathcal{D}^{\mathrm{a}}(t,x,y) =$$
$$\int_{\Delta_t} g_t(h) \left(u(x + T_u^{-\frac{1}{2}} R_u(x)h) - v(y + T_v^{-\frac{1}{2}} R_v(y)h) \right)^2 dh, \quad (11)$$

where Δ_t is a disc centered at the origin with radius proportional to the scale $t > 0$ and big enough to contain the effective support of the weighting function g_t. The distance $\mathcal{D}^{\mathrm{a}}$ provides an affine invariant distance between the patches $p_u(x)$ and $p_v(y)$. Here the patch $p_u(x) := p_u(x,\cdot)$ is defined by $p_u(x,h) := u(x+T_u(x)^{-\frac{1}{2}}h)$, with h belonging to Δ_t. We will apply it in Sect. 4 to exemplar-based inpainting. Let us also remark that formula (11) has the same complexity as the patch comparison formula (1).

The similarity measure corresponding to (11) was derived in [1] as a computationally tractable approximation of the linear case of the multiscale similarity measures introduced in [34]. There, the authors show that all scale spaces of similarity measures $\mathcal{D}(t,x,y)$ satisfying a set of appropriate axioms are solutions of a family of degenerate elliptic partial differential equations (PDE) in the variables (x,y). Images are considered in those papers as Riemannian manifolds endowed with a metric defined by a tensor field. If this tensor field is affine covariant, the resulting similarity measure is affine invariant. In this Riemannian framework $P(\cdot,\cdot)$ defines an isometry between the tangent spaces in two manifolds. The authors refer to it as the *a priori* connection, since it is related to the notion of connection appearing in parallel transport (see [34] for details).

WKB approximation method, named after Wentzel, Kramers and Brillouin, was used in [1] to find the approximate solution to a linear partial differential equation with spatially varying coefficients as a convolution with a short-time space-varying kernel.

The affine invariant patch distance (11) is used in the following section in a variational formulation for exemplar-based image inpainting and Sect. 6 will present inpainting results for both gray and color images. Let us note that for the color case we consider a generalization of (11) to multi-channel images. Let $\vec{u} : \Omega_u \rightarrow \mathbb{R}^M$ and $\vec{v} : \Omega_v \rightarrow \mathbb{R}^M$, $\Omega_u, \Omega_v \subseteq \mathbb{R}^2$ be multi-channel images (e.g., $M = 3$ for color images), then the corresponding affine invariant similarity measure is defined as

$$\mathcal{D}^{\mathrm{a},\mathrm{M}}(t,x,y) =$$

$$\int_{\Delta_t} g_t(h) \left\| \vec{u}\,(x + T_u^{-\frac{1}{2}} R_u(x)h) - \vec{v}\,(y + T_v^{-\frac{1}{2}} R_v(y)h) \right\|_2^2 dh, \quad (12)$$

where $\|\cdot\|_2$ denotes the Euclidean norm of vectors in $\mathbb{R}^M$ and the affine covariant structure tensors and neighborhoods are computed using the corresponding gray-value images associated to $\vec{u}$ and $\vec{v}$ by the iterative algorithm of Sect. 3.2.

4 Inpainting Formulation

Exemplar-based inpainting methods aim at filling-in the image so that each patch in the inpainting domain is similar to some known patch. This requires comparing known patches with partially or completely unknown patches. For this we extend the variational framework described in [17,18,20] by using the affine invariant similarity measure $\mathcal{D}^{\mathrm{a}}$ given in (11). We formulate the problem of inpainting from affinely transformed patches via the minimization of the following energy functional

$$E(u,\varphi) = \int_{\widetilde{O}} \mathcal{D}^{\mathrm{a}}\,(t,x,\varphi(x))\,dx, \quad (13)$$

where $O \subset \Omega \subset \mathbb{R}^2$ is the inpainting domain, $\hat{u} : \Omega \setminus O \to \mathbb{R}$ is the known part of the image, $\widetilde{O}$ includes all the centers of patches intersecting O and $\widetilde{O}^c$ is its complement, that is, $\widetilde{O}^c$ contains centers of fully known patches (see Fig. 4). The minimization of (13) aims at finding a visually plausible completion u of $\hat{u}$ in the unknown region O. The additional variable $\varphi : \widetilde{O} \to \widetilde{O}^c$ determines for each unknown target patch the location of a source patch from which the information will be copied.

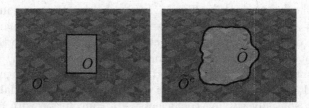

Fig. 4. Schematic representation of the sets O, O^c, $\widetilde{O}$ and $\widetilde{O}^c$.

This energy compares patches defined on elliptical regions centered at x and $\varphi(x)$. In the known part of the image, these regions are defined by the affine covariant structure tensors $T_{\hat{u}}$. Since the image is unknown inside the inpainting domain we have to estimate the structure tensors together with the image. The relationship between u and T_u introduces a complex dependency in the energy (13), which complicates its minimization. Therefore, we propose to relax it and consider instead the minimization of the energy

$$\widetilde{E}(u, \varphi, G) = \int_{\widetilde{O}} \int_{\Delta_t} g_t(h)$$
$$\left(u(x + G(x)^{-\frac{1}{2}} h) - \hat{u}(\varphi(x) + T_{\hat{u}}(\varphi(x))^{-\frac{1}{2}} R_{\hat{u}}(\varphi(x)) h) \right)^2 dh dx \quad (14)$$

where $G(x)$ is an invertible 2×2 matrix, $\forall x \in \widetilde{O}$. For now, we will not restrict the tensor field G to be given by the structure tensors T_u. Instead, we consider it as an additional variable, in principle independent of u. In this way, we do not have to deal with the complex dependency between T_u and u. In practice, due to the properties of the affine covariant structure tensors, it turns out that the $G(x)$ can be estimated from $T_u(x)$ and the additional rotation $R_u(x)$, as will be explained later in this section.

We compute a local minimum of the energy with an alternating optimization scheme on the variables u, G and φ which is summarized in Algorithm 1.

4.1 Image Update Step

In the image update step, φ and G are fixed, and the energy is minimized with respect to u. With the change of variables $z = x + G(x)^{-\frac{1}{2}} h$, the Euler-Lagrange equation leads to the following expression:

$$u(z) = \frac{1}{\varrho(z)} \int_{\widetilde{O}} g_t \left(G(x)^{\frac{1}{2}} (z - x) \right)$$
$$\hat{u} \left(\varphi(x) + T_{\hat{u}}(\varphi(x))^{-\frac{1}{2}} R_{\hat{u}}(\varphi(x)) G(x)^{\frac{1}{2}} (z - x) \right) \left| G(x)^{\frac{1}{2}} \right| dx, \quad (15)$$

where $\varrho(z)$ is normalization factor such that the sum is an average. The field G determines elliptical patches centered at each $x \in \widetilde{O}$. For each one of these patches a matching patch centered at $\varphi(x)$ is known, as well as its shape which is given by the structure tensor $T_{\hat{u}}(\varphi(x))$. The corresponding patch is then warped via the affinity

$$\widetilde{P}(\varphi(x), x) = G(x)^{-\frac{1}{2}} R_{\hat{u}}^{-1}(\varphi(x)) T_{\hat{u}}(\varphi(x))^{\frac{1}{2}}$$

and aggregated in the inpainting domain. Note that if $G(x)$ is given by $T_u(x)$, then $\widetilde{P}(\varphi(x), x)$ coincides with Eq. (10).

4.2 Affine Correspondence Update Step

Given a fixed u, the minimization of the energy with respect to (φ, G) can be performed as independent minimization of the patch distance function $\mathcal{D}^a$ for each $x \in \widetilde{O}$. This problem is very complex to solve since it is a nearest neighbor search where we also optimize for the affine transformation of the patch at x, given by G.

We will exploit the properties of the affine covariant structure tensors to obtain an approximate solution. For that, let us consider a completion candidate u and assume that a local vicinity of x on u is an affinely transformed version of a local vicinity of $\varphi(x)$ on $\hat{u}$. That is, $u(x+h) = \hat{u}(\varphi(x) + Ah)$, which is the case when x and $\varphi(x)$ do actually correspond. Setting $G(x)$ such that $G^{-\frac{1}{2}}(x)R_{\hat{u}}^{-1}(\varphi(x))T_{\hat{u}}(\varphi(x))^{\frac{1}{2}} = A$ will lead to a correct mapping and thus to the zero patch distance. On the other hand, using (10) we can find this affinity as $A = T_u(x)^{-\frac{1}{2}}RT_{\hat{u}}(\varphi(x))^{\frac{1}{2}}$ where R is some orthogonal 2×2 matrix and T_u is calculated on u. Then $G(x)$ such that $G^{\frac{1}{2}}(x) = R(x)T_u^{\frac{1}{2}}(x)$, together with $\varphi(x)$, will be global minimizers of the patch distance function $\mathcal{D}^{\mathrm{a}}$ at x. Therefore, we need to search only for $\varphi(x)$ and $R(x)$. An approximate $\varphi(x)$ can be found efficiently using our modified version of the PatchMatch algorithm [36], detailed in Sect. 5. The additional rotation $R(x)$ is determined as described in Sect. 3.1, in the same way as for the known part of an image. Notice that for notation consistency we should write $R(x) := R_u(x)^{-1}$.

Of course, if the neighborhood of x does not match any affinely transformed patch, then the estimated G might not minimize the patch distance $\mathcal{D}^{\mathrm{a}}$.

Algorithm 1. Approximate minimization of $\widetilde{E}(u, \varphi, G)$.

Input: Initial condition u^0 at O, tolerance $\tau > 0$
Output: Image completion u

repeat

    Compute affine covariant structure tensors $T_{u^{k-1}}(x)$ and rotations $R_{u^{k-1}}(x)$ for all $x \in \widetilde{O}$;

    Estimate optimal correspondences φ^k using the modified PatchMatch (see Section 5.2);

    Update image $u^k = \arg\min_u \widetilde{E}(u, \varphi^k, G^k)$, subject to $u^k = \hat{u}$ in O^c;

until $\|u^k - u^{k-1}\| < \tau$;

Another interpretation of the approximate minimization can be given by adding to the minimization of $\widetilde{E}(u, \varphi, G)$ the constraint that $G^{\frac{1}{2}}(x) = R_u(x)^{-1}T_u^{\frac{1}{2}}(x)$ for all $x \in \widetilde{O}$ and for some rotation matrix $R_u(x)$, namely,

$$\min \widetilde{E}(u, \varphi, G) \quad \text{subject to} \quad G^{\frac{1}{2}} = R_u^{-1}T_u^{\frac{1}{2}}.$$

The correspondence update step corresponds to the constrained minimization of the energy with respect to φ, G for a fixed image u. In the image update step the energy is minimized with respect to u, but without enforcing the constraint. Therefore, our approximate minimization can be seen as an alternating minimization applied to a constrained problem. The constraint is enforced only when minimizing with one of the variables (the pair φ, G). There are no theoretical guarantees for the convergence of such a scheme, although we have not yet encountered a practical case where the algorithm failed to converge.

5   Numerical Implementation

5.1   Image Update Step

The actual implementation of (15), that we use in our method, is

$$u(z) = \frac{1}{C(z)} \sum_{x \in \tilde{O}} g_t(T_u^{\frac{1}{2}}(x)(z-x)) \, m_c(x) w(x,z)$$

$$\hat{u}\big(\varphi(x) + P(x,\varphi(x))(z-x)\big) \, |T_u^{\frac{1}{2}}(x)|, \quad (16)$$

where $P(x,\varphi(x)) = T_{\hat{u}}^{-\frac{1}{2}}(\varphi(x)) R_{\hat{u}}(\varphi(x)) R_u^{-1}(x) T_u^{\frac{1}{2}}(x)$ is the estimated affinity mapping the target patch at x onto the source patch at $\varphi(x)$. The structure tensor field T_u is computed using the inpainted image u from the previous iteration.

Of course, in the discrete setting some kind of interpolation needs to be done after transforming one elliptical patch into another by $P(x,\varphi(x))$. For that we use the Nadaraya-Watson estimator [37,38] with Gaussian kernel.

The extra term m_c in (16) is a so-called *confidence mask* that takes values from 1 to 0, exponentially decreasing with the distance to the set of known pixels O^c. This mask is usual in exemplar-based inpainting, for instance, it is used in [16,20]. It helps to guide the flow of information from the boundary towards the interior of the inpainting domain, eliminating some local minima and reducing the effect of the initial condition. More precisely, we compute the confidence mask as

$$m_c(x) = (1 - c_0) \exp\left(-\frac{d(x,O^c)}{c_t} \right) + c_0,$$

where $d(x,O^c)$ is the distance from a point x to the boundary of the O^c set, such that $d(x,O^c) = 0$ when $x \in O^c$. Parameter $c_0 > 0$ defines the smallest (asymptotic) value that m_c can take and $c_t > 0$ controls the rate of decay. This confidence mask never changes during the inpainting process and can be precomputed for a given inpainting domain.

There is also another additional weighting term $w(x,z)$ in (16). In principle, all patches containing a pixel z contribute to its color value. To control the amount of contributors, we introduce the auxiliary Gaussian weight, that depends on the patch distance between a contributing patch and its corresponding known patch

$$w(x,z) = \exp\left(-\frac{\big(\mathcal{D}^a(t,x,\varphi(x)) - \min(\mathbf{D}(z))\big)^2}{2\sigma_{\text{cut-off}}^2(z)} \right), \quad (17)$$

where $\mathbf{D}(z) = \{\mathcal{D}^a(t,y,\varphi(y)) : z \in B_{T_u}(y)\}$ is a set of patch distances to known patches, computed among all patches contributing to z, and $\sigma_{\text{cut-off}}(z)$ defines a soft threshold for the patch distance values. This weight allows us to cut off contributors with low similarity (high distance) values, which in turn results in sharper reconstructions.

To compute $\sigma_{\text{cut-off}}(z)$ we begin by computing the first estimate for the cut-off distance

$$\mathcal{D}_1^a(z) = \gamma_{\text{cut-off}}(\max(\mathbf{D}(z)) - \min(\mathbf{D}(z))),$$

where $\gamma_{\text{cut-off}} \in (0,1)$ is a parameter of the method. Since distance values are usually distributed unevenly, the initial distance threshold $\mathcal{D}_1^a$ might discard too few or too many contributors. Therefore, we very roughly estimate the density of values that fall below $\mathcal{D}_1^a$ and refine the initial cut-off distance by

$$\mathcal{D}_2^a(z) = \gamma_{\text{cut-off}} \frac{\mathcal{D}_1^a \ |\mathbf{D}(z)|}{N_1},$$

where $|\mathbf{D}(z)|$ stands for the total number of elements in the set $\mathbf{D}(z)$ and $N_1 = |\{\mathcal{D}^a \in \mathbf{D}(z) : \mathcal{D}^a - \min(\mathbf{D}(z)) < \mathcal{D}_1^a(z)\}|$ is the number of distance values retained by $\mathcal{D}_1^a$. Then the final cut-off threshold is given by

$$\sigma_{\text{cut-off}}(z) = \frac{1}{6}\left(\mathcal{D}_1^a(z) + \mathcal{D}_2^a(z)\right).$$

The factor $\frac{1}{6}$ in the formula above implies that the Gaussian (17) approaches zero at the average cut-off distance between the first and second estimates. Figure 5 illustrates the cut-off distances for two different cases of distance values distribution.

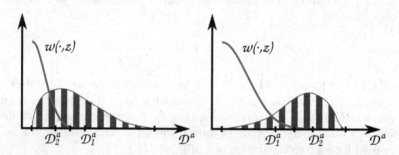

Fig. 5. Schematic depiction of two different cases of patch distance distribution. Gray filled-in curves represent histograms of distance values computed between patches overlapping $z \in O$ and their most similar known counterparts from $\widetilde{O}^c$. Weighting functions are shown in red. In both cases $\gamma_{\text{cut-off}} \approx 0.45$.

The energy (13) is non-convex and might have several local minima. As a consequence, there is a dependency on the initialization. To alleviate this dependency, we try to promote the propagation of information from the boundary towards the interior of the inpainting domain during the very first iterations of inpainting. Recall that the extended domain $\widetilde{O}$ contains the centers of all elliptical patches overlapping the inpainting domain. We enlarge $\widetilde{O}$ by a few pixels to capture a narrow stripe $\widetilde{O}^+ = (\widetilde{O} \oplus B) \setminus \widetilde{O}$ around the inpainting domain, that contains centers of completely known elliptical patches. Obviously, these elliptical patches in $\widetilde{O}^+$ do not intersect the inpainting domain. To make them contribute to the inpainting we should enlarge them first. For that we recompute them doubling the value of r. Notice that we use $r' = 2r$ only for the points within the stripe $\widetilde{O}^+$ and only in the image update step. We do not recompute the corresponding structure tensors, thus we only increase the sizes of these elliptical patches and do not modify their shapes. This additional contribution from elliptical patches, that do not depend by any means on the inpainting domain, boosts the information propagation at the boundaries of the inpainting domain. The width of the stripe $\widetilde{O}^+$ is set to 6 pixels in all our experiments.

5.2   Affine Correspondence Update Step

During the update of the correspondence map we compute an approximation of the nearest neighbor field using PatchMatch [23,36]. The PatchMatch algorithm speeds up the computation of optimal correspondences by exploiting the correlation between patches so that they can be found collectively. Since we are working with elliptical patches which might be arbitrarily rotated, we adapt the PatchMatch propagation scheme to take this into account. Let x be the current pixel and $d_1 = (\pm 1, 0)$, $d_2 = (0, \pm 1)$ be the directions of propagation. Then, the adjacent pixels $y_i = x - d_i$ ($i = 1, 2$) are tested during the propagation. Assume $i = 1$ (see Fig. 6). Pixel $\hat{y} = \varphi(y)$ is the current nearest neighbor candidate for y. The standard PatchMatch would try to propagate position $\hat{y} + d$ to pixel x (Fig. 6(a)). In contrast, we calculate the direction $\hat{d} = P(y, \hat{y})d$, where P is the a priori connection, and we try a few positions along that direction (Fig. 6(b)). This generalization gives a more meaningful propagation along edges.

Recall that to compare two elliptical patches we first transform them into discs of the same radius (see Sect. 3.3). As a result of this normalization we obtain two sets of scattered point, each of which is described by real-valued coordinates

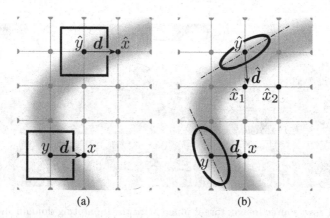

(a)   (b)

Fig. 6. Propagation directions in PatchMatch: (a) original scheme, (b) modified scheme.

and a color value. To be able to compare these sets, we use the Nadaraya-Watson estimator [37,38] with Gaussian kernel to interpolate them to a regular grid.

At early iterations of our algorithm, the inpainted image may be blurry. This is typical for iterative patch-based inpainting methods and is caused by aggregating patches that do not coincide in their overlap area at the initial iterations. As discussed in [1], the structure tensors are sensitive to blur, tending to larger elliptical patches in blurry regions. Essentially, smoothing of an image suppresses small details and produces the same effect as scaling the image down. The elliptical patches in turn capture larger areas. To compensate for this, we allow the parameter r to vary during the correspondence map estimation. That is, while $T_{\hat{u}}(\varphi(x))$ is always computed with the fixed r, say r_0 (a given parameter of the method), in the computation of $T_u(x)$ we consider a few (around five) values of r smaller than r_0 and select the one giving the smallest patch distance $\mathcal{D}^a$ between points x and $\varphi(x)$. Let us note, that to be able to compare patches, computed with different values of r, we scale the normalized patches to discs of radius one.

6   Experimental Results

In this section we present results obtained by the proposed method. For all the experiments in this section, we compare our results with the ones obtained by the multiscale NL-Means method [17,18] which we find to be a representative exemplar-based image inpainting method operating only with translations of patches. Whenever possible, we also compare against the method of [22] with a single scale and considering rotations, and the method of [26]. In both cases we use implementations provided by the authors.

As a sanity check we first test the proposed method on a synthetic example, displayed in Fig. 7. We take a textured image and create an affinely transformed

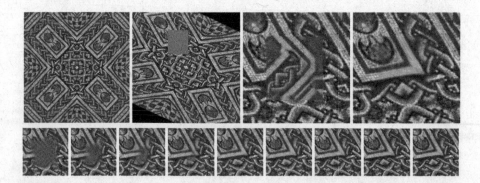

Fig. 7. First row: source image, target image with the inpainting domain shown in red, and close-ups around the inpainting area of the NL-Means result and the result of our method. Second row: evolution of the inpainting domain over iterations of our method (every second iteration).

version of it. We select a part of the transformed image as the inpainting domain. Instead of using the rest of the transformed image to copy information from, we make *the original (not transformed) image to be the source*. Let us remark that the ground truth affinity is not provided to the algorithm, hence, we test the ability of the proposed method to identify and copy affinely transformed patches. We do not show any results for [22, 26] for this experiment, since the available implementations do not support the use of a separate image as a source.

A more realistic case would be associated with a more general transformation. Since for planar objects a projective transformation can be locally approximated by an affinity, in the second example (shown in Fig. 8) we test the robustness of our method in the reconstruction of an image distorted by perspective. As usual in inpainting applications, in this experiment we use *the known part of the image as source*. We compare our method with the NL-Means method, that works only with translations, and additionally with the method of [22] in the mode when the rotations are also considered, and the method of [26]. Note that the latter method successfully determines a single plane in the image and, as expected, achieves a good reconstruction.

The third example (Fig. 9) demonstrates the reconstruction of a texture with some lens distortion applied to it. *The known part of the image is used as a source* and, like in all other experiments, just a rotation of source patches is not sufficient to obtain a good result. As in the previous case, here we compare our method

Fig. 8. First row: image with the inpainting domain shown in red. Second row: close-ups around the inpainting area of the NL-Means result, the result of [22] (considering rotations), the result of [26], and the result of our method. Third row: evolution of the inpainting domain over iterations of our method (every third iteration).

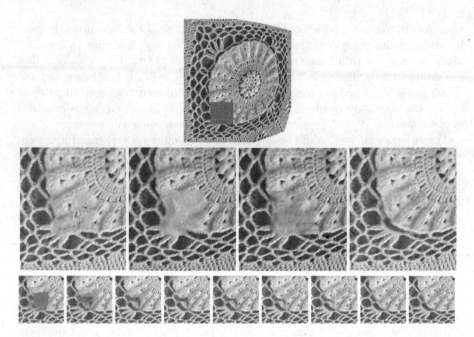

Fig. 9. First row: image with the inpainting domain shown in red. Second row: close-ups around the inpainting area of the NL-Means result, the result of [22] (considering rotations), the result of [26], and the result of our method. Third row: evolution of the inpainting domain over iterations of our method (every third iteration).

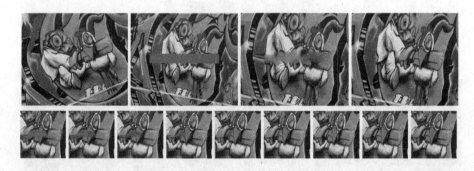

Fig. 10. First row: source image, target image with the inpainting domain shown in red, and close-ups around the inpainting area of the NL-Means result and the result of our method. Second row: evolution of the inpainting domain over iterations of our method (every second iteration).

with the NL-Means method (translations), the method of [22] (translations and rotations), and the method of [26] (projective transformation).

A final experiment, which is also potentially interesting for real applications, consists in inpainting one view of a scene using information from another view of the same scene. Figure 10 shows the results of this experiment where we have applied the proposed method to two views related by an unknown homography. As before, we compare our result with the result of the NL-Means method.

Let us note that the method of [22] also supports rotations plus scalings. However, we could not obtain meaningful results on these examples for this mode. It seems that the additional variability added by the scalings makes it easier for the algorithm to be trapped in a bad local minimum. For example, a constant region can be produced by scaling a small uniform patch.

Finally, we briefly discuss the limitations of the proposed method. Since the transformation between two patches is estimated from the surrounding texture, the method fails when there is not enough textural information (Fig. 11, first row). Severe transformations between pairs of patches may be recovered incorrectly. This can be illustrated by replacing the source image in the last experiment with a much more slanted view (Fig. 11, second row). The proposed method does not exploit the common multiscale scheme which limits the maximum possible size of the inpainting domain (Fig. 11, third row).

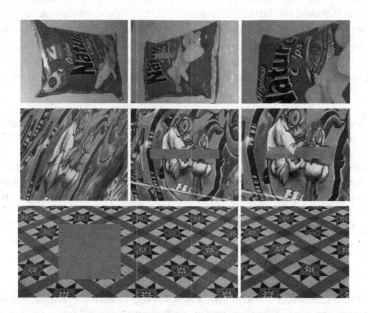

Fig. 11. Failure cases. From top to bottom: insufficient textural information, severe distortion, too big inpainting domain. More details are given in the text.

7   Conclusions

In this work we propose a new variational formulation for exemplar-based image inpainting that, for the first time, considers local full affine transformations with a tractable approximate optimization scheme. This is possible thanks to the use of an affine invariant patch similarity measure constructed from affine covariant structure tensors, both introduced in [1]. These structure tensors provide an efficient way to determine a unique affinity putting in correspondence any pair of patches. If the patches being compared are related by an affinity, then this affinity is recovered. We show that our method can be applied to complete a perspectively distorted color texture as well as to automatically inpaint one view of a scene using other view of the same scene as a source.

The problem of exemplar-based inpainting is a complex non-convex problem with many local minima. As pointed out in [5], adding transformations of patches makes it even more complex. Intuitively, the added variability makes it harder to distinguish "good" minima from other minima (a single pixel can be scaled to match a constant patch). The structure tensors are beneficial in this respect, because they constrain the number of ways in which a source patch can be transformed to match a target patch, thus eliminating some of the variability. This also allows us to design faster and more accurate minimization algorithms without the need to search the parameter space of the transformation family.

The proposed method works at a single scale. To better handle larger inpainting domains it would be desirable to develop a multiscale scheme, as is customary in the literature [17,18,20]. However, extending the multiscale approach to the problem of inpainting using affinely transformed patches is not trivial, since the filtering with an isotropic Gaussian breaks the affine invariance. Adapting multiscale inpainting approaches to this context is an interesting direction for future research.

Acknowledgements. The first, second and fourth authors acknowledge partial support by the MINECO/FEDER project with reference TIN2015-70410-C2-1-R, the MICINN project with reference MTM2012-30772, and by GRC reference 2014 SGR 1301, Generalitat de Catalunya. The second and third authors were partly founded by the Centre National dEtudes Spatiales (CNES, MISS Project), BPIFrance and Région Ile de France, in the framework of the FUI 18 Plein Phare project, the European Research Council (advanced grant Twelve Labours n246961), the Office of Naval research (ONR grant N00014-14-1-0023), and ANR-DGA project ANR-12-ASTR-0035.

References

1. Fedorov, V., Arias, P., Sadek, R., Facciolo, G., Ballester, C.: Linear multiscale analysis of similarities between images on riemannian manifolds: practical formula and affine covariant metrics. SIAM J. Imaging Sci. **8**, 2021–2069 (2015)
2. Masnou, S., Morel, J.M.: Level lines based disocclusion. In: Proceedings of IEEE ICIP, vol. 3, pp. 259–263 (1998)
3. Ballester, C., Bertalmío, M., Caselles, V., Sapiro, G., Verdera, J.: Filling-in by joint interpolation of vector fields and gray levels. IEEE Trans. Image Process. **10**, 1200–1211 (2001)
4. Masnou, S.: Disocclusion: a variational approach using level lines. IEEE Trans. Image Process. **11**, 68–76 (2002)
5. Cao, F., Gousseau, Y., Masnou, S.: Geometrically guided exemplar-based inpainting. SIAM J. Imaging Sci. **4**, 1143–1179 (2011)

6. Bertalmío, M., Sapiro, G., Caselles, V., Ballester, C.: Image inpainting. In: Proceedings of SIGGRAPH, pp. 417–424 (2000)
7. Chan, T., Shen, J.H.: Mathematical models for local nontexture inpaintings. SIAM J. App. Math. **62**, 1019–43 (2001)
8. Efros, A.A., Leung, T.K.: Texture synthesis by non-parametric sampling. In: Proceedings of the IEEE ICCV, pp. 1033–1038 (1999)
9. Foi, A., Boracchi, G.: Foveated self-similarity in nonlocal image filtering. Hum. Vis. Electron. Imaging **XVII**(8291), 829110 (2012)
10. Buades, A., Coll, B., Morel, J.M.: A non local algorithm for image denoising. In: Proceedings of the IEEE Conference on CVPR, vol. 2, pp. 60–65 (2005)
11. Gilboa, G., Osher, S.J.: Nonlocal operators with applications to image processing. Multiscale Model. Simul. **7**, 1005–1028 (2008)
12. Peyré, G.: Manifold models for signals and images. Comput. Vis. Image Underst. **113**, 249–260 (2009)
13. Pizarro, L., Mrázek, P., Didas, S., Grewenig, S., Weickert, J.: Generalised nonlocal image smoothing. Int. J. Comput. Vis. **90**, 62–87 (2010)
14. Hays, J., Efros, A.: Scene completion using millions of photographs. In: SIGGRAPH. ACM, New York (2007)
15. Demanet, L., Song, B., Chan, T.: Image inpainting by correspondence maps: a deterministic approach. Appl. Comput. Math. **1100**, 217–50 (2003)
16. Criminisi, A., Pérez, P., Toyama, K.: Region filling and object removal by exemplar-based inpainting. IEEE Trans. IP **13**, 1200–1212 (2004)
17. Wexler, Y., Shechtman, E., Irani, M.: Space-time completion of video. IEEE Trans. PAMI **29**, 463–476 (2007)
18. Kawai, N., Sato, T., Yokoya, N.: Image inpainting considering brightness change and spatial locality of textures and its evaluation. In: Wada, T., Huang, F., Lin, S. (eds.) PSIVT 2009. LNCS, vol. 5414, pp. 271–282. Springer, Heidelberg (2009). doi:10.1007/978-3-540-92957-4_24
19. Aujol, J.F., Ladjal, S., Masnou, S.: Exemplar-based inpainting from a variational point of view. SIAM J. Math. Anal. **42**, 1246–1285 (2010)
20. Arias, P., Facciolo, G., Caselles, V., Sapiro, G.: A variational framework for exemplar-based image inpainting. Int. J. Comput. Vis. **93**, 319–347 (2011)
21. Drori, I., Cohen-Or, D., Yeshurun, H.: Fragment-based image completion. In: ACM SIGGRAPH 2003 Papers, vol. 22, pp. 303–312 (2003)
22. Mansfield, A., Prasad, M., Rother, C., Sharp, T., Kohli, P., van Gool, L.: Transforming image completion. In: Proceedings of BMVC, vol. 121, pp. 121.1–121.11 (2011)
23. Barnes, C., Shechtman, E., Goldman, D.B., Finkelstein, A.: The generalized PatchMatch correspondence algorithm. In: Daniilidis, K., Maragos, P., Paragios, N. (eds.) ECCV 2010. LNCS, vol. 6313, pp. 29–43. Springer, Heidelberg (2010). doi:10.1007/978-3-642-15558-1_3
24. Pavić, D., Schonefeld, V., Kobbelt, L.: Interactive image completion with perspective correction. Vis. Comput. **22**, 671–681 (2006)
25. Huang, J.B., Kopf, J., Ahuja, N., Kang, S.B.: Transformation guided image completion. In: International Conference on Computational Photography, pp. 1–9 (2013)
26. Huang, J.B., Kang, S.B., Ahuja, N., Kopf, J.: Image completion using planar structure guidance. ACM Trans. Graph. (Proc. SIGGRAPH 2014) **33**, 129:1–129:10 (2014)
27. Gårding, J.: Shape from texture for smooth curved surfaces in perspective projection. J. Math. Imaging Vis. **2**, 327–350 (1992)

28. Gårding, J., Lindeberg, T.: Direct computation of shape cues using scale-adapted spatial derivative operators. Int. J. Comput. Vis. **17**, 163–191 (1996)
29. Ballester, C., Gonzalez, M.: Affine invariant texture segmentation and shape from texture by variational methods. J. Math. Imaging Vis. **9**, 141–171 (1998)
30. Wang, Z.: Image affine inpainting. In: Campilho, A., Kamel, M. (eds.) ICIAR 2008. LNCS, vol. 5112, pp. 1061–1070. Springer, Heidelberg (2008). doi:10.1007/ 978-3-540-69812-8_106
31. Mikolajczyk, K., Schmid, C.: Scale and affine invariant interest point detectors. Int. J. Comput. Vis. **60**, 63–86 (2004)
32. Matas, J., Chum, O., Urban, M., Pajdla, T.: Robust wide-baseline stereo from maximally stable extremal regions. Image Vis. Comput. **22**, 761–767 (2004)
33. Gårding, J., Lindeberg, T.: Direct estimation of local surface shape in a fixating binocular vision system. In: Eklundh, J.-O. (ed.) ECCV 1994. LNCS, vol. 800, pp. 365–376. Springer, Heidelberg (1994). doi:10.1007/3-540-57956-7_40
34. Ballester, C., Calderero, F., Caselles, V., Facciolo, G.: Multiscale analysis of similarities between images on riemannian manifolds. SIAM J. Multiscale Model. Simul. **12**, 616–649 (2014)
35. Lowe, D.: Distinctive image features from scale-invariant keypoints. Int. J. Comput. Vis. **60**, 91–110 (2004)
36. Barnes, C., Shechtman, E., Finkelstein, A., Goldman, D.B.: PatchMatch: a randomized correspondence algorithm for structural image editing. In: ACM SIGGRAPH 2009 Papers, pp. 1–11. ACM, New York (2009)
37. Nadaraya, E.A.: On estimating regression. Theory Probab. Appl. **9**, 141–142 (1964)
38. Watson, G.S.: Smooth regression analysis. Sankhya: Indian J. Stat. Ser. A (1961–2002) **26**, 359–372 (1964)

Real-Time Visual Odometry by Patch Tracking Using GPU-Based Perspective Calibration

Rafael F.V. Saracchini[1]([✉]), Carlos A. Catalina[1], Rodrigo Minetto[2], and Jorge Stolfi[3]

[1] Department of Simulation and Control, Technological Institute of Castilla y León, Burgos, Spain
{rafael.saracchini,carlos.catalina}@itcl.es
[2] Department of Informatics, Federal University of Technology, Curitiba, Paraná, Brazil
rminetto@dainf.ct.utfpr.edu.br
[3] Institute of Computing, State University of Campinas, Campinas, Brazil
stolfi@ic.unicamp.br

Abstract. In this paper we describe VOPT (Visual Odometry by Patch Tracking), a robust algorithm for visual odometry, which is able to operate with sparse or dense maps computed by simultaneous localization and mapping (SLAM) algorithms. By using an iterative multi-scale procedure, VOPT is able to estimate the individual motion, photometric correction and reliability tracking confidence of a set of planar patches. In order to overcome the high computational cost of the patch adjustment, we use a GPU-based least-square solver, achieving real-time performance. The algorithm can also be used as a building block to other procedures for automatic initialization and recovery of 3D scene. Our tests show that VOPT outperforms the well-known PTAMM and the state-of-art ORB-SLAM algorithm in challenging videos using the same input maps.

Keywords: 3D tracking · Augmented reality · Camera calibration · Real-time · GPU processing

1 Introduction

Visual odometry (VO) is the process of determining the motion of a video camera relative to a static environment by comparing successive video frames [30]. This technique is used in several applications in the field of robotics, entertainment and autonomous navigation of terrestrial and aerial vehicles [34]. Specifically, it is a core component in most *simultaneous tracking and mapping* (SLAM) algorithms, which acquire a 3D model of the environment as it is traversed. See Fig. 1.

In this paper we describe a robust and efficient algorithm for visual odometry, which we call VOPT (Visual Odometry by Patch Tracking). The VOPT algorithm, with a GPU-based real-time implementation, uses a sparse model of

© Springer International Publishing AG 2017
J. Braz et al. (Eds.): VISIGRAPP 2016, CCIS 693, pp. 475–492, 2017.
DOI: 10.1007/978-3-319-64870-5_23

Fig. 1. Input frame and associated visual odometry reference maps: (top) dense depth map and (bottom) sparse patch map.

the environment, consisting of an unstructured collection of salient points in 3D space (optionally with surface normals) that are associated to salient features of the video frames. The positions and normals of those tracked features can be obtained either by multiview stereo or RGBD cameras. The proposed method also uses local photometric adjustments of individual features to cope with variations in lighting, and is relatively insensitive to motion blur and image noise.

This paper is organized as follows. In Sect. 2 we review the literature on visual odometry. In Sects. 3 and 4 we describe the environment model and the algorithm. The experimental evaluation is reported in Sect. 5. Finally, in Sect. 6 we state the conclusions.

2   Related Work

Most visual odometry algorithms can be divided in three categories: *feature-based*, *appearance-based* and *patch-based*.

Feature-based algorithms [5,6] identifies a large set of distinguished *features* (such as corners and edges) in a frame, matching them with a reference set and estimating motion accordingly. By computing the positions of features with known 3D position its possible estimate the camera motion position for each captured frame. The most popular algorithm of this class and still considered its gold standard is PTAMM [5] and its variants [15,34]. This class of algorithms is relatively insensitive to lighting variations and occlusions. However, the feature detection step is rather costly, since it must allow for affine distortions of the feature's image between frames [1,2,17]. Also, they are badly affected by motion blur and noise which hinders the matching step. In order to offset the computational cost, PTAMM relies in features detected by the FAST [26] detector, which led to limited efficacy in recovery key-frame matching. State-of-the-art algorithms in this class actually use low-cost rotational and scale invariant feature descriptors such as ORB [27] in ORB-SLAM [21] and BRIEF [16].

An appearance-based or *direct* algorithm requires a 3D geometric model of the environment trying to determine the camera pose for each frame by minimizing the difference between each observed frame and the rendering of that model with the tentative camera position. This approach using a dense depth map representation was tried by the DTAM algorithm of Newcombe *et al.* [24]. Such kind of algorithms are usually associated with depth sensors to provide accurate odometry for *augmented reality* applications [20]. In order to reduce computational costs for monocular methods, Engel *et al.* [8] and Forster *et al.* [10] proposed the use of a *semi-dense* model, containing only the parts of the depth map near salient elements like edges and corners. Other approaches such as DPPTAM [7] assumes that image areas with low color gradients are planar, generating denser maps from semi-dense estimates. Engel later extended his semi-dense odometry method to a full SLAM algorithm, the LSD-SLAM [9], which is considered the state-of-art for appearance-based methods. These methods are less sensitive to motion blur and noise but are sensitive to lighting variations (although lighting variations can be alleviated by using dense maps [18]), and their monocular variants are reportedly less accurate than feature-based methods.

A patch-based algorithm follows an intermediate approach. It uses a very sparse model of the environment, consisting of a set of flat patches of the environment's surface, with known 3D positions, appearance, and normal vectors. For each frame, the camera position is adjusted until the computed projections of these surface patches match the corresponding parts of the frame image. Examples of this class are the Vachetti *et al.* [33], and the AFFTRACK algorithm of Minetto *et al.* [19]. These algorithms have lower computational cost than appearance-based algorithms because they don't have to do a full rendering of the environment model for each tentative camera pose; and also lower cost than feature-based algorithms because they can predict the position and deformation of each patch caused by the camera motion. VOPT can use either the feature-based approach or the patch-based approach, depending on the availability of surface data.

3   Data Representation

Video Frames. A video is assumed to be a sequence of N images $\mathcal{F}_0, \mathcal{F}_1,$ $\ldots, \mathcal{F}_{N-1}$, the *frames*, all with n_x columns by n_y rows of pixels. Each frame $\mathcal{F}_i$ can be viewed as a function from the rectangle $\mathcal{D} = [0, n_x] \times [0, n_y]$ of $\mathbb{R}^2$ to some color space, that is derived from the pixel values by some interpolation schema.

Camera Model. We denote by ρ the *projection function* that maps a 3D point p of $\mathcal{C}$ to a 2D point q of $\mathcal{D}$, where $\mathcal{C} \subset \mathbb{R}^3$ is the cone of visibility of the camera. We assume a pinhole camera model, so that the projection is determined by the 3×4 homogeneous *intrinsic matrix* K of the camera that defines the projection from the camera's coordinate system to the image sensor plane, and by the 4×4 homogeneous *extrinsic matrix* T that defines the current position and pose of the camera relative to the world's coordinate system. We assume K known and radial distortion removed *a priori* and that the camera has fixed focal length (zoom), so that the K matrix is the same for all video frames. We also assume that the video frames have been corrected and do not have any radial distortions.

The projection function ρ is defined by

$$q = \rho_{K,T}(p) = KT^{-1}p \tag{1}$$

where p and q are represented as homogeneous coordinate column vectors. Note that the T matrix includes the 3 coordinates of the camera's position and a 3×3 orthonormal rotation submatrix. We denote by $v(T, p)$ the unit direction vector in space from the point $p \in \mathbb{R}^3$ to the position of the camera implied by the extrinsic matrix T and the projection function $\rho_{K,T'}(p)$ which maps a point $p \in \mathbb{R}^3$ into image coordinates $q \in \mathbb{R}^2$.

Tracked Features. The essential information about the environment is a list of *features* $S = (s_0, s_1, \ldots, s_{n-1})$ on its surface. Each feature s_k has a *space position* $s_k.p$ in $\mathbb{R}^3$ and (optionally) a local *surface normal* $s_k.u \in \mathbb{S}^2$. (By convention, $s_k.u$ is $(0, 0, 0)$ if the normal is not available.) Each feature s_k is also associated to a sub-image of some reference image $s_k.\mathcal{J}$, called the *canonical projection* of the feature. We will assume that all reference images have the same domain $\mathcal{D}^* = [0, n_x^*] \times [0, n_y^*]$ and were imaged with the same intrinsic camera matrix K^*. The canonical projection of the feature is represented in the algorithm by a pointer to the reference image $s_k.\mathcal{J}$, the extrinsic camera matrix $s_k.T$ already determined for that reference image, a *bounding rectangle* $s_k.R \subset \mathcal{D}^*$, and a *mask* $s_k.\mathcal{M}$. The mask is a pixel array with values in $[0, 1]$, spanning that rectangle, that defines the actual extend and shape of the image feature on the reference image. See Fig. 2.

Key Frames an Pose Graph. In order to start the tracking, and restart it after a tracking failure, the algorithm requires a list $\mathcal{K} = (\mathcal{K}_0^*, \mathcal{K}_1^*, \ldots, \mathcal{K}_{N^*-1}^*)$ of reference images of the environment, the *key frames*. Specifically, for each key frame $\mathcal{K}_r^* \in \mathcal{K}$, the algorithm requires an extrinsic matrix T_r^* and a set S_r^* of features that are visible in it. It also requires a *pose graph* G, whose vertices are the key frames, with an edge between two key frames if their views have sufficient overlap.

This *key frame data* can be obtained from several sources: frame sequences, depth maps from devices like laser rangers and RGBD cameras, structure-from-motion, etc. The feature set and pose graph are then derived from sparse [5,8] or dense [23,35] maps computed by SLAM algorithms. The masks $s_k.\mathcal{M}$ can be determined by back-projecting the bounding rectangle $s_k.R$ into the scene

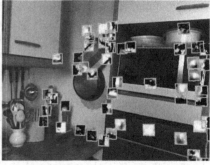

Fig. 2. Projection of features s_0 and s_1 on the image $\mathcal{K} = s_0.\mathcal{J} = s_1.\mathcal{J}$, showing their bounding rectangles $s_0.R, s_1.R$ (left) and their masks $s_0.\mathcal{M}, s_1.\mathcal{M}$ (right).

geometry, discarding regions which are not connected to its centre and computing a weight proportional with the difference of the surface normal with $s_k.\boldsymbol{u}$. In case of sparse maps, the mask is not computed and all its values are set to 1.

4   Tracking Algorithm

The VOPT algorithm is applied to each frame $\mathcal{I} = \mathcal{F}_i$ of a video sequence. Its normal case is described in Fig. 3. It takes as inputs the collection of tracked space features S, and the previous video frame $\mathcal{I}' = \mathcal{F}_{i-1}$ with its extrinsic camera matrix T'. For each feature s_k in S it also receives the *reliability weight* w'_k that measures its visibility in frame $\mathcal{I}'$ and how well it matched its canonical projection. If successful, the procedure returns the extrinsic matrix T for frame $\mathcal{I}$, and, for each feature s_k, an adjusted weight w_k. It requires an *startup procedure* [29,32] to select the features S to be tracked on the first image of the video, $\mathcal{I} = \mathcal{F}_0$, the extrinsic matrix T_0. We used the approach described in [29] in order to initialize the system and select the most likely key-frames and associated features from the pose-graph map. This procedure will be briefly described in Sect. 4.4.

The VOPT algorithm can be summarized as follows:

Procedure VOPT$(S, \mathcal{I}, \mathcal{I}', T', w')$ returns (T, w)
    1. $(d, w) \leftarrow$ DISPLACEMENTS$(S, \mathcal{I}', T', w', \mathcal{I})$
    2. $(T, w) \leftarrow$ ESTIMATECAMERAPOSITION(S, T', d, w);
    3. $(T, w) \leftarrow$ CAMERAADJUST$(S, \mathcal{I}, T, w)$

Fig. 3. The normal case of the VOPT algorithm.

The procedure DISPLACEMENTS in step 1 first computes the position c'_k of each feature point $s_k.p$ with $w'_k > 0$ on frame $\mathcal{I}'$, using the corresponding camera matrix T'. Then it uses the multiscale KLT algorithm of Kanade *et al.* [31] to determine the optical flow $\Phi : \mathcal{D} \to \mathcal{D}$ between the frames $\mathcal{I}'$ and $\mathcal{I}$. For efficiency, the flow is computed only in a subset of the domain, consisting of the union of 21×21 pixel squares centred at the points c_k'. The flow Φ is applied to the center $c_k' = \rho_{K,T'}(s_k.p)$ of each feature projection in $\mathcal{I}'$ to obtain its approximate apparent position c_k in $\mathcal{I}$. Note that each feature is independently mapped, without considering the constraints of perspective change. The procedure returns the displacement $d_k = c_k - c_k' \in \mathbb{R}^2$ between the projected feature positions on frame $\mathcal{I}'$ and their apparent positions on frame $\mathcal{I}$. The weight w_k of each feature is set to zero if $w_k' = 0$, or if any of these steps fails (e.g. if c_k' falls outside $\mathcal{D}$).

In step 2, an initial estimate of the camera matrix T is obtained from the 3D position $s_k.p$ of each feature, the camera matrix T' of the previous frame, and its apparent displacement d_k between the two frames. We use Klein's iterative M-Estimator [14] for the weighted camera motion estimation procedure ESTIMATECAMERAPOSITION, computing the *motion matrix* M such that $T = MT'$.

This motion matrix can be derived from a *motion vector* $\mu = (\mu_1, \ldots, \mu_6)$, by the exponential map $M = \exp(\mu)$ [11], where the first 3 terms of μ are the rotation and the last 3 the translation. By the observed pixel displacements d_k, one can estimate the values of μ by iteratively solving the following system

$$J^\mathsf{T} W J \mu = W d \tag{2}$$

where J is the Jacobian matrix of μ and by adjusting the diagonal weight matrix W in each iteration by the formula

$$W_{2k,2k} = W_{2k+1,2k+1} = w_k' \left(1 + \frac{r_k}{2}\right)^{-1} \tag{3}$$

where r_k, the residual of the computed and observed displacement, is the Euclidean length of the difference between the observed displacement d_k, the displacement computed from the 3D position $s_k.p$, the matrix T', and the current version of the matrix T. The M-estimator algorithm usually converges in 10 to 20 iterations. At the end, each reliability weight w_k is updated from the weight matrix W, by inverting formula (3).

In step 3, the initial estimate of the extrinsic matrix T for frame $\mathcal{I}$ (which is often imprecise, due to errors in the displacements d_k) is iteratively refined by the procedure CAMERAADJUST, simultaneously with the reliability weight w_k for each tracked feature s_k. This step is detailed in Sect. 4.1.

4.1   Camera Pose Refinement

The procedure CAMERAADJUST receives as input the list S of features that are being tracked, their current weights w_k, the next frame $\mathcal{I}$, and an initial estimate T of its extrinsic matrix. It returns the adjusted matrix T and an updated reliability weight w_k for each feature. The procedure is similar to the core of the AFFTRACK [19]. It adjusts the camera matrix T and the reliability weights w_k iteratively for each level ℓ of an image pyramid, by refining the apparent positions the features on the current frame, allowing for differences in illumination between the canonical projection and the current projection of each feature. Features that are not visible eventually get w_k set to zero, and are excluded from the computation of T. See Fig. 4.

Procedure CAMERAADJUST$(S, \mathcal{I}, T, w)$ returns (T, w)
1. Repeat while T does not converge, or until N_{ca} iterations:
2.   For each feature s_k in S,
3.   Let $P_k \leftarrow$ GETLOCALMAP(s_k, T)
4. $d_k \leftarrow (0, 0)$
5.   For each level ℓ from $N_{ms} - 1$ down to 0, do
6.   For each feature s_k with a valid map P_k, do
7.   Scale P_k, d_k to level ℓ.
8. $(d_k, w_k) \leftarrow$ FEATUREADJUST$(s_k, P_k, \mathcal{I}^{(\ell)}, d_k)$
9. $(T, w) \leftarrow$ ESTIMATECAMERAPOSITION(s, T, d, w)

Fig. 4. The iterative camera adjustment procedure.

Each iteration of CAMERAADJUST first computes a local 2D projective transformation P_k that describes the estimated position and deformation of each feature s_k from its key frame $s_k.\mathcal{J}$ to the image $\mathcal{I}$, due to camera motion between the two frames, assuming the current guess for the matrix T. Then, for each feature s_k, the procedure determines a displacement d_k from that estimated position to the position that yields the best match with the contents of frame $\mathcal{I}$. This displacement is computed by incremental adjustments at multiple scales of resolution. Finally, the displacements d_k are used to correct the matrix T, by Klein's M-Estimator. The iteration stops when the adjusted matrix T does not change significantly, or after N_{ca} iterations.

Feature Deformation Map. The projective map P_k computed in step 3 is obtained by lifting the rectangle $s_k.R$ to 3D space and projecting it onto the frame $\mathcal{I}$. More precisely, let Π_k be a plane in $\mathbb{R}^3$ that passes through the point $s_k.p$ and is perpendicular to the surface normal $s_k.u$. If the surface normal $s_k.u$ is not known, then Π is assumed to be perpendicular to $v(S, s_k.p)$. Each corner p of the rectangle $s_k.R$ is mapped to a point q of $\mathbb{R}^2$ by

$$q = \rho_{K,T}(\rho^{-1}_{K^*, S, \Pi_k}(p)) \tag{4}$$

where $S = s_k.T$ is the extrinsic matrix of the key frame of that feature, and $\rho^{-1}_{K^*, S, \Pi_k}$ is the function that back-projects that key frame onto the plane Π_k. However, if $s_k.u \cdot v(T, s_k.p) \leq 0$, or if a corner lies outside $\mathcal{D}$, the feature is considered invisible (self-occluded) in the frame $\mathcal{I}$ and the map P_k is not defined. If the projected corners fall behind the camera or outside $\mathcal{D}$ the feature is considered invisible, and P_k invalid. Partially visible features are discarded since the visible region within the frame domain $\mathcal{D}$ may be untextured or ambiguous, which may lead to incorrect adjustments with high weight values.

Multiscale Processing. In order to achieve robust and efficient discovery of large (multi-pixel) displacements, the feature adjustment in step 8 is repeated at multiple scales of resolution, using an image pyramid $\mathcal{I}^{(0)} = \mathcal{I}$, $\mathcal{I}^{(1)}$, ..., $\mathcal{I}^{(N_{ms}-1)}$, with level $\mathcal{I}^{(\ell)}$ being scaled by $1/2^\ell$ in each axis. Note that the weights w_k are recomputed from scratch at each level. Note also that the map P_k and the accumulated displacement d_k are initially scaled by $1/2^{N_{ms}-1}$ and then scaled by 2 when going from a coarse level to the next fine level. At each level, the FEATUREADJUST procedure can only adjust the displacement d_k by a few pixels. However, because of the multiscale processing, the total displacement can be of the order of $2^{N_{ms}-1}$ pixels on the original images.

4.2   Image Feature Position Adjustment

The FEATUREADJUST procedure is called for each feature s_k and scale ℓ, to refine the feature position on frame $\mathcal{I}$. The procedure receives a 2D projective map that defines the position and shape of the feature on $\mathcal{I}$, computed from the current guess of the matrix T for that frame; and a displacement d_k found by image matching at coarser scales. First, the procedure estimates the changes in the lighting of the feature between those two frames, then it applies a small correction to the displacement d_k with the Lucas-Kanade local image matching algorithm [3]. The weight w_k of each feature is defined based on the quality of the match. See Fig. 5.

Procedure FEATUREADJUST($s_k, P_k, \mathcal{I}, d_k$) returns ($d_k, w_k$)
1.   Set R_k to the bounding box of $P_k(s_k.R)$ displaced by d_k.
2.   If R_k is contained in $\mathcal{D}$ and has at least 4 rows and 4 columns,
   3. $(\alpha_k, \beta_k) \leftarrow$ PHOTOMETRICCORRECTION($s_k, P_k, d_k, \mathcal{I}$)
4.   If $\alpha_k \geq \alpha_{\min}$,
   5. $(d_k, w_k) \leftarrow$ LUCASKANADE($s_k, P_k, d_k, \mathcal{I}, \alpha_k, \beta_k$)
6.   else set $w_k \leftarrow 0$
7.   else set $w_k \leftarrow 0$

Fig. 5. The local feature position adjustment procedure.

Photometric Correction. The procedure PHOTOMETRICCORRECTION in step 3 determines the correction of pixel values under lighting effects, which is needed to compensate for changes in illumination between the two frames, including the effects of shadowing, surface orientation, glossy highlights, etc. We assume that these effects can be well approximated by an affine function $\mathcal{I}(P_k(p) + d_k) \approx \alpha_k \mathcal{K}(p) + \beta_k$, for any point p within the mask $\mathcal{N} = s_k.\mathcal{M}$; where $\mathcal{K} = s_k.\mathcal{J}$. Basically, α_k is the relative change in contrast, and β_k captures the difference of the 'black levels' of the two images around the respective projections of the feature. The values of α_k and β_k are obtained by weighted least squares, using the squared discrepancy Q defined by

$$Q(d_k, \alpha_k, \beta_k) = \sum_{p \in s'_k.R} \mathcal{N}(p)(\alpha_k \mathcal{K}(p) + \beta_k - \mathcal{I}(P_k(p) + d_k))^2 \qquad (5)$$

for the given value of the displacement d_k. If α_k is negative (meaning that darker areas of the canonical projection became lighter on the current image, and vice-versa) or zero, the procedure consider the feature lost and/or obscured by pixel noise, setting the weight w_k to zero. On the other hand, β_k can be positive or negative.

Local Feature Matching. In step 5, the Lucas-Kanade (LK) algorithm [3] is used to adjust the displacement d_k so that the pixel values of $\mathcal{I}$ are most similar to those of $\mathcal{K} = s_k.\mathcal{J}$, after mapped by P_k, displaced by d_k, and color-corrected according to α_k and β_k. The similarity is evaluated by the same Q functional (5) used for photometric correction. If the LK algorithm does not converge after N_{lk} iterations, we consider the adjustment failed, and setting w_k to zero. Otherwise we set w_k to $\exp(-z^2/2)$, where $z = Q/(\alpha_i^2\sigma^2)$ and σ is the expected standard deviation of the pixel noise.

4.3   GPU Acceleration

Steps 3 and 5 of FEATUREADJUST are responsible for most of the computational cost of the VOPT algorithm, requiring computation of sums over all pixels of a feature projection, mapped by the deformation map P_k. Compute their values for a large number of features is not feasible in real-time using CPU processing. In order to achieve real-time performance we based our implementation in the Nvidia's CUDA framework.

GPU-Based Least-Squares Adjustment. Both the photometric adjustment and LK adjustment step are weighted linear least squares problems where the quadratic error of a function $f(p)$ with its unknown coefficients $\mu = (\mu_1, \mu_2)$ in the form

$$f(p) = \sum_{i=1}^{2} \mu_i \phi_i(p) \qquad (6)$$

should be minimal compared with the observed data $Y(p)$ with weights W. The values of the coefficients μ can be determined by solving the normal equations

$$(X^{\mathsf{T}}WX)\mu = X^{\mathsf{T}}WY \qquad X_{ij} = \phi_j(p_i) \qquad (7)$$

We can generalize both the Photometric and Lucas-Kanade adjustment equations to a general form, where the functions ϕ, Y and coefficients μ take the following values

PHOTOMETRICCORRECTION   LUCASKANADE

$$\phi_1(p) = \mathcal{K}(p) \qquad\qquad \phi_1(p) = \frac{\partial}{\partial_x} \frac{(\mathcal{I}(P_k(p) + d_k) - \alpha\mathcal{K}(p) - \beta)}{2}$$

$$\phi_2(p) = 1 \qquad\qquad \phi_2(p) = \frac{\partial}{\partial_y} \frac{(\mathcal{I}(P_k(p) + d_k) - \alpha\mathcal{K}(p) - \beta)}{2} \qquad (8)$$

$$Y(p) = \mathcal{I}(P_k(p) + d_k) \qquad Y(p) = \mathcal{I}(P_k(p) + d_k) - \alpha\mathcal{K}(p) - \beta$$

$$\mu_1 = \alpha \qquad\qquad\qquad \mu_1 = d_x$$

$$\mu_2 = \beta \qquad\qquad\qquad \mu_2 = d_y$$

Since there are only two terms, we can simplify the weighted Grammian matrix $A = (X^\mathsf{T} W X)$ and the coefficients matrix $B = (X^\mathsf{T} W Y)$ from Eq. 7 as follows

$$(X^\mathsf{T} W X)_{ij} = \sum_{k=1}^{N} W_k \phi_i(p_k)\phi_j(p_k)$$
$$(X^\mathsf{T} W Y)_i = \sum_{k=1}^{N} W_k \phi_i(p_k) Y_k \qquad (9)$$

This formulation allows create a simple Least-Squares solver which is suitable to compute both adjustments by solving the system of linear equations $A\mu = B$. We call the 2D CUDA kernel LSKERNEL for each feature s_k with one thread for each pixel within the bounding box $s_k.R$. The algorithm of the least-squares CUDA 2D kernel is shown in Fig. 6.

Procedure LSKERNEL($s_k, P_k, \mathcal{I}, d_k$) returns (A, B)
Define p as kernel thread index i, j
Define $A_{--}, A_{-+}, A_{++}, B_-, B_+$ as $1 \times |s_k.R|$ arrays into kernel shared memory
1. compute $\phi_0(p), \phi_1(p)$
2. $A_{--}(p) \leftarrow W(p)\phi_0(p)^2$
3. $A_{+-}(p) \leftarrow W(p)\phi_0(p)\phi_1(p)$
4. $A_{++}(p) \leftarrow W(p)\phi_1(p)^2$
5. $B_-(p) \leftarrow W(p)\phi_0(p)Y(p)$
6. $B_+(p) \leftarrow W(p)\phi_1(p)Y(p)$
7. Synchronize threads
8. Parallel Sum Reduction [13] from $A_{--}, A_{-+}, A_{++}, B_-, B_+$ and B_-, B_+
9. if $i = 0$ and $j = 0$
10. $A_{00} \leftarrow A_{--}(ij), A_{10} \leftarrow A_{+-}(ij), A_{01} \leftarrow A_{+-}(ij), A_{11} \leftarrow A_{++}(ij)$
11. $B_0 \leftarrow B_-(ij), B_1 \leftarrow B_+(ij)$

Fig. 6. CUDA Kernel algorithm for parallel processing of weighted least squares.

In order to provide fast access to image data without using slow GPU RAM access, we store the keyframes $\mathcal{K}$, the query frame $\mathcal{I}$ and the weight masks $s_k.\mathcal{M}$ into texture memory, ensuring fast access to a pixel value from a non-discrete position p by fast cubic interpolation [28]. Storing the images into texture memory also avoids the usage of conditionals to determine if a pixel is or not within the query frame $\mathcal{I}$, reducing loss of efficiency by branching and divergence effects [12]. Since each CUDA kernel will process one single feature, we issue a call of the LSKERNEL for each feature into a CUDA stream queue. This allows process all the features in parallel. After all the features are processed, we compute the values for the coefficients μ in CPU, solving the simple least-square 2×2 systems computed by the GPU.

4.4 Initialization and Recovery

The VOPT algorithm requires an auxiliary *startup procedure* to select the features to be tracked on the first image of the video, $\mathcal{I} = \mathcal{F}_0$, and to define the extrinsic matrix T_0 for that frame. This procedure is also needed to restart the odometry if the VOPT algorithm fails, e.g. because there are not enough features that are visible in the frame (specifically, if the CAMERAADJUST procedure ends with fewer than 5 features with non-zero weight w_k). If VOPT fails when trying to go from some frame $\mathcal{I}' = \mathcal{F}_{i-1}$ to the next frame $\mathcal{I} = \mathcal{F}_i$, we give up on that frame and apply the startup procedure to frame $\mathcal{I} = \mathcal{F}_{i+1}$.

The startup procedure looks for the most similar key-frame $\mathcal{K}_r^*$ in $\mathcal{K}$ to the target image $\mathcal{I}$ by feature-matching, using the vocabulary tree technique of Nister and Stewenius [25] and extended in [29]. Afterwards, we build a list S of features to be tracked as the union of S_r^* and of all S_s^* such that $\mathcal{K}_s^*$ is adjacent to $\mathcal{K}_r^*$ in the pose graph G. The startup procedure then computes an approximate estimate T for the matrix of frame $\mathcal{I}$ using RANSAC. Finally it uses the CAMERAADJUST procedure to improve the estimated camera pose. This procedure, although very robust, is noticeably slower than the odometry procedure due the high computational cost of detecting and matching features.

5 Experiments

Test Videos. We tested the VOPT algorithm on both synthetic and real video sequences and compared its outputs with two reference algorithms: the PTAMM [5], which is the "gold standard" of visual odometry, and the ORB-SLAM [21], which is one of the state-of-art feature-based algorithms. In all tests, we set $N_{ca} = 2$, $N_{lk} = 10$ iterations, $N_{ms} = 3$ pyramid levels, $\alpha_{min} = 0$, and $\sigma = 0.1$. All the algorithms were tested in standard PC with an Intel i5 3.2 Ghz, 4 GB of RAM, and Geforce 650 GTX graphics card. The operational system was Linux Debian "wheezy" 64-bit.

The real videos were captured with a hand-held Logitech C930 camera, recording 840×480 frames at 30 fps, with fixed focal length. The intrinsic camera matrix K was determined by OpenCV calibration. One set of real videos (group M) was obtained by moving the C930 camera back and forth over a cluttered office desk. Another set of videos (group C) was taken by moving around the camera within a laboratory corridor.

The synthetic videos (group S) were generated by rendering a looped trajectory within a 3D model of an abbey with the Blender ray-tracer [4]. For each video frame $\mathcal{F}_i$, we saved the corresponding extrinsic camera matrix $T_i^{\#}$ used in the rendering. We selectively enabled challenging factors such as motion blur, localized illumination by spotlights, and occlusion by several objects. The Table 1 summarizes the video datasets used in the tests.

Table 1. Video dataset information. Notes: videos with (o) feature occlusion, (m) motion, (l) adverse lighting, (b) motion blur and (*) navigation in unmapped regions.

Video	Map	Frames	Notes
C0	C_d, C_d	1207	[o,b,m]
C1	C_d, C_d	425	[o,b,m]
C2	C_d, C_d	1015	[o]
C3	C_d, C_d	709	[o,m]
C4	C_d, C_d	666	[o,b,m,*]
M0	M_s	711	[o,m]
M1	$M1_s$	510	[o,b,m]

Video	Map	Frames	Notes
S0	S_s, S_d	241	-
S1	S_s, S_d	241	[o]
S2	S_s, S_d	241	[b]
S3	S_s, S_d	241	[o,b]
S4	S_s, S_d	241	[l]
S5	S_s, S_d	241	[l,o]
S6	S_s, S_d	241	[l,b]
S7	S_s, S_d	241	[l,b,o]

Key Frame Data. For each of the groups S, M, and C, we used the PTAMM [5] to obtain one set KPT of key frame data, including the set of key frames $\mathcal{K}$, the pose graph G, and, for each key frame $\mathcal{K}_r^*$, the extrinsic matrix T_r^*, and the set of features S_r^* using a reference video without adverse effects. This map was later converted to the internal structures of VOPT. For the ORB-SLAM algorithm, we used a separated map computed by processing the reference video its own semi-dense mapping procedure.

For the M and C groups, the reference video was acquired with the C930 camera, whether the reference video of S group was rendered with Blender, without the extra challenging factors. For each key-frame used in VOPT, we discarded redundant features, accepting at most 60, and a total limit of 180.

We also tested the algorithm with reference maps obtained from depth 3D reconstruction. For the groups S and C, we produced a set KSO of key frames containing dense data, constructing the pose-graph map with a RGBD-SLAM algorithm [29] for the C group and by depth-map rendering the 3D model of the abbey at regularly spaced intervals for the S group. The feature sets S_r^* were extracted by using a corner detector [31], discarding regions without depth, and normals computed directly from the depth map.

Algorithms and Metrics. For each test video, we ran the VOPT algorithm with the KPT key frame data set (S, M, and C groups) and with the KSO data set (S and C groups). For comparison, we also ran the visual odometry module of PTAMM (PTAMM-VO) and ORB-SLAM (ORBSLAM-VO) on each video with its original frame data with the key frame data KPT obtained by PTAMM, except that the feature sets were not trimmed. For each test run, we recorded the fraction κ of frames that were successfully calibrated (either by the startup procedure or by VOPT), the number L of tracking losses and average processing time per frame t in milliseconds. For the synthetic videos, we also computed the RMS error $\bar{e}$ between the estimated T_i and the ground-truth $T_i^{\#}$. Since the maps are not in the same scale or orientation, we aligned and re-scaled the PTAMM and ORB-SLAM maps by selecting the best correspondence between each map key-frame and applying the same transformation to the computed trajectory by the odometry algorithm. The scale was computed by computing a similarity transform between the two keyframe sets. In the case of the synthetic videos, we rescaled both maps by using the keyframe position and scale of the ground-truth map.

Results. Our results are summarized in Table 2, Figs. 7 and 8. It can be seen that VOPT was more accurate (smaller $\bar{e}$) and robust (larger κ, smaller L), out-performing PTAMM using less data and achieving best results when using dense data KSO and in both cases it performed much better than PTAMM-VO. The latter failed more often, with extreme motion blur or in absence of visible features, which is the case of video C4, and had more difficulty in recovery, in the presence of lighting variations, motion blur, and occlusions. ORBSLAM-VO showed much more stable behaviour than PTAMM, being insensitive to occlusions and managing to have less tracking loss events in the C0 and C1 videos due larger number of features tracked. Still it still it showed to be less robust than VOPT in cases of extreme blur and light variation. There was a noticeable difference between the maps generated between PTAMM and ORB-SLAM, namely, the map generated by the later had a certain degree of scaling and

Table 2. Tests with computed values of L (tracking losses), κ (successfully calibration), $\bar{e}$ (RMS error) and t (time in ms). Values in bold states the best performance in the test: error $\bar{e}$ for synthetic videos or successfully tracked frames κ for real videos.

Video	VOPT (sparse - KPT)				VOPT (dense - KSO)				PTAMM-VO				ORBSLAM-VO			
	L	κ	$\bar{e}$	t	L	κ	$\bar{e}$	t	L	κ	$\bar{e}$	t	L	κ	$\bar{e}$	t
S0	0	1.00	0.048	16.9	0	1.00	**0.008**	18.8	0	0.98	0.042	3.4	0	1.00	0.470	21.4
S1	0	1.00	0.049	18.6	0	1.00	**0.008**	20.9	0	0.98	0.043	3.2	0	1.00	0.470	20.5
S2	0	1.00	0.062	17.2	0	1.00	**0.055**	19.5	1	0.85	1.002	2.9	1	0.90	0.476	20.9
S3	0	1.00	0.061	19.2	0	1.00	**0.082**	21.6	1	0.84	0.732	2.8	3	0.88	0.515	20.4
S4	0	1.00	0.042	18.9	0	1.00	**0.009**	22.1	3	0.79	7.679	1.9	0	1.00	0.471	20.8
S5	0	1.00	0.037	19.6	0	1.00	**0.012**	23.1	3	0.48	2.897	3.0	0	1.00	0.471	20.1
S6	0	1.00	0.060	20.2	0	1.00	**0.046**	25.1	2	0.11	1.507	1.2	1	0.87	0.441	20.2
S7	0	1.00	0.064	20.7	0	1.00	**0.051**	24.8	2	0.14	4.774	1.1	1	0.83	0.439	20.6
M0	0	**1.00**	-	27.3	-	-	-	-	6	0.92	-	2.1	0	1.00	-	21.2
M1	0	**1.00**	-	30.7	-	-	-	-	13	0.67	-	3.1	2	0.99	-	21.7
C0	2	0.99	-	15.2	0	**1.00**	-	15.5	10	0.52	-	1.8	0	**1.00**	-	19.2
C1	1	0.97	-	14.9	0	**1.00**	-	18.8	2	0.88	-	2.0	0	**1.00**	-	19.6
C2	0	**1.00**	-	12.3	0	**1.00**	-	14.6	4	0.79	-	2.1	0	**1.00**	-	19.2
C3	0	**1.00**	-	13.5	0	**1.00**	-	16.4	3	0.81	-	1.7	9	0.98	-	20.8
C4	9	**0.83**	-	12.8	11	0.87	-	18.7	11	0.71	-	2.3	15	0.82	-	19.3

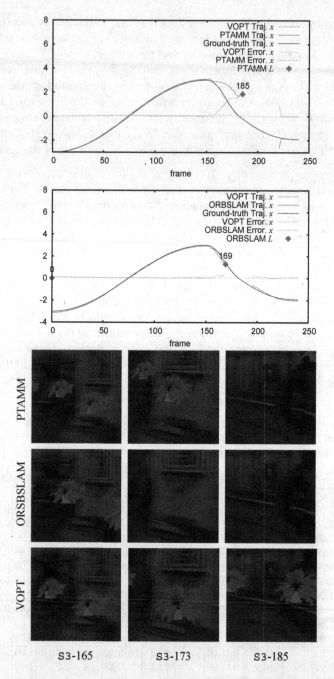

Fig. 7. Plot of x camera position of sequence S3, and AR output in the virtual scene with its respective sequence and frames. Note the increasing drift in the AR output of PTAMM. ORBSLAM has it AR output slightly displaced due the map difference, although the tracking is stable.

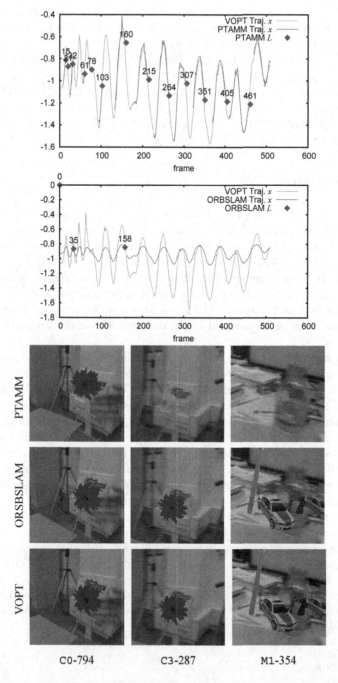

Fig. 8. Plot of x camera position of sequence M1, and AR output (plant/car) in occurrence of occlusions(a), motion blur(b) and both conditions(c) with its respective sequence and frames.

deviation in the y and z axis compared with the ground-truth trajectory. This deviation reflected to translation and scaling bias when comparing trajectories and increased error in its computed trajectory.

VOPT executed in 9–45 ms, in most cases depending on the number of iterations needed in steps 2–1 of CAMERAADJUST and in the Lucas-Kanade step 5 in FEATUREADJUST. These processing times are in par with ORB-SLAM in most of cases. However recovery after failure was much slower (50–180 ms) due the high cost of feature matching. In comparison, PTAMM processed each frame in 2–10 ms without requiring GPU processing.

6   Conclusions

Our tests show that the VOPT algorithm is slight slower than PTAMM's and ORB-SLAM's visual odometry mode, but is considerably more accurate and robust, especially in the presence of motion blur, occlusions, and localized lighting variations, combining strengths of feature-based and appearance based approaches. It owes these qualities mainly to the simultaneous adjustment of the camera matrix, the photometric correction parameters, and the reliability weights of individual features. VOPT was also used with success in indoor navigation algorithms of an assisted living device with on-site augmented reality capabilities.

Besides its robustness, the proposed algorithm still have limitations which shall be overcome in future work. Namely, it depends on an associated SLAM method to compute the features data needed for odometry. It also do not provide an efficient method to update the list of tracked features from the map after initialization. In the current implementation it allows to local tracking only, if the camera moves outside the region covered by the selected patches a tracking loss and subsequent reinitialization will follow. Finally the dependency on a GPU makes it less suitable for devices without a dedicated graphics card.

Future work will involve the development of light-weight version of VOPT aimed to mobile devices, as well the incorporation of the VOPT as effective module of a full SLAM system, overcoming the aforementioned limitations. More specifically, we aim to use it for the improvement of feature-based approaches when facing situations of extreme motion blur. We also aim for a more efficient usage of the GPU capabilities by exploiting cache and spatial locality.

Acknowledgements. This work was funded by the Ambient Assisted Living Joint Programme as part of the project Natural Communication Device for Assisted Living, ref. AAL-2010-3-116 [22].

References

1. Alcantarilla, P.F., Nuevo, J., Bartoli, A.: Fast explicit diffusion for accelerated features in nonlinear scale spaces. In: British Machine Vision Conference (BMVC) (2013)

2. Bay, H., Tuytelaars, T., Van Gool, L.: Surf: speeded up robust features. In: Leonardis, A., Bischof, H., Pinz, A. (eds.) Computer Vision - ECCV 2006. LNCS, vol. 3951, pp. 404–417. Springer, Berlin/Heidelberg (2006). doi:10.1007/11744023_32

3. Birchfield, S.: Derivation of kanade-lucas-tomasi tracking equation, December 2014. https://www.ces.clemson.edu/~stb/klt/birchfield-klt-derivation.pdf

4. Blender Online Community: Blender - a 3D modelling and rendering package. Amsterdam (2014). http://www.blender.org

5. Castle, R., Klein, G., Murray, D.: Video-rate localization in multiple maps for wearable augmented reality. In: IEEE International Symposium on Wearable Computers (ISWC), pp. 15–22 (2008)

6. Concha, A., Civera, J.: Using superpixels in monocular SLAM. In: IEEE International Conference on Robotics and Automation (ICRA), pp. 365–372 (2014)

7. Concha, A., Civera, J.: Dpptam: dense piecewise planar tracking and mapping from a monocular sequence. In: 2015 IEEE/RSJ International Conference on Intelligent Robots and Systems (IROS), pp. 5686–5693. IEEE (2015)

8. Engel, J., Sturm, J., Cremers, D.: Semi-dense visual odometry for a monocular camera. In: IEEE International Conference on Computer Vision (ICCV), pp. 1449–1456 (2013)

9. Engel, J., Schöps, T., Cremers, D.: LSD-SLAM: large-scale direct monocular SLAM. In: Fleet, D., Pajdla, T., Schiele, B., Tuytelaars, T. (eds.) ECCV 2014. LNCS, vol. 8690, pp. 834–849. Springer, Cham (2014). doi:10.1007/978-3-319-10605-2_54

10. Forster, C., Pizzoli, M., Scaramuzza, D.: SVO: fast semi-direct monocular visual odometry. In: IEEE International Conference on Robotics and Automation (ICRA), pp. 15–22 (2014)

11. Hall, B.C.: Lie Groups, Lie Algebras, and Representations: An Elementary Introduction, vol. 222. Springer, Heidelberg (2015)

12. Han, T.D., Abdelrahman, T.S.: Reducing branch divergence in GPU programs. In: Proceedings of the Fourth Workshop on General Purpose Processing on Graphics Processing Units, p. 3. ACM (2011)

13. Harris, M., et al.: Optimizing parallel reduction in CUDA. NVIDIA Develop. Technol. 2(4) (2007)

14. Klein, G.: Visual tracking for augmented reality. Ph.D. thesis, University of Cambridge (2006)

15. Klein, G., Murray, D.: Parallel tracking and mapping on a camera phone. In: IEEE International Symposium on Mixed and Augmented Reality (ISMAR), pp. 83–86, October 2009

16. Lim, H., Lim, J., Kim, H.J.: Real-time 6-dof monocular visual slam in a large-scale environment. In: 2014 IEEE International Conference on Robotics and Automation (ICRA), pp. 1532–1539. IEEE (2014)

17. Lowe, D.G.: Distinctive image features from scale-invariant keypoints. Int. J. Comput. Vis. 60(2), 91–110 (2004)

18. Maxime, M., Comport, A., Rives, P.: Real-time dense visual tracking under large lighting variations. In: British Machine Vision Conference (BMVC), pp. 45.1–45.11. BMVA Press (2011)

19. Minetto, R., Leite, N., Stolfi, J.: AFFTrack: robust tracking of features in variable-zoom videos. In: IEEE International Conference on Image Processing (ICIP), pp. 4285–4288 (2009)

20. Mohr, P., Kerbl, B., Donoser, M., Schmalstieg, D., Kalkofen, D.: Retargeting technical documentation to augmented reality. In: Proceedings of the 33rd Annual ACM Conference on Human Factors in Computing Systems, pp. 3337–3346. ACM (2015)

21. Mur-Artal, R., Montiel, J., Tardos, J.D.: Orb-slam: a versatile and accurate monocular slam system. IEEE Trans. Robot. **31**(5), 1147–1163 (2015)

22. NACODEAL: NACODEAL - Natural Communication Device for Assisted Living. European Union Project (2014). www.nacodeal.eu. Ambient Assisted Living Joint Programme ref. AAL-2010-3-116

23. Newcombe, R., Izadi, S., Hillige, O., Molyneaux, D., Kim, D., Davison, A., Kohli, P., Shotton, J., Hodges, S., Fitzgibbon, A.: KinectFusion: real-time dense surface mapping and tracking. In: IEEE International Symposium on Mixed and Augmented Reality (ISMAR), pp. 127–136 (2011)

24. Newcombe, R., Lovegrove, S., Davison, A.: DTAM: dense tracking and mapping in real-time. In: IEEE International Conference on Computer Vision (ICCV), pp. 2320–2327 (2011)

25. Nister, D., Stewenius, H.: Scalable recognition with a vocabulary tree. IEEE Comput. Soc. Conf. Comput. Vis. Pattern Recogn. **2**, 2161–2168 (2006)

26. Rosten, E., Drummond, T.: Fusing points and lines for high performance tracking. In: Tenth IEEE International Conference on Computer Vision, ICCV 2005, vol. 2, pp. 1508–1515. IEEE (2005)

27. Rublee, E., Rabaud, V., Konolige, K., Bradski, G.: ORB: an efficient alternative to SIFT or SURF. In: International Conference on Computer Vision, Barcelona (2011)

28. Ruijters, D., Romeny, B., Suetens, P.: Efficient GPU-based texture interpolation using uniform B-Splines. J. Graph. GPU Game Tools **13**(4), 61–69 (2008). http://dx.doi.org/10.1080/2151237X.2008.10129269

29. Saracchini, R.F.V., Ortega, C.C.: An easy to use mobile augmented reality platform for assisted living using pico-projectors. In: Chmielewski, L.J., Kozera, R., Shin, B.-S., Wojciechowski, K. (eds.) ICCVG 2014. LNCS, vol. 8671, pp. 552–561. Springer, Cham (2014). doi:10.1007/978-3-319-11331-9_66

30. Scaramuzza, D., Fraundorfer, F.: Visual odometry [tutorial]. Robot. Autom. Mag. IEEE **18**(4), 80–92 (2011)

31. Shi, J., Tomasi, C.: Good features to track. In: IEEE Conference on Computer Vision and Pattern Recognition (CVPR), pp. 593–600 (1994)

32. Straub, J., Hilsenbeck, S., Schroth, G., Huitl, R., Moller, A., Steinbach, E.: Fast relocalization for visual odometry using binary features. In: IEEE International Conference on Image Processing (ICIP), pp. 2548–2552 (2013)

33. Vacchetti, L., Lepetit, V., Fua, P.: Stable real-time 3D tracking using online and offline information. IEEE Trans. Pattern Anal. Mach. Intell. **26**(10), 1385–1391 (2004). http://dx.doi.org/10.1109/TPAMI.2004.92

34. Weiss, S., Achtelik, M., Lynen, S., Achtelik, M., Kneip, L., Chli, M., Siegwart, R.: Monocular vision for long-term micro aerial vehicle state estimation: a compendium. J. Field Robot. **30**(5), 803–831 (2013). http://dx.doi.org/10.1002/rob.21466

35. Whelan, T., Kaess, M., Fallon, M., Johannsson, H., Leonard, J., McDonald, J.: Kintinuous: spatially extended kinectfusion. Technical report MIT-CSAIL-TR-2012-020, MIT (2012)

Adaptive Non-local Means Using Weight Thresholding

Asif Khan and Mahmoud R. El-Sakka$^{(\boxtimes)}$

Computer Science Department, The University of Western Ontario, London, Canada
akhan644@uwo.ca, elsakka@csd.uwo.ca

Abstract. Non-local means (NLM) is a popular image denoising scheme for reducing additive Gaussian noise. It uses a patch-based approach to find similar regions within a search neighborhood and estimates the denoised pixel based on the weighted average of all pixels in the neighborhood. All weights are considered for averaging, irrespective of the value of the weights. This paper proposes an improved variant of the original NLM scheme by thresholding the weights of the pixels within the search neighborhood, where the *thresholded weights* are used in the averaging step. The threshold value is adapted based on the noise level of a given image. The proposed method is used as a two-step approach for image denoising. In the first step the proposed method is applied to generate a basic estimate of the denoised image. The second step applies the proposed method once more but with different smoothing strength. Experiments show that the denoising performance of the proposed method is better than that of the original NLM scheme, and its variants. It also outperforms the state-of-the-art image denoising scheme, BM3D, but only at low noise levels ($\sigma \leq 80$).

1 Introduction

Image denoising is the process of reducing noise artifacts from a digital image and it is one of the most fundamental problems in image processing. Noise is a random signal which affects the signal from the actual source by adding unwanted information to the signal. In digital image, noise causes random variation of brightness or color. It is usually produced during the image acquisition phase, caused by the sensors of digital cameras or scanners. Modern digital cameras have come a long way in using high quality sensors which have significantly reduced the presence of noise during image acquisition, but still noise can affect an image especially in low light conditions.

Noise in digital images can be categorized either as *additive* or *multiplicative* noise. Additive noise gets added with the image signal. It is modeled as:

$$v(i) = u(i) + \eta(i) \tag{1}$$

where $v(i)$ is the observed intensity value at pixel i, $u(i)$ is the actual raw intensity value and $\eta(i)$ is the random noise affecting pixel i. Multiplicative noise signal gets multiplied in the original image source. It is modeled as:

© Springer International Publishing AG 2017
J. Braz et al. (Eds.): VISIGRAPP 2016, CCIS 693, pp. 493–514, 2017.
DOI: 10.1007/978-3-319-64870-5_24

$$v(i) = u(i) \times \eta(i) \tag{2}$$

The main challenge of a denoising model is to reduce noise while preserving the texture, fine details and edges of an image. A model which is able to reduce significant noise artifacts but completely blurs the entire image, to a point where only minimal visual information can be extracted, is not ideal. Similarly, a denoising method which preserves the textures in the image but fails to reduce the noise to a satisfactory level is not an effective model as well.

The denoising methods can be generally categorized as either *spatial domain approaches* or *transform domain approaches*. The term spatial domain refers to the image plane itself [8] and the methods under this domain uses the raw intensity of the pixels to generate a denoised image. In transform domain approaches, the image is transformed to another domain, e.g., frequency domain using, for example, Fourier transform or wavelet transform. The transform domain decomposes smooth regions in an image into low frequencies, while edges and subtle information into high frequencies, thus making it easier to target and enhance certain regions in an image.

Among the various noise types, additive white Gaussian noise has attracted significant interest among researchers in the past few decades. Our work will focus only on this type of noise reduction. Additive white Gaussian noise is referred to noise signals with a zero-mean Gaussian distribution, having uniform power across the frequency band. Initial approaches to reduce the additive Gaussian noise included the use of basic linear filters, namely mean filter, median filter and Gaussian smoothing [8]. These filtering approaches use only the raw pixel values in a small local neighborhood around each pixels to determine the denoised image. These methods does not take into account the extent to which the neighborhood overlaps with smooth or textured regions. Thus the use of such linear filters are detrimental for edge and texture preservation, resulting in blurry denoised images. To address this problem, Perona and Malik proposed an iterative edge preserving method called Anisotropic Diffusion [16]. It attempts to determine whether a pixel is part of a smooth or a textured region and applies different degree of smoothing based on the characteristics of its locality.

Most of the earlier spatial domain denoising methods used pixel intensities within a defined local neighborhood around each pixel for estimating a denoised version of a noisy image. In recent years, Buades et al. proposed a non-local, patch based approach called *Non-local Means* (NLM) [3,4]. It takes advantage of the fact that similar local regions can be spread through out the entire image. Each of the pixels are denoised using a weighted average of all the pixels within a defined search area. The weights are assigned based on the local characteristics of the pixels used in the weighted averaging step. It uses weighted euclidean distance of the local region around the pixel being denoised, also referred to as the reference patch, and the local regions around each of the pixels within the search area. The patches with smaller euclidean distance, i.e., patches similar to the reference patch are assigned higher weights.

The concept of non-local based approach has also been applied to denoising methods in frequency domain. Dabov et al. proposed *Block Matching and 3D*

Filtering (BM3D) [6,7], using patch based concept for image denoising. It is a two-step process, where the first step groups similar patches into blocks, followed by a transform operation and hard thresholding of the transform coefficients to generate a basic estimate of the denoised image. The basic estimate is used in the second step to generate the actual denoised image. BM3D is one of the state-of-the-art approaches for denoising additive Gaussian noise.

In the field of spatial domain denoising, non-local means demonstrated significant improvement in denoising images affected with additive Gaussian noise and researchers have continued further work on the method and have proposed improvements for it. The exhaustive search nature of non-local means makes it computationally expensive. To improve the computation cost, several methods have been proposed. Tasdizen used principal component analysis (PCA) in conjunction with non-local means [19]. The image neighborhoods are projects to a lower dimension space using PCA and the reduced subspace is used for computing similarities. A similar dimension reduction approach has also been proposed by Maruf and El-Sakka [13], where the image neighborhood are projected to a lower dimension by using *t-test*.

Along with the research focused on improving the computation performance of non-local means, work has also been done on improving the denoising performance as well. Rehman and Wang proposed SSIM-based non-local means [17], utilizing structural similarity instead of euclidean distance when comparing local characteristics between patches. Chaudhury and Singer proposed *Non-Local Euclidean Medians* [5], replacing the use of mean with median. Zhu et al. proposed a two-stage non-local means approach with adaptive smoothing parameters [25]. It generates a basic denoised image by applying NLM in the first stage and the basic image is refined one more time in the second stage by using NLM but with smaller smoothing strength.

Non-local means and its variants have been used in various imaging applications such as medical imaging, including MRI brain images [10], CT scan imaging [11] and 3D ultrasound imaging [9]. It is also used in video denoising [1,22], surface salinity detection [24] and metal artifact detection [14].

Although much work has been done to improve non-local means, there are still possibilities for further improvements. In the weighted averaging step, non-local means considers all the pixels within a defined search area. The pixel patches having significantly different details than the patch of the reference pixel being denoised are likely to deviate the estimated denoised value of the reference pixel from its true noise-free pixel intensity, even with their smaller weights. In our proposed method we have thresholded the pixel weights and only the pixels with weight higher than the cut-off weight are considered for weighted averaging. The threshold is adapted based on the noise level of the given noisy image. The proposed method is applied in a two-step approach, where the first step applies the proposed method to generate a basic denoised image and in the second step the image generated from the first step is again denoised, using a smaller smoothing parameter. Experiments have illustrated better denoising performance of the proposed method compared to existing methods, e.g., the original NLM, a variant of NLM and BM3D, both in terms of objective measurements and visual image quality.

2 Background

2.1 Non-local Means (NLM)

Buades et al. proposed a non-local based approach for image denoising [3,4]. Images have redundant or similar patterns in them and the *Non-Local Means* (NLM) approach attempts to take advantage of such self-similarities to estimate the denoised gray level value of each pixel. Instead of using only a local region around each pixel for estimating the actual intensity of the pixel, NLM uses a non-local approach by searching for similar patches, within a certain search-bound, in the image. The center pixel of each patch contributes to a weighted averaging based on the similarity between the reference and search patches.

When comparing the reference patch to a search patch, a variation of the euclidean distance is measured. The euclidean distance measures the sum of squared difference between each pixel in a patch. To give more importance to pixels near the center of the patch, a Gaussian weight distribution is used, thus resulting in the final measurement being the weighted euclidean distance, $\|N(i) - N(j)\|_{2,a}^2$, where a is the standard deviation of the Gaussian kernel and $N(i)$ and $N(j)$ are the patches around pixel i and j, respectively. The weight associated with each of the search patches is based on the similarity with the reference patch. After calculating the euclidean distance between the patches, the weight is assigned using Eq. (3),

$$w(i,j) = \frac{1}{Z(i)} e^{-\frac{\|v(N_i) - v(N_j)\|_{2,a}^2}{h^2}},\tag{3}$$

where $v(N_i)$ and $v(N_j)$ are the gray values of the pixels in the patch centered on i and j respectively. $Z(i)$ is the normalizing constant as defined in Eq. (4),

$$Z(i) = \sum_j e^{-\frac{\|v(N_i) - v(N_j)\|_{2,a}^2}{h^2}}\tag{4}$$

The constant, h, controls the decay rate of the exponential weight function. Given a noisy image, the estimated value $NL[v](i)$, for pixel i, is computed as a weighted average of the center pixels of the patches in a certain search area, see Eq. (5),

$$NL[v](i) = \sum_{j \in I} w(i,j) v(j),\tag{5}$$

where $w(i,j)$ is the weight calculated based on the similarity of neighborhood around pixel i and j.

2.2 Two-Stage Non-local Means with Adaptive Smoothing Parameters

Zhu et al. [25] proposed a two-stage non-local means method with adaptive smoothing parameters. Based on the noise estimation of a given noisy image, smoothing parameter h_{basic} for the first stage is selected automatically and the basic denoised image is computed, as shown:

$$\hat{y}_{i,basic} = \sum_j w_{ij,basic} y_j, \tag{6}$$

where $w_{ij,basic}$ is the weight depending on the similarity between patches i and j and satisfies the usual conditions $0 \leq w_{ij,basic} \leq 1$ and $\sum_j w_{ij,basic} = 1$. The weight is calculated as show in Eq. (7)

$$w_{ij,basic} = exp(-\frac{1}{h^2_{basic}}\|P_i - P_j\|^2) \tag{7}$$

where P_i and P_j are the patches centered on pixel i and j and h_{basic} is the smoothing parameter which controls the decay rate of the exponential function. For the first stage the smoothing parameter is set as $h_{ij,basic} = 0.75 \times \sigma$.

Most of the image noise is removed after the first stage but for high noise levels some noise artifacts still remain in the image, thus the basic image is refined one more time. The resulting $\hat{y}_{basic}$ image of the first stage is again denoised using the non-local means method but using different smoothing parameters h_{final}. The final image is computed as shown in Eq. (8)

$$\hat{y}_{i,final} = \sum_j w_{ij,final} \hat{y}_{i,basic} \tag{8}$$

Similar to the first stage, the weights between patch i and j are calculated as:

$$w_{ij,final} = exp(-\frac{1}{h^2_{final}}\|P_i - P_j\|^2) \tag{9}$$

In the second step of the process, the smoothing parameter is defined as:

$$h_{final} = \begin{cases} \frac{\sigma^2}{100}, & \sigma < 30 \\ 0.5\sigma, & \sigma \geq 30 \end{cases}.$$

2.3 Non-local Euclidean Median

Chaudhury and Singer [5] proposed the Non-Local Euclidean Median, extending the concept of the original Non-Local Means scheme. The method is derived from the observation that the median is more robust to outliers than the mean. In the presence of noise in the image, the weights averaged over all possible patches, especially in a search-bound defined around image edges and lines, will move the resulting mean towards the outliers. The mean is the minimizer of $\sum_j w_j\|P - P_j\|^2$ over all patches P. Non-Local Euclidean Median proposed the select the patch, P, which minimizes $\sum_j w_j\|P - P_j\|$ and replace the noisy pixel value at position (x, y) in the image with the pixel value of the center of patch P.

2.4 Iterative Non-local Means

Brox and Cremers [2] proposed an iterative non-local means approach. The non-local means method is applied on an image in iterative mode. In each iteration, the similarity between patches are calculated based on the result of the previous iteration. After calculating the weights, the weighted averaging is done by multiplying the weights of a patch with it's center gray value in the noisy version of the image. The proposed method did well on regular textured image but in non-regular textured image the resulting denoising image lost texture and significant blurring is observed.

2.5 Block Matching and 3D Filtering (BM3D)

In the field of image denoising, specifically for white Gaussian noise, a popular method has emerged for tackling the problem of image noise called Block Matching and 3D Filter (BM3D). It was proposed by Dabov et al. [6,7]. BM3D demonstrated superior denoising performance compared to the existing methods and have attracted the attention of researchers working in the image of image enhancement and restoration.

BM3D is a 2-step process and has been inspired by the non-local concept first introduced in Non-Local Means (NLM). In the first step, the process starts by defining a local neighborhood, also referred to as the reference patch and searches for similar patches inside a search window, usually defined as a bounded area around the reference patch. The similarity between the search patch and the reference patch is decided based on a certain threshold. If the similarity is above the threshold, the search patch is marked as one of the similar patches. After calculating the similarity between the patches inside the bounded search area, all the similar patches are stacked together, building a 3D block. BM3D applies some 3D transformations on the block to transform from spatial domain to frequency domain. After the 3D transformation, the resulting coefficients are thresholded, called *hard thresholding*, where coefficients below a certain threshold are reduced to zero. The block coefficients are transformed back to the spatial domain, using inverse 3D transform. Next, BM3D generates a basic estimate of the denoised image from the block which has been inverse transformed. For estimating the values of the reference patch, all the patches in the 3D block are aggregated and it works by assigning different weights while estimating the pixel values in the reference patch.

In the second step of BM3D another grouping of patches are carried out, similar to that of the first step. This time however the grouping of patches into 3D blocks are done based on the basic estimate obtained from the first step. After 3D blocks are generated by selecting similar patches around the reference patch, the block is transformed using a 3D transformation. Based on the transformed coefficients the restored image coefficients are estimated using the restoration concept used in Wiener filter [21]. The Wiener shrinkage coefficients are calculated for the 3D blocks, followed by inverse transform to revert back to the spatial representation of the 3D block. Finally the Wiener shrinkage coefficients are multiplied by the 3D block to get the denoised representation of the block.

3   Non-local Means Using Adaptive Weight Thresholding

Non-local means method defines a search area of size $S \times S$ centered on the pixel, i, being denoised. The similarity of all the patches defined around each of the pixels within the search area is considered during the weighted averaging process, where higher weights are assigned to patches which are more similar, as determined by lower euclidean distance to the reference patch. The goal of the weighted averaging process is to estimate the true noise-free intensity value of pixel i, based on the similarity of the patches within the defined search area of the given noisy image. The inclusion of the center pixels of patches which are not very similar to the reference patch is likely to move the resulting estimate further from the true pixel intensity value of the noise-free image.

In our proposed method, only a subset of the available patch centers are considered for the final estimation of the denoised pixel. The patches are selected based on the similarity measure compared to the reference patch. Effectively, a cut-off weight, w_{thresh} is selected using a defined percentile position, $w_{percentile}$ among the available patch weights within the bounded search area and the weights of the patches are thresholded against w_{thresh}. All weights above w_{thresh} are unchanged and weights below w_{thresh} are reduced to zero, thus removing their pixel centers from the weighted averaging process. The selected percentile position is determined based on the noise level in a given image. In real systems, the actual amount of noise in a noisy image cannot be known beforehand. The noise can be estimated in digital image using fuzzy processing [18], image filters [15] and local variance estimate [12] methods.

For low noise levels, a higher cut-off weight, w_{thresh}, is selected for thresholding the patch weights and as the noise level of a given image increases, w_{thresh} is lowered to include more patch centers for averaging. For lower noise levels, only the patches with high similarity measure to a reference patch can be used to estimate a denoised image. The remaining patches can be considered as outliers. So, a higher cut-off threshold is selected for low noise levels. In high noise, the euclidean distance measurement may not give a true measure of patch similarity as it will end up comparing, to some extent, the noise between patches along with the structures of the patches. So, considering only the higher weighted patch centers, by keeping the threshold value high, can in fact deviate the denoised estimation from the true value. To mitigate this effect, the threshold value is lowered so that more pixels are averaged for attenuating the noise. The denoised image is calculated as shown in Eq. (10),

$$NL[v](i) = \sum_{j \in I} \hat{w}(i,j)v(j), \qquad (10)$$

where $\hat{w}(i,j)$ is the thresholded weight between patch at pixel i and patch at pixel j as shown in Eq. (11),

$$\hat{w}(i,j) = \begin{cases} w(i,j), \text{if } w(i,j) > w_{thresh} \\ 0, \qquad\qquad\qquad \text{otherwise} \end{cases} \qquad (11)$$

The proposed method is applied in two-step approach. In the first step, proposed method is used to generate a basic estimate of the denoised image. In the basic estimate, most of the noise is reduced but still some visible noise artifacts remain, especially for stronger noise levels and it is necessary to further denoise the basic image for better denoising [23]. As most of the noise is reduced in the basic image, similar regions can be identified more easily which helps to generate better denoised images in the second step. In the second step, the basic image is denoised using similar method used in the first step, but with a smaller smoothing parameters. To verify that the two-step approach is good enough, we conducted experiments to measure the improvement in the denoising performance with further steps and found the amount of improvements to be negligible, and even less in some cases.

Non-local means has two key parameters, namely the patch size and the search size. In our proposed method we have attempted to select the optimal patch and search window sizes based on the noise level in the image. We have empirically defined a model for selecting the patch size and the corresponding search window size for a noise level, σ, see Sect. 4.1.

4   Experimental Results

In this section we will report the experimental results of our proposed method. All the experiments were carried out on the standard Kodak gray-scale image set. It comprises of 24 gray-scale images of dimensions 768×512 and 512×768. The Kodak image set is shown in Fig. 1. For the purpose of our experimentation, the standard noise free image were contaminated by additive Gaussian white noise, randomly distributed throughout the image. The final intensity values were kept within the maximum intensity value of gray-scale images. The noise levels, determined by σ, ranges from 10 to 100, with a step size equals 10. The performance of our proposed method is compared with the original non-local means (NLM) [3,4], the two-stage non-local means (TS-NLM) [25] and the Block Matching and 3D Filtering (BM3D) [6,7] methods.

4.1   Parameter Selection

The primary parameters in the proposed method are:

1. Cut-off weight for thresholding
2. Patch size
3. Search window size

The parameters are determined empirically, based on experiments conducted on a set of test images. The performance measure of tuning each of the parameters are used to define the models for the parameters. The set of test of images is shown in Fig. 2. The training images are selected to address common image characteristics, including smooth regions, textured regions and fine details. The *Lena* and *Peppers* images have smooth regions, while *Barbara*, *Boats* and *Baboon* images has lot more texture and fine details. All the PSNR values reported are averaged after repeating each of the experiments 10 times.

(a) Boat2 (b) Light- (c) Woman (d) Sail (e) Statue (f) Model
house

(g) Beach (h) Bike (i) Bridge (j) Cottage

(k) Door (l) Flower (m) Raft (n) Girl

(o) Hats (p) House2 (q) Houses (r) Windows

(s) Island (t) Lake (u) Landscape (v) Lighthouse2

(w) Parrot (x) Plane

Fig. 1. Test image set (Kodak image set).

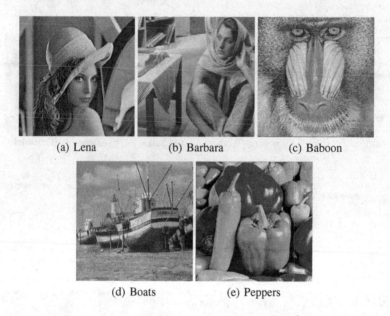

(a) Lena   (b) Barbara   (c) Baboon

(d) Boats   (e) Peppers

Fig. 2. Train image set.

4.2   Cut-Off Weight for Thresholding

The patch size and search window size are fixed and the weights are thresholded based on the cut-off weight percentile, $w_{percentile}$, determined by the model. For the purpose of determining the suitable thresholding model, the patch size is fixed at 7×7 and the search window size fixed at 11×11. Two different models are analyzed, the *linear model* and the *exponential model*.

Linear Model. The linear model for determining the cut-off weight percentile is defined as:

$$w_{percentile} = ceil(100 - a\sigma), \tag{12}$$

where $ceil()$ rounds a decimal value to the smallest following integer, coefficient a is a constant and σ is the standard deviation of Gaussian noise.

Different linear models are defined by changing the value of coefficient a. The PSNR comparison of using different linear models for weight thresholding is tabulated in Table 1. For noise levels $\sigma \leq 50$, a coefficient value of $a = 1$ generate better results and for $\sigma > 50$, the coefficient value $a = 0.5$ demonstrate better performance.

Exponential Model. The exponential model for determining the cut-off weight percentile is defined as:

$$w_{percentile} = ceil(100 \times e^{-0.01a\sigma}), \tag{13}$$

Table 1. PSNR comparison of linear models for different coefficient a.

Noise (σ)	$a = 0.25$	$a = 0.5$	$a = 1$
10	33.43	33.75	**33.87**
20	31.85	32.12	**32.25**
30	28.74	29.06	**29.15**
40	26.63	26.91	**26.98**
50	24.71	24.96	**25.07**
60	24.00	**24.28**	23.95
70	23.08	**23.33**	22.75
80	22.75	**23.14**	22.57
90	21.87	**22.28**	21.64
100	21.48	**21.71**	21.22
Average	25.85	**26.15**	25.94

where $ceil()$ rounds a decimal value to the smallest following integer, coefficient a is a constant and σ is the standard deviation of Gaussian noise.

The result of using the exponential model for determining the cut-off weight is shown is Table 2. For the exponential model, the coefficient value of $a = 1$ demonstrate better results for all noise levels. The average performance, as well as the noise-wise performance, of the exponential model, with coefficient $a = 1$, has better denoising performance compared to the linear models. So, the exponential model, with coefficient $a = 1$ is selected for determining the cut-off weight percentile for any given noise level, σ.

Table 2. PSNR comparison between exponential models for different coefficient a.

Noise (σ)	$a = 0.5$	$a = 1$	$a = 2$	$a = 4$
10	33.77	**33.89**	33.68	33.34
20	32.09	**32.22**	31.96	31.62
30	29.13	**29.36**	29.09	28.8
40	26.94	**27.11**	26.86	26.57
50	24.90	**25.13**	24.79	24.68
60	24.16	**24.35**	24.06	23.85
70	23.28	**23.48**	23.15	23.01
80	23.06	**23.24**	22.87	22.66
90	22.21	**22.34**	21.92	21.74
100	21.89	**22.05**	21.68	21.42
Average	26.14	**26.32**	26.01	25.77

4.3   Patch Size and Search Window Size

For varying the search window size, the patch size is fixed at 7×7. The weights are thresholded using the exponential model, selected in previous section. The result of varying the search window is shown in Table 3. The table shows that for $\sigma < 50$ the search window size 11×11 perform optimally and for $\sigma \geq 50$ the search size 21×21 performs best. To help determine a model for selecting the search window size, the denoising performance on the test images are again measured by varying the search window size, but this time using odd-integer window sizes only between 11×11 and 21×21, while using more noise levels.

Table 3. PSNR comparison by changing the search window size.

Noise (σ)	5×5	11×11	21×21	35×35
10	33.26	**34.19**	33.94	33.40
20	31.44	**32.48**	31.66	31.28
30	28.83	**29.35**	28.68	28.27
40	27.44	**28.12**	28.06	27.44
50	23.79	24.56	**25.05**	24.40
60	22.84	23.70	**24.27**	23.91
70	22.15	23.09	**23.73**	23.48
80	21.66	22.56	**23.41**	23.19
90	20.73	21.57	**22.88**	22.42
100	20.35	21.06	**22.52**	22.15
Average	25.25	26.07	**26.42**	25.99

The patch size and search window size for a given noise level are determined empirically, using an iterative learning approach on a training image set. At first, the patch size is fixed and the search window size is varied, for each noise levels, to select the best search window size. The noise levels, σ, ranges from 10 to 100, with a step size equals 5. Next, the patch size is varied for each noise levels, while using the best search window size for each noise as determined in the previous step. The best patch size for each noise level is used to find the corresponding optimal search window sizes one more time. This process is repeated until an iteration is reached where updating the optimal search window size for a noise level did not change the corresponding best patch size and vice versa.

To determine the patch and search window size models, the patch size is initially fixed at 7×7 and the search window size is varied. The average PSNR comparison of the various search window size is shown in Table 4.

For the next step of determining the patch size to be used in our proposed method, the patch size is varied while using the optimal search window for a given noise level, as shown in Table 3. The result of changing the patch size is

Table 4. PSNR comparison for fine tuning the search window size.

Noise (σ)	11×11	15×15	19×19	21×21
10	**34.16**	33.84	33.61	33.43
20	**32.50**	32.23	31.88	31.65
30	29.32	**29.38**	28.87	28.73
40	27.12	**27.79**	27.15	27.07
50	24.52	25.03	**25.15**	25.02
60	23.78	24.28	**24.39**	24.27
70	23.12	23.77	**23.88**	23.65
80	22.55	23.00	**23.62**	23.45
90	21.23	21.75	22.27	**22.56**
100	20.64	21.15	21.66	**22.02**
Average	25.89	26.21	**26.25**	26.19

shown in Table 5. The results indicated that the patch size 7×7 works optimally for noise level $\sigma < 85$ and for noise level $\sigma \geq 85$ the patch size 9×9 is optimal.

From our experiments, we have selected a patch size of 7×7 when the noise strength is, $\sigma \leq 80$ and for $\sigma > 80$ the patch size is increased to 9×9. For high noise levels, the larger patch size is needed to reduce the effect of noise in patch similarity measurement.

From our experiments, we also determined a model for selecting the search window size for a noise level, σ. The model used to select the search size $S \times S$ for a given noise, σ, is shown in Eq. (14),

Table 5. PSNR comparison by changing the patch size.

Noise (σ)	3×3	5×5	7×7	9×9
10	33.64	33.86	**34.12**	33.96
20	31.92	32.22	**32.45**	32.36
30	29.07	29.43	**29.61**	29.47
40	27.21	27.50	**27.74**	27.59
50	24.99	25.29	**25.46**	25.35
60	24.13	24.42	**24.60**	24.46
70	23.48	23.81	**24.02**	23.95
80	23.09	23.37	**23.55**	23.52
90	22.08	22.45	22.63	**22.82**
100	21.35	21.78	21.97	**22.41**
Average	26.10	26.41	**26.61**	26.59

$$S = round_{odd}(0.117\sigma + 9.758), \tag{14}$$

where, $round_{odd}()$ rounds a decimal value to its nearest odd integer. As the search window is centered on pixel, i, being denoised, the search window size needs to be an odd integer.

4.4   The Two-Step Approach

Using the patch and search window size models along with the exponential weight thresholding model, the proposed method was applied on the training image set as a multiple step approach. In each step, the proposed method was applied on the output image of the previous step. After each iteration, the PSNR of the resulting denoised image was measured. The change in the PSNR measurement, compared to the previous iteration, was used to determine the number of iterations which demonstrated satisfactory performance improvement due to the extra iteration. The PSNR comparison of multiple iterations of the proposed method is shown in Table 6. The 2-step has the optimal performance compared to the other multi-step approaches.

Table 6. PSNR comparison of proposed method for multiple steps for various noise levels.

Noise Level	1 step	2 step	3 step	4 step
10	34.11	**34.16**	33.71	33.56
20	32.43	**32.70**	32.31	32.23
30	29.55	**29.86**	29.47	29.34
40	27.78	**28.29**	27.92	27.83
50	25.44	**26.10**	25.77	25.69
60	24.66	**25.42**	25.18	25.08
70	23.99	**24.88**	24.73	24.66
80	23.61	**24.62**	24.48	24.30
90	22.79	23.96	**23.99**	23.85
100	22.48	23.76	**23.80**	23.68
Average	26.68	**27.37**	27.14	27.02

4.5   Performance Measure

To measure the performance of our proposed method in comparison to other existing denoising methods, we have used the Peak Signal to Noise Ratio (PSNR) and the Mean Structural SIMilarity (MSSIM) measure [20]. These measures are generally used for objective evaluation and measurement of various denoising methods. We also evaluated subject comparison between our proposed method and existing denoising methods.

Peak Signal to Noise Ratio (PSNR). The Peak Signal to Noise Ratio measures the ratio between the maximum possible power of a signal to the power of the noise which affects the quality of the original signal. The PSNR is usually expressed as the logarithmic decibel scale. A higher value in PSNR represents better reconstructed or denoised image. The PSNR is measured using Eq. (15),

$$PSNR = 10\log_{10}(\frac{MAX_I^2}{MSE}), \tag{15}$$

where MAX_I represents the maximum intensity of the image (255, for grayscale image) and MSE measures the mean squared error between the original image and the degraded image, as defined in Eq. (16),

$$MSE = \frac{1}{M \times N}\sum_{i=0}^{M}\sum_{j=0}^{N}(u_{ij} - v_{ij})^2, \tag{16}$$

where u_{ij} is the original image, v_{ij} is the degraded image and the size of the images is $M \times N$.

Mean Structural Similarity (MSSIM). One of the drawbacks of the PSNR measure is that it relies on the mean square error for calculating the ratio. Mean squared error considers only the differences between isolated data points. To evaluate the performance of a denoising method based on the degree of structural similarity between the original and the reconstructed image, the Structural SIMilarity (SSIM) measure is used. The SSIM measure provides a better assessment of an image restoration or denoising method. The SSIM between two blocks is defined in Eq. (17),

$$SSIM = \frac{(2\mu_x\mu_y + c_1)(2\sigma_{xy} + c_2)}{(\mu_x^2 + \mu_y^2 + c_1)(\sigma_x^2 + \sigma_y^2 + c_2)}, \tag{17}$$

where, x and y are two identical sized window or patch, μ_x and μ_y are the averages of x and y, σ_x^2 and σ_y^2 are the variance of x and y and σ_{xy} is the covariance. The mean SSIM (MSSIM), averaged over all SSIM, is used as for the quality measurement of a denoising method.

4.6   Performance Evaluation Using PSNR

Table 7 shows the PSNR comparison of the proposed method, the original non-local means, the variant of non-local means and BM3D, on the *Girl* image. Table 8 shows the average PSNR values over all images in the Kodak image set, for various noise levels. The performance of the proposed method is better than the original non-local means method and its variant for all noise levels. Yet, when compared to BM3D, our proposed method managed to produce better results only when $\sigma \leq 80$. The proposed method also demonstrated better performance than existing methods on the average of all the noise levels used in our experiments (Tables 7 and 8).

Table 7. PSNR comparison of the *Girl* image for the proposed method and existing methods.

Noise	NLM	TS-NLM	BM3D	Proposed
10	33.92	33.93	35.42	**35.61**
20	31.83	32.01	33.46	**33.60**
30	29.43	29.70	31.03	**31.12**
40	28.47	28.96	29.88	**30.04**
50	26.64	27.20	28.21	**28.27**
60	25.12	25.77	26.53	**26.60**
70	24.78	25.24	25.88	**26.03**
80	23.69	24.46	25.33	**25.50**
90	23.15	24.04	**24.95**	24.81
100	22.91	23.88	**24.43**	24.28
Average	26.99	27.52	28.51	**28.59**

Table 8. PSNR comparison of the proposed method with existing methods.

Noise	NLM	TS-NLM	BM3D	Proposed
10	32.61	32.63	34.05	**34.27**
20	30.77	30.94	32.25	**32.43**
30	28.58	28.83	29.80	**29.95**
40	27.02	27.47	28.19	**28.33**
50	24.88	25.54	26.07	**26.09**
60	23.93	24.66	25.38	**25.46**
70	23.24	24.02	24.74	**24.91**
80	22.90	23.56	24.46	**24.65**
90	22.21	23.18	**24.25**	24.13
100	21.98	22.83	**23.97**	23.84
Average	25.81	26.36	27.36	**27.41**

4.7 Performance Evaluation Using MSSIM

Table 9 shows the MSSIM comparison of the proposed method, the original non-local means, the variant of non-local means and BM3D, on the *Girl* image. Table 10 shows the MSSIM comparison over all images in the Kodak image set, for various noise levels. In terms of MSSIM, the performance of the proposed method is consistent with PSNR, which means it is better than the original non-local means and its variant for all noise levels. When compared to BM3D, our proposed method managed to produce better results only when $\sigma \leq 80$. On average across all noise levels, the performance of proposed method has been found to be better than existing methods.

Table 9. MSSIM comparison of the *Girl* image for the proposed methods with existing methods.

Noise	NLM	TS-NLM	BM3D	Proposed
10	0.919	0.922	0.927	**0.936**
20	0.875	0.882	0.889	**0.897**
30	0.849	0.855	0.857	**0.862**
40	0.818	0.822	0.832	**0.834**
50	0.790	0.797	0.811	**0.813**
60	0.761	0.767	0.780	**0.784**
70	0.728	0.731	0.747	**0.750**
80	0.713	0.717	0.738	**0.741**
90	0.692	0.694	**0.724**	0.720
100	0.678	0.683	**0.713**	0.708
Average	0.782	0.787	0.802	**0.804**

Table 10. MSSIM comparison of the proposed methods with existing methods.

Noise	NLM	TS-NLM	BM3D	Proposed
10	0.916	0.918	0.921	**0.932**
20	0.871	0.876	0.882	**0.891**
30	0.843	0.847	0.851	**0.857**
40	0.815	0.817	0.826	**0.829**
50	0.786	0.792	0.801	**0.806**
60	0.755	0.760	0.772	**0.778**
70	0.724	0.726	0.742	**0.744**
80	0.709	0.714	0.734	**0.735**
90	0.689	0.691	**0.712**	0.709
100	0.672	0.678	**0.708**	0.704
Average	0.777	0.782	0.795	**0.798**

4.8   Visual Quality

Figures 3 and 4 shows the visual comparison of the proposed method with the original non-local means (NLM), the two-stage non-local means (TS-NLM) and the Block Matching and 3D Filtering (BM3D) methods for noise level, $\sigma = 20$ and $\sigma = 70$ respectively. Figures 5 and 6 shows the visual comparison by zooming in on a particular region, the face. From Fig. 6, it can be noticed that the denoised output from the proposed method has fewer noise artifacts remaining when compared to the other methods. The blurring is also less in the denoised output of the proposed method compared to NLM, TS-NLM and BM3D.

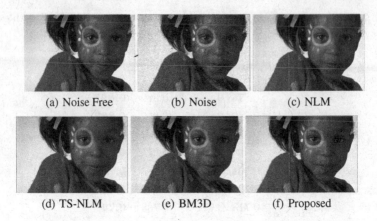

(a) Noise Free   (b) Noise   (c) NLM

(d) TS-NLM   (e) BM3D   (f) Proposed

Fig. 3. Visual comparison of proposed method with existing method ($\sigma = 20$).

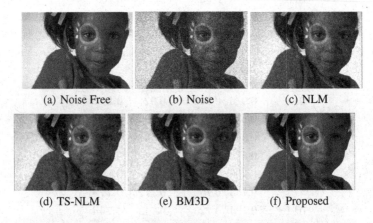

(a) Noise Free   (b) Noise   (c) NLM

(d) TS-NLM   (e) BM3D   (f) Proposed

Fig. 4. Visual comparison of proposed method with existing method ($\sigma = 70$).

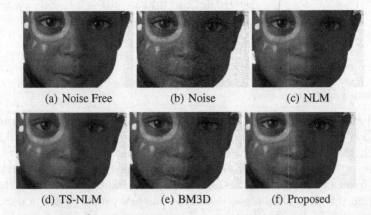

(a) Noise Free   (b) Noise   (c) NLM

(d) TS-NLM   (e) BM3D   (f) Proposed

Fig. 5. Visual comparison (zoomed) of proposed method with existing method ($\sigma = 20$).

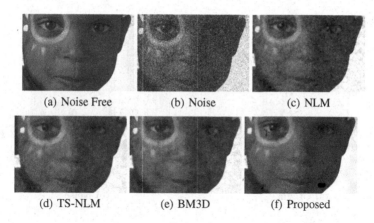

(a) Noise Free   (b) Noise   (c) NLM

(d) TS-NLM   (e) BM3D   (f) Proposed

Fig. 6. Visual comparison (zoomed) of proposed method with existing method ($\sigma = 70$).

Fig. 7. Row number 100 of *Girl* image used for generating intensity profile. Scan line shown as a black line.

4.9   Intensity Profile

The image intensity profile can help analyze how similar the profile of a denoised image is to that of the original noise-free image. Figure 7 shows the chosen horizontal scan line 100 from the *Girl* image. Figure 8 shows the intensity profiles of the true image, the noisy image at noise level, $\sigma = 70$ and the profiles of denoised images produced by the original NLM scheme, the variant of NLM, BM3D and the proposed method. The Pearson correlation coefficient between the original intensity profile and the profile of the noisy and each of the denoised images is shown in Table 11 (for $\sigma = 70$).

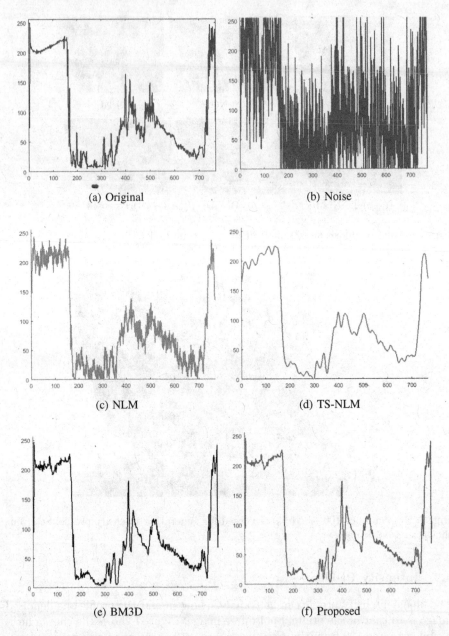

Fig. 8. Intensity profile comparison for the *Girl* image at scan line 100 ($\sigma = 70$).

The intensity profile of the proposed method shows better preservation of edges and textures, represented as sharp changes in profile graph. The original non-local means method and its variant have more noise artifacts remaining, as represented by the more jagged lines in the profile graph, closer to the origin.

Table 11. Pearson correlation coefficient comparison of the proposed method, the noisy image, the NLM method, variant of NLM and BM3D denoising scheme for noise $\sigma = 70$.

Noise	NLM	TS-NLM	BM3D	Proposed
0.680	0.975	0.980	0.988	**0.990**

When comparing the Pearson correlation coefficient, the correlation between the intensity profile of the original image and the proposed method is higher compared to those of the other existing methods. It shows that the proposed method has the closest resemblance to the intensity profile of the original image.

5 Conclusion

This paper proposed an improvement over the non-local means method, the patch-based approach for denoising additive Gaussian noise in the spatial domain. The proposed method thresholds the weights of the pixels defined around a search area of the pixel being denoised. The thresholded weights are used for weighted averaging, whereby pixels below a defined cut-off weight are ignored. The cut-off weight is determined based on the noise level estimation of an image. For a noise level, the patch and search window size are determined by a model, which is empirically defined through a learning approach. The proposed method is applied in a two-step approach for image denoising. The proposed method has demonstrated better objective and subjective denoising performance, compared to the original non-local means algorithm and its variant. When compared to BM3D, the state-of-the-art approach for image denoising, the proposed method demonstrated better results when $\sigma \leq 80$.

References

1. Basavaraja, V., Bopardikar, A., Velusamy, S.: Detail warping based video super-resolution using image guides. In: International Conference on Image Processing (2010)
2. Brox, T., Cremers, D.: Iterated nonlocal means for texture restoration. In: Sgallari, F., Murli, A., Paragios, N. (eds.) SSVM 2007. LNCS, vol. 4485, pp. 13–24. Springer, Heidelberg (2007). doi:10.1007/978-3-540-72823-8_2
3. Buades, A., Coll, B., Morel, J.: A non-local algorithm for image denoising. In: Computer Vision and Pattern Recognition (2005)
4. Buades, A., Coll, B., Morel, J.: A review of image denoising algorithms, with a new one. Multiscale Model. Simul.: SIAM Interdisc. J. 4, 490–530 (2005)
5. Chaudhury, K.N., Singer, A.: Non-local euclidean medians. IEEE Signal Process. Lett. 19, 745–748 (2012)
6. Dabov, K., Foi, A., Katkovnik, V., Egiazarian, K.: Image denoising by sparse 3-D transform-domain collaborative filtering. IEEE Trans. Image Process. 16, 2080–2095 (2007)

7. Dabov, K., Foi, A., Katkovnik, V., Egiazarian, K.: Image denoising with block-matching and 3D filtering. In: SPIE Electronic Imaging (2006)
8. Gonzalez, R.C., Woods, R.E.: Digital Image Processing. Prentice Hall, New Jersey (2008)
9. Hu, S., Hou, W.: Denosing 3D ultrasound images by non-local means accelerated by GPU. In: International Conference on Intelligent Computation and Bio-Medical Instrumentation (2011)
10. Iftikhar, M., Rathore, S., Jalil, A., Hussain, M.: A novel extension to non-local means algorithm: application to brain MRI de-noising. In: International Multi Topic Conference (2013)
11. Kelm, Z., Blezek, D., Bartholmai, B., Erickson, B.: Optimizing non-local means for denoising low dose CT. In: International Symposium on Biomedical Imaging: From Nano to Macro (2009)
12. Lim, J.S.: Two-Dimensional Signal and Image Processing. Prentice Hall, New Jersey (1990)
13. Maruf, G.M., El-Sakka, M.R.: Improved non-local means algorithm based on dimensionality reduction. In: Kamel, M., Campilho, A. (eds.) ICIAR 2015. LNCS, vol. 9164, pp. 43–50. Springer, Cham (2015). doi:10.1007/978-3-319-20801-5_5
14. Mouton, A., Megherbi, N., Flitton, G.T., Bizot, S., Breckon, T.P.: A novel intensity limiting approach to metal artefact reduction in 3D CT baggage imagery. In: International Conference on Image Processing (2012)
15. Mouton, A., Megherbi, N., Flitton, G.T., Bizot, S., Breckon, T.P.: Gaussian noise estimation in digital images using nonlinear sharpening and genetic optimization. In: Instrumentation and Measurement Technology Conference (2007)
16. Perona, P., Malik, J.: Scale-space and edge detection using anisotropic diffusion. IEEE Trans. Pattern Anal. Mach. Intell. 12, 629–639 (1990)
17. Rehman, A., Wang, Z.: SSIM-based non-local means image denoising. In: Proceedings of Image Processing (2011)
18. Russo, F.: Noise estimation in digital images using fuzzy processing. In: International Conference on Image Processing (2001)
19. Tasdizen, T.: Principal Components for non-local means image denoising. In: International Conference on Image Processing (2008)
20. Wang, Z., Bovik, A.C., Sheikh, H.R., Simoncelli, E.P.: Image quality assessment: from error visibility to structural similarity. IEEE Trans. Image Process. 13, 600–612 (2004)
21. Wiener, N.: The Interpolation Extrapolation and Smoothing of Stationary Time Series. MIT Press, New York (1949)
22. Xu, Q., Jiang, H., Scopigno, R., Sbert, M.: A new approach for very dark video denoising and enhancement. In: International Conference on Image Processing (2010)
23. Zhang, L., Dong, W., Zhang, D., Shi, G.: Two-stage image denoising by principal component analysis with local pixel grouping. Pattern Recogn. 43, 1531–1549 (2010)
24. Zhao, Y., Liu, Y.: Patch based saliency detection method for 3D surface simplification. In: International Conference on Pattern Recognition (2012)
25. Zhu, S., Li, Y., Li, Y.: Two-stage non-local means filtering with adaptive smoothing parameter. Int. J. Light Electron Opt. 125, 7040–7044 (2014)

How Good Can a Face Identifier Be Without Learning?

Yang Zhong[(✉)], Anders Hedman, and Haibo Li

Computer Science and Communication,
KTH Royal Institute of Technology, 100 44 Stockholm, Sweden
{yzhong,ahedman,haiboli}@kth.se

Abstract. Constructing discriminative features is an essential issue in developing face recognition algorithms. There are two schools in how features are constructed: hand-crafted features and learned features from data. A clear trend in the face recognition community is to use learned features to replace hand-crafted ones for face recognition, due to the superb performance achieved by learned features through Deep Learning networks. Given the negative aspects of database-dependent solutions, we consider an alternative and demonstrate that, for good generalization performance, developing face recognition algorithms by using hand-crafted features is surprisingly promising when the training dataset is small or medium sized. We show how to build such a face identifier with our Block Matching method which leverages the power of the Gabor phase in face images. Although no learning process is involved, empirical results show that the performance of this "designed" identifier is comparable (superior) to state-of-the-art identifiers and even close to Deep Learning approaches.

Keywords: Face recognition · Controlled scenario · HD Gabor Phase · Block Matching · Learning-free · Deep Learning

1 Introduction

Automatic human face recognition is a well-defined research problem in the fields of computer vision and pattern recognition. The technical core is to define a distance to measure the similarity between two given face images X and Y. The simplest way to define the distance is using the $l2$ metric on the whole raw images as:

$$d = \|X - Y\|_2 . \tag{1}$$

Besides $l2$ form, other forms like $l0$, $l1$, etc., are also widely used. A distance metric is believed good if d is small when both X and Y are from the same person and large when they are from two different persons (so-called small intra-personal variations and large inter-personal variations). Unfortunately, it is hard to directly employ the raw face images for similarity measurement in practice. This is because human face images exhibit significant appearance variations in

© Springer International Publishing AG 2017
J. Braz et al. (Eds.): VISIGRAPP 2016, CCIS 693, pp. 515–533, 2017.
DOI: 10.1007/978-3-319-64870-5_25

scale, pose, lighting, background, hairstyle, clothing, expression, color saturation, image resolution, focus, etc., as they occur in real world applications.

To distinguish persons from their faces, a more effective and efficient way is to represent face images using visual features so face images can be projected into a feature space and classified. Then the similarity between two images X and Y is measured with the following distance metric:

$$d = \sum_i \|x_i - y_i\|_2, \qquad (2)$$

where x_i and y_i are features extracted from two face images X and Y.

The power of using features for face recognition comes from, not only the construction of visual features, but importantly from the flexibility and possibility of **weighting** visual features for classification. With weighting, the similarity is measured by calculating the distance metric:

$$d = \sum_i w_i \|x_i - y_i\|_2, \qquad (3)$$

where w_i is the weight received by feature i. The intuition of giving weights is that for each face image point (in a high-dimension space) such a metric should make face image points from the same person closer than points from different persons.

In face recognition, one of the most technically challenging issues is how to construct suitable facial features for face classification. The facial features constructed by conventional approaches are so-called "hand-crafted features", i.e. features are constructed mathematically or engineered. Commonly used mathematical tools include Wavelet and Gabor filtering. The two most remarkable engineering features used in face recognition are SIFT [28] and LBP [1]. An entirely different way to construct facial features is through learning from face image data, i.e. learning to extract facial representations from training sets. The classical Eigenface approach is about how to extract principal facial components from training data sets for classification. Since the principal components are learned from training sets, the extracted facial features are called learned features. Another well-known algorithm for feature learning is Linear Discriminant Analysis (LDA). Today, with the rise of Deep Learning networks, almost all facial features used in face recognition are learned features.

Although it is easy to see that hand-crafted features and learned features are two different approaches, few realize that they are from two different facial feature constructing schools and there is consequently little debate around this topic. Successful stories of deep neural networks have led us to believe that learning is king! The unspoken assumption is: hand-crafted features are out of date, and only approaches using learned features are viable. The consequence is that we have become blind to their inherent problems. Solutions that (over) learn from training sets (particularly Deep Learning) are becoming increasingly database-dependent, even worse, it is hard to distinguish cases where general progress is made in face recognition from just good solutions to particular problems defined over specific databases.

In this paper, we argue that in the interest of making fundamental progress in face recognition, we ought to adequately study how to develop database-independent face recognition algorithms. We are interested in how good a modern face recognition system can be without learning. We consider face identification mainly due to two reasons: the problem itself is more challenging than face verification; it has been a research topic for quite some time and there are extensive experimental results available for comparison. The scientific methodology we employ here is to construct a face identifier and test and compare with state-of-the-art identifiers to explore empirically the question of how good a face identifier can be without learning.

We propose a method that merely leverages the power of the Gabor phase to address the problem of face identification in controlled scenarios. A slim filter bank of only two Gabor filters is applied to extract the Gabor phase information and explicit phase code matching is performed on the quantized phase map via our Block Matching scheme [55]. Different from other elastic matching schemes, the Block Matching scheme not only cancels the patch-wise spatial shift in the phase map but also simultaneously evaluates the patch-wise utility during the *learning-free matching process*. Combining the matching scheme with phase codewords enables the employment of high-definition phase information (4 times higher than [47]) from the 2 utilized Gabor filters. Thus, the proposed approach can significantly bring up the algorithmic efficiency without sacrificing the recognition accuracy. Furthermore, it is totally comparable to state-of-the-art Gabor solutions and even Deep Learning based solutions.

The disposition of our paper is as follows: we first briefly review related Gabor based approaches in Sect. 2; our approach is then described in Sect. 3 followed by comparative experiments presented in Sect. 4 where we also compare the performance between our approach and Deep Learning solutions. Finally, we discuss our work as a whole and offer our conclusions.

2 Related Work

Gabor filtering enables the employment of rich low-level, multi-scale features by transforming images from the pixel domain to the complex Gabor space. A Gabor face is obtained by filtering a face image with the Gabor filter function, which is defined as:

$$\psi_{u,v}(z) = \frac{\|k_{u,v}\|^2}{\sigma^2} e^{(-\|k_{u,v}\|^2\|z\|^2/2\sigma^2)}[e^{ik_{u,v}z} - e^{-\sigma^2/2}], \quad (4)$$

where u and v define the orientation and scale of the Gabor kernels respectively, and the wave vector is defined as:

$$k_{u,v} = k_v e^{i\phi_u}, \quad (5)$$

where $k_v = k_{max}/f^v$, $\phi_u = u\pi/8$; k_{max} is the maximum frequency, σ is the relative width of the Gaussian envelop, and f is the spacing factor between kernels in the frequency domain [27]. The discrete filter bank of 5 different spatial frequencies ($v \in [0, \cdots, 4]$) and 8 orientations ($u \in [0, \cdots, 7]$) is mostly exploited to filter face images to facilitate multi-scale analysis for face recognition.

In the complex Gabor transformed space, most state-of-the-art face recognition approaches utilize the amplitude of Gabor filtered image for face representation and facial feature construction. As in [51], the LGBP feature is extracted from the amplitude spectrum. One of the motivations is because the amplitude varies slowly with spatial shifts, making it robust to texture variations caused by dynamic expressions and imprecise alignment.

By constructing LBP type features from the amplitude and adopting different learning techniques, many Gabor filtering based approaches have shown remarkable advantages over pixel feature based methods: the identification rate in benchmark evaluations has been improved by more than 20% (reaching around 90%) thanks to the so-called "blessing of dimensionality" [17] (but with a high cost of less computational efficiency [8, 29]).

The Gabor phase is robust to light change. It has been well-known that phase is more important than amplitude for signal representation and reconstruction [30]. It is reasonable to believe that the Gabor phase should have played a more important role in face identification. However, the use of the Gabor phase in face recognition is far from common and it has often been unsuccessful with worse or nearly the same performance as the amplitude in comparative experiments [6, 16, 47, 53].

This is largely due to two challenging issues: (1) the Gabor phase is a periodic function and a hard quantization occurs for every period; (2) the Gabor phase is very sensitive to spatial shifts [45, 53], which imposes a rigid requirement on face image alignment. The first issue was partly solved by introducing the phase-quadrant demodulation technique [14], but the second issue is still far from being solved. The state-of-the-art Gabor phase approach (LGXP in [47]) extracts varied LBP from the phase spectrum. Since the combination of phase and LBP is also sensitive to spatial shifts, the power of the Gabor phase has not been demonstrated in face identification.

Fusing other features that are independent of the local Gabor features can also lead to better performance: [38, 42, 52] fuse the global (holistic) features with local ones at feature level; [47] proposes a fusion of the Gabor phase and amplitude on score and feature levels; [8] fuses real, imaginary, amplitude and phase data. Alternatively, attaching an illumination normalization step and weighting the local Gabor features is shown to be helpful as well [6].

3   Our Learning-Free Face Matching Approach

In this section, we first introduce the philosophy of our proposed approach in Sect. 3.1, and then demonstrate how it is used in face identification to achieve competitive performance with respect to the state-of-the-art.

3.1 Overview

Repeatable features extracted from small face portions are known as good discriminative traits for identifying persons. In addition, such local features are less likely to be influenced than the holistic features by pose changes and facial expressions. Thus, it is natural to divide face images into blocks and perform similarity measurements between them. Practically, in most face recognition methods, the matching process compares spatially corresponding patches after face alignment.

But such a matching process implies that the spatially corresponding features are the best match. This implication is hardly true even after face alignment. Because of the movement of facial components, head pose variability and imprecise alignment, the spatially corresponding patches easily become dislocated (see Fig. 2 in [58]). It is nearly impossible to achieve reasonable face alignment by using similarity transformations applied holistically to images.

In our approach, facial components are aligned individually by our Block Matching algorithm. Our Block Matching segments a face image into non-overlapping blocks and treats individual blocks as features explicitly. Given a pair of face images X and Y, the core of the algorithm is to use a given block (feature) x_i of image X to search for the corresponding block y_i in image Y. Then we measure the distance of two blocks as $\|x_i - y_i\|_2$, which is used to form the similarity between two face images as in Eq. 2. This is a direct application of the Elastic Matching concept [45] in face recognition.

Moreover, since not all blocks contribute to face identity equally, it is natural to weight the face blocks during the matching process as shown in Eq. 3. By computing proper weight factors, we can expect larger distance values for patches from different persons and smaller distances for patches from the same persons. The key is how to acquire the weight factor w_i.

Without doubt, we can learn weights from the training sets using metric learning techniques as in [12,18], but the developed algorithm will be database dependent. To have good generalization performance, we developed an efficient on-line learning step to calculate the weight factor w_i during matching the face pair at hand in our Block Matching approach, which is introduced in the following.

3.2 Algorithm

We designed the algorithm based on the observation that a face can be distinguished by its unique feature(s) which is more informative than its surrounding one(s), e.g. scars, moles, nasolabial folds, etc. This means that in an Elastic Matching context, if a segmented patch is discriminative, it gives very small distance when a good match is found and the distance varies dramatically if it is matched to surrounding locations. By considering both the minimum matching distance and the variation of the matching distance, we can evaluate the discrimination power of the local patches.

Specifically, given a face-matching pair, a probe image P (denoted by pb) and a gallery image G (denoted by gl), we first segment the probe image into N non-overlapping patches that are denoted by $\{f_n^{(pb)}\}_0^{N-1}$. (The local features are simply formed by the corresponding patches, e.g. by pixels from a patch for gray-scale images.)

For each probe patch $f_n^{(pb)}$ centered at image coordinate (x_n, y_n) (denoted by $f^{(pb)}(x_n, y_n)$), it searches its best matching block within the corresponding search window and yields a patch-wise distance vector d_n where:

$$d_n = \{d_n^i\}, i \in [0, L-1] \tag{6}$$

where L is the number of candidate gallery patches within the $(2R+1) \times (2C+1)$ search window, i.e. $L = (2R+1) \cdot (2C+1)$ when applying an "exhaustive" search method, R and C stand for the search offset in vertical and horizontal directions respectively. Each element in d_n is computed as:

$$d_n^i = \left\| (f^{(pb)}(x_n, y_n) - f^{(gl)}(x_i, y_i)) \right\|_2, \tag{7}$$

where the patch-wise distance metric is the $l2$-norm of element wise distance of local features (patches) and $f^{(gl)}(x_i, y_i)$ denotes the patch that centered at image coordinate (x_i, y_i) within the search window on the gallery face image so that

$$x_i = x_n + \Delta x, \Delta x \in [-C, C], \; y_i = y_n + \Delta y, \Delta y \in [-R, R]. \tag{8}$$

We then calculate the slope k_n of the linear fitting of the first 5 ascendingly sorted values of d_n for normalization of the patch wise distance for each patch, such that the weight factors for each local feature w_n is calculated as:

$$w_n^* = k_n / d_n^*, \tag{9}$$

where $d_n^* = min(d_n)$. w_n^* is then normalized by its $l1$-norm as:

$$w_n = w_n^* / \sum_{n=0}^{N-1} w_n^*. \tag{10}$$

Finally, the distance between a matching pair of probe and gallery face images is the weighted sum of d_n^* as:

$$dist^{(pb,gl)} = \sum_{n=0}^{N-1} w_n \cdot d_n^*. \tag{11}$$

More details of our Block Matching approach are given in [55].

3.3   Gabor Phase Block Matching (GPBM)

It is known that the features constructed from pixels are vulnerable to lighting and pose variations. To further improve the recognition performance, one way is to construct more robust features. Another effective way is increasing the dimensionality of features to raise recognition rate dramatically thanks to the "blessing of dimensionality". Traditionally, the most popular way is to exploit the Gabor features via Gabor transformation, which normally increases dimensionality of image representations by 40 times [27].

The Gabor transformation enables the employment of rich low-level multi-scale features by transforming images from the pixel domain to the complex Gabor space. Different strategies of using either Gabor magnitude or Gabor phase, or a hybrid of both magnitude and phase have been proposed to construct features. One reasonable option for many state-of-the-art approaches has been to utilize Gabor amplitude for face representation and feature construction.

But high dimensional features lead to high cost and create difficulties for training, computation, and storage (as pointed out in [8,29]). To build a practical solution, patch-based approaches and dimensional reduction techniques, such as PCA or LDA, and rotated sparse regression, are commonly used to learn a subspace to reduce intra-class variation and expand inter-class variations. Since the learning process has to be involved and training datasets are needed (e.g. for LDA, the leading eigenvectors of the covariance matrix are needed to calculate over training image pairs), the advantage of using hand-crafted features to achieve generalization performance is diminished.

Can we remain learning-free (to promise generalization) in our face matching approach and also further improve the recognition performance without suffering from heavy computational load brought about by high-dimension representations? We focus on the **Gabor Phase**, since it better reconstructs signals than amplitude [30]. We combine the Gabor Phase face representation with our Block Matching approach introduced in Sect. 3.2, and demonstrate that increasing the signal dimension is not the only way to boost the recognition performance.

Specifically, we filter faces with only a single-scale Gabor filter pair and calculate the phase of the filtered face. That is, for each face image, only two demodulated Gabor phase spectra are used as in the input of our Block Matching method, see Fig. 1. We first segment the probe phase spectrum into N

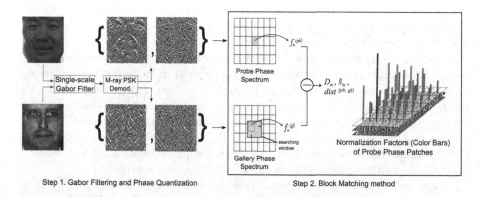

Step 1. Gabor Filtering and Phase Quantization   Step 2. Block Matching method

Fig. 1. The matching process of our Gabor Phase Block Matching (GPBM) approach.

non-overlapping patches and the patches $\{f_n^{(pb)}\}_0^{N-1}$ are simply formed by the *raw phase* codes of the patches. Then the Block Matching approach is utilized to calculate the distance of the two faces. The only difference is that when calculating the phase distance, each element in d_n is computed by performing an explicit matching over the raw demodulated phase as:

$$d_n^i = \left\| \text{XOR}(f^{(pb)}(x_n, y_n), f^{(gl)}(x_i, y_i))_{decimal} \right\|_2, \tag{12}$$

where the patch-wise distance metric is the $l2$-norm of element wise Hamming distance in decimal. More technical details are provided in [56].

4   Experiment

4.1   Database Selection

There are a variety of large-scale datasets available for benchmark evaluation of different face recognition approaches, such as the FERET [33], FRGC2.0 [32] and the LFW [21] datasets. Since we focus on face recognition in controlled scenarios in this paper, the FERET database—the most commonly used face identification benchmark—is selected to evaluate and compare our method with state-of-the-art face identification approaches. In addition, the CMU-PIE [36] dataset is selected to evaluate our GPBM against variations of pose, expression and illumination.

4.2   Experimental Setup

Face images were first normalized (aligned) based on the positions of both eyes as in [47]. A central facial area of 150×136, which maintained the same aspect ratio (1.1:1) as in [47,54], was segmented from the face image and used for our experiments.

Due to our Block Matching scheme, the Gabor phase information with a higher definition can be utilized in our approach. We found that a single-scale Gabor filter pair with two orientations is sufficient for face identification. In our implementation, the selected Gabor filters had the following parameters: $v = 0$, $u \in \{2, 6\}$, $f = \sqrt{2}$, $k_{max} = \pi/2$, $\sigma = 2\pi$.

One can see that the chosen Gabor filters have broad high-frequency coverage. These high-frequency components correspond to facial texture variations and are insensitive to the factors of lighting, pose, and aging. Accordingly, to retain high phase definition and to be tolerant to potential phase change caused by texture shift, a Gray-coded 16-PSK demodulator was used for phase demodulation and the constellation is shown in Fig. 2. Compared to the quadrature phase demodulation used in [47,52], 4 times the phase information can be utilized thanks to the employment of our Block Matching approach.

To provide a thorough answer to "How good can a face identifier be without learning", we compare our Block Matching approach [55] and our GPBM to other methods on both image domain and Gabor transformed space in the following.

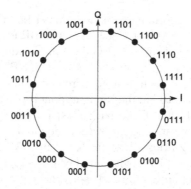

Fig. 2. 16-PSK demodulator constellation.

4.3   Evaluations on the CMU-PIE Database

The CMU-PIE database contains 41368 images of 68 subjects. Images with pose labels 05, 07, 09, 27, and 29 under 21 illuminations (Flash 2 to 22) of all the 68 persons are selected as the probe set.

When applying the Block Matching method, the most important parameters are the block size (H and W) and searching offset (R and C). Our empirical tests on other datasets indicate that it makes sense to divide a central facial area into 5×7 patches, which semantically correspond to components of human faces, like eyes, nose, etc. Thus, for a facial area of 150×136, a reasonable size of a block is 30×20. In our implementation, our block size was 29×19 (we prefer odd block sides) in the block search. To have a good coverage while keeping low computational complexity, the search offset was set to around a quarter of the block size and we selected the search offset of $R = 7$, $C = 6$ pixels in our experiments. To test how sensitive the performance was to the selected parameters, we selected the first 2000 probe images on the CMU-PIE to evaluate the performance with the chosen parameters and other parameters randomly selected around them. The evaluation results in Fig. 3 show that the recognition performance is rather insensitive to parameter selections.

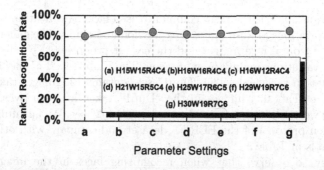

Fig. 3. Recognition rates under different parameters on the CMU-PIE dataset.

We then conducted experiments on the CMU-PIE probe set and compared our GPBM with G_LBP and G_LDP [54]. The G_LBP is the Gabor version LBP and the G_LDP is a type of improved Gabor amplitude Local Binary Pattern. The G_LDP achieved equivalent performance as LGXP (Gabor Phase pattern) on the FERET evaluations so it is a good reference for comparison. Since the Gabor phase is inadequate to handle the non-monotopic illumination variations, we employed the Difference of Gaussian (DoG) image to equalize the illumination on the face. The corresponding results is denoted by $GPBM + DoG$. The comparative rank-1 recognition rates are listed in Table 1. It can be seen that even with use of the raw pixels without any photometric processing for face matching, the Block Matching approach performs comparably to the LBP and LDP approaches where hand-crafted features were employed. Similarly, for our GPBM, it is slightly better than the G_LDP, even though LDP extracts much more complicated Gabor amplitude patterns. The results indicate that with the environment of dramatic illumination and pose variation, our Block Matching approaches have equal recognition power to the hand-crafted features.

Table 1. Comparative rank-1 recognition rates on the CMU-PIE database.

Method	Accuracy
Unweighted LBP*	58%
Best of LDP, 3rd order*	61%
Block Matching [55]	**60%**
G_LBP*	71%
Best of G_LDP, 3rd-order*	79%
GPBM	**82%**
GPBM + DoG	**85%**

* The recognition rates are estimated from Fig. 12a in [54].

4.4   Evaluations on the FERET Database

The FERET database is the most commonly used face identification benchmark. It contains variations in illumination, expression and aging. The gallery set "Fa" contains 1196 frontal face images and the easiest probe set "Fb" contains 1195 images with variations mainly in expression. The probe set "Fc" has 194 images with illumination variations. The "Dup1" set contains 722 images taken later in time than the "Fa". 234 images in the "Dup1" taken at least 1 year after the "Fa" session were selected to form the hardest "Dup2" set. We faithfully followed the evaluation protocol of the FERET dataset and compare with other feature based methods in Table 2.

It is easy to observe that when recognizing faces in the image domain, the Block Matching approach outperforms the LBP and LDP (first 3 rows in

Table 2. Comparative rank-1 recognition rates on the FERET database.

Method	Fb	Fc	Dup1	Dup2
Unweighted LBP [2]	93%	51%	61%	50%
LDP 3rd-order [54]	90%	88%	63%	61%
Block Matching [55]	**93%**	**83%**	**66%**	**64%**
LGBP_Pha [53]	93%	92%	65%	59%
HGPP [52]	97.5%	99.5%	79.5%	77.8%
LGXP [47]	98%	100%	82%	83%
LGXP+BFLD [47]	99%	100%	92%	91%
S[LGBP+LGXP] [47]	99%	100%	94%	93%
LN+LGXP [6]	99.9%	100%	94.7%	91.9%
PCANet-2 [9]	99.6%	100%	95.4%	94%
GPBM	**99.4%**	**100%**	**95.3%**	**94.9%**

Table 2). If Gabor transformation is employed, in a fair comparison (non-learning component was involved and only Gabor phase was utilized for face matching), our GPBM is almost 12% better than LGXP on the hardest "Dup2". Even in unfavorable comparisons, where pre-processing, training, and fusion methods were exploited by LN+LGXP and S[LGBP Mag+LGXP], our GPBM still excels. To our best knowledge, the method S[LGBP_ Mag+LGXP]—aided by the Gabor amplitude and training procedures—was the state-of-the-art Gabor phase based method in terms of performance on the hardest FERET "Dup2", and our GPBM is entirely comparable.

Our approach also has comparable performance to the Deep Learning based PCANet-2 [9]. It firmly confirms again that in a controlled scenario by weighting the image-wise distance via our Block Matching process, we can achieve equally effective face identification as the state-of-the-art. Our results indicate that feature design and high-dimensional signal representation might be less important than commonly believed.

We further compare our GPBM with other state-of-the-art approaches based on other techniques on the FERET in Table 3. From the table one can see that all these approaches are based on Gabor features, which indicates that the Gabor filter is a very effective tool for signal representation. Our GPBM method outperforms all the other approaches on the hardest "Dup2" set and it features three advantages: (1) it enables high definition Gabor phase to be utilized for face identification; (2) a single-scale Gabor filter with two orientations is sufficient to generate an effective face image representation, with 1/20 of the computational complexity of other methods that utilized 40 Gabor filters; (3) further to this, it is not a learning-based face identification method and, therefore, promises good generalization.

We also evaluated our approach under pose variations using the pose probe sets "bd", "be", "bf", "bg" and gallery set "ba" on the FERET dataset. These

Table 3. Comparative summary of recent state-of-the-art face identification approaches.

Methods	Image size	Gabor feature space Gabor filter bank	Training data independent	Rank-1 Rate on FERET Dup2
LGXP[a]	88 × 80	Phase 5 × 8	No	83%
LGBP+LGXP[b]	88 × 80	Amplitude + Phase 5 × 8	No	93%
GOM[c]	160 × 128	Amplitude + Phase 5 × 8	No	93.1%
LN+LGXP[d]	251 × 203	Phase 5 × 8	No	91.9%
LN+LGBP[e]	251 × 203	Amplitude 5 × 8	No	93.6%
SLF-RKR_$l2$[f]	150 × 130	Amplitude 5 × 8	No	94.4%
GPBM, ours	150 × 136	Phase (explicit matching) 1 × 2	**Yes**	**94.9%**

[a][47], [b][47], [c][8], [d][6], [e][6], [f][48]

sets correspond to pose angles of $+25°$, $+15°$, $-15°$, $-25°$, and $0°$ to the camera, and each of these sets contains 200 persons. The comparative performance of our GPBM is listed in Table 4. It can be observed that our GPBM approach is significantly more accurate than the non-learning feature-based approaches (*LGBP* and *LGXP*) when probe faces have relatively large pose angles. In addition, it outperforms the learning-based methods on all probe sets and is even comparable to the recent Deep Learning approach *SPAE* on "be" and "bf" sets.

Table 4. Comparative rank-1 recognition rates on the FERET dataset with the posed probe sets.

Method	bd	be	bf	bg
Pose angle	$-25°$	$-15°$	$+15°$	$+25°$
LGBP [6]	86.5%	98%	97.5%	88.5%
LGXP [6]	73.5%	95.5%	96%	65.5%
StackFlow [3]	89%	96%	94%	92%
DAE [4]	93%	96%	96%	94%
LN//LGXP [6]	97.5%	99%	99.5%	96%
SPAE [23]	98%	99%	99%	99%
GPBM	**96%**	**100%**	**100%**	**95%**

The computational complexity is always a big concern. As in Table 4 of [29], under the image size of $128 × 128$ with a $5 × 8$ Gabor filter bank, the histogram extraction of LGBP takes around 0.45 s, S[LGBP_ Mag+LGXP] takes 0.99 s. Extracting GOM feature takes 0.7 s [8]. However, the "feature extraction" time in our method is **0** s since only the raw phase is used for matching; the demodulation is the only on-line computation of the probe face, thus, it is extremely fast. Our Matlab implementation executes the matching of a face pair in 0.05 s on average (Gabor filtering included) on a 3.4 GHz Intel CPU. We can therefore safely conclude that our GPBM outperforms the best Gabor-phase based approach (S[LGBP_ Mag+LGXP]) in efficiency with a big margin. We can also infer that the other methods in Table 3 could hardly be more efficient than our GPBM due to higher image resolution, Gabor face dimensions, and additional photometric processing. Here we should mention that our GPBM needs to run block matching. Right now, we used an "exhaustive search" strategy. Since we have just a few blocks per probe image, matching is still fast. In future work, we could also incorporate fast-search strategies from the video compression field to speed up face matching.

4.5 Deep Learning for Face Recognition in Controlled Scenarios

Before we conclude this paper, it would be interesting to investigate *how good Deep Learning can be for face recognition in controlled scenarios*. To answer this, we trained several CNNs with well-known architectures of AlexNet [24], VGG-net [37], Google's InceptionNet [40] and FaceNet [35]), and evaluated them on the most difficult probe set "Dup2" of the FERET database.

For fair comparisons between different architectures, layers after the last spatial pooling in our implementation of the InceptionNet and the FaceNet were replaced by two concatenated Fully Connected (FC) layers, and Softmax was selected as the lost function. We used the WebFace dataset [50] to train our networks. WebFace is a face image collection with half a million instances of around 10000 celebrities. Since the FERET dataset was formed with non-celebrity people, all the trained nets were fine-tuned carefully with the FERET gallery images.

To illustrate how architecture choice affects the recognition performance, we investigated how rank-1 accuracy varies under different sizes of the FC layers. The results are listed in Table 5. We can see that the architecture (length of FC) does influence the recognition accuracy. Explicitly inherit network architectures designed for one image classification task (e.g. networks with $FC - 4096$ layers work well on the ImageNet) may not perform well in a novel face recognition task. Investigations on suitable deep feature representations must be made correspondingly and here we found that $FC - 1024$ is a good choice which is also verified in [31]. On the other hand, the performance strongly correlates to architecture in general: even with $FC - 1024$, the InceptionNet outperformed others.

Table 5. Rank-1 accuracy of several well-known CNN architectures on FERET Dup2 (input image size to CNNs was 120×120).

Architectures	Length of the FC-4096	Last 2 FCs FC-1024
AlexNet	91.9%	94.4%
VGG-13 layers	93.6%	94.9%
VGG-16 layers	93.2%	97.0%
InceptionNet	95.3%	98.7%
FaceNet	94.9%	98.3%

Apparently, CNNs can outperform the proposed approach for around 4%, but such advantage is not statistically significant for the test: the best CNN correctly identified 9 more probe faces than our proposed approach which made 222 correct answers out of 234 probes on the "Dup2" set. We can see that even with the most advanced CNNs and trained with a massive face dataset, deep-learning doesn't solve the face identification problem defined over the FERET set significantly better than our non-learning approach. At least, one can conclude that for an unseen face identification task, the approach developed without learning could be a promising solution.

5   Discussion

Although without learning, we have shown that a combination of Block Matching with Gabor phase could work as a very good identifier. Unlike in most state-of-the-art face identifiers where a highly engineered design of facial features or high dimensional features ("blessing of dimensionality") are "must-have", the proposed face identifier has no designed features. It just uses the blocks of raw face images, two orientation channels of a single scale Gabor filter to construct the phase features for face recognition. Our experimental results show that the form and dimensionality of features are of course important, but not the key in building a good identifier. The key lies in how to handle the factors causing unlimited variations of facial textures. The crucial issue in constructing features is knowing how the features affect the relationship between within-class variability and between-class variability. Face recognition can be performed reliably only when the between-class variability is larger than the within-class variability. Why does the proposed identifier work? It is due to that spatial shift caused by camera, pose, expression, scale, and aging between two face images is effectively tackled by our Block Matching approach. In addition, the 2D Gabor filtering family uniquely achieves the theoretical lower bound on joint uncertainty over spatial position and frequency. These properties are particularly useful in characterizing facial textures. Leveraged by our Block Matching technique, the Gabor phase demonstrates its power in handling lighting factors and detailing of local facial texture changes.

Just like others, the performance of our face identifier was also tested with standard database sets. However, there is a clear difference between the evaluations. Since it is not learned from the training data, our identifier is of good generalization. One can expect a similar performance when applying to other databases. Of course, according to the *No Free Lunch Theorem*, if we are interested solely in the generalization performance, there is no reason to prefer one identifier over another. Certain prior knowledge about the problem or a concrete application is always used explicitly or implicitly, for example, choice of operating parameters in our case. In fact, our identifier itself is of a good technical platform where learning can be well integrated. For example, instead of being computed on-line, the weights can be learned from database sets. Thus, our approach is becoming the so-called metric learning:

$$d = (\mathbf{x} - \mathbf{y})^T W (\mathbf{x} - \mathbf{y}), \tag{13}$$

where $\mathbf{x}$ and $\mathbf{y}$ are the feature representations of X and Y, W is a weight matrix, typically a symmetric positive definite matrix. A typical example is the Mahalanobis metric. The weight matrix can be learned from either sets of labeled image pairs or just sets of labeled images with an objective of finding a matrix such that positive pairs have smaller distances than negative pairs. Of course, once learning is involved the developed identifier will be more database-dependent and less apt for generalization.

Face recognition has been developed for over more than three decades and three-order of magnitude improvement in recognition rate has been achieved. One has to realize that such an achieved performance increase was only on the selected databases.

Due to the popularity of machine learning, we don't know how to measure the real progress that has been made in face recognition. Taking the "LFW benchmark" as an example, so many advanced CNN networks have been trained to do face recognition and some of the best can achieve 99.5% recognition accuracy in benchmark evaluations. But when such networks were practically deployed in a real-world application, it was found that they were still far from usable mostly due to the divergence between the training dataset and the real-world data [57]. Though Deep Learning made a big stride in solving challenging face recognition tasks, it is still early to confirm that it is the only right way to go. It is wise to include diverse solutions using "hand-crafted" features and/or features learned from data. This is why in this paper we take on a radical approach to see how far face recognition can go without learning.

6   Conclusions

In this paper we argue strongly that it makes sense to study how good a face identifier can be without learning, particularly today when Deep Learning is very commonly used. We have shown how to construct such an identifier that simply uses Block Matching technique over Gabor phase codes to achieve state-of-the-art performance. We have demonstrated that engineered feature designs or those adhering to the slogan "blessing of dimensionality" are not essential ingredients for building a good identifier. The key issue in constructing features is to achieve between-class variability larger than within-class variability. Since it is not learned from the training data, our identifier lends itself well for generalization. One can expect similar performance when applying it to other databases. This is very important for developing algorithms that constitute real progress in face recognition.

References

1. Ahonen, T., Hadid, A., Pietikäinen, M.: Face recognition with local binary patterns. In: Pajdla, T., Matas, J. (eds.) ECCV 2004. LNCS, vol. 3021, pp. 469–481. Springer, Heidelberg (2004). doi:10.1007/978-3-540-24670-1_36

2. Ahonen, T., Hadid, A., Pietikainen, M.: Face description with local binary patterns: application to face recognition. IEEE Trans. Pattern Anal. Mach. Intell. **28**(12), 2037–2041 (2006)
3. Ashraf, A.B., Lucey, S., Chen, T.: Learning patch correspondences for improved viewpoint invariant face recognition. In: IEEE Conference on Computer Vision and Pattern Recognition, CVPR 2008, pp. 1–8. IEEE (2008)
4. Bengio, Y.: Learning deep architectures for AI. Found. Trends Mach. Learn. **2**(1), 1–127 (2009)
5. Belhumeur, P.N., Hespanha, J.P., Kriegman, D.: Eigenfaces vs. fisherfaces: recognition using class specific linear projection. IEEE Trans. Pattern Anal. Mach. Intell. **19**(7), 711–720 (1997)
6. Cament, L.A., Castillo, L.E., Perez, J.P., Galdames, F.J., Perez, C.A.: Fusion of local normalization and Gabor entropy weighted features for face identification. Pattern Recogn. **47**(2), 568–577 (2014)
7. Cao, X., Wei, Y., Wen, F., Sun, J.: Face alignment by explicit shape regression. Int. J. Comput. Vis. **107**(2), 177–190 (2014)
8. Chai, Z., Sun, Z., Mendez-Vazquez, H., He, R., Tan, T.: Gabor ordinal measures for face recognition. IEEE Trans. Inf. Forensics Secur. **9**(1), 14–26 (2014)
9. Chan, T.-H., Jia, K., Gao, S., Lu, J., Zeng, Z., Ma, Y.: PCANet: a simple deep learning baseline for image classification? IEEE Trans. Image Process. **24**(12), 5017–5032 (2015)
10. Chen, D., Cao, X., Wang, L., Wen, F., Sun, J.: Bayesian face revisited: a joint formulation. In: Fitzgibbon, A., Lazebnik, S., Perona, P., Sato, Y., Schmid, C. (eds.) ECCV 2012. LNCS, vol. 7574, pp. 566–579. Springer, Heidelberg (2012). doi:10.1007/978-3-642-33712-3_41
11. Chan, C.H., Tahir, M.A., Kittler, J., Pietikainen, M.: Multiscale local phase quantization for robust component-based face recognition using kernel fusion of multiple descriptors. IEEE Trans. Pattern Anal. Mach. Intell. **35**(5), 1164–1177 (2013)
12. Cui, Z., Li, W., Xu, D., Shan, S., Chen, X.: Fusing robust face region descriptors via multiple metric learning for face recognition in the wild. In: 2013 IEEE Conference on Computer Vision and Pattern Recognition (CVPR), pp. 3554–3561. IEEE (2013)
13. Dalal, N., Triggs, B.: Histograms of oriented gradients for human detection. In: IEEE Computer Society Conference on Computer Vision and Pattern Recognition, CVPR 2005, vol. 1, pp. 886–893. IEEE (2005)
14. Daugman, J.: How iris recognition works. IEEE Trans. Circ. Syst. Video Technol. **14**(1), 21–30 (2004)
15. Daugman, J.: Probing the uniqueness and randomness of iriscodes: results from 200 billion iris pair comparisons. Proc. IEEE **94**(11), 1927–1935 (2006)
16. Gao, Y., Wang, Y., Zhu, X., Feng, X., Zhou, X.: Weighted Gabor features in unitary space for face recognition. In: 7th International Conference on Automatic Face and Gesture Recognition, FGR 2006, p. 6. IEEE (2006)
17. Givens, G.H., Beveridge, J.R., Lui, Y.M., Bolme, D.S., Draper, B.A., Phillips, P.J.: Biometric face recognition: from classical statistics to future challenges. Wiley Interdisc. Rev.: Comput. Stat. **5**(4), 288–308 (2013)
18. Hu, J., Lu, J., Tan, Y.-P.: Discriminative deep metric learning for face verification in the wild. In: 2014 IEEE Conference on Computer Vision and Pattern Recognition (CVPR), pp. 1875–1882, June 2014
19. Hua, G., Akbarzadeh, A.: A robust elastic and partial matching metric for face recognition. In: 2009 IEEE 12th International Conference on Computer Vision, pp. 2082–2089. IEEE (2009)

20. Huang, G., Lee, H., Learned-Miller, E.: Learning hierarchical representations for face verification with convolutional deep belief networks. In 2012 IEEE Conference on Computer Vision and Pattern Recognition (CVPR), pp. 2518–2525 (2012)
21. Huang, G.B., Ramesh, M., Berg, T., Learned-Miller, E.: Labeled faces in the wild: a database for studying face recognition in unconstrained environments. Technical report 07-49, University of Massachusetts, Amherst (2007)
22. Jia, Y., Shelhamer, E., Donahue, J., Karayev, S., Long, J., Girshick, R., Guadarrama, S., Darrell, T.: Caffe: Convolutional architecture for fast feature embedding. arXiv preprint. arXiv:1408.5093 (2014)
23. Kan, M., Shan, S., Chang, H., Chen, X.: Stacked progressive auto-encoders (spae) for face recognition across poses. In: 2014 IEEE Conference on Computer Vision and Pattern Recognition (CVPR), pp. 1883–1890. IEEE (2014)
24. Krizhevsky, A., Sutskever, I., Hinton, G.E.: Imagenet classification with deep convolutional neural networks. In: Pereira, F., Burges, C., Bottou, L., Weinberger, K. (eds.) Advances in Neural Information Processing Systems, vol. 25, pp. 1097–1105. Curran Associates Inc. (2012)
25. Lades, M., Vorbruggen, J.C., Buhmann, J., Lange, J., von der Malsburg, C., Wurtz, R.P., Konen, W.: Distortion invariant object recognition in the dynamic link architecture. IEEE Trans. Comput. **42**(3), 300–311 (1993)
26. Li, H., Hua, G., Lin, Z., Brandt, J., Yang, J.: Probabilistic elastic matching for pose variant face verification. In: 2013 IEEE Conference on Computer Vision and Pattern Recognition (CVPR), pp. 3499–3506. IEEE (2013)
27. Liu, C., Wechsler, H.: Gabor feature based classification using the enhanced fisher linear discriminant model for face recognition. IEEE Trans. Image Process. **11**(4), 467–476 (2002)
28. Lowe, D.G.: Distinctive image features from scale-invariant keypoints. Int. J. Comput. Vis. **60**(2), 91–110 (2004)
29. Mu, M., Ruan, Q., Guo, S.: Shift and gray scale invariant features for palmprint identification using complex directional wavelet and local binary pattern. Neurocomputing **74**(17), 3351–3360 (2011)
30. Oppenheim, A.V., Lim, J.S.: The importance of phase in signals. Proc. IEEE **69**(5), 529–541 (1981)
31. Parkhi, O.M., Vedaldi, A., Zisserman, A.: Deep face recognition. In: Proceedings of the British Machine Vision Conference (2015)
32. Phillips, P.J., Flynn, P.J., Scruggs, T., Bowyer, K.W., Chang, J., Hoffman, K., Marques, J., Min, J., Worek, W.: Overview of the face recognition grand challenge. In: IEEE Computer Society Conference on Computer Vision and Pattern Recognition, CVPR 2005, vol. 1, pp. 947–954. IEEE (2005)
33. Phillips, P.J., Moon, H., Rizvi, S.A., Rauss, P.J.: The feret evaluation methodology for face-recognition algorithms. IEEE Trans. Pattern Anal. Mach. Intell. **22**(10), 1090–1104 (2000)
34. Pinto, N., DiCarlo, J.J., Cox, D.D.: How far can you get with a modern face recognition test set using only simple features? In: IEEE Conference on Computer Vision and Pattern Recognition, CVPR 2009, pp. 2591–2598. IEEE (2009)
35. Schroff, F., Kalenichenko, D., Philbin, J.: Facenet: A unified embedding for face recognition and clustering. In: Proceedings of the IEEE Conference on Computer Vision and Pattern Recognition, pp. 815–823 (2015)
36. Sim, T., Baker, S., Bsat, M.: The CMU pose, illumination, and expression (PIE) database. In: Fifth IEEE International Conference on Automatic Face and Gesture Recognition, Proceedings, pp. 46–51. IEEE (2002)

37. Simonyan, K., Zisserman, A.: Very deep convolutional networks for large-scale image recognition. CoRR, abs/1409.1556 (2014)
38. Su, Y., Shan, S., Chen, X., Gao, W.: Hierarchical ensemble of global and local classifiers for face recognition. IEEE Trans. Image Process. **18**(8), 1885–1896 (2009)
39. Sun, Y., Wang, X., Tang, X.: Deep learning face representation from predicting 10,000 classes. In: Proceedings of the IEEE Conference on Computer Vision and Pattern Recognition, pp. 1891–1898 (2013)
40. Szegedy, C., Liu, W., Jia, Y., Sermanet, P., Reed, S., Anguelov, D., Erhan, D., Vanhoucke, V., Rabinovich, A.: Going deeper with convolutions. In: 2014 IEEE 12th International Conference on Computer Vision (2014)
41. Taigman, Y., Yang, M., Ranzato, M., Wolf, L.: Deepface: closing the gap to human-level performance in face verification. In: Proceedings of the IEEE Conference on Computer Vision and Pattern Recognition, pp. 1701–1708 (2013)
42. Tan, X., Triggs, B.: Fusing Gabor and LBP feature sets for kernel-based face recognition. In: Zhou, S.K., Zhao, W., Tang, X., Gong, S. (eds.) AMFG 2007. LNCS, vol. 4778, pp. 235–249. Springer, Heidelberg (2007). doi:10.1007/978-3-540-75690-3_18
43. Turk, M.A., Pentland, A.P.: Face recognition using eigenfaces. In: IEEE Computer Society Conference on Computer Vision and Pattern Recognition, Proceedings of CVPR 1991, pp. 586–591. IEEE (1991)
44. Vedaldi, A., Lenc, K.: Matconvnet-convolutional neural networks for matlab. arXiv preprint arXiv:1412.4564 (2014)
45. Wiskott, L., Fellous, J.-M., Kuiger, N., Von Der Malsburg, C.: Face recognition by elastic bunch graph matching. IEEE Trans. Pattern Anal. Mach. Intell. **19**(7), 775–779 (1997)
46. Wright, J., Yang, A.Y., Ganesh, A., Sastry, S.S., Ma, Y.: Robust face recognition via sparse representation. IEEE Trans. Pattern Anal. Mach. Intell. **31**(2), 210–227 (2009)
47. Xie, S., Shan, S., Chen, X., Chen, J.: Fusing local patterns of Gabor magnitude and phase for face recognition. IEEE Trans. Image Process. **19**(5), 1349–1361 (2010)
48. Yang, M., Zhang, L., Shiu, S.-K., Zhang, D.: Robust kernel representation with statistical local features for face recognition. IEEE Trans. Neural Netw. Learn. Syst. **24**(6), 900–912 (2013)
49. Yi, D., Lei, Z., Li, S.Z.: Towards pose robust face recognition. In: 2013 IEEE Conference on Computer Vision and Pattern Recognition (CVPR), pp. 3539–3545. IEEE (2013)
50. Yi, D., Lei, Z., Liao, S., Li, S.Z.: Learning face representation from scratch. arXiv preprint arXiv:1411.7923 (2014)
51. Zhang, W., Shan, S., Gao, W., Chen, X., Zhang, H.: Local Gabor binary pattern histogram sequence (LGBPHS): a novel non-statistical model for face representation and recognition. In: Tenth IEEE International Conference on Computer Vision, ICCV 2005, vol. 1, pp. 786–791. IEEE (2005)
52. Zhang, B., Shan, S., Chen, X., Gao, W.: Histogram of Gabor phase patterns (HGPP): a novel object representation approach for face recognition. IEEE Trans. Image Process. **16**(1), 57–68 (2007)
53. Zhang, W., Shan, S., Qing, L., Chen, X., Gao, W.: Are Gabor phases really useless for face recognition? Pattern Anal. Appl. **12**(3), 301–307 (2009)
54. Zhang, B., Gao, Y., Zhao, S., Liu, J.: Local derivative pattern versus local binary pattern: face recognition with high-order local pattern descriptor. IEEE Trans. Image Process. **19**(2), 533–544 (2010)

55. Zhong, Y., Li, H.: Is block matching an alternative tool to LBP for face recognition? In: 2014 IEEE International Conference on Image Processing (ICIP), pp. 723–727 (2014)
56. Zhong, Y., Li, H.: Leveraging Gabor phase for face identification in controlled scenarios. In: Proceedings of the 11th Joint Conference on Computer Vision, Imaging and Computer Graphics Theory and Applications, pp. 49–58 (2016)
57. Zhou, E., Cao, Z., Yin, Q.: Naive-deep face recognition: touching the limit of LFW benchmark or not? arXiv preprint. arXiv:1501.04690 (2015)
58. Zou, J., Ji, Q., Nagy, G.: A comparative study of local matching approach for face recognition. IEEE Trans. Image Process. 16(10), 2617–2628 (2007)

Object Tracking Guided by Segmentation Reliability Measures and Local Features

Cristian M. Orellana[1] and Marcos D. Zuniga[2(✉)]

[1] Department of Computer Science, Universidad Técnica Federico Santa María,
Av. España 1680, Valparaíso, Chile
[2] Electronics Department, Universidad Técnica Federico Santa María,
Av. España 1680, Valparaíso, Chile
marcos.zuniga@usm.cl

Abstract. Real world applications need to cope with unreliable data sources that affect negatively the performance of visual systems, adding error to the whole process. Existing solutions focus their efforts on decreasing the probability of making errors, but if an error occurs, there is no mechanism to deal with it. This work focuses in dealing with this problem by modelling the quality of the segmentation phase in order to apply control mechanisms to mitigate negative effects in later stages. Our control mechanism is based on determining the reliability of local features to discard the less reliables. Local features are characterized using colour, texture, and an illumination reliability model to quantify the quality of illumination. The use of local features enables us to deal with partial occlusion problems by determining the global object position via local features consensus. Experiments were performed, showing promising results in object position estimation under poor illumination conditions.

Keywords: Multi-target tracking · Feature tracking · Local descriptors · Background subtraction · Surveillance tracking · Reliability measures · Quality segmentation

1 Introduction

Real problems often lack on the possibility of obtaining manual initialisation for properly obtaining a reliable first model of an object. Many tracking algorithms require a robust initial object model to perform tracking, often obtained with manual procedures [1,2]. These methods often fail in dealing with problems as severe illumination changes or lack of contrast, or perform expensive procedures to keep the coherence of tracking in these complex situations. Also, these tracking approaches are focused on moving camera applications, so they neglect the utilisation of background subtraction to determine the regions of interest in the scene.

© Springer International Publishing AG 2017
J. Braz et al. (Eds.): VISIGRAPP 2016, CCIS 693, pp. 534–554, 2017.
DOI: 10.1007/978-3-319-64870-5_26

A wide variety of applications can be solved utilising a fixed camera setup (e.g. video-surveillance, health-care at distance, behaviour analysis, traffic monitoring). This kind of setup allows the consideration of inexpensively utilising background subtraction approaches to detect potential regions of interest in the scene. This work focuses on this kind of applications, focusing in solving the problem of robust tracking of multiple unknown (uninitialised) objects, independently of the scene illumination conditions, in real-time. Then, tracking is performed without manual intervention.

Segmentation is commonly the early stage of any vision system, prior to tracking and higher level analysis stages, where regions of interest are extracted from the video sequence. Background subtraction approaches present several issues as: low contrast, poor illumination, gradual and sudden illumination changes, superfluous movement, shadows, among others [3]. Any error emerging from this stage would be propagated to the subsequent stages. A way to deal with these issues is to determine the *quality of the segmentation process* in order to activate control mechanisms to mitigate those errors on later stages.

Assuming that we do not know the model of objects present in the scene, we initially use a bounding box representation extracted from segmented blobs using background subtraction methods. This representation is general enough to track any object in real-time, and serves as the initial region of interest for applying more complex object models. Nevertheless, as the segmented blobs are obtained from background subtraction, they are sensitive to changes in contrast and illumination. This sensitivity affects the object tracking process incorporating noise (in terms of false positive and negative) to the system.

In order to control the effect of noisy information in tracking, we propose a local feature tracking approach, which reinforces the tracking of the bounding box associated to the object. We extract a contrast map from segmentation, to obtain reliability measures which allow us to characterise the local features in terms of illumination and contrast conditions. The local descriptors are obtained from a multi-criteria approach, considering colour (through HSV histograms), structural (through a binary descriptor), and segmentation region (through foreground mask and contrast maps) features. Then, the most reliably tracked local features are utilised, together with the tracked bounding box and the foreground information associated to the tracked object in the current frame, to adjust the estimation of the bounding box in the current frame.

This paper is organised as follows. First, Sect. 2 presents the state-of-the-art in order to clearly establish the contribution of the proposed approach. Then, Sect. 3 performs a complete description of the approach. Next, Sect. 4 presents the results obtained on several benchmark videos. Finally, Sect. 5 presents the conclusion and future work.

2 State of the Art

In the context of segmentation quality measures, the most recent approach is presented in [4]. The authors propose a metric to quantify the segmentation quality for remote sensing segmentation, in terms of over-segmentation and under-segmentation. In order to detect under or over-segmentation, they use a similarity function to evaluate the quality of the segmentation. A good segmentation is obtained if a segment is well separated from its neighbouring segments. Errors can occur, like splitting a segment in two similar segments (over-segmentation) or merging two distinct segments (under-segmentation). Using the similarity function, the authors are able to measure over-segmentation and under-segmentation for each segment in the image. That information then is utilised to improve the segmentation applying the corresponding mechanisms to the erroneous segment (e.g. splitting a segment with under-segmentation problem).

In [5] the authors make a review of video segmentation quality. They identify that quality measurements can be object-based (individually) or globally (as meaning of overall segmentation). These measurements can also be classified as *relative*, when the segmentation mask is compared with ground-truth or as *stand-alone*, when the evaluation is made without using a reference image. Other classifications are subjective evaluation using human judgement or objective evaluation, using a set of a priori expected properties. For our scope, we are interested on a *individual stand-alone objective* quality measurement. In the same article, the features describing this kind of measures are intra-object metrics such as shape regularity, spatial uniformity, temporal stability and motion uniformity; or inter-object metrics like local contrast or neighbouring objects feature difference. The authors propose measures for each two classes of content, the stable content and the moving content. The first one is temporally stable and has regular shape, while the second one has strong and uniform motion. These measures take into account the characteristics of each content to make an unique quality value for the object.

In [6] the authors proposed three disparity metrics: local bound contrast, temporal color histogram difference and motion difference along object boundary. The local bound contrast is focused on determining the quality of the bounds by

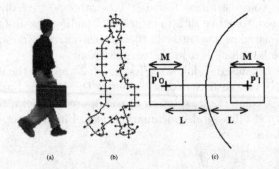

Fig. 1. Spatial color contrast along boundary metric from [6]. (a) image: Object detected, (b) image: Boundary with normal lines, (c) image: A zoom-in of a normal line where each cross represents a pixel inside (P_I) or outside the object (P_O).

comparing internal features (inside of the object) with external features (outside of the object). The next image depicts this metric:

To determine the quality of the boundary, a pixel P_I from the object is compared with a pixel of its neighbourhood P_O, both at distance L of the boundary. The comparison considers the average color in the square of size M, centered in the pixel P_* as shown in the Fig. 1(c). In this sense, good quality segmentation is achieved when there is a high difference between internal and external features. Special care must be taken with the meaning of the value, because a good boundary can be represented by a high quality value, but a high quality value does not necessarily mean a good quality boundary. The second metric tries to measure the temporal stability of color histogram distribution by comparing current object histogram with a smoothed version generated as an average of k previous histograms. A good temporal color stability is obtained if both histograms are similar. The third metric models the quality of the movement by estimating how the points P_* change from one frame to another. The movement metric considers the difference of motion vectors from both points (P_i and P_O) and a reliability factor defined as the precision of the estimation compare the measurement and the color consistency of the points in the square. The authors proposed a combined metric to determine the quality of the object segmentation. As well, they can determine if a particular segment of the boundary has poor quality using a combination of local bound contrast and motion metrics. If the combined value is higher than a predefined threshold, the related segment is considered as low quality. This threshold is obtained as a factor of the standard deviation of the mean object quality.

In the context of, local descriptor-based trackers, some similar approaches are presented in the literature. In [7] a reliable appearance model (RAM) that uses local descriptor (HOG) to learn the object shape and histogram is proposed. This appearance model effectively incorporate color and edge information as discriminative features. However, it is necessary to get a reliable first model to perform the training of the Adaboost learner, leaving this approach as semi-automatic, as well as many other approaches [1,2,8–10].

In [8] the authors proposed a weighted histogram that gives a higher weight to foreground pixel in order to make target features more prominent. The weighted component is based on the pixel's degree of belonging to the foreground. The way of producing the weighted histogram is very similar to our weighted histogram from Eq. (4), but it does not incorporate the reliability of illumination $R_i(y)$, that defines how illumination affect color-based features.

The authors in [9] combine a local descriptor (SIFT) with a global representation (PCA). In contrast to classical PCA, where pixels are weighted uniformly, they add a higher weight to pixels close to SIFT descriptor's position. The tracking phase depends on how reliable are the descriptors matching. This reliability is obtained based on how well the descriptor has been matched previously. Also the amount of reliable descriptors is used to determine if the occlusion is present in the frame. There are three modes of tracking, (1) if there are enough descriptor matched and they are reliable, then the tracking is perform by approximating the affine matrix that described the movement of the previous frame's descriptors with the current descriptors. (2) if there are reliable matched descriptor but they are scarce, a translation model (position and velocity) is calculated instead. (3) is there no reliable matches, previous information is used to estimate the object's movement. In our case, the reliability of the descriptors comes from the reliability map, but the idea of use previous information when there is no reliable match of the descriptors remains. Another tracker that uses reliability is presented in [11]. In this case, the reliability is based on self-incorporated object detector (that is trained off-line). In order to get a good tracking performance, it is necessary to weight properly the information of tracking history and the classifier, otherwise drifting problems may arise.

Fragtrack is proposed in [10]. It uses local patches to avoid partial occlusion problems. If a patch is occluded, other patches can be used to predict the bounding box position (they assume that at least 25% of patches are visible). Each of this patches has associated a histogram and the relative position of its bounding box. The estimation of the bounding box in the next frame is done by a voting scheme. Each patch's histogram is searched in a neighbourhood and votes for a possible position of the bounding box. So, the estimated bounding box's position is whose has more votes. As the method rely heavily on the use of histogram, they use integral matching to perform real time tracking. This also allows search in different scales at without increasing so much the computational cost.

We summarise the contributions of the proposed approach as:

- A reliability model for background subtraction methods (or methods with similar behaviour: background modelling, comparing current frame with background model and applying a threshold to classify pixels into foreground or background). This is a pixel-level reliability model, which we refer as *reliability map*.
- An illumination reliability model to quantify the effects of illumination on color-based features.
- A way to convert a *reliability map* to *attribute-level reliability*. The attributes depend on the object representation. In our case, we will use a 2D bounding box and local features as object representation.
- A multi-target tracking approach incorporating attribute-level reliability measures for weighting the contribution of detected local features to the object model. The idea is to prevent the incorporation of information that could negatively affect the estimation of the object model, and focus on the most reliable information to reduce the effect of noise.

3   Reliable Local Feature Tracking

The proposed tracking approach is depicted in Fig. 2.

For each new frame of the video sequence, a background subtraction algorithm is applied for obtaining the foreground mask, the reliability map (see Sect. 3.1, for details), and the regions of interest (ROI), represented as a set of

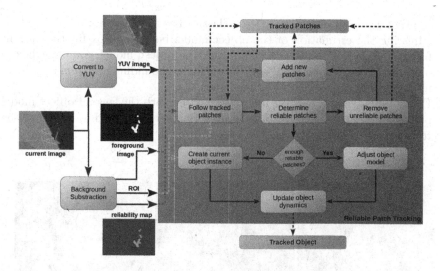

Fig. 2. General schema of the proposed tracking approach.

bounding boxes, using a connected components algorithms. Also, the new frame is converted to YUV color space.

For the first frame where a new object appears (new bounding box not associated to any other previously tracked object), a set of tracked patches is initialised, according to the procedure described in Sect. 3.2.

For the next frames, a ROI (or merge of partial ROIs), determined with a Multi-Hypothesis Tracking (MHT) algorithm [12], is associated to the object as input to the robust patch tracking approach, and the following procedure is applied:

1. If a patch is considered unreliable in terms of positioning. Then, an optimal association to the patch is searched in the current frame considering the information of the ROI displacement and dimension change, compared to the previously associated ROI. This optimal association is determined using a global reliability measure, which integrates temporal coherence, structural, colour, and contrast measures (see Sect. 3.4). If a set of patches has been reliably tracked from previous frames, this information is utilised to determine the displacement of all the patches for the current frame, according to the procedure detailed in Sect. 3.3.
2. Then, according to the global reliability measure calculated at the previous step, the highest reliability patches can be classified as *highly reliable*, the patches with low reliability are classified as *unreliable* and marked for elimination (see Sect. 3.3, for details).
3. Next, *unreliable* patches are eliminated and new patches are added in positions not properly covered by the remaining tracked patches. The construction of these patches follows the same procedure as the patch initialisation phase (Sect. 3.2).

4. If a significant number of patches is classified as reliable, they are utilised for adjusting the estimation of the object model bounding box for the current frame. If this number is not significant, the object model bounding box is obtained from the input ROI and the estimated bounding box from the object model dynamics (see Sect. 3.5, for details).
5. Finally, the dynamics object model is updated with the current object model bounding box (see Sect. 3.5, for details). Bottom image of Fig. 3 depicts the result of the tracking process.

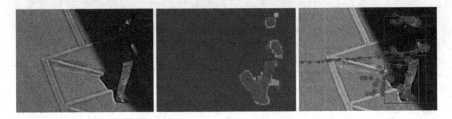

Fig. 3. Top figure shows the current frame. Center figure depicts the reliability map, with a thermal map, where high reliability is red. Bottom figure shows the result of the tracking process; red boxes represent the bounding boxes from segmentation, the blue box represents the estimated bounding box of the tracked object, the dots represent the tracked patches coloured according to reliability in thermal scale, and blue segments represent the object trajectory. (Color figure online)

3.1 Reliability Map from Background Subtraction

The key factor for a good tracking is how distinguishable is the object of interest from its surroundings. If we are working in a background subtraction scheme, we are going to interpret the *surrounding* of the object as the background model and *how distinguishable is* as the degree of difference between the current image and the background model. If we have a significant difference, we have certain margin of error on defining the threshold and the segmentation algorithm will still be able to perform a good classification. Nevertheless, if that difference is low, we have to accurately define the threshold value to avoid a misclassification. In this sense, the last example is less *reliable*, because it is more prone to make a wrong classification.

Based on the previous idea, we propose a method that can model the reliability of any background subtraction technique through the following steps:

1. Generate a pixel-level difference value D between current image pixels and background model pixels.
2. Define a range $[inf, sup]$ for the difference value D. We are interested in generating a reliability image representation with different degrees of reliability. If we consider all the range, sometimes it can generate a binary image (just low and high reliability) that is not useful for our interest. This range can be

defined as the neighbourhood of the threshold value, for example using the range $[inf = \alpha \times Threshold,\ sup = \beta \times Threshold]$, with $0 < \alpha < 1$ and $1 < \beta$.

3. Apply the scaling function from Eq. (1), to every difference value D generated in Step 1, to convert difference values into reliability measures:

$$S(D) = \begin{cases} 0\% & if \quad D < inf \\ f(x) & if \ inf \leq D \leq sup, \\ 100\% & if \quad D > sup \end{cases} \tag{1}$$

where D is the differnece value, inf and sup are values defined in Step 2 and $f(x)$ is a increasing function (we use a linear function).

At the end of these steps we generate a pixel-level representation of the reliability which we named as *reliability map*. This map is internally represented as a grayscale image, but for proper visualisation we transform it into thermal scale, as shown in Fig. 4.

Usually, several post-processing functions are applied to the segmentation mask in order to reduce the noise. This operation also should be applied to the reliability map to maintain the coherence of its representation with the foreground mask. Figure 5 is an example of applying morphology operations to the foreground image and the reliability map (considering gray-scale morphological operators).

Fig. 4. Reliability map visualization. Left image: current image frame, right image: thermal scale reliability map. Blue color means a low difference between modeled background and current frame. Red color means a high difference. (Color figure online)

We illustrate how this method works using naive background subtraction [13]: This model performs difference of current image with a background subtraction image (image without any object interest). Our implementation uses the sum of square differences as distance value before applying the classification threshold.

The sum of square difference, shown in the Eq. (2), is a common metric to measure the distance between current pixel and background pixel in a RGB color space:

$$D = (R_{bg} - R_i)^2 + (G_{bg} - G_i)^2 + (B_{bg} - B_i)^2, \tag{2}$$

where subindex $(\cdot)_i$ refers to current image pixel and $(\cdot)_{bg}$ refers to background pixel.

The classification is performed applying the threshold value as in the Eq. (3):

$$fg_{mask} = \begin{cases} foreground \ if \ D > \tau^2 \\ background \ if \ D < \tau^2 \end{cases}, \tag{3}$$

where τ is the classification threshold.

Applying the proposed scheme to this method using a range of $[0.1 \times \tau^2, 2.0 \times \tau^2]$ with $\tau = 13$ and using a morphology window of size 7×7, we can obtain image shown in Fig. 6.

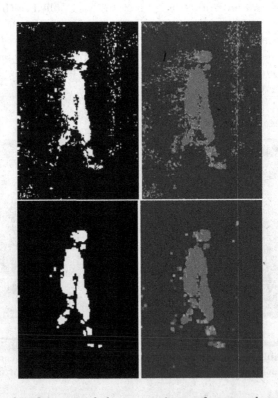

Fig. 5. Example of applying morphology operations to foreground mask and reliability map. The top images show the foreground mask and the reliability map with noise. The bottom images show the results after applying the morphological operation (binary morphology for foreground mask and gray-scale morphology for reliability map).

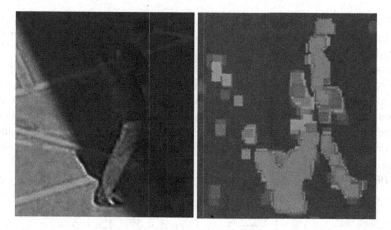

Fig. 6. Reliability map using naive background subtraction. Left image: current image, right image: reliability map from naive background subtraction.

3.2   Patch Initialisation Phase

The first step is to find patches of size $patchSize \times patchSize$ in the contour of the object (defined by the foreground mask) in such way that any two patches do not overlap between each other. Then, the strongest point inside of the patch, obtained by FAST algorithm [14] from the Y-channel of the current frame converted to YUV color space, is added as a new patch position if no other existing patch is near this position.

Then, each candidate patch stores the following information:

- The central patch position (x, y).
- The 512 bits FREAK descriptor [15], generated using the reliability map, representing the structural information of the patch.
- A normalised colour histogram, using chroma channels U and V from the YUV current frame, considering only pixels belonging to the foreground mask in the analysed patch. Considering $H_{UV}(i, j)$ as the bin of a 2D histogram of the UV channels, with $i, j \in [0..BinsNumber]$, The Eq. (4) represents the way this histogram is calculated.

$$H_{UV}(i,j) = \frac{\sum\limits_{p \in Q} F(p) R_m(p) R_i(Y(p))}{\sum\limits_{p \in P} F(p) R_m(p) R_i(Y(p))}, \qquad (4)$$

with

$$Q = \left\{ p \in P : \left\lfloor \frac{U(p)}{binSize} \right\rfloor = i \ \wedge \ \left\lfloor \frac{V(p)}{binSize} \right\rfloor = j \right\}, \qquad (5)$$

where $Y(p)$, $U(p)$, and $V(p)$ correspond to the channel level in $[0..255]$ in pixel position p of the current frame in YUV color space, P is the set of pixel positions inside the analysed patch, and Q is the set of patch positions, where values $U(p)$ and $V(p)$ fall inside the bin $H_{UV}(i,j)$. For each pixel a weighted value is added, where: $F(p) = 1$ if the pixel p corresponds to the foreground, and 0 otherwise; $R_m(p) \in [0;1]$ is the reliability map value in position p, where a value of 1 corresponds to maximum contrast reliability (see Sect. 3.4, for details); and $R_i(Y(p))$ corresponds to the illumination reliability, accounting the pertinence of colour information given different illumination levels, according to the gray-scale level in channel $Y \in [0..255]$ at pixel position p. The reliability measure R_i considers maximum reliability near 128 value (medium illumination) and decays to 0 near the extremes of the interval. Eq. (6) formulates this reliability and Fig. 7 depicts the reliability function.

$$R_i(Y) = \begin{cases} 0 & \text{if } Y \leq 128 - \gamma \\ \frac{Y+\gamma-128}{\beta} & \text{if } 128 - \gamma < Y < 128 - \alpha \\ 1 & \text{if } 128 - \alpha \leq Y \leq 128 + \alpha \\ \frac{128+\gamma-Y}{\beta} & \text{if } 128 + \alpha < Y < 128 + \gamma \\ 0 & \text{otherwise} \end{cases} \qquad (6)$$

where α and β are predefined parameters, and $\gamma = \alpha + \beta$.

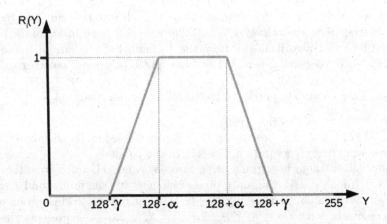

Fig. 7. Illumination reliability function.

- A colour histogram reliability measure accounting for the reliability of colour information (Eq. (7)).

$$R_{colour} = \left(\sum_{p \in P} F(p) R_m(p) R_i(Y(p)) \right) / N_{pix},\qquad (7)$$

where N_{pix} is the number of foreground pixels in the patch.

- A normalised gray-scale histogram of *NumBins* bins, accumulating channel Y of the current image in YUV color space, for those pixels inside the patch which belong to the foreground.

All this information is utilised to properly characterise the patch, in order to match with potential patches in future frames. These patches then initialise patch tracking buffers for future processing.

3.3 Patch Tracking Phase

Given a set of patches S from the previous frame, the patch tracking process follows the process described bellow:

- Consider S_H as the set of tracked patches considered as *highly reliable* from the previously processed frame. A reliably tracked frame is a frame of high reliability, which has a coherent movement with the mobile object and high contrast, colour, and structural accumulated reliabilities (as described in Sect. 3.4). Then, these patches are considered able to estimate the behaviour of less reliable patches near to them. For this reason, tracking becomes more exhaustive for these patches, but in a reduced region. Then, the reliable patches are tracked in the following way:
  1. Displacement vector (dx, dy) is determined from the displacement vector inferred from their associated patch tracking buffer.
  2. Search window is determined from the accumulated difference (x_d, y_d) between the accumulated object center movement vector with the accumulated movement vector of the patch, considering all the patches in the tracking buffer. The window is centered in $(x_W, y_W) = (x_p + dx, y_p + dy)$, where (x_p, y_p) is the position of the patch in the previous frame.
  3. Then, the patch position with minimal global distance D_{global} to the previous patch is associated to the current reliable patch position, following the Eq. (8).

$$(x^*, y^*) = \arg \max_{(x,y) \in W_H} D_{global}(p_t(x, y), p_{t-1}), \qquad (8)$$

with

$$W_H = \{(x, y) : |x - x_W| \le x_d \wedge |y - y_W| \le y_d\}, \qquad (9)$$

where $p_t(x, y)$ is the current patch at position (x, y), and p_{t-1} is the patch at previous frame. The distance measure D_{global} globally calculates the patch distance, considering the structural, colour, segmentation and gray-scale information. This measure is described in detail, in Sect. 3.4.
- If the patch buffer has been built just in the previous frame (previous initialisation step) or the patch is not *highly reliable*, the positioning of the patch is determined in the following way:
  1. If set S_H size is adequate, the displacement vector (dx, dy) for the patch is determined from the displacement vectors of *highly reliable* patches, each weighted by the position of the *highly reliable* patch to the analysed patch in the previous frame and the R_{global} reliability measure.

2. The window is determined in a similar way as for *highly reliable* patches, but, as the patch is less reliable, it would normally have a bigger search window. For this reason, FAST algorithm is applied to the search window for candidate positions.

3. Then, maximal reliability patch is determined in a similar way as in Eq. (8), but from the set of FAST points detected on the window.

- Then, according to the global reliability measure R_{global}, the tracked patches are classified as *highly reliable* if they pass a high threshold T_H. Patches with reliability below a low threshold T_U are classified as unreliable and eliminated.

- As the object can be represented by less patches, new patches are added in positions not properly covered by the remaining tracked patches, using the same procedure described in Sect. 3.2.

3.4   Patch Distance and Reliability Measures

To match two patches, the distance between them in terms of their different attributes must be calculated. We propose the distance measure D_{global}, described in Eq. (10).

$$D_{global} = \frac{w_{st}D_{st} + w_{fg}D_{fg} + w_{co}D_{co} + w_{gs}D_{gs}}{w_{st} + w_{fg} + w_{co} + w_{gs}}, \tag{10}$$

with

$$D_{st}(p_1, p_2) = \frac{\|Freak[p_1]; Freak[p_2]\|_H}{512}, \tag{11}$$

$$D_{fg}(p_1, p_2) = \frac{|\#FG[p_1] - \#FG[p_2]|}{\max(\#FG[p_1], \#FG[p_2])}, \tag{12}$$

$$D_{co}(p_1, p_2) = D_{rcol}(p_1, p_2) \|H_{UV}[p_1]; H_{UV}[p_2]\|_B, \tag{13}$$

$$D_{rcol}(p_1, p_2) = |R_{colour}(p_1) - R_{colour}(p_2)|, \tag{14}$$

$$D_{gs}(p1, p2) = \|H_Y[p1], H_Y[p2]\|_B, \tag{15}$$

where $\|\cdot; \cdot\|_H$ is the distance of Hamming for binary descriptors, and $\|\cdot; \cdot\|_B$ is the Bhattacharyya distance [16] for histograms. $Freak[p]$ corresponds to the FREAK descriptor, $\#FG[p]$ is the number of foreground pixels, $H_{UV}[p]$ is the colour histogram, and $H_Y[p]$ is the gray-scale histogram, of patch p. $D_{rcol}(\cdot, \cdot)$ accounts for the difference in R_{colour}, considering that histograms are more comparable under similar conditions in terms of illumination and contrast reliability.

It has been previously discussed that we need a measure to account for the reliability of the tracked patches in the scene, in order to determine the usefulness of the patch information on contributing to a more robust object tracking. This reliability measure is R_{global}, described in Eq. (16), considering a tracked patch buffer $B_p = \{p_1, .., p_N\}$, where p_1 is the current patch, and N is the buffer size, and the object bounding box buffer $B_I = \{I_1, .., I_N\}$, where I_j is the bounding box in buffer position j.

$$R_{global}(B_p) = \frac{R_{pos}(B_p) + R_c(B_p) + R_g(B_p)}{3}, \tag{16}$$

with

$$R_{pos}(B_p) = \frac{\sum_{i=1}^{N-1}(N-i)\,\|c[p_i] - c[p_{i+1}]; c[I_i] - c[I_{i+1}]\|_M}{\sum_{i=1}^{N-1}(N-i)}, \tag{17}$$

$$\|c1; c2\|_M = |x[c1] - x[c2]| + |y[c1] - y[c2]|, \tag{18}$$

$$R_c(B_p) = \frac{\sum_{i=1}^{N}(N-i+1)C(x[p_i], y[p_i])}{\sum_{i=1}^{N}(N-i+1)}, \tag{19}$$

$$C(x_p, y_p) = \frac{\sum_{x=x_p-\frac{L}{2}}^{x_p+\frac{L}{2}} \sum_{y=y_p-\frac{L}{2}}^{y_p+\frac{L}{2}} G(x - x_p, y - y_p)FG(x, y)R_m(x, y)}{\sum_{x=x_p-\frac{L}{2}}^{x_p+\frac{L}{2}} \sum_{y=y_p-\frac{L}{2}}^{y_p+\frac{L}{2}} G(x - x_p, y - y_p)FG(x, y)}, \tag{20}$$

$$R_g(B_p) = 1 - \frac{\sum_{i=1}^{N-1}(N-i)D_{global}(p_i, p_{i+1})}{\sum_{i=1}^{N-1}(N-i)}. \tag{21}$$

The three components of R_{global} are calculated weighting by the novelty of the information. R_pos is the position coherence reliability, which takes into account the displacement coherence between the history of the patch (measured as the displacement vector of the patch centers $c[p_i] - c[p_{i+1}]$) and the history of the central position of the object model bounding box ($c[I_i] - c[I_{i+1}]$), using the Manhattan distance between displacement vectors at the different frames. R_c accumulates the contrast reliability measure $C(x, y)$, which accumulates the values of the reliability map R_m, weighted by a Gaussian function G centred at (x, y) and only accumulating foreground pixels (considering $FG(x, y)$ as the foreground image, with value 1 for foreground pixels and 0 for background). R_g accumulates the reliability on the similarity of the patches in the buffer.

3.5   Adjustment of Object Model

Finally, if the current input bounding box is significantly different in dimensions compared to the previous frame or several reliable patches present a low contrast reliability for the current frame and a relevant change on patch mean illumination from previous frame (inferred from Y channel), the bounding box is recalculated based on the information provided by the remaining reliable patches. The displacement of each bound of the bounding box (Left, Right, Bottom, Top) is obtained from the weighted mean of patches displacement from the previous frame, weighted by the distance to the bound and the reliability of the patches.

If no reliable patches are available, the bounding box projected from the object dynamics model is considered as input. We utilise a dynamics model similar to Kalman Filter [12]. If the current input bounding box is similar in size to the previous frame, this bounding box is considered as the object model for the current frame. Then, the dynamics model is updated with the current object model.

4   Experimental Validation

The visual coherence of the estimation has been first tested in three short sequences of diverse contrast. The results are shown in Fig. 8.

A qualitative test has been performed using a section of the sequence Light_Video032 of dataset Alov-300[1]. In this sequence a man walks from a good to a poor illumination region (from left to right). Figure 9 depicts the evolution of the patches through the last five frames:

As we can see in the Fig. 9, each patch can track its next position even with structural deformation (e.g. row num. 0) or some changes in color (e.g. row num. 4) due to illumination changes.

The performace of the approach is compared with a reduced version (without local patches nor reliability measurements). A compacted view of the sequence is shown in the Fig. 10, considering key frames for the analysis.

Initially both methods perform similar, but when the object enters into the poor illumination region, our approach outperforms the basic method. However, at the end of the sequence, when the object is completely immerse in the shadow region, both methods are unable to keep tracking the object. Neverthless, when reliable local information is available, our approach is able to correctly infer the global position of the object and unrealible local features enable the approach to properly characterize regions with poor illumination conditions. This characterization will allow us in the future to properly handle poor illumination situations when they are detected. Figure 11 presents the evolution of the illumiantion conditions for the tested sequence.

To quantify the improvement of the approach, two videos of changing contrast situations have been tested. Both videos have ground-truth segmentation, in order to obtain the ideal track of the analysed objects. The first video consists

---

[1] http://www.alov300.org/.

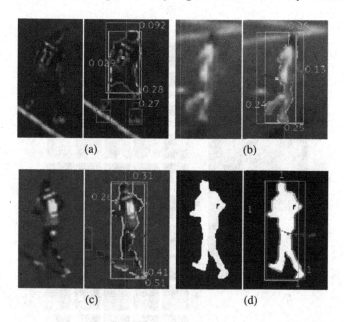

Fig. 8. Resulting tracking for three soccer player sequences with different levels of contrast. Figures (a), (b), and (c) show the result for low, medium, and high contrast situations, respectively. Figure (d) is a control case for ground-truth segmentation. The segmentation blob bounding boxes are colored red, the merged bounding box for the object hypothesis colored yellow, and the estimated bounding box from the dynamics model colored cyan. The central object position trajectory is depicted with blue squares. (Color figure online)

in a single football player sequence (27 frames), where a player goes from a light to a dark zone of the pitch. This video is a zoomed short sequence extracted from the Alfheim Stadium dataset[2]. The second video consists in a sequence (51 frames) where a rodent is exploring a confined space with better illumination in the center. The sequence is part of a set of sequences provided by the Interdisciplinary Center of Neuroscience of Valparaiso[3]. This sequences are intended to study the behavior of the degu, a rodent which commonly presents the Alzheimer disease.

The experiment consists in performing object tracking using the new dynamics model with and without considering the proposed reliability measures, and compare the obtained tracks with the ideal tracks obtained from the ground-truth segmentation. The results were summarised in Table 1.

The results for the first experiment are exemplified in Fig. 12. Figure 12(b) and (c) show the core motivation of this work: the effect of considering different measures for tracked attributes allows a finer control of the trade off between

---

[2] Open dataset extracted from Alfheim Stadium, the home arena for TromsøIL (Norway). Available from: http://home.ifi.uio.no/paalh/dataset/alfheim/.

[3] Interdisciplinary Center of Neuroscience of Valparaiso, Chile. http://cinv.uv.cl/en/.

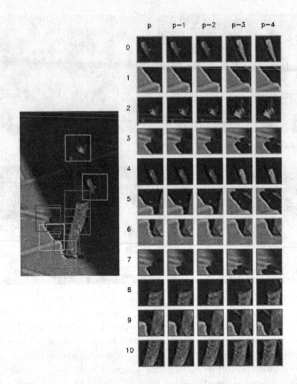

Fig. 9. The local patches are shown in this figure. The patches are sorted from current patches in column p, to older patches (columns $p - i$). (Color figure online)

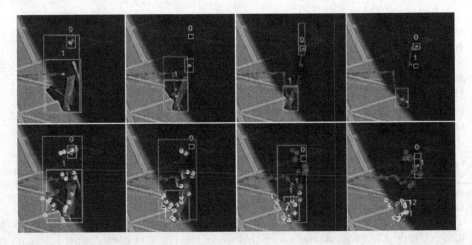

Fig. 10. Comparison of the approach using local patches and reliabilty measurements. First row shows the performance without utilisation of local patches nor reliabilty measurements. Second row depicts our approach. In the second row, red circles represent matched patches and blue circles represent new patches. (Color figure online)

Table 1. Results for evaluation sequences with respect to ground-truth sequences. The column Imp.% is the percent of improvement utilising the proposed approach.

Sequence	Distance (pixels)		
	No rel.	Patch rel.	Imp.%
Football (T = 15)	602.2	579.5	3.8%
Football (T = 20)	640.7	570.8	10.9%
Rodent (T = 10)	600.4	581.6	3.1%
Rodent (T = 15)	506.7	491.4	3.0%
Rodent (T = 20)	1086.8	1011.5	6.9%
Rodent (T = 25)	1071.1	1023.0	4.5%

the estimated state and the measurement in the update process. In the example, the patch tracking algorithm was able to properly weight unreliable data to not affect considerably the dynamics model, and the legs of the player were not lost (Fig. 12(c)).

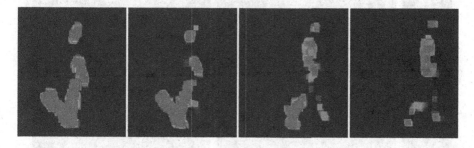

Fig. 11. Reliability map of the object from video Light_Video032. The reliability of the object is properly related to illumination, thus it can be used to detect bad segmentation in order to activate control mechanisms.

For the second experiment, the challenge is to follow a rodent of quick acceleration changes and not homogeneous illumination conditions. Also, poor segmentation occurs due to the sudden changes of speed. The sequence was tested for different segmentation thresholds ($T \in \{10, 15, 20, 25\}$). From these results, we are able to state that a more robust tracking can be achieved utilising the patch reliability measure, with an improvement higher than a 3% in precision. Examples of these results are depicted in Fig. 13.

All the results generated by qualitative and quantitative experiments can be found in: http://profesores.elo.utfsm.cl/~mzuniga/videos/.

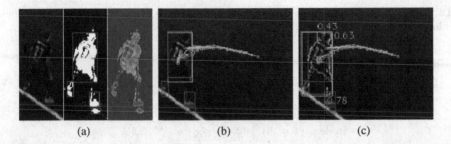

(a)   (b)   (c)

Fig. 12. Example of the effect on utilising the patch reliability on the tracking process ($T = 20$). Figure (a), from left to right, shows the current, segmentation, and contrast map images, respectively. Figure (b) shows the tracking result without considering the patch reliability measures (every reliability is set to 1). Figure (c) shows the result of using the patch reliability measure. Note the difference in tracking bounding box, where the feet of the player are more properly incorporated to the object. The boxes are colored the same way as previous images. The central object position trajectory is depicted with green squares, the ground-truth positions in cyan squares, and the distance between them is represented with a yellow line. (Color figure online)

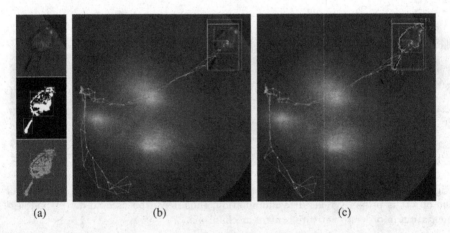

(a)   (b)   (c)

Fig. 13. Example of the effect on utilising the patch reliability on the tracking process ($T = 25$). Figure (a), from top to bottom, shows the current, segmentation, and contrast map images, respectively. Figures (b) and (c) show the tracking result not considering and considering the patch reliability measures, respectively.

5   Conclusions

For addressing real world applications, computer vision techniques must properly handle noisy data. In this direction, we have proposed a new tracking schema considering local features and reliability measures which have shown promising results for improving the dynamics updating process of the tracking phase. The reliability measures were utilised to control the uncertainty in the obtained information, through a direct interpretation of the criteria utilised by the segmentation phase to determine the foreground regions. In this sense, this approach can be applied to other segmentation algorithms to improve the tracking phase in the same way.

In particular, the proposed global patch reliability measure, considering a diverse range of features, has shown one of the many possible ways of integrating segmentation phase data to object modelling. In the present work, no a priori knowledge has been considered about the objects to be tracked. The integration of the data from the segmentation phase with more complex object models can also improve the tracking phase, by better determining the objects of interest for a context or application. At the same time, these reliability measures can help these object models to better determine their parameters, subject to noisy measurements.

The preliminary evaluation obtained promising results both in robust tracking and quick processing. Nevertheless, extensive testing is required for fully validating the approach.

This work can be extended in several ways: the approach can be tested for different types of detectors of interest points and local feature detectors. Also, the algorithm can be tested for different background subtraction approaches. However an extensive parameter sensitivity evaluation is still needed. As local features are utilised for partial occlusion, this approach could be naturally extended to deal with dynamic occlusion situations. As previously mentioned, one of the most important extension of this work will be the development of more sophisticated control mechanisms in case of local patches describing the presence of poor illumination conditions.

Acknowledgements. This research has been supported, in part, by Fondecyt Project 11121383, Chile.

References

1. Kalal, Z., Matas, J., Mikolajczyk, K.: Tracking learning detection. IEEE Trans. Pattern Anal. Mach. Intell. **34**, 1409–1422 (2011)
2. Yang, F., Lu, H., Yang, M.: Robust superpixel tracking. IEEE Trans. Image Process. **23**, 1639–1651 (2014)
3. Toyama, K., Krumm, J., Brumitt, B., Meyers, B.: Wallflower: principles and practice of background maintenance. In: Proceedings of the International Conference on Computer Vision (ICCV 1999), pp. 255–261 (1999). doi:10.1109/ICCV.1999. 791228
4. Troya-Galvis, A., Gancarski, P., Passat, N., Berti-Equille, L.: Unsupervised quantification of under- and over-segmentation for object-based remote sensing image analysis. IEEE J. Sel. Top. Appl. Earth Obs. Remote Sens. **8**, 1936–1945 (2015)
5. Correia, P.L., Pereira, F.: Objective evaluation of video segmentation quality. IEEE Trans. Image Process. **12**, 186–200 (2003)
6. Erdem, Ç.E., Sankur, B., et al.: Performance measures for video object segmentation and tracking. IEEE Trans. Image Process. **13**, 937–951 (2004)

7. Lee, S., Horio, K.: Human tracking using particle filter with reliable appearance model. In: 2013 Proceedings of SICE Annual Conference (SICE), pp. 1418–1424 (2013)
8. Wang, L., Yan, H., Wu, H.Y., Pan, C.: Forward-backward mean-shift for visual tracking with local-background-weighted histogram. IEEE Trans. Intell. Transp. Syst. **14**, 1480–1489 (2013)
9. Sun, L., Liu, G.: Visual object tracking based on combination of local description and global representation. IEEE Trans. Circ. Syst. Video Technol. **21**, 408–420 (2011)
10. Adam, A., Rivlin, E., Shimshoni, I.: Robust fragments-based tracking using the integral histogram. In: 2006 IEEE Computer Society Conference on Computer Vision and Pattern Recognition, vol. 1, pp. 798–805 (2006)
11. Breitenstein, M., Reichlin, F., Leibe, B., Koller-Meier, E., Van Gool, L.: Robust tracking-by-detection using a detector confidence particle filter. In: 2009 IEEE 12th International Conference on Computer Vision, pp. 1515–1522 (2009)
12. Zuniga, M.D., Bremond, F., Thonnat, M.: Real-time reliability measure driven multi-hypothesis tracking using 2D and 3D features. EURASIP J. Adv. Signal Process. **2011**, 142 (2011). doi:10.1186/1687-6180-2011-142
13. McIvor, A.: Background subtraction techniques. In: Proceedings of the Conference on Image and Vision Computing (IVCNZ 2000), Hamilton, New Zealand, pp. 147–153 (2000)
14. Rosten, E., Drummond, T.: Machine learning for high-speed corner detection. In: Proceedings of the IEEE European Conference on Computer Vision (ECCV 2006), vol. 1, pp. 430–443 (2006)
15. Alahi, A., Ortiz, R., Vandergheynst, P.: Freak: fast retina keypoint. In: Procedings of the IEEE Conference on Computer Vision and Pattern Recognition (CVPR 2012), pp. 510–517 (2012)
16. Bhattacharyya, A.: On a measure of divergence between two statistical populations defined by probability distributions. Bull. Calcutta Math. Soc. **35**, 99–110 (1943)

Affordance Origami: Unfolding Agent Models for Hierarchical Affordance Prediction

Viktor Seib[✉], Malte Knauf, and Dietrich Paulus

Active Vision Group (AGAS), University of Koblenz-Landau,
Universitätsstr. 1, 56070 Koblenz, Germany
{vseib,mknauf,paulus}@uni-koblenz.de
http://agas.uni-koblenz.de

Abstract. Object affordances have moved into the focus of researchers in computer vision and have been shown to augment the performance of object recognition approaches. In this work we address the problem of visual affordance detection in home environments with an explicitly defined agent model. In our case, the agent is modeled as an anthropomorphic body. We model affordances hierarchically to allow for discrimination on a fine-grained scale. The anthropomorphic agent model is unfolded into the environment and iteratively transformed according to the defined affordance hierarchy. A scoring function is computed to evaluate the quality of the predicted affordance. This approach enables us to distinguish object functionality on a finer-grained scale, thus more closely resembling the different purposes of similar objects. For instance, traditional methods suggest that a stool, chair and armchair all afford sitting. However, we additionally distinguish sitting without backrest, with backrest and with armrests. This fine-grained affordance definition closely resembles individual types of sitting and better reflects the purposes of different chairs. We report evaluation results of our approach on publicly available data as well as on real sensor data.

Keywords: Affordance · Affordance prediction · Visual affordances · Affordance hierarchies · Object recognition

1 Introduction

Since Gibson's work on affordances [1] a lot of effort was put into the theoretical investigation of affordances [2,3] and their applications in other fields. When it comes to classification in computer vision, many approaches struggle with large intraclass appearance variations. The reason is at hand: classes are defined by the functionality of objects, rather than their visual appearance. By describing action possibilities between an agent and an object, affordances allow to detect similarities on a functional level, rather than solely rely on the object's appearance. Thus, approaches exploiting affordances were shown to augment the classification process [4,5].

© Springer International Publishing AG 2017
J. Braz et al. (Eds.): VISIGRAPP 2016, CCIS 693, pp. 555–574, 2017.
DOI: 10.1007/978-3-319-64870-5_27

This is why reasoning about an object's purpose has become an important area in today's research in robotics. While shape features are often acquired locally (i.e. around salient points) and might therefore be misleading, detecting a functionality of an object facilitates categorization. Additionally, predicting affordances of objects instead of the object classes allows objects and tools to be applied even without the precise knowledge of the class the object belongs to. Even objects of different classes can be applied according to a certain affordance required by the agent. For example, if an agent (e.g. a robot) needs to hammer, it would pick a heavy object providing enough space for grasping and a hard surface to hit on another object. This works without knowing the category *hammer* or having a hammer available by e.g. using a stone instead.

While in robotics humans play an active role in teaching affordances to robots, e.g. by interaction [6,7] or by imitation [8–10], the vision community follows other approaches. Some approaches completely omit the interacting agent and propose to derive object descriptors by physical simulation [4], by data from additional sensors (e.g. kinematic data [11]) or purely from visual sensors [12]. Other approaches create or "imagine" human models in the environment [13]. These human models are exploited to propose comfortable poses for sitting [5], to learn human-relative placement of objects [13] or to explore action possibilities in human workspaces [14]. In contrast to approaches in robotics we do not record kinematic data of an agent, neither do we detect affordances by interaction. Nowadays, it can be expected that visual perception is mostly common in robots and it is thus plausible to rely on that data. Thus, the approach proposed in this paper relies on visual data only.

In our approach we employ the *observer's* view on affordances as introduced by Şahin et al. [15]. While the environment is being observed by a robot equipped with certain sensors, the system is looking for affordances that afford actions to a predefined model. In our case this predefined model is an anthropomorphic agent representing a humanoid. In recent work [5,13] this *observer's* view is often referred to as *hallucinating interactions*.

In the proposed method we focus on the complementary nature of a humanoid agent and its environment. In our previous work we proposed detecting fine-grained or hierarchical sitting affordances with a simulated anthropomorphic agent [16]. Given an agent and an affordance model, the agent's joints are transformed from a start to a goal pose. Figuratively speaking, the agent model is unfolded from an initial pose to a functional pose that corresponds to predicted affordances (Fig. 1). The final state of the individual joints determines the predicted fine-grained affordances in the hierarchical affordance model.

We use indoor or home environments that are considered as environments specifically designed to suit the needs of humans. Therefore, the complementary agent to the investigated environment is an anthropomorphic, i.e. human, body. Thus, for the purpose of this work affordances shall be informally defined as *action possibilities that the environment offers to an anthropomorphic agent*.

In this work we extend our previous work by refining our affordance model and including more action possibilities: sitting and lying. The refined model

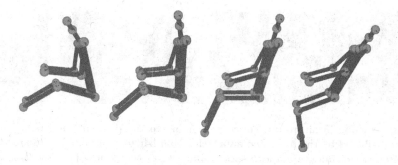

Fig. 1. Unfolding the agent model for the sitting affordance hierarchy.

allows to handle complex scenes in contrast to individual objects of our previous approach. We further provide a formal model for hierarchical affordance prediction and show its applicability on large datasets of furniture objects.

Related approaches in the literature [4,17,18] distinguish affordances on a coarse scale. The considered affordances often include sitting (chairs), support for objects (tables) and liquid containment (cups). We propose looking closely at the individual affordances and distinguishing their functional differences an a fine-grained scale. We already introduced the concept of *fine-grained* affordances in [19] to closely resemble the functional differences of related objects. Although good results could be obtained, our previous work was a proof-of-concept with several limitations.

In the presented work, we concentrate on fine-grained affordances derived from the affordance *sitting* and *lying*. We present a new algorithm for fine-grained affordance prediction that differentiates between 4 typical functionality characteristics of the *sitting* affordance. We divide the coarse affordance *sitting* into the fine-grained affordances *sitting without backrest, sitting with backrest, sitting with armrest* and *sitting with headrest*, whenever the sitting functionality is supported by additional environmental properties that can be exploited by the considered agent. Further, we give an outlook on different subaffordances of the course affordance *lying*.

A system that is able to find affordances either encounters only those objects that were specifically designed to support the affordance in question or environmental constellations that afford the desired action. Our algorithm takes point clouds e.g. from a RGB-D camera as input. The input data is directly searched for affordances (and thus functionalities) without prior object segmentation. In the core of the algorithm, the agent model is unfolded and checked for collisions with the environment. Specific goal configurations of the agent model represent different types of fine-grained affordances. The encountered affordances are segmented from the input point cloud. This segmentation can serve as an initial segmentation for a subsequent object classification step (not further explored in this work). Since the found affordances (especially on a fine-grained scale) provide many hints on the possible object class, categorization can be performed with fewer training objects or simpler object models. The presented fine-grained

Fig. 2. Example furniture objects corresponding to the different fine-grained affordances predicted by the presented approach. From left to right: stool (*sitting without backrest*), two chairs (*sitting with backrest*) and three armchairs (*sitting with backrest* and *sitting with armrest*). Additionally, the rightmost chair also supports the *sitting with headrest* affordance.

affordances correspond to objects such as a stool, chair, armchair and a chair with head support (Fig. 2). Specifically, an affordance-based categorization system can be exploited as outlined in the following. Affordances enable the detection of *sittable* objects even without knowing object classes as *stool, chair* or *couch*. Following the idea of fine-grained affordances, a stool standing close to a wall can even provide both affordances: sitting with and without backrest (in the former case the back is supported by the wall). This intuitively corresponds to the way a human would utilize an object to obtain different functionalities.

The remainder of this work is structured as follows. Related work on affordances in robotics is presented in Sect. 2. Section 3 introduces the model definitions applied in our algorithm and Sect. 4 explains our approach for fine-grained affordance prediction in detail. The proposed algorithm is evaluated in Sect. 5. Finally, a discussion is given in Sect. 6 and Sect. 7 concludes the paper and gives an outlook to our future work.

2   Related Work

Hierarchies in affordances have been explored mainly in design theory to reason about functional parts of objects [20,21]. Their goal is to divide objects into different functional parts that represent different affordances. This allows a designer to identify desired and undesired affordances in early stages of product design. Note however that this hierarchical affordance modeling is conceptually different from the fine-grained affordances applied in this paper. We do not separate objects in different parts with different affordances. Rather, our object independent approach separates an affordance (in this case the *sitting* and *lying* affordances) into different subaffordances on a fine-grained scale.

Other approaches like the work of Hinkle and Olson [4] use physical simulation to predict object functionality. The simulation consists of spheres falling onto an object from above. A feature vector is extracted from each object depending on where and how the spheres come to rest. The objects are classified as cup-like, table-like or sitable.

Research especially focusing on sitting affordances has been conducted over the past years. Office furniture recognition (chairs and tables) is presented by Wünstel and Moratz [12]. Affordances are used to derive the spatial arrangement of the object's components. Objects are modeled as graphs, where nodes represent the object's parts and edges the spatial distances of those parts. The 3D data is cut into three horizontal slices and within each slice 2D segmentation is performed. The segmentation results are classified as object parts and matched to the object models. Wünstel and Moratz' approach detects sitting possibilities also on objects that do not belong to the class *chair*, but intuitively would serve a human for sitting. Unlike the approach of Wünstel and Moratz, we encode the spatial information needed for affordance prediction in an anthropomorphic agent model and affordance models, rather than creating explicit object models.

Hierarchical classification of object parts has been explored in [22]. Complex object models were proposed to identify object parts and thus infer the subcategory of an object type. Affordances were not mentioned explicitly, though. In [23] a human agent model is used to classify objects. Again, affordances were not explicitly mentioned here. Contrary to our approach they needed a segmented object as input for classification.

Our algorithm takes 3D data and detects affordances inside these data. In our approach, individual objects exposing these affordances are subsequently segmented based on the detection result. We propose a hierarchical affordance model and the detection of fine-grained affordances by unfolding an anthropomorphic agent model and fitting the agent to functional object parts. By applying our agent and affordance models we do not need to create complex object models as opposed to [22] and do not have any constraints on the environment (e.g. segmented objects) as in [23]. In our case, the segmented part of the scene is a result of the detected affordances on the input data.

More recently, Grabner et al. [5] proposed a method that learns sitting poses of a human agent to detect sitting affordances in scenes to classify objects. For training, key poses of a sitting person need to be placed manually on each example training object. In detecting chairs, their approach achieves superior results over methods that use shape features only. However, as pointed out by Grabner et al. their approach has difficulties in detecting stools, since they were not present in the training data. Consequently, the approach of Grabner et al. does not allow to find affordances per se, but rather affordances of trained object class examples.

In the present paper we follow a different approach. Our goal is to directly predict sitting affordances in input data, independently of any possibly present object classes. Further, if a sitting affordance is hypothesized, it will be categorized on a fine-grained scale according to the characteristics of the input data at the position where the affordance is assumed. Our approach does not rely on examples of sitting furniture, but only on the agent and affordance models. Our fuzzy function formulation encodes expert knowledge to connect the input data with the desired functionality with respect to the given agent model. Still, our models remain simple and also work even if important parameters are changed.

We show this generality by varying the size of the applied agent model during our evaluation (Sect. 5). Additionally and similar to Grabner et al., our approach suggests a pose how the detected object can be used by the agent.

Note that our approach is ignorant of any object categories. However, our fine-grained affordance formulation allows for a more precise object categorization as a consequence of affordance prediction. Due to the fine-grained scale on which affordances are predicted, object categories can be easily linked to the prediction result (e.g. if a backrest could be detected or not). Our approach thus suggest as which kind of object the detected object exhibiting the affordance can be used. However, the detailed analysis of detected objects and their classification is left for future work.

3   Affordance Modeling

Usually, affordances are defined as relations between an agent and its environment [1,15,24]. Since these two entities are crucial for affordances, we start with their definitions. Then, a definition of fine-grained affordances is provided.

3.1   Environment and Agent

Contrary to our previous work [19], we do not need an explicit environment model. Our algorithm is designed to work on point clouds (e.g. from an RGB-D camera). Thus, a point cloud $P = \{p_i\}, p_i \in \mathbb{R}^3$ defines the environment in this setting. There are no further models or assumptions involved in our environment definition except the two necessary constraints when working with affordances. Firstly, the environment must correspond to the body-scale metrics of the agent. Secondly, the agent and environment must share a common coordinate frame (i.e. common ground plane and up-vector).

The affordances applied in this approach correspond to functional properties offered to humanoid agents. The agent H is modeled as a directed rooted tree $H = (V_H, E_H)$ with vertices V_H and edges $E_H \subseteq V_H \times V_H$ representing a scene graph. In this graph, nodes represent joints in a human body and edges represent parameterized spatial relations between these joints. The spatial relations correspond to average human body proportions. The nodes contain information on how the joints can be revolved while maintaining an anatomically plausible state (i.e. without harming a real human if the same state would be applied).

In the refined approach that we present here, the agent H is modeled according to average human body size and proportions as reported in a statistical investigation [25] and in [26]. Each edge $e \in E_H$ in the graph H represents parameterized body parts of the agent and is attributed with a length $l = \|e\|$ to reflect the dimensions of the human body. When fitting the agent into the environment during affordance prediction, the edges of the graph are approximated by cylinders for collision detection. Further, each vertex $v \in V_H$ represents movable joints in the broader sense and is attributed with an angle θ. This angle θ defines the current state of the joint and describes the rotation relative to the parent joint in the graph around the lateral axis. Note that this simple model does not reflect all possible degrees of freedom of a human body. However, this simplified human model is sufficient for our purposes.

3.2 Affordances and Affordance Hierarchy

An affordance is an action opportunity offered to an agent by its environment. We suggest to consider affordances on a fine-grained scale. This means that an affordance is a generalization of similar action opportunities and thus can be divided into fine-grained affordances or subaffordances. For instance, the affordance *sitting* is a generalization of more precise relations that an agent and its environment can take. We demonstrate our ideas by distinguishing between the fine-grained affordances *sitting without backrest, sitting with backrest, sitting with armrest* and *sitting with headrest*. Further, we give an outlook on the *lying* affordance which can be seen as a generalization of lying with *elongated* or *raised body* and lying with *stretched* or *raised legs*. As is obvious from the specializations of the *lying* affordance, subaffordances can be mutually exclusive.

In the context of this paper we define an affordance A as a set $A = \{F_0 \ldots F_j\}$ of fine-grained affordances F_i. For a given environment P, affordance A and initial agent configuration H_A the function Aff : $H_A \times P \times A \rightarrow \{(F, p, H_g)_i\}$ determines a set of tuples. Each tuple contains $F \subseteq A$, a set of fine-grained affordances present at position p in the environment with a goal agent configuration H_g. The algorithm described in the next Section is an implementation of the above function Aff.

For each affordance A, an initial pose H_A of the agent needs to be defined. Thus, every fine-grained affordance F_i specializing the same affordance A has the same initial pose. In this work we use a separate initial pose for sitting and one for lying. The initial pose refers to the joint states of the simulated agent prior to any transformations and collision tests.

The fine-grained affordances F_i of each affordance A are organized in an affordance hierarchy (Fig. 3). In this work we examine two affordance hierarchies: $\mathcal{A}_{sitting}$ for sitting affordances and $\mathcal{A}_{lying}$ for lying affordances.

An affordance hierarchy is defined as a directed rooted tree $\mathcal{A}$. Each node $F \in \mathcal{A}$ corresponds to a fine-grained affordance and is a tuple $F = (J_F, E_F)$. Here, $J_F = \{(\theta_1, \theta_2, \theta_d, \mathbf{x}, v)_i\}$ defines constraints on valid angles $\theta_1 \leq \theta \leq \theta_2$ for a vertex $v \in V_H$ around axis $\mathbf{x} \in \mathbb{R}^3$. The angle θ_d defines a default pose that is used for stability checks (see Sect. 4). These angle constraints are chosen to reflect a broad range of valid poses for the corresponding affordance. Further, $E_F \subseteq E_H$ define affordance specific edges in the agent model H. The agent model edges E_F are checked for collision while the agent is unfolded and the corresponding vertices are transformed from θ_1 to θ_2.

After processing a node $F \in \mathcal{A}$ either a collision occurred or not (i.e. the fine-grained affordance F_i was found or not). The edges in the affordance hierarchy graph $\mathcal{A}$ are annotated with a constraint $c \in \{true, false\}$. The constraint c indicates whether child nodes of F_i are processed in case the affordance was found (*true*) or not (*false*). An affordance A is found if each subtree of the root

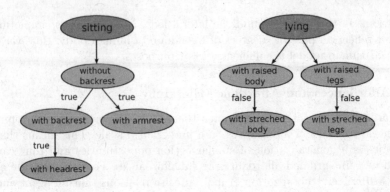

Fig. 3. The presented affordance hierarchies specialize the *sitting* and the *lying* affordance into fine-grained affordances. The arrows indicate the dependencies between the fine-grained affordances.

node in the affordance hierarchy $\mathcal{A}$ has at least one valid subaffordance F (i.e. at least one subaffordance where the agent H has contact with the environment).

Note that some of the fine-grained affordances inside an affordance hierarchy $\mathcal{A}$ depend on others. For example, if the environment affords *sitting with backrest* to the agent it must necessarily afford *sitting without backrest* as well, because the agent can choose not to use the backrest while seated. The dependencies as defined in our models are depicted in Fig. 3.

4   Predicting Fine-Grained Affordances

The algorithm for fine-grained affordance detection is essentially based on dropping an agent model in its default pose into the scene at appropriate positions. These positions need to be found beforehand. The joints of the model are then transformed to achieve maximum contact with the point cloud. Only joints relevant for a certain affordance are considered. The initial pose of the agent, as well as the joint transformations are determined by the affordance models. Further, only the agent model and the affordance models determine the current functionality of the detected object. This means that the presented approach also finds objects that might not have been designed to fulfill a certain functionality. However, based on visual information and their position in the scene they afford the desired actions. We confined the evaluation to an agent representing an average human adult and to fine-grained affordances derived from the affordance *sitting* and *lying*.

4.1   Extracting Positions of Interest

Before unfolding the agent into the scene, the search space needs to be reduced to the most promising positions. We therefore create a height map of the scene (Fig. 4(b)). The point cloud is subdivided into cells. In our experiments a size of 0.05 m provided a good balance between precision and calculation time. The highest point per cell determines the cell height. We decided in favor of the highest point instead of the average to avoid implausible values at borders of objects, where a cell may contain parts of the object and e.g. the floor.

Subsequently, a circular template, approximating the agent's torso, is moved over the height map to test whether a cell is well suited to provide support for the agent. The diameter of this template corresponds to the width of the agent as defined in the model. The decision for each cell is based on fuzzy sets as introduced by Zadeh [27]. We define 3 membership functions: discontinuity, roughness and height (Fig. 5). Discontinuity is a measure defined in percent of invalid cells or holes within the current position of the circular template. Roughness is the standard deviation of the height of all cells within the circular template. Finally, the membership function height is used to include only cells in a certain height that allow comfortable sitting with bent knees, while the feet still touch the ground. Note that we use the same height function for lying, since in home environments positions where a person would lie down also afford sitting. However, this function can be disabled in the algorithm configuration to allow for valid positions on the ground or on higher planes like tables.

One single rule is enough to decide whether a position is a valid hypothesis for further processing. We use the intersection of these membership functions to obtain the following rule: *IF roughness is low AND discontinuity is low AND height is comfortable THEN the position is a valid hypothesis*. Of course, with more affordances, more rules will be needed. The fuzzy value obtained from these functions is defuzzyfied on the function depicted in Fig. 5(d) using the *first of maximum* rule. The test is performed for both fuzzy sets of this rule, obtaining a crisp value for *available* and *not available* and deciding in favor of the fuzzy set with the higher crisp value. For each position, a possible agent orientation is obtained by considering the height gradient descent in the height map. The orientation for *sitting* affordances is parallel to the gradient vector, while the orientation for *lying* affordances is orthogonal to this vector. The positions obtained in this manner are used as possible positions in further algorithm steps (Fig. 4(c)).

4.2 Initial Agent Fitting

In the next step, each hypothesis position is checked to provide enough space for the agent model. We test several agent model orientations in this step since the initial circular template was an approximation of the agent's torso. However, in this step also the corresponding rotation needs to be found to provide enough room for e.g. the agent's legs in case of the *sitting* affordance. To reduce the amount of tested orientations we use the hypothesis orientation obtained in the previous step (Sect. 4.1) and test a few rotations in a certain range around that orientation.

For this tests, the agent is put into a default pose (defined by the angles θ_d) and is positioned above the hypothesis position. We use the FCL library [28] for collision detection between the scene P and the agent H. FCL detects collisions between 2 objects and returns the exact position at which the collision occurred.

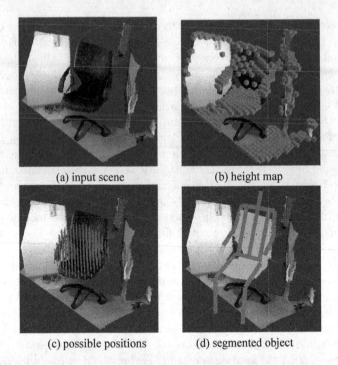

(a) input scene　　　　　　　　(b) height map

(c) possible positions　　　　　(d) segmented object

Fig. 4. Illustration of different algorithm steps. The input scene is shown in (a) and the corresponding height map in (b). Image (c) shows the possible positions for sitting affordances found by our fuzzy set formulation. The length of the red arrows corresponds to the defuzzyfied value from the availability function. The final agent pose as well as the object segmentation is shown in (d) for the fine-grained affordances *sitting with backrest* and *sitting with armrest*.

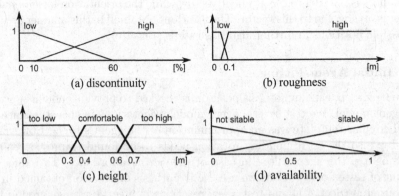

(a) discontinuity　　　　　　　　(b) roughness

(c) height　　　　　　　　　　　　(d) availability

Fig. 5. Membership functions used to find valid positions for sitting and lying affordances. The functions in (a), (b) and (c) are used to evaluate the rule, while the function in (d) is used for defuzzyfication to determine the possible presence of the affordance.

As input for FCL we convert the point cloud of the scene to the OctoMap representation [29] and approximate the individual body parts of the agent by cylinders.

If a collision occurs before any lowering of the model, the current orientation at that position is discarded. If there is no collision (i.e. the scene provides enough free space at that position), the agent is lowered until a contact with the scene occurs. The edges E_F belonging to the most general affordance $F \in \mathcal{A}$ are subsampled and the distance d of each sample to the scene is determined and a stability score obtained by using an unnormalized Gaussian

$$s = \exp \left(-\frac{(\frac{1}{n} \sum_i d_i)^2}{2\sigma_d^2} \right),$$
(1)

where σ_d is a threshold and n the number of sampled distances d. This score ensures a stable positioning of the agent and avoids that only a small part of the agent collides with the scene. The orientation with the best stability score s at a hypothesis position is kept for further steps. All stable positions qualify for the next algorithm step.

4.3   Unfolding the Agent Model

The affordance hierarchy $\mathcal{A}$ is iterated from the root node. Each node F corresponds to a fine-grained affordances and is checked at the given position. The agent is transformed to its initial pose, defined by the θ_1 parameters in each agent node $v \in V_H$. Subsequently, all vertices in the set J_F of F are iteratively transformed from θ_1 to θ_2, while the edges $e \in E_F$ are checked for contacts with the scene P.

For instance, the fine-grained affordance *sitting with backrest* is detected during the transformation of the agent's torso, comparable to the agent's movement of leaning backwards against a backrest. The affordance is detected if a contact with the scene is encountered during the transformation. If a joint reaches its maximum goal pose without a collision the algorithms assumes that the current fine-grained affordance is not present.

The subaffordance F is detected if a collision occurs at an angle $\theta_c \in [\theta_1, \theta_2]$ during this transformation. The resulting angle θ of the vertex v is then determined as

$$\theta = \begin{cases} \theta_c, & \text{if a collision occurs,} \\ \theta_d, & \text{otherwise.} \end{cases}$$
(2)

In the next step, all child nodes of F in $\mathcal{A}$ are processed if the associated constraint c of the outgoing edge of F matches the collision state in F.

Note that in contrast to normal affordances, a fine-grained affordances might depend on the existence of another fine-grained affordance (Fig. 3). Taking the *sitting* affordance as an example, the *sitting without backrest* subaffordance is checked first, as other affordances depend on it. *Sitting with backrest* and *sitting with armrests* are checked subsequently. The *sitting with headrest* affordance is checked as the last one, since it depends on the presence of a backrest. The output of this step is the final pose of the agent (position p and joint states H_g), as well as a set of predicted fine-grained affordances F. The resulting joint states represent the suggested body pose of the agent H, specifying how a hypothetical object could be used exhibiting the predicted affordance at the given position.

4.4 Combining Evidence for Affordance Presence

In our previous approach [16] the score for a detected affordance was determined by the number of detected contacts over all processed edges $e \in \bigcup_i E_{Hi}, \forall F_i \in \mathcal{A}$. However, this score did not produce meaningful results in some situations.

Our goal here is to compute the score for each affordance $F \in \mathcal{A}$ and to combine each individual score in a meaningful way. To achieve this, we follow the approach proposed in [22]. It is important to distinguish between a *local* score for all transformed edges $e \in E_F$ defined by the transformation of v in J_F and the *accumulated* score over all processed $F \in \mathcal{A}$. The local score is determined by all edges $e \in E_F$. If the processing of an edge e results in a lower score than the local score so far, the total local score should be lowered accordingly. On the other hand, when combining local scores over different affordances F_i the total accumulated score should increase, whenever there is evidence for a present fine-grained affordance F_i. The *T-norm* and *T-conorm* operators [30] have been shown to work best for the desired properties of local and accumulated scores [31].

The local score r for one transformed edge $e \in E_F$ is determined as

$$r = \begin{cases} \exp\left(-\frac{(\theta_d - \theta_c)^2}{2\sigma_\theta^2}\right), & \text{if a collision occurs,} \\ \epsilon, & \text{otherwise} \end{cases} \tag{3}$$

where σ_θ is a threshold value and $\epsilon > 0$ is a low default score. We use the *T-norm* operator

$$T(r, q) = rq \tag{4}$$

to combine local scores r and q over all $e \in E_F$ of an affordance F. At initialization, $r = 1$, however, at later steps r is the previously computed local score, whereas q is the next local score computed to be combined with r. Further, the *T-conorm* operator

$$S(r, q) = r + q - rq \tag{5}$$

is used to accumulate all local scores of all detected affordances $F \in \mathcal{A}$. For this operator, $r = 0$ at initialization, or, at later steps, the accumulated score, whereas q is the next score to be accumulated.

4.5 Object Hypothesis Segmentation

We select the pose with the best score according to Eq. 5 for each hypothesis position. All neighboring hypothesis positions around the selected pose that lie within the agent model radius are omitted, since we do not want to obtain intersecting agent goal poses H_g.

After obtaining the affordances and highest rated poses, the partition of the scene exhibiting that affordance is segmented. We use a region growing algorithm where the position of the detected affordance serves as seed point. Each point below a certain Euclidean distance is added to the segmented scene part. A low value is well suited to close small gaps in the point cloud, but at the same time limit the segmentation result to one object. Further, points close to the floor are ignored. The segmentation result is shown in Fig. 4(d).

5 Evaluation

This section describes the different datasets used for the evaluation of our approach. Further, we present the experiments and the obtained results in this section.

5.1 Datasets

Our approach was evaluation on 3 datasets. These datasets are described in the following.

Real-world Dataset. These data was acquired in our lab. Data acquisition was performed with an RGB-D camera (Kinect version 1) that was moved around an object and roughly pointed at that object's center. In total, we acquired data from 17 different chairs and 3 stools to represent the fine-grained affordances. From these data, we extracted 248 different views of the chairs and 47 different views of the stools. Example views of these objects are shown in Fig. 2. Additionally, negative data (i.e. data without the fine-grained affordances) from 9 different furniture objects was obtained and 109 views of these objects extracted. Negative data includes objects like desks, tables, dressers and a heating element. Example views of negative data are presented in Fig. 6. The whole dataset contains 404 scene views with 295 positive and 109 negative data examples. This data is provided online[1].

Warehouse Dataset. We collected 3D models from Google Warehouse with 431 objects in total. This dataset contains 323 models with sitting affordances (stools, chairs, benches, sofas) and 108 negative examples (other furniture models without sitting affordances).

Grabner Dataset. This dataset is used for comparison with other approaches. It was used in [5] to augment chair recognition by adding affordance cues. This dataset consists of 890 objects, 110 chairs, 720 non-chairs and 60 other sittable objects.

---

[1] *Real-World* dataset available at http://agas.uni-koblenz.de/data/datasets/furniture _affordances/uni-koblenz_kinect_v1.tar.gz.

Fig. 6. Example scenes without sitting affordances in the evaluation dataset.

5.2   Experiments and Results

All models in the datasets were annotated with fine-grained affordances. Object points representing sitting surfaces, backrests, armrests and headrests were assigned different labels corresponding to fine-grained affordances of the sitting affordance hierarchy $\mathcal{A}_{sitting}$. We evaluate the ability of our algorithm to distinguish sittable objects and also to detect fine-grained affordances as defined in $\mathcal{A}_{sitting}$. Similar to [5] we split the evaluation with their dataset in 2 parts. In the first part only the chair and non-chair objects are used, whereas the second part uses the sittable objects to test the generalization of the approach. To further show the general validity of our approach, we perform each evaluation using 3 differently parameterized agent models corresponding to humans with the body sizes of 1.85 m, 1.75 m and 1.65 m.

Examples of affordance predictions on the *Real-World* dataset are shown in Fig. 7 and some sample poses of the agent on artificial data in Fig. 8. Additionally to the evaluated affordance hierarchy $\mathcal{A}_{sitting}$ we modeled a hierarchy $\mathcal{A}_{lying}$ for lying. The different fine-grained affordances here are lying with or without raised back and with or without raised legs. Examples for this affordances are shown in Fig. 8 (right).

Table 1 presents the results on the *Real-World* dataset. The evaluation results for the *Warehouse* dataset can be found in Table 2. Finally, the results on the *Grabner* dataset are reported in Table 3 (chairs vs. non-chairs) and in Table 4 (other sittable objects). Note that these results were achieved without any training. All the knowledge required for detection is encoded in the simple agent and affordance models.

Fig. 7. Resulting agent poses for some scenes from the *Real-World* dataset.

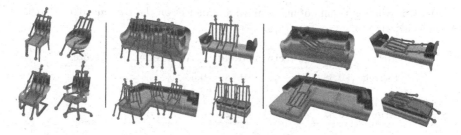

Fig. 8. Sample poses of predicted affordances for sitting on chairs (left), sitting on other furniture (center) and lying (right) on artificial data from the *Warehouse* and the *Grabner* datasets.

Table 1. Prediction results of fine-grained affordances on the *Real-World* dataset.

Agent size	1.85 m			1.75 m			1.65 m		
Metric	f-score	Precision	Recall	f-score	Precision	Recall	f-score	Precision	Recall
Sitting	0.95	1	0.90	0.95	1	0.91	0.95	1	0.91
Backrest	0.86	0.98	0.77	0.86	0.97	0.77	0.86	0.98	0.77
Armrest	0.72	0.66	0.79	0.71	0.64	0.79	0.69	0.64	0.76
Headrest	0.41	0.80	0.28	0.44	0.63	0.34	0.60	0.71	0.52

Table 2. Prediction results of fine-grained affordances on the *warehouse* dataset.

Agent size	1.85 m			1.75 m			1.65 m		
Metric	f-score	Precision	Recall	f-score	Precision	Recall	f-score	Precision	Recall
Sitting	0.95	1	0.90	0.95	1	0.91	0.95	1	0.91
Backrest	0.86	0.98	0.77	0.86	0.97	0.77	0.86	0.98	0.77
Armrest	0.72	0.66	0.79	0.71	0.64	0.79	0.69	0.64	0.76
Headrest	0.41	0.80	0.28	0.44	0.63	0.34	0.60	0.71	0.52

The results on the *Real-World* dataset are promising. Our algorithm is able to find almost all sitting possibilities, while making only little mistakes, as indicated by the results for the *sitting without backrest* and the *sitting with backrest* affordance. The ability of our algorithm to detect these two specialized affordances at the presented high rates speaks in favor of the presented approach. Further, there is almost no difference between the 3 agent model body sizes that were tested.

The results for the fine-grained affordances involving an armrest and a headrest are below the aforementioned ones. F-scores of armrests and headrests indicate that our algorithm successfully differentiates between closely related object functionalities and is able to detect the corresponding fine-grained affordances in RGB-D data. However, the low values indicate that the agent model might need more degrees of freedom during collision detection to better find differently shaped chairs.

Table 3. Prediction results of fine-grained affordances on the *Grabner* dataset (chairs vs. non-chairs).

Metric	f-score	Precision	Recall	Grabner et al. [5] (f-score)
Sitting	0.88	1	0.78	0.53
Backrest	0.75	1	0.60	-
Armrest	0.53	0.54	0.52	-

Table 4. Prediction results of fine-grained affordances on the *Grabner* dataset (other sittable objects).

Metric	f-score	Precision	Recall	Grabner et al. [5] (f-score)
Sitting	0.88	1	0.78	0.53
Backrest	0.75	1	0.60	-
Armrest	0.53	0.54	0.52	-

In contrast to the *Real-World* dataset, the *Warehouse* dataset contains full 3D models. According to the results in Table 2 our approach is again able to distinguish sitting and non-sitting objects on a fine-grained scale. This is supported by the high f-score for *sitting* and *backrest* affordances.

The *armrest* affordance has lower f-score which we attribute to the limited degrees of freedom in the agent model. This drawback will be addressed in future work. The results for most fine-grained affordances are similar along the different agent model sizes, confirming the validity of our approach. However, the *headrest* affordance has significant deviations. We attribute this to the large ambiguity of the presence of a headrest. Depending on where the agent is seated when leaning back (closer or farther away from the backrest) a normal backrest can also serve as a headrest. Additionally, the significantly higher f-score for the smallest agent model indicates that the objects in the dataset are of small size.

The results for the *Grabner* dataset were very similar across different agent sizes. We therefore report their average in Table 3 and in Table 4. Further, we omit the evaluation of the *headrest* affordance, since only very few objects with this affordance were present in the dataset. For the comparison with [5], we report their f-score corresponding to the same recall that we obtained. Considering the results reported in Table 3, our approach seems to fail. However, in [5] sitting affordances were used as a cue for object recognition, where the main goal was to tell apart chairs from other objects. For a fair evaluation, Grabner et al. scaled these other objects to the typical size of chairs, which was completely justified for their evaluation. However, since our approach is detecting sitting affordances independently of underlying object categories, many of the non-chair objects are recognized as sittable (low precision in Table 3). Indeed, e.g. a huge object with a flat and stable surface would also be considered as sittable by real humans. Still, if all objects were true to scale many of the non-chair objects would have been identified as not providing a sitting affordance by our approach. Although only trained to recognize chairs, the approach of Grabner et al. was shown to generalize well for other sitting objects due to the additional affordance cues. Our algorithm outperforms [5] on other sittable objects, since it detects sitting affordances per se (Table 4).

6   Discussion

We have shown in this paper that our algorithm is able to differentiate affordances on a fine-grained scale without prior object or plane segmentation. Thus, the presented approach is more general and can be applied to the input data directly. To our best knowledge, no similar approaches exist in the literature that are able to differentiate affordances on a fine-grained scale. This makes it hard (if not impossible) to assess the quality of our approach and compare it to related work. The comparison made with the work of Grabner et al. [5] can only serve as an approximate comparison, since their work and ours had a different goal. We therefore want to give a discussion on certain properties of our algorithm and give a detailed outlook to our ongoing work in that field.

Apart from introducing the notion of fine-grained affordances the biggest difference to related work is that we detect affordances directly. In contrast, e.g. in [5] affordances are learned as properties of objects which allows to augment the classification ability of object recognition. However, our approach is ignorant of any object categories.

While we believe that our approach will also benefit from machine-learning techniques (e.g. by learning the membership functions for the fuzzy sets), at this point we have completely omitted the learning step. This comes at the cost of manually defining "reasonable" values for the fuzzy sets (low effort) and a deformable human model (medium effort). Additionally, this raises the question on the extensibility of the approach. An initial agent pose needs to be provided for any new affordance that is included. However, if an agent model is already available (as for *sitting* affordance) new poses can be added by simply transforming joint values in the corresponding configuration file. As a second step, the joints of interest that are involved in the new affordance description, need to be provided with a minimum and maximum angle for transformation. In total, a new affordance model can be added to the algorithm with moderate effort as we have shown with the *lying* affordance hierarchy.

A more complex extension of the algorithm would be to include a different agent, e.g. a hand for grasping. While the hand itself can be modeled again as an directed rooted tree of joints, the initial hypotheses selection step must be changed completely. Instead of finding potential sitting or lying positions in the height map, for a hand a different hypotheses selection needs to be applied (e.g. finding small salient point blobs). However, as soon as these hypotheses are found, the rest of the algorithm is the same: unfolding joints of the agent and evaluating a cost function that reflects the quality of the predicted affordance. We thus believe that the presented approach is generalizable and well suited for extension.

7   Conclusion and Outlook

In this paper we have refined the term *fine-grained affordances* to better distinguish similar object functionalities. We have presented a novel algorithm that is based on fuzzy sets to detect these affordances modeled in hierarchies of subaffordances. The algorithm has been evaluated on 4 specializations of the *sitting* affordance and examples for predicting 4 specializations of the *lying* affordance were given. We have thus shown that the presented approach is able to differentiate affordances on a fine-grained scale. Since this approach detects affordances independently of underlying object categories it can be regarded as complementary to current state of the art approaches which mostly use affordances only as an additional cue for object recognition.

We believe that an object independent affordance detector could be beneficial in existing object recognition pipelines. The segmented object that results from the affordance prediction is constrained to object classes that provide the detected affordance. Where algorithms for 3D object recognition tend to detect false positive objects, these objects could be discarded due to missing affordances that the recognized object classes should posses. If this object needs to be classified, it does not have to be matched against the whole dataset, but only against object classes exhibiting the found affordance. Thus, we are currently working on combining our 3D object recognition pipeline with the presented approach for affordance detection.

The presented algorithm is ignorant of any object classes, since our goal is to detect affordances. This is evident from the leftmost image in Fig. 7, where the agent is sitting with a backrest although the object it is sitting on does not have one. Clearly, here the environmental constellation (object and wall) provided the detected affordance. This demonstrates a strength of the concept of fine-grained affordances that we will further explore in our future work.

Further, we will investigate how an anthropomorphic agent model can be exploited to detect more fine-grained affordances from other body poses. Fine-grained affordances can also be defined for other agents, e.g. a hand. In that case, *grasping with the whole hand* and *grasping with two fingers* could be distinguished, e.g. for grasp planning for robotic arms. Additionally, fine-grained affordances for grasping actions can include drawers and doors that can be *pulled open* or *pulled open while rotating* (about the hinge). We are currently looking for more examples for fine-grained affordances for different agents, to generalize our approach of fine-grained affordances.

References

1. Gibson, J.J.: The ecological approach to visual perception. Routledge, Abingdon (1986)
2. Chemero, A.: An outline of a theory of affordances. Ecol. psychol. **15**, 181–195 (2003)
3. Turvey, M.T.: Affordances and prospective control: an outline of the ontology. Ecol. psychol. **4**, 173–187 (1992)

4. Hinkle, L., Olson, E.: Predicting object functionality using physical simulations. In: 2013 IEEE/RSJ International Conference on, Intelligent Robots and Systems (IROS), pp. 2784–2790. IEEE (2013)
5. Grabner, H., Gall, J., Van Gool, L.: What makes a chair a chair? In: 2011 IEEE Conference on Computer Vision and Pattern Recognition (CVPR), pp. 1529–1536 (2011)
6. Montesano, L., Lopes, M., Bernardino, A., Santos-Victor, J.: Learning object affordances: from sensory-motor coordination to imitation. IEEE Trans. Robot. **24**, 15–26 (2008)
7. Ridge, B., Skocaj, D., Leonardis, A.: Unsupervised learning of basic object affordances from object properties. In: Computer Vision Winter Workshop, pp. 21–28 (2009)
8. Stark, M., Lies, P., Zillich, M., Wyatt, J., Schiele, B.: Functional object class detection based on learned affordance cues. In: Gasteratos, A., Vincze, M., Tsotsos, J.K. (eds.) ICVS 2008. LNCS, vol. 5008, pp. 435–444. Springer, Heidelberg (2008). doi:10.1007/978-3-540-79547-6_42
9. Kjellström, H., Romero, J., Kragić, D.: Visual object-action recognition: inferring object affordances from human demonstration. Comput. Vis. Image Underst. **115**, 81–90 (2011)
10. Lopes, M., Melo, F.S., Montesano, L.: Affordance-based imitation learning in robots. In: IEEE/RSJ International Conference on Intelligent Robots and Systems IROS 2007, pp. 1015–1021. IEEE (2007)
11. Castellini, C., Tommasi, T., Noceti, N., Odone, F., Caputo, B.: Using object affordances to improve object recognition. IEEE Trans. Auton. Ment. Dev. **3**, 207–215 (2011)
12. Wünstel, M., Moratz, R.: Automatic object recognition within an office environment. In: CRV, vol. 4, pp. 104–109. Citeseer (2004)
13. Jiang, Y., Saxena, A.: Hallucinating humans for learning robotic placement of objects. In: Desai, J.P., Dudek, G., Khatib, O., Kumar, V. (eds.) Experimental Robotics, vol. 88, pp. 921–937. Springer, Heidelberg (2013). doi:10.1007/978-3-319-00065-7_61
14. Gupta, A., Satkin, S., Efros, A., Hebert, M., et al.: From 3D scene geometry to human workspace. In: 2011 IEEE Conference on, Computer Vision and Pattern Recognition (CVPR), pp. 1961–1968. IEEE (2011)
15. Şahin, E., Çakmak, M., Doğar, M.R., Uğur, E., Üçoluk, G.: To afford or not to afford: a new formalization of affordances toward affordance-based robot control. Adapt. Behav. **15**, 447–472 (2007)
16. Seib, V., Knauf, M., Paulus, D.: Detecting fine-grained sitting affordances with fuzzy sets. In: Magnenat-Thalmann, N., Richard, P., Linsen, L., Telea, A., Battiato, S., Imai, F., Braz (eds.) Proceedings of the 11th Joint Conference on Computer Vision, Imaging and Computer Graphics Theory and Applications. SciTePress (2016)
17. Sun, J., Moore, J.L., Bobick, A., Rehg, J.M.: Learning visual object categories for robot affordance prediction. Int. J. Robot. Res. **29**, 174–197 (2010)
18. Hermans, T., Rehg, J.M., Bobick, A.: Affordance prediction via learned object attributes. In: International Conference on Robotics and Automation: Workshop on Semantic Perception, Mapping, and Exploration (2011)
19. Seib, V., Wojke, N., Knauf, M., Paulus, D.: Detecting fine-grained affordances with an anthropomorphic agent model. In: Agapito, L., Bronstein, M.M., Rother, C. (eds.) ECCV 2014. LNCS, vol. 8926, pp. 413–419. Springer, Cham (2015). doi:10.1007/978-3-319-16181-5_30

20. Maier, J.R., Ezhilan, T., Fadel, G.M.: The affordance structure matrix: a concept exploration and attention directing tool for affordance based design. In: ASME 2007 International Design Engineering Technical Conferences and Computers and Information in Engineering Conference, pp. 277–287. American Society of Mechanical Engineers (2007)

21. Maier, J.R., Mocko, G., Fadel, G.M., et al.: Hierarchical affordance modeling. In: DS 58–5: Proceedings of ICED 09, the 17th International Conference on Engineering Design, vol. 5, Design Methods and Tools (pt. 1), Palo Alto, CA, USA, 24–27 08 2009 (2009)

22. Stark, L., Bowyer, K.: Function-based generic recognition for multiple object categories. CVGIP: Image Underst. **59**, 1–21 (1994)

23. Bar-Aviv, E., Rivlin, E.: Functional 3D object classification using simulation of embodied agent. In: BMVC, pp. 307–316 (2006)

24. Chemero, A., Turvey, M.T.: Gibsonian affordances for roboticists. Adapt. Behav. **15**, 473–480 (2007)

25. GESIS: Wie groß sind Sie? allgemeine bevölkerungsumfrage der sozialwissenschaften allbus 2014 (2015). http://de.statista.com/statistik/daten/studie/278035/umfrage/koerpergroesse-in-deutschland/. Accessed 26 Jan 2016

26. Bogin, B., Varela-Silva, M.I.: Leg length, body proportion, and health: a review with a note on beauty. Int. J. Environ. Res. Public Health **7**, 1047–1075 (2010)

27. Zadeh, L.A.: Fuzzy sets. Inf. Control **8**, 338–353 (1965)

28. Pan, J., Chitta, S., Manocha, D.: FCL: a general purpose library for collision and proximity queries. In: 2012 IEEE International Conference on Robotics and Automation (ICRA), pp. 3859–3866 (2012)

29. Hornung, A., Wurm, K.M., Bennewitz, M., Stachniss, C., Burgard, W.: OctoMap: an efficient probabilistic 3D mapping framework based on Octrees. Autonomous Robots (2013). http://octomap.github.com

30. Bonissone, P.P., Decker, K.S.: Selecting uncertainty calculi and granularity: an experiment in trading-off precision and complexity. Uncertainty in Artificial Intelligence (1985)

31. Stark, L., Hall, L.O., Bowyer, K.: Investigation of methods of combining functional evidence for 3-D object recognition. In: Intelligent Robots and Computer Vision IX: Algorithms and Techniques (1991)

From Occlusion to Global Depth Order, a Monocular Approach

Babak Rezaeirowshan, Coloma Ballester, and Gloria Haro$^{(\boxtimes)}$

Department of Information and Communication Technologies,
Universitat Pompeu Fabra, Barcelona, Spain
babak.re.r@gmail.com, {coloma.ballester,gloria.haro}@upf.edu

Abstract. Estimating 3D structure of the scene from a single image remains a challenging problem in computer vision. This paper proposes a novel approach to obtain a global depth order of objects by incorporating monocular perceptual cues such as T-junctions and object boundary convexity, which are local indicators of occlusions, together with physical cues, namely ground contact points. The proposed combination of these local cues complement each other and creates a more thorough partial depth order relationship. The different partial orders are then robustly aggregated using a Markov random chain approximation to obtain the most plausible global depth order. Experiments show that the proposed method excels in comparison to state of the art methods.

Keywords: Monocular depth · Ordinal depth · Depth layering · Occlusion reasoning · Convexity · T-junctions · Boundary ownership · 2.1D

1 Introduction

Depth perception in humans enables a robust 3D vision even in the presence of a single view stimulus. Such a system is desirable in computer vision mainly due to its many applications and the abundance of monocular cameras. Human vision harnesses monocular cues to resolve inherent ambiguity caused by 3D to 2D projection in the image formation process and creates a sensible 3D perception. Monocular depth perception cues consist of dynamic cues and static cues. Dynamic cues, such as motion occlusion and motion parallax require multiple frames and motions in the scene as stimuli which are out of scope of this work. In this proposal, the focus is on static cues, namely, convexity and T-junctions; other cues in this category are perspective, relative dimensions, lighting and shadow.

While physiological aspects of these cues have been widely studied in the literature of psychophysics and vision, there is only a handful of research works that test these theories in a practical scenario using computer vision methods. Most of the work related to depth estimation in computer vision focuses on stereo disparity or motion parallax, both of which use triangulation to compute depth.

© Springer International Publishing AG 2017
J. Braz et al. (Eds.): VISIGRAPP 2016, CCIS 693, pp. 575–592, 2017.
DOI: 10.1007/978-3-319-64870-5_28

While triangulation-based methods provide absolute depth, which is desirable in many applications, they require two or more views. Monocular static cues on the other hand, can be combined to create a depth perception in the absence of binocular and dynamic monocular cues or as a complement to improve existing depth perception in a much wider domain.

The goal of this paper is, given a single image from an uncalibrated camera and its decomposition in shapes (that are assumed to represent the projection of the 3D objects on the image plane; e.g., a segmentation), to create a globally consistent depth order of these shapes that constitute the image scene. For this purpose, occlusion cues between objects, namely T-junctions and convexities are used. Additionally, we use physical cues such as ground contact points. Following the underlying assumption for extracting depth from occlusion cues, we assume that the image is composed of objects that are fronto-parallel to the camera. This is also referred to as the *dead leaves model*, a term coined by Matheron [1], which constitutes a model for image formation where the image is made by objects falling on top of each others and partially occluding them. The reason for making such assumption is that in the presence of non fronto-parallel objects in the image, e.g. floor, occlusion does not translate to depth order (see Fig. 1).

Fig. 1. Dead leaves model (DLM) and correctness of convexity cues. The left image follows the DLM while the right one doesn't. Arrows indicate the occluding object suggested by convexity cues. Bright arrows indicate a correct depth order while dark arrows indicate a wrong one.

Given an image that satisfies the dead leaves model, the occlusion cues provide a depth order among neighbouring regions. However, we require a global order to establish a rough 3D model of the scene, which is understood here as obtaining a consistent global order from a number of partial orders, which may contain some discrepancies. This problem is in general referred to as rank aggregation and it has been dealt with in several fields of computer science [2,3]. This ordering problem appears whenever there are multiple operators providing partial orders with transitive relations. The goal is to use the transitivity to obtain a global robust order as consistent as possible with the partial orders. Transitivity between orders can be stated as the following property: if we have a partial order indicating A < B and another one indicating B < C, thus we can infer the global order A < B < C. Our approach stems from the fact that transitivity of local orders can be utilized to obtain a global order using rank aggregation.

Our main contributions in this paper are (i) a depth ordering system based on monocular perceptual cues that allows reasoning without need for camera calibration, multiple frames, or motion, (ii) a novel general convexity cue detector that assigns a local depth order based on convexity and which is based on the convex hull of a shape, and (iii) the extraction of a global depth order by a robust integration of the partial orders.

2 Related Work

3D modeling has received a significant attention from the computer vision community, with studies focusing on various aspects of 3D perception. Due to the vastness of the literature in this field, we will focus on studies conducted on monocular static cues. Computational methods for depth extraction from a single image can be categorized into supervised methods and Gestalt-based methods. Alternatively, other approaches have been suggested in the literature that use human perception and vision as the basis from which to attempt to infer a computational model simulating the known processes of human vision. Our work falls in the latter category. Thus, we focus on the use of T-junctions and convexity cues for establishing a depth order. The role of T-junctions as a cue for recovering surface occlusion geometry was introduced by [4], and later stressed by [5,6]. Moreover, through the Gestalt school of thought in psychology, T-junctions were described as a basis of monocular depth perception by the work of Kanizsa [7]. Later on, more computational works demonstrated the capability of T-junctions for depth estimation; to the best of our knowledge, one of the first attempts at depth ordering methods using T-junctions was performed by [8]. Later on, an inspiring work of Nitzberg et al. [9,10] proposed the so-called 2.1D sketch through a joint segmentation and depth estimation model. More recently, studies have been conducted using energy minimization approaches which use either explicit [11,12] or implicit [13] junction detection algorithms.

In addition to T-junctions, convexity is considered to be one of the most dominant cues for figure-ground organization [7]. A computational model for utilizing convexity has been developed for figure-ground organization in the recent past [14]. Moreover, works on occlusion reasoning using Gestalt-based methods have used convexity as a complementary cue to T-junctions for a more robust relative depth estimation [15–17]. While it has been suggested that convexity affects human depth perception and is coded explicitly in the brain [18], the literature in computational models that use convexity is divided in this sense. In the works [15,17], convexity is explicitly detected and coded, while in [16] this is done implicitly. The proposed approach shares with [15–17] the use of convexity and T-junctions cues. In order to integrate the partial depth orders suggested by the monocular depth cues we use a graph-based approach. Previous works [17,19] also use a graph representation but need to reduce it to an acyclic graph and remove conflicts among different cues. In contrast, our work can directly handle conflicting transitive orders in the graph by using a rank-aggregation-based method [3], and obtain a globally consistent depth order. Here, transitive order is the order established by a path in the graph involving more than two nodes using the transitivity property mentioned in Sect. 1. A very recent work on depth layering using occlusion cues is the work of [20] where convexity, T-junctions and a ground contact cue is used to obtain a depth order of the image. An energy minimization scheme is used to find the correct depth order which makes their method more complex and time consuming than our proposed method. Moreover, they have to make more restrictive assumptions to obtain the correct ground contact cue which limits their method to a smaller domain. As the method proposed by [20] shows promising results and performs superior to other similar methods [17,21], it has been used as a benchmark for evaluation of our proposed method. A comparative evaluation using the experimental setup in [20] is presented in Sect. 4.

3   Proposed Method

We propose a method to extract a global depth order from a single image from an uncalibrated camera. The idea is motivated by studies showing human vision capability to integrate monocular depth cues to create a sensible depth perception. Given an input image, let us consider the set of its (segmented) shapes - the notion of shape used in this paper will be clarified in Sect. 3.1. Then, a global depth order can be obtained following the steps below:

1. Determine a local depth order between each pair of adjacent shapes by analysing the convexity of their common boundaries.
2. Detect T-junctions and use a multi-scale feature to determine a local depth order between the shapes that meet at each T-junction.
3. Establish a global depth order by rank aggregation of the previous partial local orders.
4. Refine the order using ground contact cue.

Each step of the proposed method is detailed in the following sections. Figure 2 illustrates the different steps of the algorithm.

3.1   Local Depth Cues Detection

Local depth cues are extracted to establish a local depth order between neighbouring objects. In this work, convexities (L-junctions) and T-junctions are used for this purpose. We use a segmentation of the image as an input to the cue detection mechanism. In order to compute a local depth order in a manner that follows the human perception based on psychophysics studies [7,18,22], T-junctions and convexities must be treated in a different manner. Thus, an explicit detection of such depth cues is required. In the following, we explain how we detect both kind of junctions.

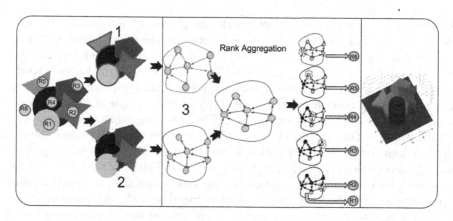

Fig. 2. Diagram of the proposed method. From left to right: The segmented image; detected convexities cues (above) and T-junctions cues (below) and the local depth order from each cue inferred by the local cues (green areas are in front of red ones, whereas yellow indicates an inconclusive cue); global depth order extraction by rank aggregation on a graph whose nodes represent the different shapes and the directed edges indicate local depth orders; final result with global depth order illustrated as a depth map, where warmer color values indicate closer objects to the camera. (Color figure online)

Convexity Cue. In this paper we propose a global convexity decision about each connected boundary between any two adjacent (segmented) regions in the image. The aim of this step is to determine which side of the boundary is the occluder and which side is the occluded, thus establishing a local depth order. Given the dead leaves model assumption, this cue can be used to infer the local depth order of the shapes that share a boundary.

To find the occluding region, we propose a method to determine which side of the boundary is closer to a convex shape. Figure 3 illustrates this process. Initially the segmented image is used to obtain the set of all the common boundaries

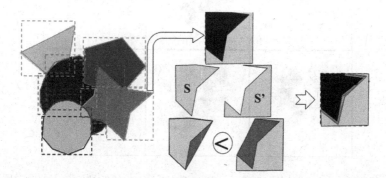

Fig. 3. Illustration of the main steps of the convexity cue detector and the estimated local order where green areas are estimated to be in front of red areas. (Color figure online)

between any two regions or objects in the image (Fig. 3, left image). For each connected common boundary, we consider its bounding box (shown in Fig. 3, middle-up). A connected common boundary divides the bounding box into two shapes (denoted by S and S' in Fig. 3). The shape whose area is closer to the area of its convex hull (i.e. smaller red area in Fig. 3, middle-down) is considered more convex and assigned as the occluder (S in the example of Fig. 3). On the other hand, the complement shape (S' in Fig. 3) is assigned as the occluded.

Let us notice that there is the possibility that a given boundary does not provide a conclusive depth cue. In other words, the convexity cue does not provide enough information to clarify which side is the occluder and which side is the occluded. This phenomenon appears, for instance, when the common boundary is either a straight line or a sinusoidal curve. To deal with such cases we define a criteria based on a threshold on our proposed global convexity measure of the connected boundary between two adjacent regions. This criteria is derived from the absolute difference between the convexity defect areas (red areas in Fig. 3) of the shapes (S and S'). If this value is not significant enough (i.e. it is lower than a prescribed threshold thr_{CX}) then these boundaries are considered inconclusive and will have no effect on the result. We define this threshold as $thr_{CX} = L \cdot \pi \cdot thr$, where L is the length of the boundary and thr is a tuning parameter that controls the sensitivity of the criteria and is independent of the length of the boundary. Examples of such inconclusive boundaries for different values of thr can be found in Fig. 4; namely, the figure displays examples for a smaller value of $thr = 0.0$ and a bigger value of $thr = 0.6$. In order to study the effect of this parameter, both on the local and global depth ordering, we present in Sect. 4 some experiments where the threshold thr is modified in the range of $[0.0, 0.6]$ with step size of 0.05.

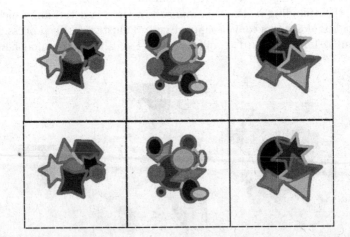

Fig. 4. Illustration of modifying the value of thr_{CX} through the parameter thr. Top row, $thr = 0.5$; bottom row, $thr = 0.15$. Decreasing thr leads to accepting more conclusive boundaries (less inconclusive boundaries in yellow). (Color figure online)

T-Junctions Cue. One of the pivotal depth cues used in this paper are T-junctions. T-junctions appear at the meeting points of three shapes boundaries and are related to occlusion configurations (see Fig. 2). Two of the three regions present in the T-junction are separated by the stem of the T; these two regions are perceived to be partly occluded by the region which presents a larger section or angle. The latter region is then in front of the other two. Moreover, the angle of each object forming the junction must satisfy some criteria to be classified as a T-junction.

In this paper, we compute T-junctions using the method in [23] where the authors gave a definition of T-junction which overcomes the difficulty of computing angles in a discrete image. They proposed an efficient algorithm which is mainly based on thresholding and computes junctions directly on the image without previous preprocessing or smoothing. The segmented image is used as an input to this method and the output is the locations of T-junctions.

The definition is based on the topographic map of an image $u : \Omega \in \mathbb{R}^2 \to \mathbb{R}$ (in our case, the segmented image), that is, the family of the connected components of the so-called, level sets of u, $[u \geq \lambda] := \{x \in \Omega : u(x) \geq \lambda\}$, and on its boundaries, the so-called level lines. Here, λ represents the gray level of the segmented image u. The set of level sets is invariant to monotonic non-decreasing illumination changes, a classical requirement in image processing and computer vision [24], and the level lines contain the boundaries of the parts of the physical objects projected on the image plane. In practice, the algorithm computes the T-junctions as all the pixels p where two level lines meet and such that the area of the connected component of each of the bi-level sets $[u \leq \alpha]$, $[\alpha < u < \beta]$, $[u \geq \beta]$, with $\alpha < \beta$, meeting at p is big enough.

After detecting the location of T-junctions, for establishing a local depth ordering one could use some angle or area of the regions meeting at the T-junction, both of which have been used in the literature [17,25]. Problems arise when certain configurations of the cue lead to an inaccurate computation. One of the problems is related to the scale at which the depth cue is obtained.

Noise in the image can also lead to incorrect cues, so one could use larger scales but they are less discriminative in depth. To avoid these issues, we stem from the work by [16] to create a reliable multi-scale measure to establish a local depth order (according to human vision) at the located cues. To this end, features are formulated using the curvature of the level lines of the distance function of each connected component in the segmented image at different scales. The features are computed for each scale s by adding the contribution from each connected component using the following formula:

$$E_s(x) = \sum_{c=1}^{nc} (e^{\beta_s |K_{c,s}(x)|^{\gamma_s}} - 1), \tag{1}$$

where $K_{c,s}$ is the curvature of the level lines of the distance function to the connected component c at scale s, nc is the number of connected components at scale s, γ_s and β_s are scale-related parameters which are fixed as proposed in [16]. In order to keep these features local and avoid overlapping with other boundaries, the distance function is clipped at a distance 5. In order to generate a multi-scale local feature we combine the local features according to (1) by computing an average of the normalized features at several scales, as in [16]. In this work, we integrate the features from scales 1 to 5. Figure 5 illustrates with an example the behaviour of this multi-scale features. As it can be seen in Fig. 5 right, the part of the cue that is perceptually closer to the observer has a higher multi-scale feature value.

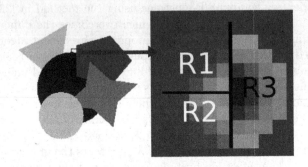

Fig. 5. Multi-scale features obtained after averaging the features E_s (1) of the first five scales.

Finally, to estimate the local depth order induced by a certain T-junction, first a representative value of the multi-scale depth features is computed for each region (e.g. R1, R2, and R3 in Fig. 5 right) in the neighborhood of the T-junction given by a disk of radius 5. The representative value is computed by applying either the median or max operators on the features of the respective region (i.e. R1, R2, R3). In Sect. 4 we compare the performance of both operators. The region with a higher representative local feature value is assigned to be in front of the other two neighboring regions (R3 in front of R1 and R2 in the example of Fig. 5).

Ground Contact Cue. The physical restriction of the real world imposes that every non-flying, non-wall-mounted, object should be connected to the ground. A reasonable extension of this phenomena suggests that every object in the scene is either directly connected to the ground or is occluded by other objects, and at least one of which is connected to the ground. Making reasonable assumption that the ground plane can be estimated, or manually segmented, the lowest point of contact of each object to the ground can be easily used as an extra cue for depth order. As ground contact cues can establish an order between non-adjacent objects they can complement the occlusion relations computed from T-junction and convexity. This kind of cue was previously used in [20].

3.2   From Local to Global Depth

In order to establish a global depth order given by the local cues we use an approximation of rank aggregation [2] similar to the one used in [3] for photosequencing. To do so, we construct a weighted graph $G(U, E)$ to represent the partial order between pairs of shapes (objects), which are represented in the graph by the nodes in U. The graph is constructed by placing a directional edge $e(i, j) \in E$ connecting the node i to node j if the local cues relating the objects suggest that object i is in front of object j (represented here by $i \gg j$). The weight of the edge gathers up the local depth order cues. Each convexity cue indicates a depth order relation between two nodes (e.g. $i \ll j$) and each T-junctions indicates a relation between three nodes using two edges (e.g. $i \ll j$, $i \ll k$). The weight of the edge $e(i, j)$ between nodes i and j is proportional to the number of local cues indicating the local order $i \gg j$, which can be interpreted as proportional to the number of votes for the local order $i \gg j$. This weight corresponds to the probability that $i \gg j$. In such a graph, a random walk after a sufficient time (in the steady state) will reach the sink of the graph (or of a sub-graph) which represents the object (or objects) perceptually furthest from the viewer. Repeating this process iteratively while in each iteration removing the sink node (or nodes) from the previous iteration will provide us with the global depth order. In particular the iteration number in which a set of nodes is removed reveals the global order of this set of nodes. For illustration of this process see Fig. 2 – step 3.

The steady state can be computed using an eigenvector analysis of M, the transition state matrix associated to the graph. The elements of M are the probabilities of moving from one state (node) to another. To construct the matrix M with non-negative entries, we initially form a matrix V collecting the votes, where the rows and columns indices correspond to the index of each associated connected component. Thus, an image with N shapes will produce an $N \times N$ matrix V. The i, j-th element of matrix V, $V(i, j)$, collects the number of votes (local cues) that agree with the partial ordering $i \gg j$.

Once the matrix V is filled, we compute the matrix M which specifies the probability that $i \gg j$. Firstly, the cycles of length two which may have been introduced by conflicting cues are removed. We follow the method proposed in [3] to resolve these conflicts. In particular, $M(i, j) = 1 - \frac{V(j,i)}{V(i,j)}$, and $M(j, i) = 0$ if $V(i, j) > V(j, i)$. The rest of the cycles do not need to be removed since the rank aggregation method automatically solves them. Finally, the rows of M are normalized to 1 in order to get transition probabilities.

After an initial depth order is computed from occlusion cues, it is possible to refine the depth for those segments which are assigned the same depth level due to lack of occlusion relations. To this end, a ground contact cue is used. For each set of objects placed at the same depth, the lowest ground contact point is computed as the point with lowest y-coordinate of common boundary between objects and the ground plane (which has been pre-segmented). Subsequently, the order is refined so that objects whose ground contact is closer to the lower border of the image are assigned a higher (i.e. closer) depth level.

4   Experimental Results

This section presents three different experiments with different kind of data designed to evaluate and illustrate various aspects of the proposed method. An initial experiment is first presented as a proof of concept using synthetic images with the following parameters: $thr = 0.15$ for convexity cue detection, and median as T-junction feature operator. The goal of the second experiment is twofold: first, to present an experimental study of different parameter settings to find the best performance and fix the parameter values for the rest of the experiments and, second, to provide a quantitative comparison of the proposed method and the most recent state-of-the-art methods [17,20,21]. This experiment is done using a dataset of 52 images proposed by [20]. For both the first and second experiments the ground truth segmentation is available, whereas in the third experiment the segmentation is done using an interactive tool [26].

Figure 6 illustrates the results of applying the proposed method to a small set of synthetic images. The first row shows the input images and the second row the global depth order images with convexity and T-junctions cues superimposed on them, respectively. The local depth order is illustrated in each cue, where green indicates the section perceived to be closer to the observer. As for global depth order, the grey values indicate global depth order, particularly the brighter areas are closer to the observer. As it can be seen all T-junction cues indicate a correct local depth order, whereas some of the convexity cues are incorrect or inconclusive (marked as yellow). However, the T-junctions cues are able to compensate these errors and create a globally consistent depth order that complies with human depth perception.

In the first part of the second experiment the proposed method is evaluated under different parameter settings with the dataset proposed by [20]. Figures 7 and 8 illustrate these results. The horizontal axis denotes the parameter thr

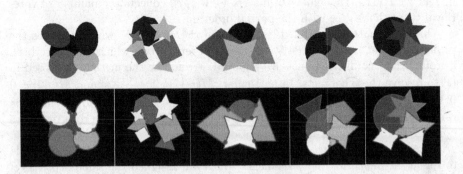

Fig. 6. Experiments with synthetic images: estimated global depth ordering (brighter gray levels indicate closer objects). The automatically detected local depth cues, convexities and T-junctions establish a local depth order (green areas are estimated to be in front of red areas). Inconclusive convexity cues are marked in yellow. (Color figure online)

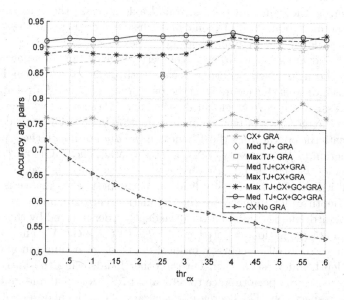

Fig. 7. Accuracy of local depth order between adjacent pairs of shapes.

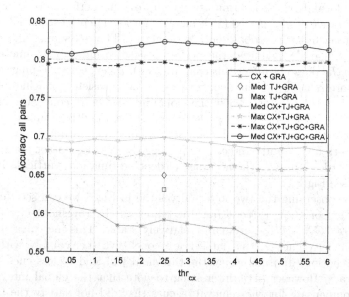

Fig. 8. Accuracy of global depth order between all pairs of shapes.

that defines the threshold $thr_{CX} = L \cdot \pi \cdot thr$ applied to the difference of defect areas. Then, as the values in the horizontal axis increase the threshold thr_{CX} increases and more boundaries become inconclusive, meaning that the sensitivity for detecting global convex boundaries decreases (see also Fig. 4). In Fig. 7 the vertical axis indicates the accuracy as the percentage of pairs of adjacent shapes which have been assigned a correct local depth order. Whereas in Fig. 8 the vertical axis indicates the accuracy as the percentage of pairs of all shapes which have been assigned a correct global depth order. These accuracy measures are identical to the measures of performance evaluation in [20]. The legend of Figs. 7 and 8 indicate the operator for the T-junction (*median* or *max*), the type of local cue (T-junction (TJ), convexity (CX), ground contact cue (GC) or combination of these), and whether or not a global rank aggregation was used ("*GRA*" or "*No GRA*"). Both Figs. 7 and 8 indicate that the best performance is achieved when the depth order induced by the T-junction cues is computed using the *median* operator and is combined with the depth order induced by the convexity cue using rank aggregation, denoted as "*Med TJ+CX+GRA*". Thus, achieving a performance of 91.49% accuracy in local depth order estimation and 69.94% in global depth order estimation. On the other hand, using the *max* operator slightly decreases the performance to 89% and 67.8% for local and global depth estimation, respectively. The decrease in accuracy of the global order with respect to the local order can be explained by the fact that the proposed method can only infer depth relations between objects connected by a path in the graph. It should also be noted that the performance of the *max* operator is slightly less stable. Further, it can be seen that the contribution of T-junctions is significant for both global and local depth estimation as they improve the performance compared to when only convexities are used (16% increase for local depth estimation and 19% increase in accuracy of global depth estimation). The red square and diamond in the Figs. 7 and 8 highlight the performance of using only T-junctions (the parameter *thr* does not affect this computation). As expected, T-junctions seem to be a more reliable cue than convexities as they consistently achieve a higher accuracy. Figure 7 illustrates how the global integration of convexity cues using rank aggregation improves the performance of local depth estimation between adjacent pairs of shapes, namely, the performance increases from 59% to 75%. Further, both Figs. 7 and 8 indicate that using the ground contact point to refine the depth order improves the accuracy of depth estimation, particularly among all pairs (see Fig. 8).

Finally, observing the two lower curves the in Fig. 7 we can see that, while the accuracy of "*CX + No GRA*" decreases as the threshold increases, the accuracy of "*CX + GRA*" remains relatively stable. This indicates that most of the convexity cues in the dataset are conclusive (i.e. comply with human depth perception) and increasing the threshold will lead to less cues and thus less accuracy. However, it is interesting to note that the global integration is able to compensate for the removal of cues that did not satisfy the threshold and stabilize the performance. It can be seen that the best operation point for the threshold of the global convexity is the mid-range value $thr = 0.25$, where the average of the two accuracy measures is the highest. While the effect of *thr* is not significant it leads to a slight increase in the performance of the global depth estimation (see Fig. 8).

In the second part of the second experiment, the proposed method is compared with the state-of-the-art [17,20,21] with the accuracy measures presented in [20]. According to the results obtained in the previous analysis, we fix the parameters to the following values: $thr = 0.25$ and *median* as the operator in the depth order estimated from the T-junctions. To this end, we follow the experimental setup suggested by [20] on their proposed depth ordering dataset. The results in Table 1 show that using a combination of T-junction, convexity and ground contact (GC) cues achieves the highest performance. As it can be seen, the proposed method, both with and without the ground contact cue, outperforms all of the state-of-the-art methods in the adjacent pairs case and, in the

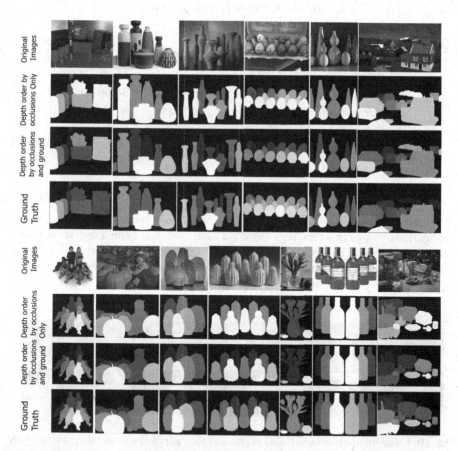

Fig. 9. Depth ordering results using the proposed method (using occlusion cues only and both occlusion and ground contact cues) on near-view scenes from the dataset by [20].

all pairs case, the proposed method performs superior to [17,21] and very close to [20]. If we do not use the ground contact cue, our methods falls short of [20] in the all pairs case. This is mainly due, as previously noticed, to the fact that our proposed method cannot infer depth relations between objects that are not connected with a path in our graph i.e. there are no transitive relations to be used to infer a global depth order. In contrast, when we add the ground contact cue, as in [20], as a post-processing of the depth order it helps to infer new depth relations when the other cues (T-junction, convexity) are not present. The improvement is illustrated in Table 1 where the accuracy increases from 69.94 to 82.59 in the all pairs case. A qualitative comparison of the best performance of the proposed method can bee seen in Fig. 9. The improvements of the depth refinement step using the ground contact point are evident in here.

Table 1. Depth order accuracy.

	Adj. pairs	All pairs
Jia et al. [21]	79.84	29.88
Palou and Salembier [17]	43.85	43.56
Zeng et al. [20]	82.66	**84.60**
Our method: CX + GRA	74.83	59.21
Our method: TJ + GRA	84.39	65.3
Our method: TJ + CX + GRA	**91.49**	69.94
Our method: TJ + CX + GRA + GC	**92.11**	**82.59**

Finally, to show how the proposed method may be used as a real world application, the interactive segmentation tool [26] has been used to segment some images from the Berkeley dataset [27] and the global depth order of the segmented objects is estimated with the proposed method. As it can be seen in Fig. 10 the order of the segmented objects is correct in most of the cases.

Fig. 10. Using interactive segmentation [26] and the proposed method to create a depth ordering of objects in the scene.

5   Limitations and Assumptions

Estimating depth from a single image is a very challenging and under-determined problem. It is necessary to make suitable assumptions to make the problem tractable. Our first assumption is that a good segmentation is available where the boundaries of the segmentation regions coincide with the actual object boundaries. As the method is based on a convexity cue defined on boundaries and T-junctions (which are points at the intersection of boundaries), a deficient segmentation leads to significant depth artifacts in the estimated depth order. A second limitation may be noticed in one of the examples in Fig. 11: the one in box

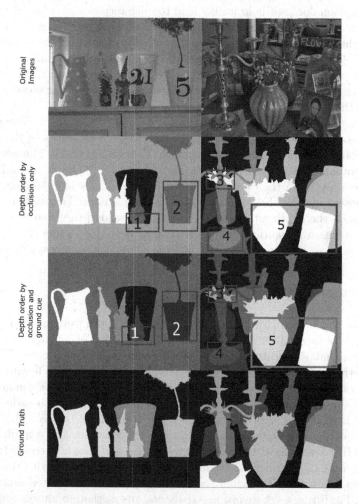

Fig. 11. Due to some limitations of the proposed approach, the violation of certain assumptions leads to errors in the estimated depth order which have been delimited with red boxes (see Sect. 5 for a detailed explanation of these problems). (Color figure online)

1 of the left image. The T-junction and convexity cues that are detected on the ground contact of the object indicate incorrect depth order. In some cases, there exist other cues that compensate for these mistakes, either directly or indirectly using the transitivity property of the graph. However, this is not the case in the aforementioned example. Another limitation is inaccuracies in our convexity detector which can be seen in Fig. 11 box 2, a misinterpretation of convexities in cases where a long narrow shape is next to two concavities. Figure 11 box 3 shows the bias of the proposed method to interpret small convex objects to be in front of their neighbouring shapes (this may happen also in visual holes, such as windows or arch bridges). A more general limitation is that objects in the scene should be approximated with fronto-parallel planes to the camera. When this assumption does not hold it may lead to misinterpretation of local cues and thus misestimation in the order of objects. An example of this can be found in Fig. 11 right, box 4. In this case, since the two objects sharing the same border cannot be approximated with fronto-parallel planes, the algorithm misestimates the depth order. Another limitation marked in the box 5, illustrating that, if the ground cue is not used or is not available, the non-adjacent objects can be placed in the same depth level.

Fortunately, in some cases there are solutions to deal with the aforementioned limitations. The problem illustrated in box 5 can be resolved using the ground contact point as shown in the third row right column of Fig. 11. The non-fronto parallel problem can be resolved by ground separation in simple cases. In cases where there are more than one non-fronto parallel planes in the image, a geometric context method, based for instance on surface normal extraction, may be used to guide the depth estimation. The problem related to visual holes can be addressed using a semantic labelling method that is able to identify the visual hole; for example by classifying areas like the sky which are always in the back.

6   Discussion and Conclusion

Inspired by the human vision capability to perceive depth using monocular cues, we proposed a method for the detection and integration of T-junction and convexity cues that is able to obtain a globally consistent depth order. The proposed method computes partial depth orders using multi-scale features, then, integrates them using a rank aggregation method that resolves conflict. This allows to simultaneously compensate for incorrect partial depth orders introduced by invalid cues and also harnesses the transitivity between the cues to obtain a global order from partial orders. The proposed method is applicable to any scene that complies with the dead leaves model and does not require training. In the presense of a segmented ground plane, the contact point with the ground can be used as an additional cue to refine the depth order estimated using occlusion cues. For future work we propose to extend the method to images containing non fronto-parallel objects using other monocular and binocular cues that may be integrated in the rank aggregation step as additional votes for partial depth orders.

Acknowledgements. The authors acknowledge partial support by the MINECO/ FEDER project with reference TIN2015-70410-C2-1-R, the MICINN project with reference MTM2012-30772, and by GRC reference 2014 SGR 1301, Generalitat de Catalunya.

References

1. Matheron, G.: Modèle séquentiel de partition aléatoire. Technical report, CMM (1968)
2. Dwork, C., Kumar, R., Naor, M., Sivakumar, D.: Rank aggregation methods for the web. In: Proceedings of the 10th International Conference on World Wide Web, pp. 613–622. ACM (2001)
3. Basha, T., Moses, Y., Avidan, S.: Photo sequencing. In: Fitzgibbon, A., Lazebnik, S., Perona, P., Sato, Y., Schmid, C. (eds.) ECCV 2012. LNCS, vol. 7577, pp. 654–667. Springer, Heidelberg (2012). doi:10.1007/978-3-642-33783-3_47
4. Guzmán, A.: Decomposition of a visual scene into three-dimensional bodies. In: Proceeding AFIPS 1968 (Fall, part I) (1968)
5. Malik, J.: Interpreting line drawings of curved objects. Int. J. Comput. Vis. **1**(1), 73–103 (1987)
6. Rubin, N.: Figure and ground in the brain. Nat. Neurosci. **4**, 857–858 (2001)
7. Kanizsa, G.: Organization in Vision: Essays on Gestalt Perception. Praeger, New York (1979)
8. Marr, D.: Vision: A Computational Approach. Freeman & Co., San Francisco (1982)
9. Nitzberg, M., Mumford, D.: The 2.1-D sketch. In: Proceedings of Third International Conference on Computer Vision, pp. 138–144. IEEE (1990)
10. Nitzberg, M., Mumford, D., Shiota, T.: Filtering, Segmentation and Depth. LNCS, vol. 662. Springer, Heidelberg (1993)
11. Gao, R.-X., Wu, T.-F., Zhu, S.-C., Sang, N.: Bayesian inference for layer representation with mixed markov random field. In: Yuille, A.L., Zhu, S.-C., Cremers, D., Wang, Y. (eds.) EMMCVPR 2007. LNCS, vol. 4679, pp. 213–224. Springer, Heidelberg (2007). doi:10.1007/978-3-540-74198-5_17
12. Palou, G., Salembier, P.: Occlusion-based depth ordering on monocular images with binary partition tree. In: 2011 IEEE International Conference on Acoustics, Speech and Signal Processing (ICASSP), pp. 1093–1096. IEEE (2011)
13. Esedoglu, S., March, R.: Segmentation with depth but without detecting junctions. J. Math. Imaging Vis. **18**, 7–15 (2003)
14. Pao, H., Geiger, D., Rubin, N.: Measuring convexity for figure/ground separation. In: 1999 The Proceedings of the Seventh IEEE International Conference on Computer Vision, vol. 2, pp. 948–955. IEEE (1999)
15. Dimiccoli, M., Morel, J.M., Salembier, P.: Monocular depth by nonlinear diffusion. In: Sixth Indian Conference on Computer Vision, Graphics & Image Processing, ICVGIP 2008, pp. 95–102. IEEE (2008)
16. Calderero, F., Caselles, V.: Recovering relative depth from low-level features without explicit t-junction detection and interpretation. Int. J. Comput. Vis. **104**, 38–68 (2013)
17. Palou, G., Salembier, P.: Monocular depth ordering using t-junctions and convexity occlusion cues. IEEE Trans. Image Process. **22**, 1926–1939 (2013)

18. Burge, J., Fowlkes, C., Banks, M.: Natural-scene statistics predict how the figure-ground cue of convexity affects human depth perception. J. Neurosci. **30**, 7269–7280 (2010)

19. Dimiccoli, M., Salembier, P.: Hierarchical region-based representation for segmentation and filtering with depth in single images. In: 2009 16th IEEE International Conference on Image Processing (ICIP), pp. 3533–3536 (2009)

20. Zeng, Q., Chen, W., Wang, H., Tu, C., Cohen-Or, D., Lischinski, D., Chen, B.: Hallucinating stereoscopy from a single image. In: Computer Graphics Forum, vol. 34, pp. 1–12. Wiley Online Library (2015)

21. Jia, Z., Gallagher, A., Chang, Y., Chen, T.: A learning-based framework for depth ordering. In: 2012 IEEE Conference on Computer Vision and Pattern Recognition (CVPR), pp. 294–301. IEEE (2012)

22. McDermott, J.: Psychophysics with junctions in real images. Perception **33**, 1101–1127 (2004)

23. Caselles, V., Coll, B., Morel, J.: Topographic maps and local contrast changes in natural images. Int. J. Comput. Vis. **33**, 5–27 (1999)

24. Serra, J.: Introduction to mathematical morphology. Comput. Vis. Graph. Image Process. **35**(3), 283–305 (1986)

25. Dimiccoli, M., Salembier, P.: Exploiting t-junctions for depth segregation in single images. In: Acoustics, Speech and Signal Processing, pp. 1229–1232 (2009)

26. Santner, J., Pock, T., Bischof, H.: Interactive multi-label segmentation. In: Kimmel, R., Klette, R., Sugimoto, A. (eds.) ACCV 2010. LNCS, vol. 6492, pp. 397–410. Springer, Heidelberg (2011). doi:10.1007/978-3-642-19315-6_31

27. Martin, D., Fowlkes, C., Tal, D., Malik, J.: A database of human segmented natural images and its application to evaluating segmentation algorithms and measuring ecological statistics. In: Proceedings of 8th International Conference on Computer Vision, vol. 2, pp. 416–423 (2001)

Infinite, Sparse 3D Modelling Volumes

Eugen Funk[⊠] and Anko Börner

Department of Information Processing for Optical Systems,
German Aerospace Center (DLR), Institute of Optical Sensor Systems,
Berlin, Germany
{eugen.funk,anko.boerner}@dlr.de

Abstract. Modern research in mobile robotics proposes to combine localization and perception in order to recognize previously visited locations and thus to improve localization as well as the object recognition processes recursively. A crucial issue is to perform updates of the scene geometry when novel observations become available. The reason is that a practical application often requires a system to model large 3D environments at high resolution which exceeds the storage of the local memory. The underlying work presents an optimized volume data structure for infinite 3D environments which facilitates (i) successive world model updates without the need to recompute the full dataset, (ii) very fast in-memory data access scheme enabling the integration of high resolution 3D sensors in real-time, (iii) efficient level-of-detail for visualization and coarse geometry updates. The technique is finally demonstrated on real world application scenarios which underpin the feasibility of the research outcomes.

1 Introduction

Research on autonomous vehicles has become eminent in recent years. A significant driving force is the vision of autonomous transport. Cars or pillars, which can operate autonomously 24 h a day are highly attractive for logistics and public or private transport [1]. This vision has lead to intensive research that is also emphasized by the European Commission [7].

Another fruitful area of research is the automation of production sets, where robots optimized for a single-task are carefully separated from people [4]. The future of automation lies in flexible factory floors and quick burst manufacturing processes, which can provide complex, short-life-cycle products without investing into reconfiguration of the production set.

Both research disciplines have a particular aspect in common: When it comes to simultaneous application of multiple robots, lifelong world modelling or accurate localization using optical 3D sensors becomes a critical task. Moreover, both research disciplines heavily rely on 3D sensors such as stereo cameras or laser scanners. These are the low-level interfaces between the algorithmic data analysis and the physical world. The goal is to integrate each 3D measurement from the environment, whether a recognized object, its state or its geometrical 3D

© Springer International Publishing AG 2017
J. Braz et al. (Eds.): VISIGRAPP 2016, CCIS 693, pp. 593–605, 2017.
DOI: 10.1007/978-3-319-64870-5_29

shape, into a global and consistent database. Such a database aims at supporting all other mobile platforms in navigation and scene understanding. Only when a mobile robot can localize itself with respect to walls or other static obstacles, it is able to move and to approach targeted locations.

Common 3D sensors such as laser scanners, stereo or *time of flight* cameras provide samples in $\mathbb{R}^3$ of the environment. The set of all given samples is usually referred to as a *point cloud*, since no structural information is provided by the sensors. The goal in perception robotics is further to process the point clouds to meaningful information, such as 3D maps, obstacles and its positions, or any other objects of interest. In fact, autonomous vehicles are able to avoid obstacles, to navigate or to pick up load only when the 3D samples have successfully been processed to an application specific model. The information processing challenge is aggravated by the circumstance that the sensors deliver large 3D point datasets. A stereo camera, working at VGA resolution (640×480) at 10 frames per second delivers 3 million 3D points per second. Therefore, in order to integrate all measurements over multiple days or even years, a highly efficient 3D database is required. Only then it is possible to deploy a long-term operating robot capable of surface extraction, or object recognition.

Today, intensive research is undertaken by the robotics community focusing on the 3D perception. Issues such as strong noise, huge data sets and restricted computation resources make the task particularly challenging. Furthermore, 3D modelling via stereo cameras is remarkably difficult since variations in light or object surface properties lead to non-gaussian errors and hamper the modelling processes. Stereo cameras for autonomous vehicles are, however, favourable since no interference between other sensors and no a priori infrastructure in the application domain is required. Improvements in this domain are expected to enable autonomous vehicles to operate in factories and in public environments with a strong impact on the production efficiency, traffic safety, logistics and public transportation.

When processing 3D points from range sensor in general, it is an accepted practice to group the 3D samples into small cubic volumes, *voxels*. This enables redundant data to be removed and the memory layout to be structured efficiently since the voxel resolution is defined by the user a priori.

The underlying work presents a voxel-based database approach which makes it possible to create and to store 3D models and maps of unlimited size and to access voxels efficiently by a given 3D coordinate (x, y, z). In contrast to standard approaches, where the modeling volume is bounded, the presented technique enables the extension of the volume dynamically. This is of particular interest for all 3D robotics applications, where the size of the environment the robot needs to operate upon is now known a priori.

Section 2 introduces details from the state of the art research on 3D modelling via voxels, aligned to the proposed approach. Section 3 states the research objectives investigated in this work. In Sect. 4 the proposed methodology is presented, which clarifies the novel data structure (Sect. 4.1) and its implication on the voxel query speed using 3D coordinates (Sect. 4.2). Section 5 introduces the

3D modelling framework. A standard approach is discussed critically with respect to its limitations and an improved technique is proposed. Section 6 demonstrates the application of the voxel database and the 3D fusion technique from range (RGB-D) camera images. Finally, concluding remarks and aspects for further research are given in Sect. 7.

2   Literature Review

Random access of a 3D point by a given coordinate (x, y, z) is a difficult task, since the search complexity usually depends on the number of samples in the database. Grouping the data to cubic cells on a regular grid and storing them in a 3D array enables very fast access and has been the state of the art technique for many years. The reason is, when the resolution r of a volume is known, e.g. $r = 1$ cm, then the access to the corresponding voxel coordinate $x = 13$ m is performed by computing its index i in the storage array $i = x/r = 1300$. The drawback of this approach is that the memory requirements grow with the cube of the space size. Representing a dense volume $100 \times 100 \times 100$ m^3 at 1 cm resolution, would require 3.7 TB of memory when using 32 bit data inside each cell. A common approach to this issue is to structure the occupied cells with a hierarchical *octree*, where each node (cube) contains eight cubes of smaller sizes [9, 12]. When searching a voxel given a coordinate (x, y, z), the tree is traversed starting from the largest top node, as shown in Fig. 1. This means, that the number of hierarchy levels increases the number of path checks and thus directly affects the access speed of a voxel. Each time an arbitrary voxel is addressed, either when iterating or performing random access, it is necessary to traverse the full height of the octree. In short, the search technique in octrees suffers from the access complexity $\mathcal{O}(d)$ with d being the octree-depth.

Teschner [18] proposed to apply a hash-map to achieve constant time access $\mathcal{O}(1)$ to sparse voxels. In principal, a coordinate (x, y, z) is encoded to a hash index which is used for direct data access. Another benefit of hash based databases, is that the amount of data can be theoretically infinite. This further allows to store data at nearly arbitrary resolution enabling huge models to be managed which is only limited by the capacity of the physical memory. However, generating unique hash values is a difficult if not an impossible task. The goal is to avoid generating a hash value representing different coordinates [18].

Nießner et al. [17] approached the *collision issue* by storing also the coordinate of a voxel and by grouping voxels with same hash values to *buckets* (See Fig. 2). When a voxel is found by a hash value, which does not correspond to the searched coordinate, the next element in the bucket is accessed.

However, direct voxel hashing does not allow to apply level of detail (LOD) visualization or multi scale 3D modelling which is favourable when low and high resolution processing is required. In fact, LOD data structure enables to perform coarse rendering depending on the distance to the virtual camera and coarse 3D modelling depending on the expected error of a measurement [8]. Practically speaking, when it is known that the covariance of a 3D sample covers several

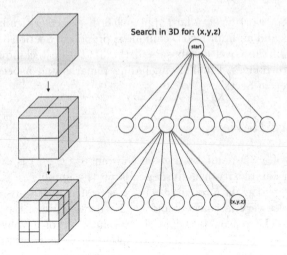

Fig. 1. Illustrated octree structure and the data access path (red), when searching for a voxel at coordinates (x, y, z). (Color figure online)

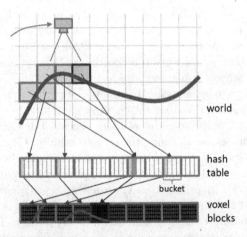

Fig. 2. Nießner et al. [17] applied a hash table for voxels. Voxel coordinates leading to the same hash key are grouped to buckets.

meters, it is not necessary to perform high resolution modelling on a 1 cm grid. The efficiency of the hash maps and the LOD capability of octrees motivated the development of a hybrid hashed octree, which is discussed next. We combine the hashing technique from Teschner [18], approach the collision issue similarly to Nießner but reference entire octrees by a hash index instead of small voxels.

3   Research Objectives

The first research objective is the development of voxel database to enable storage of geometrical models or 3D maps of unlimited size. The hash search collision issues need to be addressed in order to guarantee correct voxel access given a coordinate query.

The second objective is the integration of multiple levels of detail. Since direct voxel hashing does not enable to query larger spatial groups of voxels, the objective is to use the hashing technology for entire octrees covering larger volumes. This would combine efficient data access known from hash tables and the favourable LOD data access scheme known from octrees.

The third research objective is the development of a 3D modelling technique applied upon the proposed voxel database. Motivated by the application in autonomous vehicles, successive 3D measurements need to be used for successive model updates. That means that each new point cloud from a camera frame is expected to update an existing geometrical model. Thus, a recursive technique is required for this task to be accomplished in real-time. A crucial aspect is the suppression of noise and outliers from the input data.

Section 4 presents the developed octree hashing technique, and the hash table applied for storage and search of hash indices. In Sect. 5 the proposed voxel database is applied for successive 3D modelling. In contrast to the work from [17], the proposed work incorporates an adaptive noise suppression technique enabling low cost and low power range sensors such as stereo cameras to be applied.

4   Methodology

The hashing technique, inspired by [18] is extended to octrees. The fundamental part of the technique is the storage and search of arbitrary data elements indices via a hash map [10]. Finally, the performance of the developed hashed octree framework is compared to a standard octree approach and the sparse octree technique from [12].

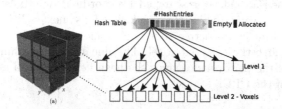

Fig. 3. Multiple octrees are stored independently referenced by a hash. (Color figure online)

4.1   Hash Indexing

We propose to combine octrees with a hash table (Fig. 3) leading to sparse voxel representation. The hash table is used to access the top level root nodes which further contain an octree in itself. Since the internal octrees are constructed with low depth (e.g. $d = 2$), this significantly decreases the access time compared to standard octrees. To access the voxel at central index coordinates (x, y, z), we begin by computing the rootKey

```
int rootKey[3] = {x&~((1<<d)-1),
    y&~((1<<d)-1),
    z&~((1<<d)-1)};
```

where d is the depth of the internal octree, & and ~ denote, respectively, bit-wise AND and NOT operations. At compile time, this reduces to just three hardware AND instructions. The shift by d makes sure that the coordinates (x, y, z) are represented by coarser values. For instance, applying an internal octree of depth $d = 3$ with $2^d = 8$ subdivision nodes in each dimension the space is divided in coordinate ranges $\{[0, 7], [8, 15], \cdots\}$. This process is illustrated in Fig. 4. For example, the operation (1<<3)-1 results in first three bits set to one. Negating the same, leads to five one-bits and three zero bits, as shown in the illustration of ~((1<<3)-1). Similar to [18], the rootKey is further processed to a hash using large prime numbers.

```
unsigned int rootHash=((1 << N)-1)&
    (rootKey[0] * 73856093^
    rootKey[1] * 19349663^
    rootKey[2] * 83492791);
```

Here, N is the constant bit-length of the hash, the three constants are large prime numbers, ^ is the binary XOR operator and << is the bit-wise left shift operator. Since the hash is imperfect, collisions occur so that multiple different coordinates are mapped by the same hash. The size of the hash map N influences the collision probability. In our experiments N was set to 32 bit leading to 70cpm (collisions per million of distinct coordinates). A drawback of the hash is that it is not well suited for negative coordinates. Thus, when negative values in (x, y, z) are processed to a hash, the number of collisions increases by up to 50%. Such high rates require countermeasures which are undertaken by additional octree place holders (green cells in Fig. 3). Each octree reference stores also its coordinate. When an octree is searched by coordinate given by the user, the resulting octree is validated. If the validation fails, next cell in the reference list is checked. Finally, an octree root node enclosing the searched coordinate is traversed to give the targeted voxel. The linear search within a hash block slightly reduces the performance since the hash keys are small and fit into the Level-1 cache of the CPU.

4.2   Hashed-Octree Performance

We have compared the performance of the proposed hashed-octree with a standard octree implementation and the recent work from [12]. The second column in Table 1 shows the achieved random access times for each approach. The third column contains the best achievable resolution applying the corresponding technique. For example, using an octree with depth $d = 16$, the maximum number of voxels is 32768^3 which corresponds to $(327\,\text{m})^3$ when each voxel represents a box volume of $1 \, cm^3$. While octrees are usually limited, the hashed octree approach is not. In theory, geometrical models of infinite size can be represented by the proposed technique. However, direct storage of the voxels in the computer memory is not practical and streaming out of core techniques [2] need to be considered.

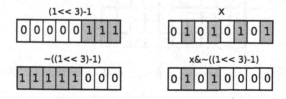

Fig. 4. Illustration of the key generation steps.

Table 1. Octree access time comparison.

Method	Access time	Max. resolution
Octree ($d = 16$)	6.43 μs	32768^3 (327 m)^3 @ 1 cm
Octree [12] ($d = 16$)	2.55 μs	32768^3 (327m)^3 @ 1 cm
Hashed-Octree ($d = 2$)	0.45 μs	∞

5 Application to 3D Modelling

The final goal of the presented research work is the application of the developed hashed octree techniques for successive 3D modelling. When a new measurement, a 3D point cloud, becomes available from the range sensor, the goal is to update the existing 3D model as fast as possible. Several years ago Curless and Levoy [6] proposed a volumetric update approach, applying the Nadaraya Watson regression technique [16] for successive volume updates from streaming range measurements. In contrast to standard surface modelling techniques with polygons, the surface is represented by a zero level set. In principle, each voxel contains a positive or a negative scalar value indicating its location inside or outside of an object. Figure 5 outlines this concept. When a surface is observed by a range camera, the volumetric model divides the full space into interior and exterior areas. The goal is further to assign correct values to the voxels around the surface in order to approximate the shape as accurate as possible.

During the reconstruction process each pixel in the camera image is processed to a 3D point $p \in \mathbb{R}^3$. The ray between the camera centre and the 3D point is traversed updating the implicit value of each voxel lying on the ray. In order to increase the computation speed and to reduce the memory overhead, only voxels within a small distance away from p are updated. Figure 6a shows the updated neighbouring voxels in red and the sample p as a dark red circle. As mentioned before, it is assumed that voxels inside an object receive a negative

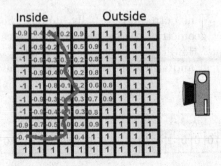

Fig. 5. Implicit shape representation by the zero level set of voxels.

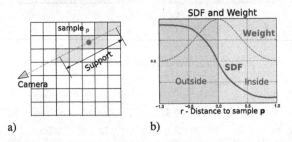

a)   b)

Fig. 6. (a) Range of influenced voxels from a 3D sample p. (b) The signed distance function applied and its weight over the distance away from the sample p.

and outside-voxels a positive scalar value. The distribution of the positive and negative values is described by a *signed distance function* (SDF) in Fig. 6b. The goal is finally to incorporate the SDF into the voxel model updating existing values.

As illustrated in Fig. 6(b), each voxel receives a weight decreasing with the distance r away from the sample p. This represents the certainty of the SDF value f_i around p. Given the SDF function values f_i and its weight w_i, the existing voxels along the ray and around the sample p are updated following Eq. (1). The new state $(k+1)$ of the i-th voxel is computed from its previous weight w_i^k and its previous implicit value f_i^k also incorporating the novel measurements w_i and f_i.

$$f_i^{k+1} = \frac{f_i^k \cdot w_i^k + f_i(r) \cdot w_i(r)}{w_i^k + w_i}$$

$$w_i^{k+1} = w^k + w_i(r) \tag{1}$$

where r is representing the distance away from p. The weights follow the Gaussian distribution

$$w_i(r) = w_i^{sample} = e^{-\lambda r} \tag{2}$$

where λ is set to $\lambda = \frac{-r_{max}}{\ln(0.1)}$ which gives a weight of 0.1 at the boundaries of the SDF support. Figure 7a shows a simple synthetic scene, where a range camera faces a wall. A ray is traversed through each camera pixel and its intersection point with the plane is integrated into the 3D model as a sample. After the fusion operation (1) is applied, the voxels in front of the plane receive positive and voxels behind the wall receive negative values. This is shown in Fig. 7b. Because of the simplicity of this technique, it is being applied in several 3D modelling frameworks [5,13,17]. However, this technique does not consider the sample sparsity and scale of the measurement errors. In cases when a wall is far away from the camera, its sample distribution is very sparse. In such a case, two neighbouring pixels in a range camera represent samples which are far away from each other (see point cloud in Fig. 7c). This leads to holes in a volume grid and in the reconstructed surface, which are not recovered by the algorithm (Fig. 7e). Thus, consistent 3D modelling of surfaces is not possible when the standard technique is applied.

In order to prevent this, the technique from Curless and Levoy is extended to cones. The width of the cone is small close to the camera and large when the distance is increased (Fig. 7d). Furthermore, the weights w_i are extended to

$$w_i^{full} = w_i^{sample} \cdot w_i^{cone}$$
$$w_i^{cone} = e^{-\lambda_c r_c} \tag{3}$$

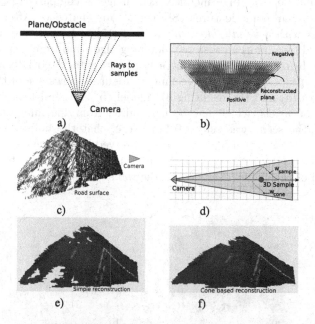

Fig. 7. (a) Camera faces a wall in a synthetic setup, (b) resulting voxel values, (c) point cloud from a camera observing a road, (d) the proposed cone fusion concept, (e) mesh reconstruction of the road segment with the standard Curless & Levoy technique, (f) mesh reconstruction with the proposed cone fusion technique.

with r_c as the distance of the i-th voxel orthogonal to the ray. Similar to (2) λ_c is set to give 0.1 at the boundary of the cone, which is however not a critical parameter.

This extension is the main difference of the presented 3D modelling approach compared to the state of the art methods. Figures 7e–f show the effect of cone fusion on 3D samples acquired from a road surface. While the application of the standard fusion technique is likely to produce holes caused by sparse samples and noise, the presented approach still achieves consistent surfaces.

Note, that the recursive nature of the update process has a linear time complexity and is not affected by the size of the 3D model. Moreover, the voxel updates (1) inferred by each sample can be performed in parallel which further increases the computation efficiency of the technique.

Section 6 discusses the application of the technique on realistic datasets from a multi view high resolution UAV-set-up and a mobile stereo system mounted on a vehicle.

6   Experiments

The cone based 3D fusion technique with hashed octrees is demonstrated on two different applications. In the first application a UAV with a high resolution camera flew around a chapel. The images have been processed by a multi view software similar to [19]. In principle, each image is compared with all other images and similar point features (SIFT [14]) are matched. After estimating the trajectory with a *bundle block adjustment* [15] technique, the images have been processed by a multi view stereo matching algorithm from [11]. Finally, the obtained depth images for each camera frame are integrated into a global 3D model via the proposed 3D fusion technique. Figure 8 shows one of the acquired camera images (a) and the resulting 3D model (b). The full model consists of 167 millions of voxels, which has been acquired from 450 image frames. The resolution of the scene was set to 0.1 m. Note, that the holes are caused by occlusions and areas which have not been observed by the UAV camera during the flight. These often relates to the ground under the trees, or the ground in the backyard of the chapel occluded by the walls.

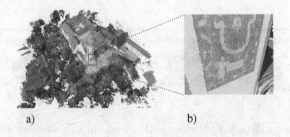

a)   b)

Fig. 8. (a) Obtained 3D model from UAV at 1 cm resolution, (b) enlarged view on the chapel tower.

The second application uses a stereo camera system in combination with an inertial measurement unit (IMU). This enables to obtain the six degrees of freedom (6dof) pose of the camera in real time. This set-up is of particular interest for a wide range of indoor applications such as inspection, autonomous transport or logistics. More details about the hardware and software of the real time localization system can be found in [3]. Again, the stereo images are processed to dense disparity images [11]. The trajectory provided by the IMU+stereo system and the disparity images are directly used for 3D fusion. Figure 9 shows a point cloud (a) and the resulting 3D model (b) when the cone based 3D fusion using the hashed octree is applied.

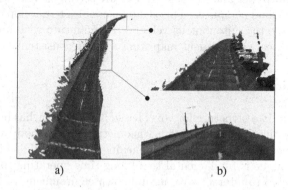

a)   b)

Fig. 9. (a) Point cloud from stereo, (b) modelled surface of the road.

The presented results clearly show that the developed technique is capable of handling large data sets and to process them to simplified 3D models in linear time depending on the number of 3D samples. It has been observed that the multi view 3D reconstruction point clouds suffer from less noise and errors than the real time stereo depth images. The reason is that when multiple images from a single object are available, each pixel in each depth image contains multiple depth hypotheses. This enables the optimization of the depth consistency and to increase the overall 3D reconstruction quality dramatically. As for the stereo data, the standard 3D fusion technique [6] lead to a high number of holes and artefacts in the final model. Only the cone fusion approach achieved smooth and consistent surfaces.

When the standard 3D fusion technique from [6] is applied, the algorithm achieves a runtime performance of 500 ms for a single VGA (640 × 480) depth image on a standard desktop PC with 16 cores. After extending the algorithm to the cone fusion approach, the support width of the SDF depends on the distance between the camera and centre and p. This increases the runtime by about 20%. The worst case run time (in seconds) for a single frame via cone fusion can be estimated a priori via

$$t_{frame} = \frac{n_{res} \cdot p_{sup} \cdot r_{sup}^2}{4} \cdot t_v \tag{4}$$

with n_{res} as the resolution of the image (e.g. $640 \times 480 = 307200 = n_{res}$), p_{sup} is the width of the support around a sample (see w_{sample} Fig. 7d), and r_{sup} as the maximal width of the cone in voxels (Fig. 7d). The value $t_v = 0.45 \cdot 10^{-6}$ represents the time required to access a single voxel in the database as shown in Table 1.

Compared to the recently proposed GPU driven 3D fusion technique from Nießner et al. [17], the runtime performance is significantly lower. Nießner reported processing times around 15 ms, which is possible when the data is cached in the internal GPU memory. In contrast to this, our method focuses on multi-threading and distributed computing which enables to obtain 3D models even when low power and low cost sensors are set-up on vehicles and the raw data is sent to a cloud computer. This strategy enables to develop a server-client architecture where the autonomous vehicles communicate with a central server and update the local environment map from a single consistent source.

7 Conclusion and Outlook

A highly efficient data structure for voxel based 3D geometry has been presented. The approach enables to model arbitrary geometries and to modify them dynamically, for instance when new 3D measurements become available from a mobile robot. The technique is fundamental to all long time operating robotic systems which are expected to interact with an unknown environment.

An advanced 3D modelling technique has been presented and applied on the 3D voxel database. The application enables very large environments to be modelled and to create high resolution maps.

Future research will focus on the integration of loop-closing localization algorithms. Another aspect of future developments will cover the extension of the presented framework to a cloud computing architecture. Low level communication with sensors, consistent global mapping and client based visualization will be targeted. The mid term goal of the project is an ubiquitous framework focusing on 3D object detection, 3D mapping and visualization of huge 3D scene.

References

1. Andreasson, H., Bouguerra, A., Cirillo, M., Dimitrov, D., Driankov, D., Karlsson, L., Lilienthal, A., Pecora, F., Saarinen, J., Sherikov, A., Stoyanov, T.: Autonomous transport vehicles: where we are and what is missing. IEEE Robot. Autom. Mag. **22**(1), 64–75 (2015)
2. Baert, J., Lagae, A.: Dutré, P.: Out-of-core construction of sparse voxel octrees. In: Proceedings of the 5th High-Performance Graphics Conference, HPG 2013, pp. 27–32. ACM, New York (2013)
3. Baumbach, D.G.D., Zuev, S.: Stereo-vision-aided inertial navigation for unknown indoor and outdoor environments. In: Proceedings of the International Conference on Indoor Positioning and Indoor Navigation (IPIN), 2014. IEEE (2014)
4. Bekris, K., Shome, R., Krontiris, A., Dobson, A.: Cloud automation: precomputing roadmaps for flexible manipulation. IEEE Robot. Autom. Mag. **22**(2), 41–50 (2015)

5. Chajdas, M.G., Reitinger, M., Westermann, R.: Scalable rendering for very large meshes. In: International Conference on Computer Graphics, WSCG 2014 (2014)
6. Curless, B., Levoy, M.: A volumetric method for building complex models from range images. In: Proceedings of the 23rd Annual Conference on Computer Graphics and Interactive Techniques. SIGGRAPH 1996, pp. 303–312. ACM, New York (1996)
7. EUCp: FP7-Transport, Research supported by the European Commission (2015). http://bit.ly/1btLACw. Accessed 22 Sept 2015
8. Floater, M.S., Hormann, K.: Surface parameterization: a tutorial and survey. In: Dodgson, N.A., Floater, M.S., Sabin, M.A. (eds.) Advances in Multiresolution for Geometric Modelling. MATHVISUAL, pp. 157–186. Springer, Heidelberg (2005). doi:10.1007/3-540-26808-1_9
9. Frisken, S.F., Perry, R.N., Rockwood, A.P., Jones, T.R.: Adaptively sampled distance fields: a general representation of shape for computer graphics. In: Proceedings of the 27th Annual COnference on Computer Graphics and Interactive Tehniques, pp. 249–254. ACM Press/Addison-Wesley Publishing Co. (2000)
10. Google inc. Google sparse hash, ver1.5 (2015). http://goog-sparsehash.sourceforge.net/. Accessed 26 Sept 2014
11. Hirschmuller, H., Scharstein, D.: Evaluation of stereo matching costs on images with radiometric differences. IEEE Trans. Pattern Anal. Mach. Intell. **31**(9), 1582–1599 (2009)
12. Hornung, A., Wurm, K.M., Bennewitz, M., Stachiss, C., Burgard, W.: OctoMap: an efficient probabilistic 3D mapping framework based on octrees. Auton. Robots **34**, 189–206 (2013)
13. Izadi, S., Kim, D., Hilliges, O., Molyneaux, D., Newcombe, R., Kohli, P., Shotton, J., Hodges, S., Freeman, D., Davison, A., Fitzgibbon, A.: Kinectfusion: real-time 3d reconstruction and interaction using a moving depth camera. In: ACM Symposium on User Interface Software and Technology. ACM, October 2011
14. Lowe, D.G.: Distinctive image features from scale-invariant keypoints. Int. J. Comput. Vis. **60**(2), 91–110 (2004)
15. Moulon, P., Monasse, P., Marlet, R.: Global fusion of relative motions for robust, accurate and scalable structure from motion. In The IEEE International Conference on Computer Vision (ICCV), December 2013
16. Nadaraya, E.A.: On estimating regression. Theory Probab. Appl. **9**(1), 141–142 (1964)
17. Nießner, M., Zollhöfer, M., Izadi, S., Stamminger, M.: Real-time 3d reconstruction at scale using voxel hashing. ACM Trans. Graph. (TOG) **32**, 169 (2013)
18. Teschner, M., Heidelberger, B., Mueller, M., Pomeranets, D., Gross, M.: Optimized spatial hashing for collision detection of deformable objects. In: Proceedings of Vision, Modeling, Visualization (VMV 2003), pp. 47–54 (2003)
19. Visualsfm, C.: A visual structure from motion system (2011). http://ccwu.me/vsfm/. Accessed 30 Sept 2015

Author Index

Arias, Pablo 454

Bak, Peter 287
Ballester, Coloma 454, 575
Bischof, Horst 353
Black, Michael J. 175
Börner, Anko 593
Breidt, Martin 175

Catalina, Carlos A. 475
Chaudhuri, Parag 135
Cristani, Marco 3

Egger, Bernhard 95
El-Sakka, Mahmoud R. 493
Ertl, Thomas 220, 242

Facciolo, Gabriele 454
Fan, Haoli 72
Fedorov, Vadim 454
Feldmann, Dirk 113
Feng, Jie 35, 72
Fleming, Reuben 175
Fratarcangeli, Marco 153
Fraundorfer, Friedrich 353
Fujita, Shu 439
Fukushima, Norishige 439
Funk, Eugen 593

Goldblatt, Ran 287
Gupta, Heena 135

Haag, Florian 242
Hajder, Levente 395
Haro, Gloria 575
Hedman, Anders 515
Holzmann, Thomas 353

Jähne, Bernd 329
Jänicke, Stefan 199
John, Markus 220

Kaufmann, Dinu 95
Khan, Asif 493

Knauf, Malte 555
Koch, Steffen 220
Kolingerová, Ivana 51
Kraft, Dirk 418
Krüger, Norbert 418
Krüger, Robert 242

Li, Haibo 515
Liu, Zhihong 72
Lohmann, Steffen 220
Lothe, Pierre 329

Machado, Penousal 310
Minetto, Rodrigo 475
Miranda, Catarina Runa 374
Mittelstädt, Sebastian 287
Mohammadi, Sadegh 3
Mohler, Betty J. 175
Murino, Vittorio 3

Nguyen, Hoa 264

Omer, Itzhak 287
Orellana, Cristian M. 534
Orvalho, Verónica Costa 374

Patel, Priyanka 135
Paulus, Dietrich 555
Perina, Alessandro 3
Pohl, Melanie 113
Polisciuc, Evgheni 310
Prantl, Martin 51
Pusztai, Zoltán 395

Rezaeirowshan, Babak 575
Romero, Javier 175
Rosen, Paul 264
Roth, Volker 95
Rumman, Nadine Abu 153

Saracchini, Rafael F.V. 475
Scheuermann, Gerik 199
Schönborn, Sandro 95
Schreck, Tobias 287

Seib, Viktor   555
Setti, Francesco   3
Shao, Lin   287
Stolfi, Jorge   475

Thomsen, Mikkel Tang   418

Váša, Libor   51
Vetter, Thomas   95
von Schmude, Naja   329

Witt, Jonas   329
Wörner, Michael   220

Xiong, Xiaoliang   35, 72

Zhong, Yang   515
Zhou, Bingfeng   35, 72
Zuniga, Marcos D.   534